New Insights into Structural Interpretation and Modelling

Special Publication reviewing procedures

The Society makes every effort to ensure that the scientific and production quality of its books matches that of its journals. Since 1997, all book proposals have been refereed by specialist reviewers as well as by the Society's Books Editorial Committee. If the referees identify weaknesses in the proposal, these must be addressed before the proposal is accepted.

Once the book is accepted, the Society has a team of Book Editors (listed above) who ensure that the volume editors follow strict guidelines on refereeing and quality control. We insist that individual papers can only be accepted after satisfactory review by two independent referees. The questions on the review forms are similar to those for *Journal of the Geological Society*. The referees' forms and comments must be available to the Society's Book Editors on request.

Although many of the books result from meetings, the editors are expected to commission papers that were not presented at the meeting to ensure that the book provides a balanced coverage of the subject. Being accepted for presentation at the meeting does not guarantee inclusion in the book.

Geological Society Special Publications are included in the ISI Index of Scientific Book Contents, but they do not have an impact factor, the latter being applicable only to journals.

More information about submitting a proposal and producing a Special Publication can be found on the Society's web site: www.geolsoc.org.uk.

It is recommended that reference to all or part of this book should be made in one of the following ways:

Nieuwland, D. A. (ed.) 2003. *New Insights into Structural Interpretation and Modelling*. Geological Society, London, Special Publications, **212**.

Rutten, K.W. & Verschuren, M.A.J. 2003. Building and unfaulting fault-horizon networks. *In*: Nieuwland, D.A. (ed.) 2003. *New Insights into Structural Interpretation and Modelling*. Geological Society, London, Special Publications, **212**, 39–57.

GEOLOGICAL SOCIETY SPECIAL PUBLICATION NO. 212

New Insights into Structural Interpretation and Modelling

EDITED BY

D. A. Nieuwland

Vrije Universiteit Amsterdam, The Netherlands

2003
Published by
The Geological Society
London

THE GEOLOGICAL SOCIETY

The Geological Society of London (GSL) was founded in 1807. It is the oldest national geological society in the world and the largest in Europe. It was incorporated under Royal Charter in 1825 and is Registered Charity 210161.

The Society is the UK national learned and professional society for geology with a worldwide Fellowship (FGS) of 9000. The Society has the power to confer Chartered status on suitably qualified Fellows, and about 2000 of the Fellowship carry the title (CGeol). Chartered Geologists may also obtain the equivalent European title, European Geologist (EurGeol). One fifth of the Society's fellowship resides outside the UK. To find out more about the Society, log on to www.geolsoc.org.uk.

The Geological Society Publishing House (Bath, UK) produces the Society's international journals and books, and acts as European distributor for selected publications of the American Association of Petroleum Geologists (AAPG), the American Geological Institute (AGI), the Indonesian Petroleum Association (IPA), the Geological Society of America (GSA), the Society for Sedimentary Geology (SEPM) and the Geologists' Association (GA). Joint marketing agreements ensure that GSL Fellows may purchase these societies' publications at a discount. The Society's online bookshop (accessible from www.geolsoc.org.uk) offers secure book purchasing with your credit or debit card.

To find out about joining the Society and benefiting from substantial discounts on publications of GSL and other societies worldwide, consult www.geolsoc.org.uk, or contact the Fellowship Department at: The Geological Society, Burlington House, Piccadilly, London W1J 0BG: Tel. +44 (0)20 7434 9944; Fax +44 (0)20 7439 8975; E-mail: enquiries@geolsoc.org.uk.

For information about the Society's meetings, consult *Events* on www.geolsoc.org.uk. To find out more about the Society's Corporate Affiliates Scheme, write to enquiries@geolsoc.org.uk.

Published by The Geological Society from:
The Geological Society Publishing House
Unit 7, Brassmill Enterprise Centre
Brassmill Lane
Bath BA1 3JN, UK

(*Orders*: Tel. +44 (0)1225 445046
         Fax +44 (0)1225 442836)
Online bookshop: http://bookshop.geolsoc.org.uk

British Library Cataloguing in Publication Data
A catalogue record for this book is available from the British Library.

ISBN 1-86239-133-5

Project management by Swales & Willis, Exeter, UK
Typeset by Lucid Digital, Honiton, UK
Printed by Alden Press, Oxford, UK

Distributors
USA
  AAPG Bookstore
  PO Box 979
  Tulsa
  OK 74101-0979
  USA
Orders: Tel. + 1 918 584-2555
         Fax +1 918 560-2652
         E-mail bookstore@aapg.org

India
  Affiliated East-West Press PVT Ltd
  G-1/16 Ansari Road, Daryaganj,
  New Delhi 110 002
  India
Orders: Tel. +91 11 327-9113
         Fax +91 11 326-0538
         E-mail affiliat@nda.vsnl.net.in

Japan
  Kanda Book Trading Co.
  Cityhouse Tama 204
  Tsurumaki 1-3-10
  Tama-shi
  Tokyo 206-0034
  Japan
Orders: Tel. +81 (0)423 57-7650
         Fax +81 (0)423 57-7651
         E-mail geokanda@ma.kcom.ne.jp

Contents

Introduction: new insights into structural interpretation and modelling

D. A. NIEUWLAND

*Faculty of Earth and Life Sciences, Tectonics, Vrije Universiteit, De Boelelaan 1085,
1081HV Amsterdam, The Netherlands (e-mail: dick.nieuwland@falw.vu.nl)*

This book is a follow-up to a Geological Society of London conference with the same title, held in Burlington House, London, in February 2000. Following the conference it was decided to compile a Special Publication based in part on papers presented at the conference and in part on papers that were added later to form a balanced content.

The content of *New Insights into Structural Interpretation and Modelling* presents a balanced overview of what the title promises. It is intended as a book that will serve the experienced professional as well as more advanced students in Earth sciences with a wide range of topics described in high quality publications. Some chapters are by leaders in the field, other chapters are written by young authors with fresh ideas.

Structural interpretation and modelling go hand in hand in structural geology. Interpretation needs to be based on data. These data can be found on many scales and involve direct observations of natural phenomena based on fieldwork, and indirect observations of natural data by such means as seismic or stress measurements. Interpretation is no longer limited to drawing lines along fault planes in cross-sections or to constructing horizon maps. New methods and techniques have developed in many fields of acquisition processing and interpretation. Modelling relies on two complementary approaches: analogue models using physical materials as analogues to rocks, and numerical models that are based on mathematical algorithms to mimic rocks and deformation processes with computer programs.

These approaches to solving complex structural problems have their specific advantages and limitations. Significant progress in structural geology is therefore to be expected due to improved integrated methods. In this book a collection of papers has been compiled with the intention to provide the reader with a comprehensive overview of state-of-the-art approaches and techniques in structural interpretation and modelling. The chapters are organized following the view that the topics that are closest to the natural world appear first, fol-lowed by gradually more 'artificial' approaches. Although the multidisciplinary nature of many papers makes a straightforward order of appearance according to topic difficult, the chapters dealing with natural geological data appear in principle early, interpretation precedes modelling and analogue modelling precedes numerical modelling.

Since the publication of the predecessor of this Special Publication (Buchanan & Nieuwland 1996) much progress has been made. This has been primarily thanks to the continuously increasing computing speed and computer memory capacity, which has positively affected all fields in structural interpretation, seismics and modelling, directly or indirectly.

Interpretation

Natural data sets have the obvious advantage that they are real, however strange they may appear to be. The data result directly from natural deformation processes that have deformed natural rocks in natural time. However, data from outcrops also have problems associated with them. It is often difficult or even impossible to observe and measure a feature (fault, fold or fracture) completely in 3D. Commonly some elements need to be estimated because they cannot be observed. Another problem with natural data is the time factor. Deformation processes that took place over geological time-scales cannot be observed in a human lifetime. What we see in outcrop is a static end result of a very dynamic process. The 3D problem with natural data is in part resolved when 3D seismic can be applied. Unfortunately the method is costly and therefore almost exclusively used for commercial purposes, preferably on offshore locations where the acquisition is so much better and easier than in the mountainous areas where field geologists often like to work because of the abundance of outcropping rocks. The quality of 3D seismic has improved significantly over the last ten years or so. It has become a standard tool for geologists and

From: NIEUWLAND, D. A. (ed.) *New Insights into Structural Interpretation and Modelling.* Geological Society, London, Special Publications, **212**, 1–5. 0305-8719/03/$15

is no longer the exclusive domain of the trained geophysicist/seismic interpreter.

The section on interpretation begins with a chapter by Cello *et al.* The work relies on a large database collected in the Apennines. The project was based on collaboration between industry and academia, a combination that may become more common in the near future. Industry has decreased its own research efforts and relies more heavily on university projects, focused on geological problems that are of interest to the industry. The chapter by Cello *et al.* illustrates that for fracture systems, fieldwork is still the most important source of information.

Details of complex fault structures can be interpreted very well from high-resolution 3D seismic, as demonstrated by the examples in the chapter by Schroot & de Haan. However, faults with an offset smaller than some 20 m cannot be interpreted reliably and tension fractures cannot be observed at all. Nine-component seismic may be able to identify tension fracturing (joints), but cannot distinguish between a few large fractures and many small fractures. Orientations of tension fracture systems derived from seismic data are also uncertain because conjugate fractures tend to become averaged along the bisectrix of the acute angle. Nevertheless, without 3D seismic there would be no hope of interpreting the vast and complex subsurface structures that form reservoirs for natural resources such as oil and gas. The message is again, integration of fieldwork, modelling and seismics, to combine full 3D and intricate fault geometries at all scales is the key to reliable structural interpretation.

Next to fault geometries, the kinematics of faulting processes is just as important, presenting a problem in another dimension – time. One approach to the kinematics problem is that of palinspastic reconstruction, which presents problems of its own. Traditionally, palinspastic reconstruction (or section balancing) can only be performed on dip sections of strata that were deformed in plain strain. As many deformations take place in 3D, a palinspastic reconstruction requires a 3D approach also. In order to reconstruct faults correctly in 3D, one needs to know the complete strain-path. Preservation of line and area is necessary, but not sufficient to perform a mechanically correct palinspastic reconstruction; the strain path needs to be incorporated as well.

Rutten & Verschuren have worked out a 3D approach to this problem. They apply their technique on the scale of a delta. To attack this problem on the much larger scale of mountain belts, Hindle has developed a technique that can be seen as similar to that used by Rutten & Verschuren in that it is also based on displacement paths. However, Hindle's approach is specifically suitable for the reconstruction of displacements on a much larger scale. In his example he uses displacement fields rather then vectors to reconstruct movements on the scale of mountain belts rather than along individual faults.

Projects in the oil and gas industry and in civil engineering routinely require accurate structural interpretations of complicated fault systems and the related stress fields. Accurate definition of the present-day tectonic stress is a fundamental requirement to perform such tasks correctly and reliably. The reliability of the results depends to a large extent on the quality assurance of the data. The requirements for this quality assessment and implementation have been laid down in a program called TENSOR. The program is the basic tool for the World Stress Map but also has other applications in the field of stress inversion. Delvaux & Sperner explain more about this widely used program. The World Stress Map provides the data in the form of a freely accessible global database as described in the chapter by Sperner *et al.*

A different field of structural interpretation with relevance for modern society is that of earthquake mechanics. Predicting earthquakes has been attempted for decades if not longer, but so far predictions have not been very successful. In a new unconventional approach to earthquake prediction, Volti & Crampin present an application of seismic shear-wave splitting, based on a long-term field study. The first chapter on this topic describes the background, preliminary observations, and analysis of shear-wave splitting in Iceland. The second chapter discusses potential applications of this approach including monitoring hydrocarbon production, stress-forecasting earthquakes and some volcanic eruptions.

Modelling

In this time of massive computing power a commonly asked question with regard to analogue modelling (AM) is, does it still have a place in science, and why not solve all our problems with computers? This simple question immediately brings forward the strengths and weaknesses of both modelling techniques. In the modelling section a brief treatment of some fundamental aspects of analogue modelling and numerical modelling (NM) is presented.

In analogue modelling the small size of the models directly raises questions about scaling. (Mandl 1988; McClay 1990; Brun *et al.* 1994). The results of computer models are seemingly free of a scaling problem, especially when model results are given dimensions that are of the same order as in natural examples. This is misleading in that, for example, in the much used finite element (FE) technique, the size of the mesh elements is very large compared to the smallest elements in natural deformation processes. For comparison, a typical analogue model contains about 30 000 000 sand grains. In order to reach the same resolution as such an AM experiment, a FE mesh should also be built of 30 000 000 elements. FE meshes of such dimensions are still unthinkable. The sand grains in analogue models are not scaled; in a typical model 1:10 000 to 1:100 000 sand grains with an average diameter of 0.1 mm would resemble boulders of roughly 1 m to 10 m in size. However, the resolution of analogue models in terms of definition of individual fault planes and intricate fault patterns is much greater than in numerical models. Without going into the mathematics of scaling, a good argument in support of the validity of scaled analogue models comes from nature itself. Field observations of faults of all types on almost all scales, indicate that fractal properties can be applied to faulting and fracturing (Mandelbrot 1989). Self-similarity has as a consequence that small fault systems created in analogue models have a similar geometry as larger scale systems.

Whereas numerical models rely on mathematical algorithms to mimic natural processes and materials, analogue models work with natural materials to mimic rocks; however, the deformation processes are all natural, not imitated by algorithms. A disadvantage of using analogue materials to simulate rocks is that the choice of materials is limited compared to the variation in natural rocks. For most brittle rocks this does not pose a problem. The reason is that the most important mechanical parameter that governs deformation geometries and processes is the internal friction angle ϕ, which is about 30° for most brittle rocks and is also about 30° for sand. The main difference between sand and rocks is strength; however, strength governs only the magnitude of the stress that is required to drive a deformation process. The resulting structural geometries are independent of the stress magnitude. For both methods (AM and NM) scaling is a real but manageable problem: not everything can be exactly scaled, but most of the important parameters can be scaled satisfactorily.

The main goal of analogue models is to study the geometries and kinematics of tectonic deformation processes in manageable space and time. The duration of the deformation process (strain rates), high resolution and free formation of faults are of fundamental importance to this method.

Stress orientations and magnitudes are much more difficult to acquire from AM experiments than geometric and kinematic information. So far only directional point measurements are possible.

Such a stress measurement has an advantage of being continuous in time, but it is limited to being a point in 3D space. Numerical models lack much in the field of geometry and to a lesser extent kinematics because free-growing faults cannot yet be generated. However, concentrations of deformation in shear bands can give useful information: especially in the early deformation stages, stresses calculated with numerical methods are much more detailed than the AM point measurements even though the NM stress measurements are not continuous in time.

So far deformation in the brittle domain has been discussed. When viscous or ductile materials start to play a role, the material properties become more important. In numerical models material properties can be varied more easily than materials in analogue models. However, the viscous material silicone putty in a variation of qualities can be modified to suit many desired mechanical properties and stepping engines provide the required control over strain rates. Temperature effects are still a major problem in analogue experiments. Numerical models can treat this aspect better by simply varying the input parameters in algorithms that are hopefully realistic enough to model the processes that we are trying to understand.

Analogue modelling

The contribution by Dooley *et al.* is a good illustration of an integrated approach in which field data (from seismic) and modelling have been brought together to interpret and understand a complex multiphase fault system. The example also makes clear the importance of stratigraphy for reliable modelling and interpretation of deformation processes. The mechanical stratigraphy is the basis for a good model, be it analogue or numerical. In the chapter by Schellart & Nieuwland the emphasis lies on the kinematics of a growing complex fault system, and the related changes in the local stress field. The results make clear that with an unchanging regional stress field, the local stresses within a structure can change dramatically. The work makes intensive use of computed tomograph (CT) scanning to generate 3D data volumes. CT scans permit non-destructive acquisition of 3D data volumes of analogue models. The resulting series of 3D data volumes through time forms a powerful 4D database for detailed structural interpretation that can be performed on 3D seismic interpretation systems. This implies that advances made in 3D seismic software are also of immediate use for the analysis and interpretation of analogue models that have had the benefit of CT scanning. Tentler & Temperley have focused their analogue modelling on a large scale of deformation including the asthenosphere, which they approximate with carefully chosen silicon polymers of two different viscosities and densities. The model highlights the strength of analogue modelling in representing the development of intricate 3D structures in time, and at the same time demonstrates the complexity of making the appropriate analogue materials for experiments that include layers with brittle and ductile or viscous behaviour.

Mulugeta & Sokoutis have applied analogue modelling to gain more insight into the mechanics of thrust tectonics. Their chapter describes the use of analogue experiments for studying the dynamic and rheologic control of hanging wall accommodation in ramp–flat thrust models. The complexity of 3D deformation processes in time becomes more apparent as strain rate and rheology are also used as variables. Much can be learned from such analogue models about the strain path of complicated structures. The study highlights the usefulness of modelling tectonic processes with analogue materials and at the same time draws attention to the practical difficulty involved with such models. Variations in material are much more easily achieved in numerical models, but here the algorithms that need to be applied form another challenge.

Numerical modelling

Following the observations and conclusions of the analogue models, the next logical step is to find out how numerical techniques can be applied to find answers to questions that cannot be solved by analogue modelling. In their chapter Burov & Poliakov use a numerical approach to model basin dynamics. The work is exemplary for numerical modelling. In addition to the main topic of the paper, erosional forcing of basin dynamics, they present advances in the mathematics of 3D basin modelling. The work also demonstrates the quantitative insights that can be gained from this approach and that are such a welcome complementary result to the more qualitative analogue experiments. Burov & Poliakov and also Robin *et al.* work with a variable that has not yet been successfully controlled in analogue models – temperature. In analogue models of the brittle upper crust, temperature effects are normally not a factor that needs to be considered; however, as soon as the deeper lithosphere or asthenosphere is involved in the modelling, temperature effects become important. Robin *et al.* integrate a wealth of geological data with their numerical work.

The chapter by Hobbs *et al.* deals with 3D seismic data, a field in which a lot of progress has been made over the last five years. Hobbs *et al.* have written an important contribution on seismic processing of difficult data with severe topography

effects and velocity problems. Their approach has now been applied to an interesting geoscience problem, that of imaging an axial magma chamber on the East Pacific Rise, but could well be modified for application to economic targets as well.

The next chapter on geomechanical modelling by Heidbach & Drewes deals with a large-scale problem – tectonic processes in the Eastern Mediterranean. Although their approach is to a large extent based on geomechanics with an elasto-visco-plastic 3D finite element model at the core, important boundary conditions still need to be provided as input. The model does not generate faults; however, on the scale of the problem individual faults are not so important. 3D geomechanical finite element models are still in a development stage and this paper marks a significant step forward.

The next two chapters, by Cornu *et al.*, avoid the problem of 3D geomechanics by using a kinematic approach to large scale geometrical problems. A disadvantage of this is that input of fault geometries is required, but an advantage is that relatively complicated structural geometries can be modelled. Cornu *et al.* describe their approach and apply it to three cases: basin deformation, and compressional and extensional tectonics.

In the last two chapters, Skar & Beekman and van Wees *et al.* apply geomechanical finite element modelling to problems on smaller scales than entire basins and limit themselves to the brittle upper crust. The method is possible in the case worked on by Skar & Beekman because it is not necessary to grow faults; they are part of the starting geometry and derived from well defined geological cross-sections. The complicated structural cross-section that has been used as the basis for their work could be used because the model is 2D. The work demonstrates that in many cases, where it is valid to approximate the deformation history with plain strain, there is a definite advantage in using 2D instead of 3D. As the lithosphere is not involved, the rheological model can be elasto-plastic, a viscous element is not required. Van Wees *et al.* also base their calculations on a wealth of geological and geophysical data. They integrate these in a geomechanical 3D finite element model and successfully model the quantitative effects of hydrocarbon depletion from reservoirs.

From the chapters on numerical modelling it can be concluded that numerical modelling and analogue modelling are indeed complementary in many ways. Both methods continue to evolve and to produce new approaches and techniques. Geomechanical numerical models cannot yet generate free-growing faults; however, in very large scale problems individual faults are not so important and geomechanical numerical models can produce realistic results. On scales where individual faults become important the technique that is most suitable depends on the nature of the problem. Whether a kinematic or a mechanical approach is chosen depends on the question that needs to be answered.

Acknowledgements

The editor wishes to thank the Geological Society of London for giving the opportunity to organize the conference which inspired this volume and their subsequent support during the compilation of this book. The book would not have appeared without the help of the referees who spent considerable time reviewing the manuscripts. The following referees are gratefully acknowledged: G. Bertotti; T. den Bezemer; J.C. Blom; P. v.d. Boogaert; S. Cloetingh; T. Dooley; H. Doust; T. van Eck; R. Gabrielsen; O. Heidbag; D. Hindle; F. Kets; Y. Leroy; I. Moretti; F.J. Nieuwland; J. Reijs; P. Richard; K. Rutten; W. Schellart; A. Seihl; J.D. van Wees; B.F. Windley and R. Zoetemeijer.

References

BRUN, J., SOKOUTIS, D. & VAN DEN DRIESSCHE, J. 1994. Analogue modelling of detachment fault systems and core complexes. *Geology,* **22,** 319–322.

BUCHANAN, P. G. & NIEUWLAND, D. A. (eds) 1996. *Modern Insights into Structural Interpretation, Validation and Modelling.* Geological Society, London, Specal Publications, **99.**

MANDELBROT, B. 1989. Multifractal measures, especially for the geophysicist. *Geophysics,* **1/2,** 5–42.

MANDL, G. 1988. Mechanics of tectonic faulting. *In:* ZWART, H. J. (ed.) *Developments in Structural Geology.* Elsevier, Amsterdam.

McCLAY, K. R. 1990. Extensional fault systems in sedimentary basins. A review of analogue model studies. *Marine and Petroleum Geology,* **7,** 206–233.

Geometry, kinematics and scaling properties of faults and fractures as tools for modelling geofluid reservoirs: examples from the Apennines, Italy

G. CELLO[1], E. TONDI[1], J. P. VAN DIJK[2], L. MATTIONI[1], L. MICARELLI[1,3] & S. PINTI[1]

[1]Dipartimento di Scienze della Terra, Università di Camerino, Via Gentile III da Varano, 62032 Camerino (MC), Italy (e-mail: giuseppe.cello@unicam.it)
[2]Eni Spa, Divisione Agip, San Donato Milanese (MI), Italy
[3]Institut Français du Pétrole (IFP), Rueil-Malmaison, France

Abstract: The study of some major fault zones in the Apennines was mostly focused on the acquisition of quantitative data *in situ* aimed at deriving input parameters for modelling faulted rock volumes. Structural data collected in the Monte Alpi area (southern Italy) and in the central Apennines allowed us to: (i) estimate the fractal dimension characterizing the geometric complexity and size distribution of different fault and fracture patterns; (ii) assess the appropriate parameters defining the overall architecture, anisotropy and related permeability structure of the mapped fault zones; and (iii) constrain the scaling properties of some of the attributes (i.e. length, spacing, map pattern, fracture density, etc.) of both fault-related and regional fracture sets.

The results of our work suggest that the above data are appropriate for extrapolating field based information at different scales, and for producing 3D models of fault and fracture networks.

Systematic studies of fault and fracture characteristics are fundamental for approaching the many problems arising from both research work (i.e. think of the role of fluids in faulting) and from the many possible applications in geofluid reservoirs modelling. In the management of fractured hydrocarbon reservoirs, for example, the need for integrating surface geological data into a decision-making process aimed at optimizing a reservoir model for production purposes, the above statement is obvious. In this case, the advantage of using geostructural data from an exposed analogue is that specific parameters (such as the typology of fracture swarms and damage zones, as well as the three-dimensional relationships between faults and fractures) can be directly measured in the field and then used to supply additional information to available subsurface data sets. As a result, one may tie together information derived from outcrop exposures to those acquired over a scale range of a few centimetres (data from well cores) to tens of metres (data from seismics). The integration of field-based information with subsurface data may greatly improve the estimates of the following parameters: (i) the spatial distribution of fractures, their density and orientation variations throughout a geological structure; (ii) the matrix block size (which is related to the shape and size of the rock blocks in between effectively permeable fracture surfaces); (iii) the efficiency of connectivity within fracture networks (for evaluating drainage volumes); and (iv) fracture porosity.

In order to illustrate a few case studies of fault and fracture analysis focused on obtaining quantitative structural data for modelling faulted rock volumes, we report on some examples from the Apennines, where the Neogene fold and thrust belt structure is dissected by Quaternary faults that are part of regional networks including active structures that are the sources for shallow and intermediate earthquakes.

Geological framework

In this study we report on the results of quantitative geostructural analyses performed in two sectors of

From: NIEUWLAND, D. A. (ed.) *New Insights into Structural Interpretation and Modelling.* Geological Society, London, Special Publications, **212**, 7–22. 0305-8719/03/$15

the Apennines where different tectonostratigraphic units can be recognized (Fig. 1). In the Southern Apennines, these include (Cello & Mazzoli 1999, and references therein):

(1) remnants of a Mesozoic–Palaeogene passive margin succession consisting of (i) shallow-water carbonates of the Apenninic Platform, and (ii) basinal deposits of the Upper Triassic–Lower Cretaceous Lagonegro basin which are made up of cherty limestones (Calcari con selce Formation), siliceous shales (Scisti silicei Formation) and marls and

argillites (Galestri Formation) (Scandone 1972; D'Argenio 1988);

(2) synorogenic (foredeep and satellite thrust-top basin) Miocene successions, unconformably overlying the above units and themselves involved in the thrust structure (Pescatore 1978; Patacca & Scandone 1989; Cello *et al.* 2000).

The main structural features in the area include thrust-related folds and associated fabrics involving the Mesozoic Lagonegro basin units, and a major thrust sheet of Apenninic Platform carbonates and

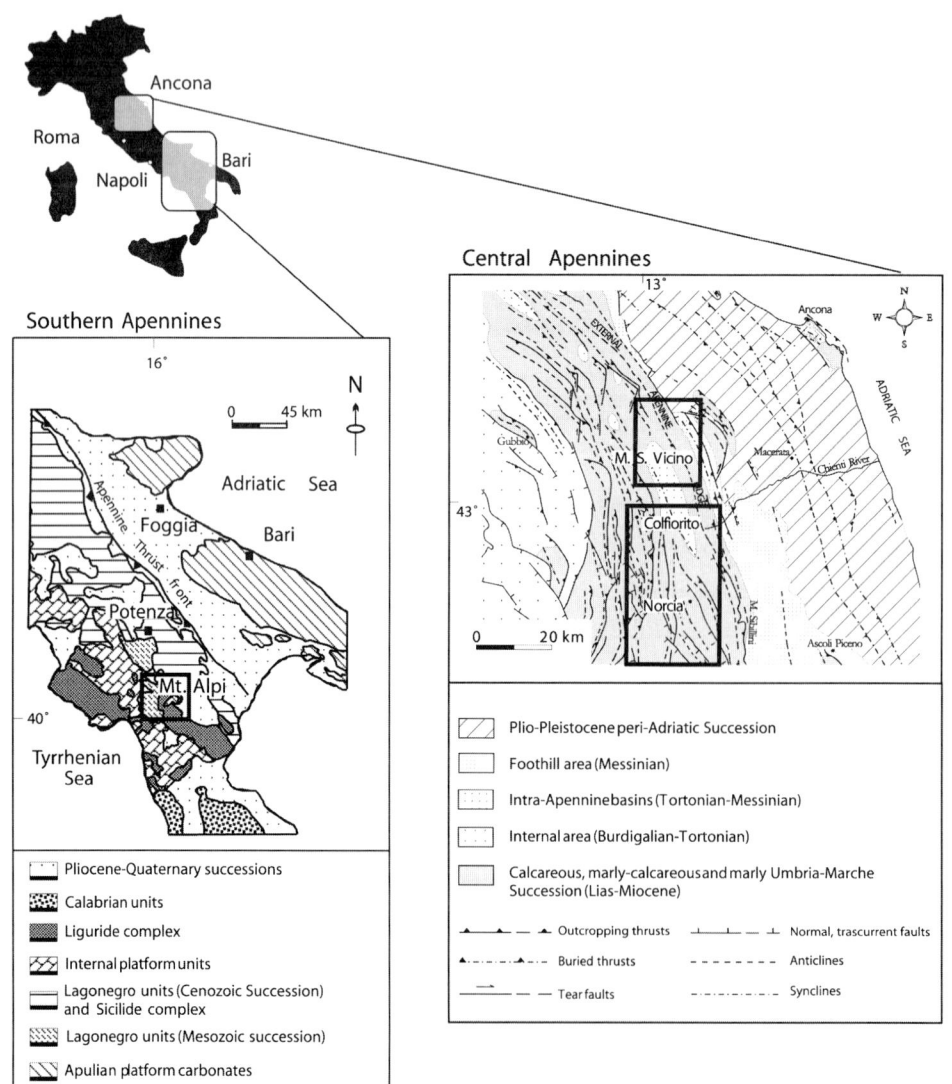

Fig. 1. Simplified geological maps showing the locations of field study areas in the Southern Apennines (after Cello *et al.* 2000) and the Central Apennines (after Deiana & Pialli 1994).

the overlying Miocene marls and clastics, which rests tectonically above the Lagonegro units (Cello *et al.* 2000). The outcropping fold-and-thrust-related features developed during the various stages of thrust belt evolution which, in the area, were active since Middle Miocene times (Cello & Mazzoli 1999, and references therein).

In the Monte Alpi area (Lucanian Apennines), Late Pliocene to Pleistocene faults (i.e. the Monte Alpi Fault System, MAFS) accommodate most of the brittle strain stored during the latest stages of evolution of this sector of the peri-Mediterranean orogen, and control the fragmentation pattern of the crustal volumes undergoing deformation within the current stress regime. The MAFS dissects a kilometre-scale backthrust unit consisting of about 2000 m thick Jurassic–Cretaceous carbonate platform sequence unconformably covered by Upper Miocene terrigenous clastics and pelitic sequences. The backthrust structure developed in Late Pliocene times and is now dissected by various faults which have been active as transtensional and transpressional features up to recent times (Van Dijk *et al.* 2000).

In the Central Apennines (Fig. 1), the fold and thrust structure of the mountain belt is made up of several thrust sheets consisting of different portions of the Umbria–Marche succession (Deiana & Pialli

1994), which is part of a passive margin sequence including (from bottom to top): (i) evaporites (Anidriti di Burano Formation, Upper Trias); (ii) shallow water limestones (Calcare massiccio Formation, Lower Lias); and (iii) pelagic limestones and marls of Jurassic to Middle Miocene age.

In the investigated area of the Central Apennines, thrust-related deformation developed since Middle Miocene–Early Pliocene times (Calamita *et al.* 1994; Deiana & Pialli 1994) and the whole structure of the belt is now dissected by a Late Quaternary fault system (the Central Apennine Fault System, CAFS; Cello *et al.* 1997) which is characterized by geometrically complex fault zones of different size (Fig. 2). Some of the main faults bound intramontane basins infilled with Pleistocene–Holocene continental deposits, and trend mostly from NW–SE to roughly north–south. Long-term displacements recorded across basin-bounding faults are in the range of a few hundred metres, and their kinematics vary from normal motion, on roughly NW–SE trending planes, to transtensional and strike-slip motion on NNW–SSE and north–south oriented faults, respectively. Fault slip data from the main fault planes cutting through middle Pleistocene–Holocene deposits suggest that the present-day stress field is characterized by a NE–SW extension and by a maximum compression

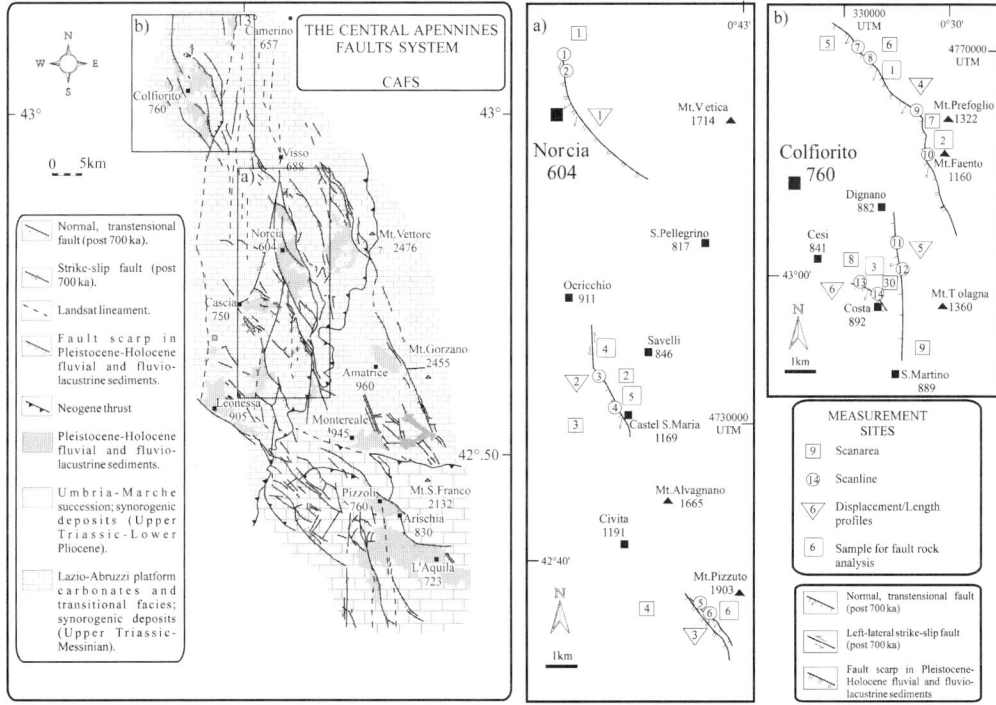

Fig. 2. Geological sketch map of the Colfiorito-Norcia areas (modified after Cello *et al.* 2001*a*).

which is sub-horizontal and oriented roughly NW–SE (Cello et al. 1995, 1997).

In this sector of the Central Apennines, we performed detailed structural analyses in two areas: the Monte San Vicino and Colfiorito–Norcia areas (Fig. 1).

Structural analysis

Methodology

Our study was mostly based on the systematic analysis of fracture and fault characteristics *in situ*.

From a methodological viewpoint, we used fractal statistics (Mandelbrot 1983) as a tool for defining the validity range within which the different attributes of a fault zone may be considered to be fractal (i.e. scale invariant), and quantitative structural analysis for assessing both the deformation history of the investigated areas and the overall properties of fault zones, including the permeability structure of different fault sections (Fig. 3). This latter property is of outstanding importance for evaluating different portions of the hydraulic characteristics of a fault zone (Fig. 4).

Following Caine et al. (1996) it may be assumed

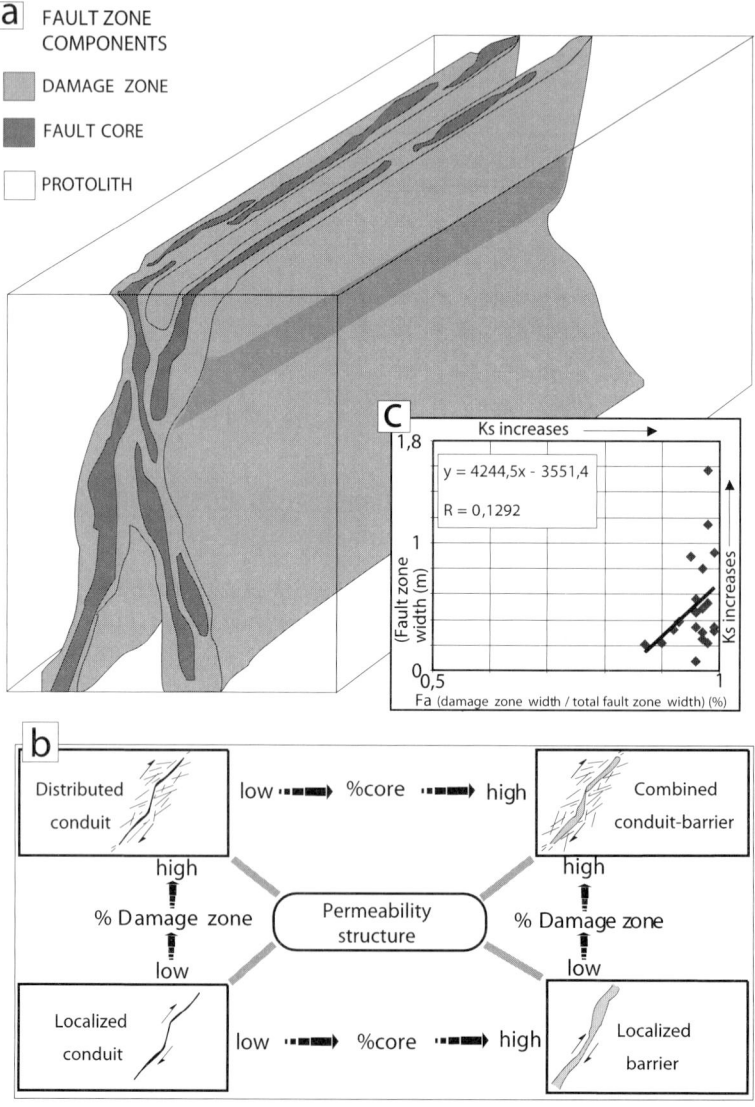

Fig. 3. Fault volume characteristics: (**a**) fault zone components; (**b**) hydraulic properties of conduit-type faults; (**c**) inferred permeability structure (modified after Caine et al. 1996).

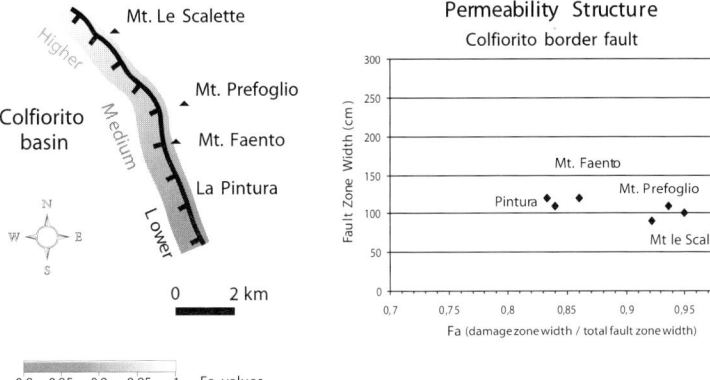

Fig. 4. Fault segmentation model of the Colfiorito basin border fault (Central Apennines) based on the hydraulic properties derived for each fault section (from Cello *et al.* 2001*a*).

that, within a fault zone, the core generally behaves as a barrier to fluid flow, due to grain size reduction and geochemical processes which tend to decrease permeability, whereas within the fault damage zone, permeability is higher and fluid flow is greatly enhanced, due to the occurrence of high density fracture populations. Based on this observation, structural data can be used to assess the ratio between fault core and damage zone thickness and to derive the architectural indices needed to evaluate the permeability structure of fault zones and sections (Fig. 3a).

Another important parameter to be evaluated in order to furnish information on fault and fracture permeability is connectivity. Fluid flow within fracture networks depends on their geometrical properties (i.e. aperture, length, spacing, wall morphology, etc.) and on the degree of interconnectiveness. This latter property can be evaluated semi-quantitatively by assessing the degree of physical connection among fractures within a network (Fig. 5a) and by representing the results of the analysis in a triangle diagram of the type shown in Figure 5b. There, the ratio of the number of fractures which are isolated (type I), simply connected (type II), or multiply connected (type III) over the total number of fractures in the network may be shown (Ortega & Marret 2000).

As concerns fractal statistics, it has been shown by several authors (e.g. Barton & La Pointe 1995; Cowie *et al.* 1996) that the size distribution of fault and fracture populations, as well as their attributes (i.e. length, displacement, thickness, spacing, etc.), are often described by power-law relations of the type: $N(L) = kL^{-D}$, where N is the number of features of size L, k is a constant, and D is the fractal dimension (Mandelbrot 1983). Map patterns also show fractal properties, and their 2D geometric complexity can be analysed by means of a box-

a)

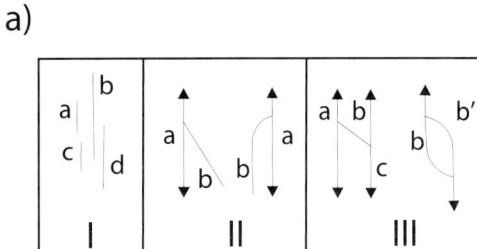

b)

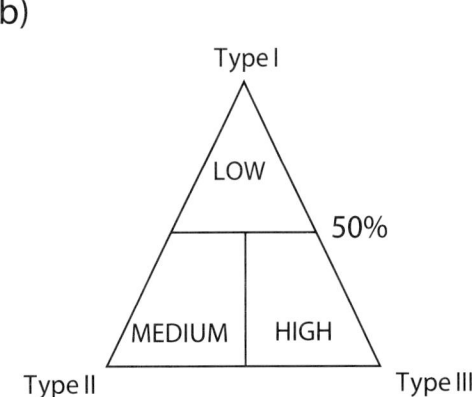

Fig. 5. (**a**) Fracture connectivity diagram (type I – isolated; type II – singly connected; type III – multiply connected fractures); (**b**) evaluation of the degree of connectivity of fracture networks (slightly modified from Ortega & Marret 2000).

counting technique which allows to draw log–log diagrams from which *D* values can be obtained together with their validity range (Walsh & Watterson 1993; Cello 1997). Consequently, systematic analysis of fault and fracture populations allows one to derive (and model) the appropriate scale relations between the different attributes of fault zones and systems.

The main steps in our analysis of a few study areas in the Apennines included: (i) remote sensing studies and high resolution field mapping of fault zones and fault rocks (leading to the production of fault maps where all the outcropping segments with a trace length of more than 100 metres are shown); and (ii) the acquisition of fracture data along both *scan lines* and *scan areas*.

Remote sensing data and fault maps based on field work at different scales allowed us to assess the relative chronology of deformation in the study areas and to discriminate between regional and fault-related structural features. Furthermore, they furnished an appropriate image of the 2D geometry of the fault systems from which we could evaluate their degree of complexity, as derived from box-counting analysis. Additional parameters, collected within fault zone volumes, are: (i) orientation data relative to lower rank features (i.e. Y and Riedel shears, pressure solution cleavages, stylolites, etc.); (ii) occurrence and type (i.e. grain size, texture, etc.) of fault rocks; (iii) azimuth and angle of plunge of mechanical striae/shear fibres; (iv) sense and magnitude of displacement; and (v) fault

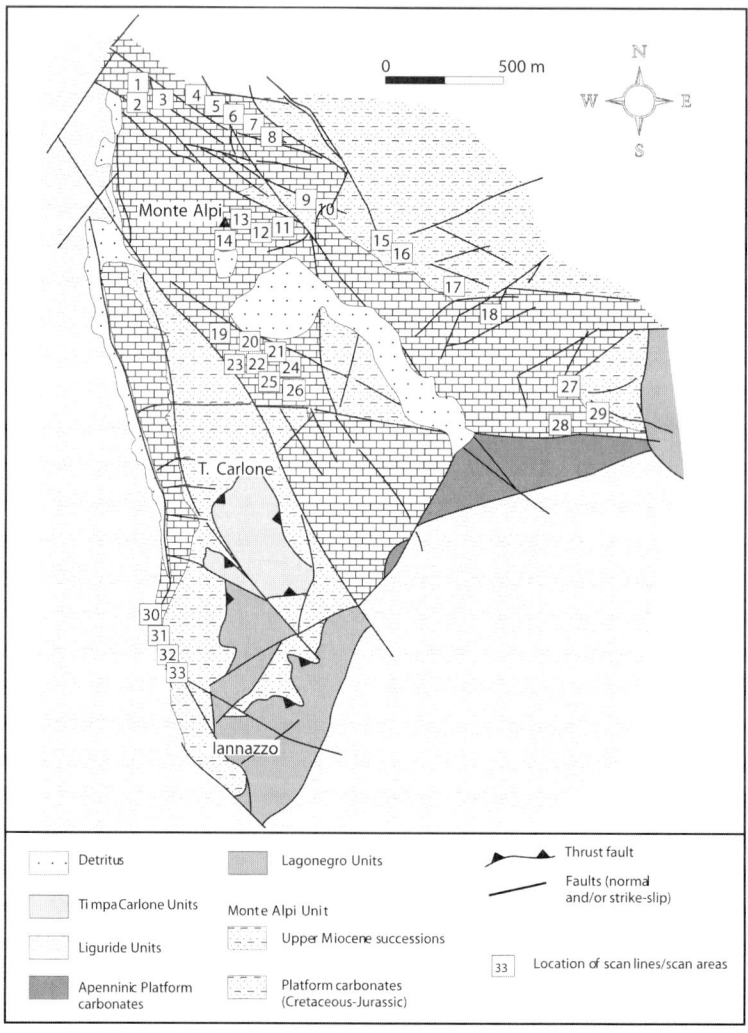

Fig. 6. Geological map of Monte Alpi, showing *scan area* sites (after Van Dijk *et al.* 2000, simplified).

length. The last two parameters, when combined together with fault rock analysis, give information on fault growth processes (Scholz & Gupta 2000; Cowie & Roberts 2001), as well as on the deformation mechanisms operating during fault zone evolution. Fault rock studies are even more effective if combined with fluid inclusion analysis aimed at assessing the P–T conditions of deformation during fluid entrapment. This type of analysis has been used by Cello *et al.* (2001*b* to discriminate among shallow and deep-rooted faults in the axial zones of the Central Apennines.

Scan line data collected across fault zones and *scan area* data from the different lithological units cropping out in the study areas include the following parameters: (i) spatial position; (ii) attitude (dip direction and angle of dip); (iii) typology (minor fault, fracture, cleavage plane, extension vein); (iv) width; (v) occurrence and type of filling; (vi) morphology of discontinuity surface (ranging from very rough to smooth). Additional information from *scan areas* include the 2D fracture pattern images and the length and spacing distribution of the fracture networks. Three corrections were routinely performed on this dataset: length weighting, the Terzaghi correction (removal of sample orientation bias), and tilt removal.

Fault and fracture data

The Monte Alpi area (Southern Apennines) In the Monte Alpi area, structural information on fault and fracture networks were collected from *in situ* analysis of discontinuity features exposed at 33 selected sites (Fig. 6). The Monte Alpi area was chosen as a case study because it is thought to represent a possible analogue of the nearby Val d'Agri oil reservoir; for this reason, a field-based study of the MAFS-related features (Fig. 7) was carried out, for exploration and development purposes, in order to integrate these data with information obtained from *in situ* analysis of subsurface data (including 3D seismics and well log analysis). The resulting database was inserted and calibrated in a 3D GIS-CAD (Geographic Information Systems and Computer Aided Design) environment consisting of three-dimensional geographic framework comprising topographic maps, DTM (Digital Terrain Model), geological maps, aerial photograph lineament maps, and a 3D reconstruction of major fault and fracture surfaces (Fig. 8). From this database, it was concluded that (Van Dijk *et al.* 2000): (i) the analysed fault and fracture sets are part of the same structural fabric which was generated prior to Plio-Quaternary tectonics and then passively rotated; (ii) fracture and fault length distributions (Fig. 9) show that small and large scale features, within the network, form part of one

a) *Scan area* fractures

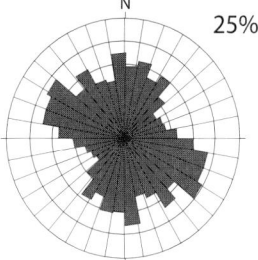

b) Aerial photograph lineaments

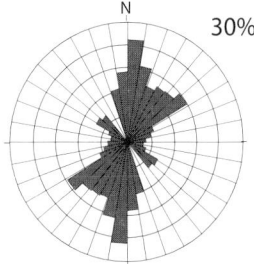

c) Mesostructural faults

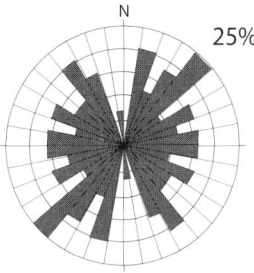

d) Mapped faults

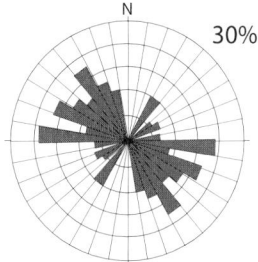

Fig. 7. Rosette diagrams showing the statistical distribution of (**a**) fractures, (**b**) aerial photograph lineaments, (**c**) mesostructural faults and (**d**) major faults (after Van Dijk *et al.* 2000, modified).

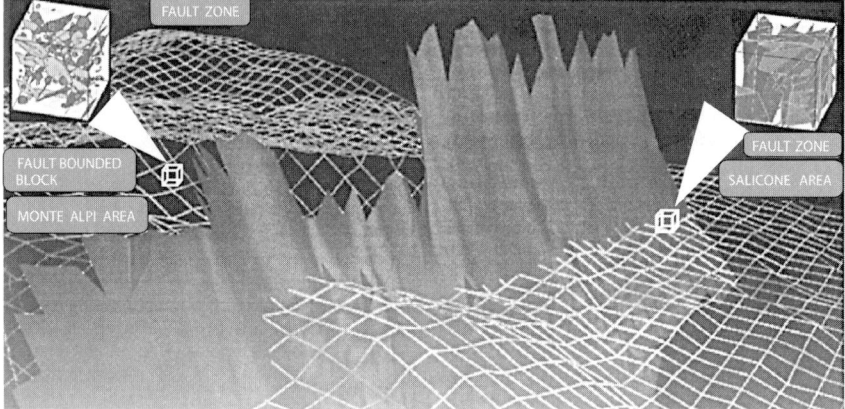

Fig. 8. Discrete fracture network model (slightly modified from Van Dijk *et al.* 2000).

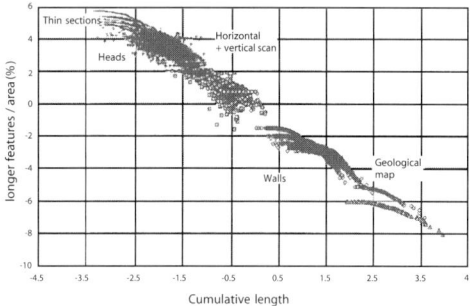

Fig. 9. Fracture frequency diagram. The diagram was constructed by integrating all the information available from both surface and subsurface data (after Van Dijk *et al.* 2000, modified).

scale-invariant system with fractal properties; (iii) no scale gap is present in the data set, which comprises about 30 000 measurements in the 'subseismic' scale range; (iv) the 3D geometry of the analysed network can be described by a fractal dimension of 2.9.

Monte San Vicino–Colfiorito–Norcia areas (Central Apennines) Fault and fracture data collected by means of *scan line* and *scan area* analysis in the Monte San Vicino–Colfiorito–Norcia areas, are illustrated below.

For the Monte San Vicino area (Fig. 10), a few examples from our database are shown in Figures 11–19 and will serve to discuss the main points of the study.

In Figure 11, it is shown that the total fracture field recorded along the selected *scan line* is composed of discontinuity sets which are spatially

associated with specific faults and show frequency values well above 2%, whereas regional sets never exceed this background value. Frequency–distance diagrams of the type shown in Figure 11 are therefore particularly useful for discriminating between fault-related fractures and regional features which developed during thrust belt evolution. Furthermore, this type of analysis furnishes the appropriate data to estimate the extent of the fault damage zone width, which coincides with a drop, in fracture density, to the level of background frequency values.

In the selected examples of *scan line* data of Figure 12, one can see that fault-related discontinuities within the Calcare Massiccio Formation (Lower Lias) are arranged in such a way as to show analogies and differences between fracture patterns belonging to fault segments with different trends and kinematic characters. For example, in the central diagram of Figure 12 it can be seen that the most abundant fractures are Y shears paralleling the main faults trending roughly north–south, whereas subordinate sets trending roughly NW–SE represent pre-existing regional features related to folding during thrust belt evolution. In the diagrams of Figure 12, which refer to NW–SE trending transtensional faults, the most abundant fracture sets include mostly extensional features, whereas the pre-existing Apenninic trend is represented by a few NW–SE oriented fractures and minor faults.

Available data on fault zone thickness in the Monte San Vicino area are shown in Figure 13; they emphasize a power law relation over at least one order of magnitude. In Figure 14 the values of the architectural indices derived from the ratio between fault core and damage zone thickness have

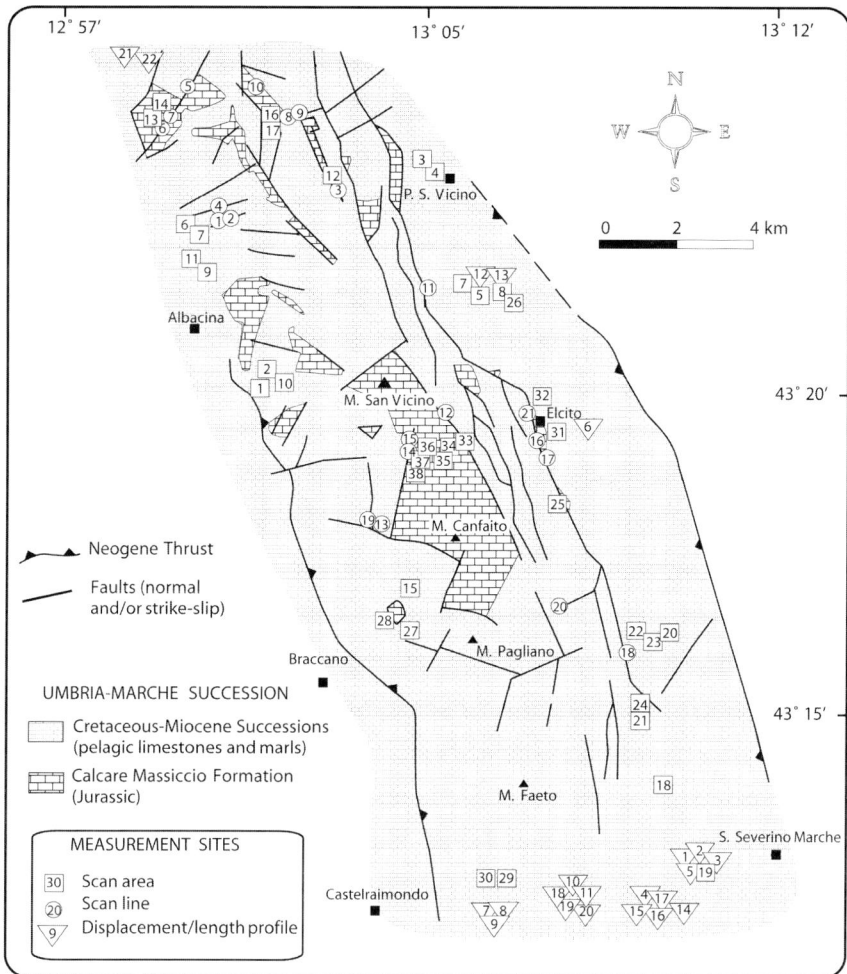

Fig. 10. Geological map of the Monte San Vicino area, showing the locations of survey sites.

been linearly correlated in order to show that all the analysed faults behave as typical conduit-type structures.

Selected examples of *scan area* data collected from different lithological units cropping out in the study areas, are shown in Figure 15; they indicate that most of the extensional fractures may be grouped into two main trends: one oriented roughly north–south, and the other about NW–SE. In the same figure, cumulative length and spacing distributions of the analysed fracture sets are also shown, together with the box-counting curves of the fracture trace map of each *scan area*. From the latter diagrams, one may derive the appropriate power-law relation and the fractal dimension (*D*) of each map pattern.

As concerns the geometrical complexity of

major fault zones, as expressed by the fractal dimension computed from box-counting analysis of fault trace maps at different scales, it can be seen, from Figure 16, that *D* values are quite stable around 1.3. Since box-counting data also give information on the degree of maturity of a given structure (Cello 1997), it is suggested that the calculated *D*-values may be due to a low degree of maturity of the faults belonging to the CAFS.

In Figure 17, the degree of connectivity within fracture networks affecting the shallow water limestones of the Calcare Massiccio Formation and the pelagic lithological units is shown by means of appropriate triangle diagrams. As can be seen, the data relative to the pelagic units indicate that the network is characterized by a high percentage of multiply connected fractures (type III), hence sug-

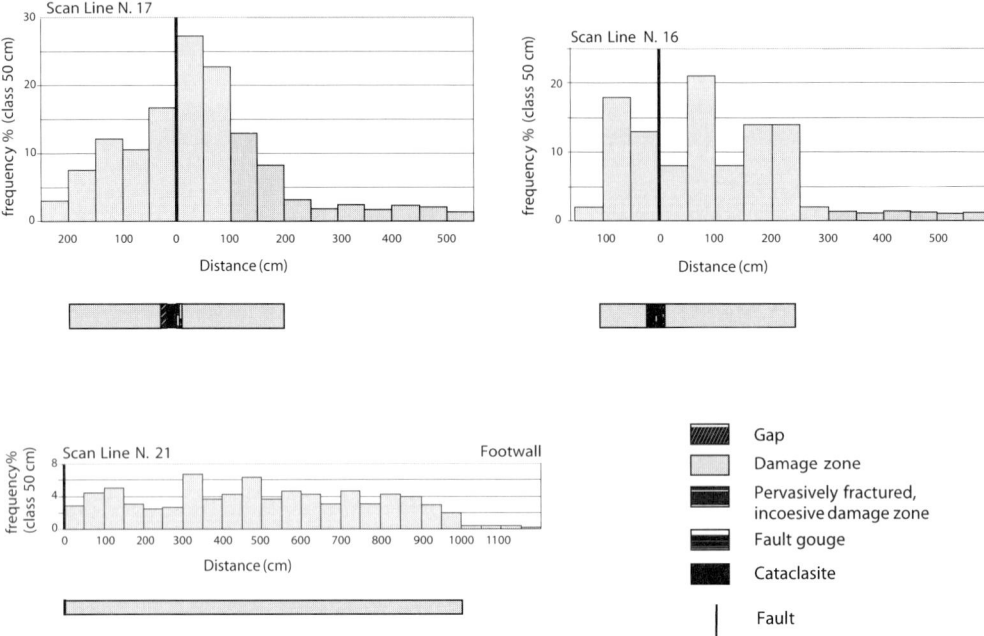

Fig. 11. Spatial distribution of discontinuity surfaces associated with the main faults in the Monte San Vicino area. The horizontal bar records the characteristics and position of the fault rocks. For the location of the *scan line* refer to Figure 10.

gesting that permeability is also high, whereas the data points relative to fracture linkage within the Calcare Massiccio Formation show that the network can be classified as type II.

In Figure 18, a few examples of displacement profiles obtained from outcrop analysis across normal/transtensional faults are shown. As can be seen, except for the case of Figure 18b, most of the profiles are asymmetric, hence suggesting a high degree of interaction between the fault segments cropping out at Monte San Vicino.

As concerns fault-displacement profiles along major (regional) fault segments, a few examples are shown in Figure 19. They were constructed by using both stratigraphic markers and microtopographic measurements across the bounding faults of the Norcia and Colfiorito basin areas (Cello *et al.* 1995). It may be seen that, in the case of the Norcia fault zone, the displacement profiles are all asymmetric, hence indicating strong interaction among the outcropping fault segments of S. Scolastica, Castel S. Maria, and Le Piatenette. In the Colfiorito basin area, however, the overall symmetry of the displacement profiles relative to the border fault segment (Monte Le Scalette–La Pintura fault) suggests that it behaves as a single structure, as

opposed to the other two faults (the Costa and Tolagna faults) which may be interpreted as growing segments interacting with other minor features within the basin (Cello 2000).

Fault rocks associated with CAFS structures were investigated in the field and classified according to the scheme proposed by Sibson (1977); a detailed microstructural analysis was also carried out by means of transmission optical microscopy (Cello *et al.* 2001*a*). In general, rock samples from fault cores in carbonate rocks range from protocataclasite to cataclasite with little calcite cement, 50–60% of matrix, and grain size below 0.5 mm. Samples from damage zones within the Calcare Massiccio Formation consist of highly fractured rocks and fault breccia; this confirms that permeability within damage zones should be quite high, whereas fault cores, which are made up of crush breccia to fine crush breccia consisting of micrite and microspar clasts within a fine-grained matrix, behave essentially as seals to fluid flow. As a rule, fault rocks within damage zones appear to be not well cemented; however, calcite precipitation along veins and microvein networks is observed in some thin sections. These veins can be classified as: (i) extensional veins developed within

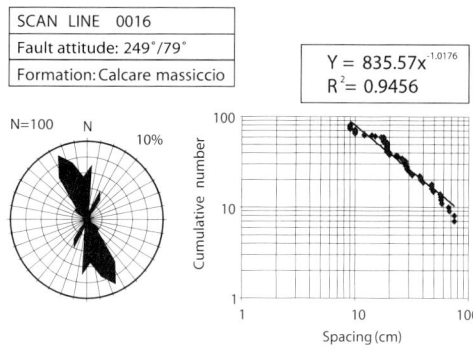

SCAN LINE 0016
Fault attitude: 249°/79°
Formation: Calcare massiccio

$Y = 835.57x^{-1.0176}$
$R^2 = 0.9456$

N=100

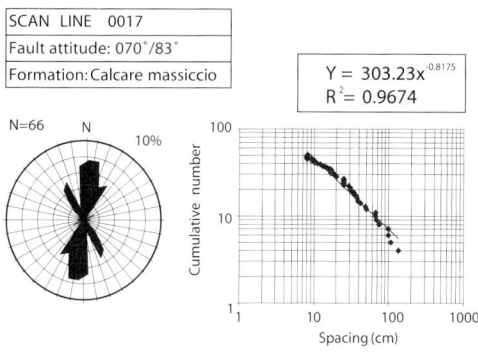

SCAN LINE 0017
Fault attitude: 070°/83°
Formation: Calcare massiccio

$Y = 303.23x^{-0.8175}$
$R^2 = 0.9674$

N=66

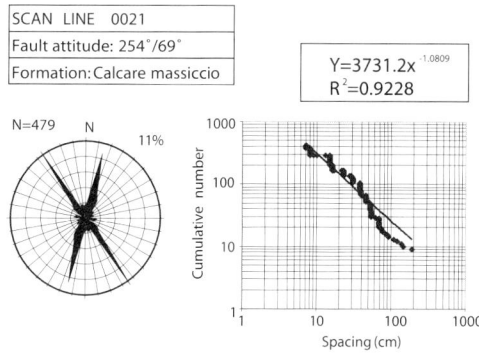

SCAN LINE 0021
Fault attitude: 254°/69°
Formation: Calcare massiccio

$Y = 3731.2x^{-1.0809}$
$R^2 = 0.9228$

N=479

Fig. 12. Orientation data and cumulative spacing diagrams of fault-related discontinuity surfaces in the Monte San Vicino area. For the location of the *scan line* refer to Figure 10.

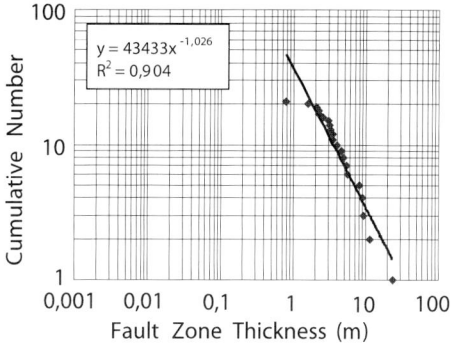

$y = 43433x^{-1.026}$
$R^2 = 0.904$

Fig. 13. Fault zone thickness distribution in the Monte San Vicino area.

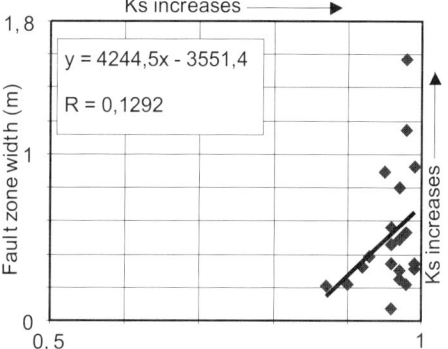

$y = 4244.5x - 3551.4$
$R = 0.1292$

Fig. 14. Permeability structure of some mesoscopic fault zones in the Monte San Vicino area.

Discussion and conclusions

The results of this work confirm that the integration of subsurface and field-based data on fault and fractures is fundamental for modelling faulted crustal volumes and for the management of geofluid resources (oil fields, geothermal and water reservoirs). A few examples from the Apennines are used to illustrate this through the discussion of the following points (which are relevant for possible applications to geofluid reservoir modelling):

(i) the nature and spatial distribution of fault-related fractures and their degree of connectivity;
(ii) the spacing and length characteristics of structural discontinuities;
(iii) the fractal properties of structural networks;

the damage zone; and (ii) veins developed in the fault core. Veins of the first type are filled with calcite blocky crystals showing little internal deformation. The veins in the fault core show a 'pinch and swell' geometry that is thought to be due to fluid flow along ridges and furrows on channelling fault surfaces (Cello *et al.* 2001a).

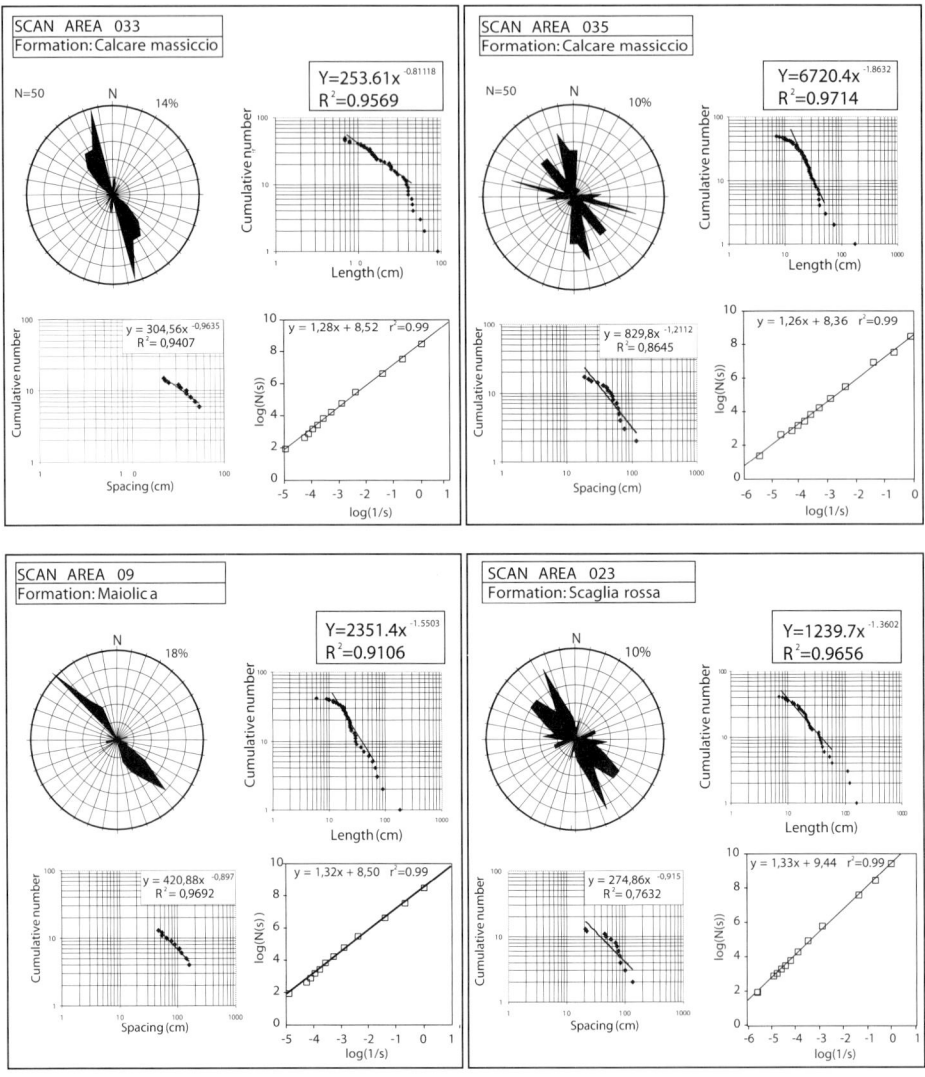

Fig. 15. Orientation data, cumulative length, spacing, and box-counting curves from a *scan area* site located in different lithological units cropping out in the Monte San Vicino area. For the location of the *scan area* refer to Figure 10.

(iv) the length–displacement profiles of multi-scalar faults.

As concerns point (i), we collected information on the orientation, clustering, and interconnectiveness of fracture sets associated with specific fault zones. This dataset, combined with fault rock analysis and information on fault core and damage zone thickness, allowed us to assess the anisotropic properties and the architectural indices characterizing the permeability structure of some of the main faults exposed in the study areas. The results of our work indicate that fault damage zones in carbonate rocks respond mostly as conduits to fluid flow, whereas fault cores behave essentially as hydraulic seals.

As concerns point (ii) we derived the power-law relations to be used for modelling the spacing and length properties of fractures. Our results suggest, however, that these attributes also depend on lithology and bed thickness; consequently, power-law relations show a limited range of validity and their

CAFS

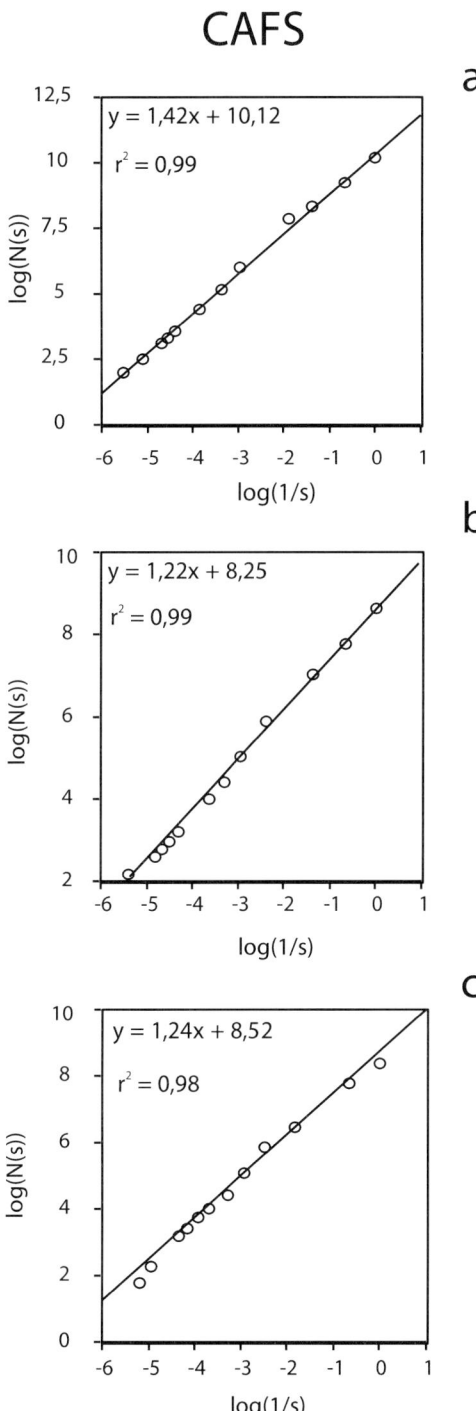

Fig. 16. Box-counting curves from surface trace maps at different scales: (**a**) the whole CAFS, (**b**) the Colfiorito fault array, and (**c**) the Norcia–L'Aquila set (after Cello *et al.* 2001*a*).

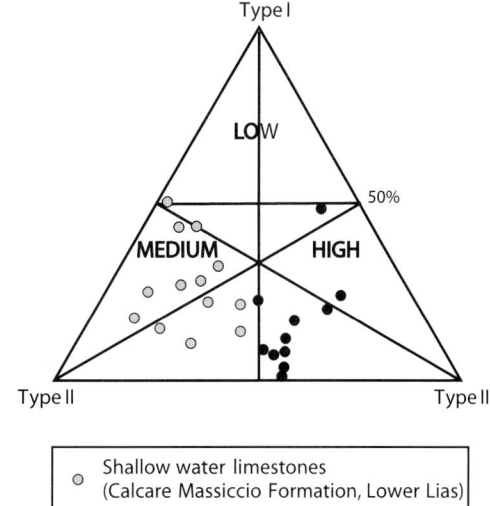

Fig. 17. Fracture connectivity diagram relative to two different lithologies. The analysed 2D patterns of the fracture networks were derived from image analysis of *scan areas*.

use must be integrated with information on the sedimentary successions affected by fracturing.

As concerns point (iii) we analysed information on the size distribution of fault and fracture populations, as well as their geometric complexity, as shown in various maps constructed at different scales (from outcrop to regional scales). Our results, obtained by means of 2D fractal statistics, suggest that small and large scale features in the Apennines are part of several fractal systems, each characterized by a specific fractal dimension (*D*) and by a different degree of evolution. This information, which can be expressed in terms of specific power-law relationships, is extremely useful for modelling fault and fracture distributions within a given rock volume.

As concerns point (iv) we established the appropriate relations between slip on a fault and its length; data were collected at both the mesoscale and the regional scale. The results obtained from displacement profiles confirm the general notion that slip tends to be larger at the centre of a fault and to decrease towards the tip zones. The distribution pattern of slip data, however, can be either symmetric or asymmetric; in the first case, the displacement profile records the evolution of single isolated faults (Monte Scalette–La Pintura fault),

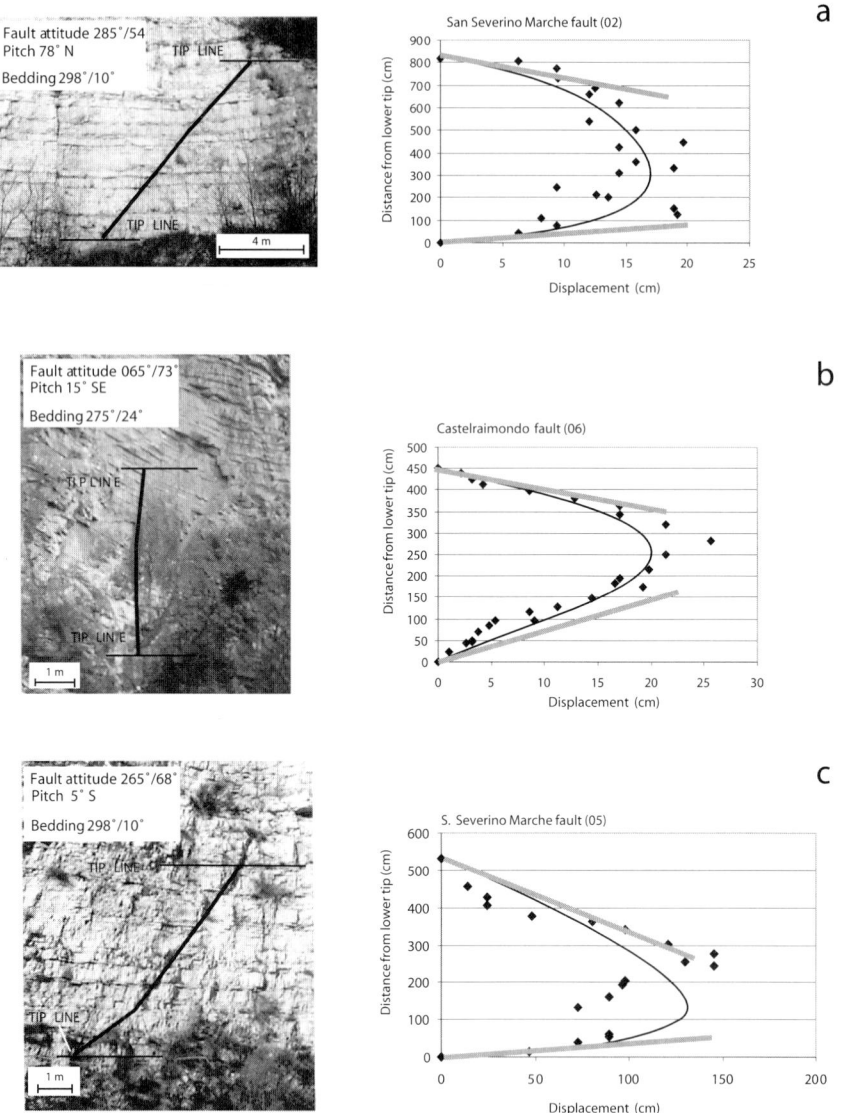

Fig. 18. Length–displacement profiles across mesoscopic faults in the Monte San Vicino area.

whereas asymmetric patterns record the growth of interacting fault segments (S. Scolastica, Castel Santa Maria, and Le Piatenette faults).

In conclusion, based on our results and observations, we emphasize that: (1) systematic geostructural analysis of well exposed fault and fracture networks, in the Apennines, yielded the appropriate information needed to separate regional fracture patterns related to thrust belt evolution from fault-related discontinuity sets which developed in response to the stress conditions operating in the study areas since Quaternary times; (2) *in situ* analysis of the geometric and kinematic characteristics, as well as of the spatial and scaling properties of faults and fractures, integrated with fault rock studies, provided the basic parameters needed for assessing the overall architecture and permeability structure of fault zones, and for modelling fault-dominated crustal volumes by means of power-law based fracture generators.

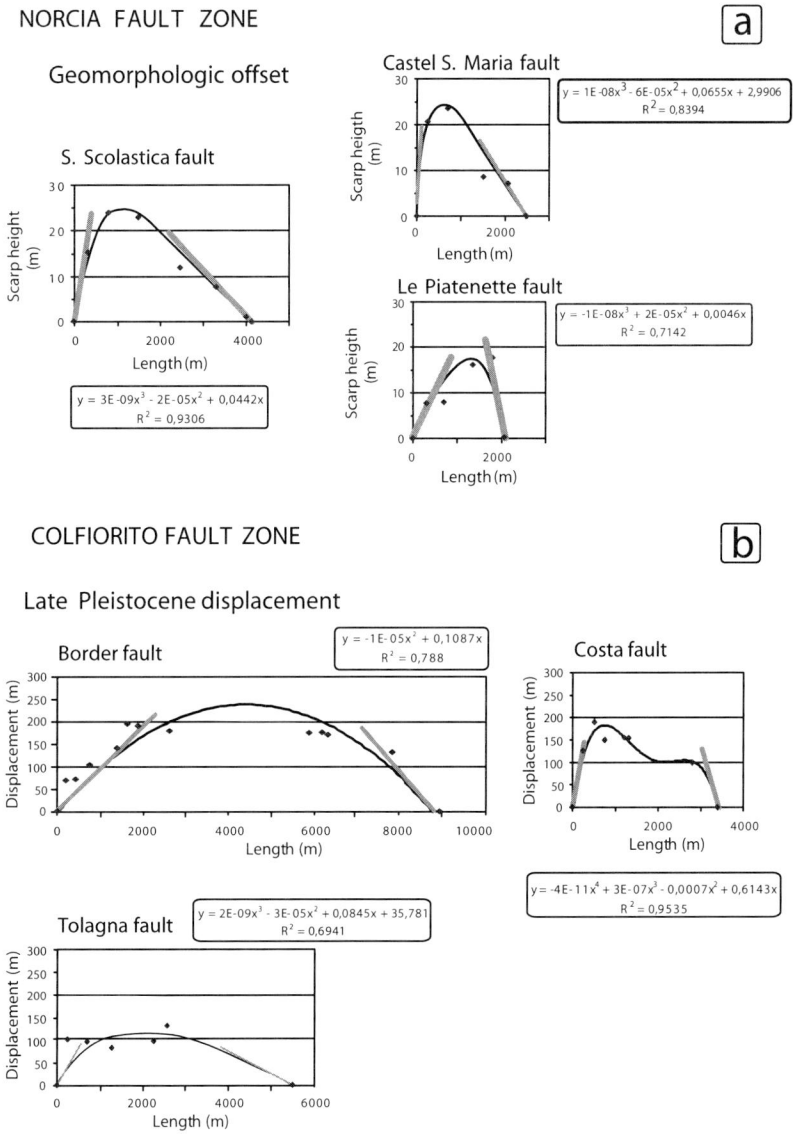

Fig. 19. Length–displacement profiles across the active faults of the (**a**) Norcia and (**b**) Colfiorito basin areas (after Cello *et al.* 2001*b*).

References

BARTON, C. A. & LA POINTE, P. R. 1995. *Fractals in the Earth Sciences*. Plenum Press, New York.

CAINE, J. S., EVANS, J. P. & FORSTER, C. B. 1996. Fault zone architecture and permeability structure. *Geology*, **11**, 1025–1028.

CALAMITA, F., CELLO, G., DEIANA, G. & PALTRINIERI, W. 1994. Structural styles, chronology rates of deformation, and time–space relationships in the Umbria–Marche thrust system (central Apennines, Italy). *Tectonics*, **13**, 873–881.

CELLO, G. 1997. Fractal analysis of a Quaternary fault array in the central Apennines, Italy, *Journal of Structural Geology*, **19**(7), 945–953.

CELLO, G. 2000. A quantitative structural approach to the study of active fault zones in the Apennines (peninsular Italy). *In*: CELLO, G. & TONDI, E. (eds) *The Resolution of Geological Analysis and Models for Earthquake Faulting Studies, Journal of Geodynamics*, **29**(3/5), 265–292.

CELLO, G. & MAZZOLI, S. 1999. Apennine tectonics in southern Italy: a review. *Journal of Geodynamics*, **27**, 191–211.

CELLO, G., MAZZOLI S., TONDI, E. & TURCO, E. 1995. Tettonica attiva in Appennino centrale ed implicazioni per l'analisi della pericolosità sismica del settore assiale della catena umbro-marchigiana-abruzzese. *Studi Geologici Camerti,* **13,** 115–138.

CELLO, G., MAZZOLI, S., TONDI, E. & TURCO, E. 1997. Active tectonics in the central Apennines and possible implications for seismic hazard analysis in peninsular Italy. *Tectonophysics,* **272,** 43–68.

CELLO G., GAMBINI R., MATTIONI L., MAZZOLI S., READ A., TONDI E. & ZUCCONI V. 2000. Geological analysis of the High Agri Valley (Lucania Apennines, Southern Italy). *Memorie della Società Geologica Italiana,* **55,** 149–155

CELLO, G., INVERNIZZI C., MAZZOLI, S. & TONDI, E. 2001*a.* Fault properties and fluid flow patterns from quaternary faults in the Apennnines, Italy. *Tectonophysics,* **336,** 63–78.

CELLO, G., TONDI, E., MICARELLI, L., & INVERNIZZI, C. 2001*b.* Fault zone fabrics and geofluid properties as indicators of rock deformation modes. *Journal of Geodynamics,* **32,** 543–565

COWIE, P. & ROBERTS, G. 2001. Constraining slip rates and spacings for active normal faults. *Journal of Structural Geology,* **23,** 1901–1915.

COWIE, P. A., KNIPE, R. J. & MAIN, I. G. (eds) 1996. Scaling laws for fault and fracture populations: Analyses and applications. *Journal of Structural Geology,* Special Issue, **18**(2/3), 135–383.

D'ARGENIO, B. 1988. L'Appennino campano lucano. Vecchi e nuovi modelli geologici tra gli anni sessanta e gli inizi degli anni ottanta. *Memorie della Società Geologica Italiana,* **41,** 3–15.

DEIANA, G. & PIALLI, G. 1994. The structural provinces of the Umbro–Marchean Apennines. *Memorie della Società Geologica Italiana,* **48,** 473–484.

MANDELBROT, B. B. 1983. *The Fractal Geometry Of Nature.* Freeman, New York.

ORTEGA, O. & MARRET, R. 2000. Prediction of macrofracture properties using microfracture information, Mesaverde Group sandstones, San Juan basin, New Mexico. *Journal of Structural Geology,* **22**(5), 571–588.

PATACCA, E. & SCANDONE, P. 1989. Post-Tortonian mountain building in the Apennines: the role of the passive sinking of a relic lithospheric slab. *In*: BORIANI, A., BONAFEDE, M., PICCARDO, G. B. & VAI, G. B. (eds) *The Lithosphere in Italy.* Accademia Nazionale dei Lincei, Advances in Earth Sciences Research, **80,** 157–176.

PESCATORE T. 1978. Evoluzione tettonica del Bacino Irpino (Italia Meridionale). *Bollettino della Società Geologica Italiana,* **97,** 783–805.

SCANDONE P. 1972. Studi di geologia Lucana: carta dei terreni della serie calcareo-silico-marnosa e note illustrative. *Bollettino della Società dei Naturalisti in Napoli,* **81,** 225–300.

SCHOLZ, C. H. & GUPTA, A. 2000. Fault interactions and seismic hazard. *In*: CELLO, G. & TONDI, E. (eds) *The Resolution of Geological Analysis and Models for Earthquake Faulting Studies, Journal of Geodynamics,* **29**(3/5), 317–323.

SIBSON, R. H. 1977. Fault rocks and fault mechanisms. *Journal of Geological Society of London,* **133,** 191–231.

WALSH, J. J. & WATTERSON, J. 1993. Fractal analysis of fracture patterns using the standard box-counting technique: valid and invalid methodologies. *Journal of Structural Geology,* **15**(12), 1509–1512.

VAN DIJK, J. P., BELLO, M., TOSCANO, C., BERSANI, A. & NARDON, S. 2000. Tectonic Model and 3D Fracture Network Analysis of Monte Alpi (Southern Apennines). *Tectonophysics,* **324,** 203–237

An improved regional structural model of the Upper Carboniferous of the Cleaver Bank High based on 3D seismic interpretation

B. M. SCHROOT & H. B. DE HAAN

Netherlands Institute of Applied Geoscience TNO – National Geological Survey, PO Box 80015, 3508 TA Utrecht, The Netherlands (e-mail: b.schroot@nitg.tno.nl and h.dehaan@nitg.tno.nl)

Abstract: The use of 4500 km^2 of amalgamated 3D seismic surveys allowed for an improved intra-Carboniferous seismic interpretation of the Dutch Cleaver Bank High, which is part of the Southern North Sea Carboniferous Basin. The observations of faults that were active during the Late Carboniferous are reviewed in the context of what is described in literature about the regional structural framework of the basin. The high quality seismic data show at least three distinct fault trends, namely east–west, NE–SW and NW–SE, active before Rotliegend times. All of these trends are inherited from older existing zones of weakness, and furthermore, all three trends have been reactivated again during the Mesozoic or Cenozoic to some extent. The interpretation of major controlling east–west shear zones that have been reactivated in different senses throughout geological history is the result of careful examination of the data. The dominance of such systems is not obvious on simple fault maps. Their role in the regional plate tectonics fits the model of escape tectonics of the North Sea–Baltic plate. The distinction of different styles of Late Carboniferous normal fault systems results in a better understanding of the different tectonic phases during the period from Westphalian D to Autunian.

The structural geology of Carboniferous rocks in NW Europe is described in detail in the literature, mainly based upon outcrops in the UK and Germany. However, much less is known about the characteristics and the structural development of the extension of the Carboniferous in the Southern North Sea Carboniferous Basin (SNSCB), where these rocks are buried at a depth of some 3500–4000 m. All knowledge is derived from scattered boreholes and from seismic surveys. Obviously, the quality of the latter is crucial when it comes to unravelling the structural development and style of the sequences. Until the early 1990s seismic resolution of the Upper Carboniferous in the basin was rather limited, and as a result the structural models were simple. However, when higher quality 3D data became available, it became obvious that the structural complexity is much higher than previously assumed.

In this paper we review the regional structural models for the Carboniferous of the SNSCB, and show examples of the intra-Carboniferous structural style from a local study area of about 4500 km^2 of amalgamated 3D surveys in the Cleaver Bank High. The aim is to unravel the pre-Permian

tectonic framework of the area. It will be demonstrated that the use of 3D seismic data leads not only to a higher degree of detail in the structural models, but also to some significantly different interpretations. Some implications to the regional structural model will then be discussed.

Exploration history of the area

The area of the Dutch Cleaver Bank High (CBH) in the northern part of the central SNSCB, consisting of (parts of) the Dutch quadrants D, E, J and K and parts of the UK quadrants 43 and 44, was one of the first areas which was specifically explored for Upper Carboniferous gas reservoirs since the beginning of the 1980s. In the area the sequences subcropping the Base Permian unconformity (BPU) are of Westphalian A–Stephanian age. Despite the relatively high exploration risk, this area north of the Rotliegend fairway was attractive for a number of reasons (Besly 1998). First of all the Carboniferous source rocks are gas-prone. Secondly, the Rotliegend shales and Zechstein evaporites constitute an excellent seal. Finally, in a large part of this area the deformation

From: NIEUWLAND, D. A. (ed.) *New Insights into Structural Interpretation and Modelling.* Geological Society, London, Special Publications, **212**, 23–37. 0305-8719/03/$15
© The Geological Society of London 2003.

of the overburden (Permian and younger) has been relatively mild, thus giving seismic imaging a good chance, and limiting the risk of migration paths through the seal. At first sight it also seemed that in this stable platform setting the structural deformation of the Upper Carboniferous rocks had only been mild, something which would have made the intra-Carboniferous correlation along seismic profiles easier. As a matter of fact, this later turned out to be a severe underestimation of the internal structural complexity of the sequences, due to the inability to properly image the rocks at some 3500 m depth, underneath thick Zechstein salt, with the 2D seismic method of the time.

The first wells were drilled in the 1980s on base Zechstein structural highs, easily identifiable on 2D seismic data. The Permian structural style almost exclusively controlled the exploration concepts, simply because not much could be reliably correlated within the pre-Permian sequences. A number of discoveries (mostly on the UK side of the median line) were made, and stimulated further exploration. The quality of 2D seismic surveys started to improve, and in some cases allowed for the interpretation of intra-Carboniferous reflectors across a field or a licence block. However, it was still extremely difficult to make a reliable intra-Carboniferous structural interpretation consistent over a larger area.

A major change occurred in the early 1990s, when 3D seismic surveys were acquired over the area. The tremendous success of the 3D seismic method elsewhere persuaded the operators to rapidly apply this new technique to the Carboniferous play of the Cleaver Bank High. By 1995 most of the area of interest on the CBH had been covered by 3D seismic surveys. It is now possible to make more detailed and more reliable intra-Carboniferous structural interpretations based on seismic data.

Regional tectonic setting

The Carboniferous rocks of the Southern North Sea Basin underwent a long and complex history of multiphase structural deformation. What can be observed today on seismic data and in outcrops is the cumulative result of superposed tectonic events which affected the area since Carboniferous times. Since it is generally assumed (e.g. Bartholomew et al. 1993; Coward 1993) that the reactivation of structural features inherited from pre-Carboniferous times is a major factor determining the distribution of later faults, the tectonic events of the Early Palaeozoic and even the Proterozoic are considered to be relevant too.

Whereas much of the structural history of the Late Palaeozoic and later episodes can be derived directly from seismic data, the events which took place before the Late Palaeozoic can only be deduced in a general sense from plate tectonic reconstructions. Most authors (Ziegler 1978; Coward 1993; Besly 1998) agree that the dominant events during the Palaeozoic, responsible for the overall tectonic framework, are the Caledonian (550–400 Ma) and Variscan (400–300 Ma) orogenies. Coward (1993) also suggests that already Middle and Late Proterozoic tectonic events, namely the Laxfordian and Svecofennian (1800–1600 Ma), and the Grenville (1100–900 Ma) episodes respectively, influenced the later tectonic framework, leaving structural features in the basement rocks. Here we will focus on the Caledonian and Variscan orogenies.

Caledonian belt

The Late Silurian to Early Devonian Caledonian orogeny resulted in a 'Northern Continent' (called Laurasia), which was a suture of the North American–Greenland plate with the Fennoscandian–Russian plate (Ziegler 1978). The main Caledonian orogenic belt links Scotland to Norway, trending SW–NE. The Caledonian orogeny can be summarized as a process of closure of an early Atlantic (Iapetus) Ocean by the accretion of magmatic island arcs and old continental fragments onto the North American continental craton, in a dominantly NW direction. Sinistral strike-slip movements took place. In addition to the main orogenic belt, a branch running from southern Norway through the central Southern North Sea (SNS) connects to the Caledonides of north Germany and Poland, and a third branch further south, called the Mid-European Caledonides, links the German Caledonides and the Ardennes to the British Isles (Ziegler 1978). From such a plate-tectonic model it would thus appear that two Caledonian structural grains (NE–SW and north–south to NW–SE respectively) are relevant to the later SNSCB. This seems to be confirmed by the pre-Carboniferous modelling of gravity and magnetic data of the southern North Sea by Horscroft et al. (1992), which revealed the same trends. When considering the Carboniferous basin-bounding faults and their underlying basement grain, Corfield et al. (1996) distinguished three different structural domains in and around the British Isles, each of them having their own local dominant basement lineament. The Southern North Sea would be within their Tornquist Domain; with a dominant NW–SE structural grain inherited from the Lower Palaeozoic thrust and fold belt.

The branch of the Caledonian orogenic belt linking the Mid-European Caledonides to the British Isles probably constitutes the low-grade metamorphic basement to the later Carboniferous deposits in the SNSCB. It was intruded by numerous post-orogenic late-Caledonian plutons (Ziegler 1978). One of these granites has been drilled in Dutch well A17-1. That pluton was given an age of 350 Ma, somewhat younger than Caledonian, a difference which might be explained by the occurrence of a mild intra-Carboniferous thermal event (Leeder & Hardman 1990).

Devonian–Early Carboniferous tectonics

The period of Caledonian orogenesis and the collision of plates was followed by large-scale wrench tectonics in the North Atlantic during the Devonian. Sinistral displacements along faults such as the Great Glen fault of hundreds of kilometres are reported (Harland & Gayer 1972). By the end of the Devonian these movements had ceased, and the area north of the future Variscan deformation front (including the SNSCB) showed the characteristics of a relatively stable plate. The next phase was the purely extensional rifting during the Late Devonian and Dinantian. It has been observed (Leeder & Hardman 1990) that many of the structural highs of this rifting phase seem to have one of the Late Caledonian plutons as a core. Figure 1 shows the schematic configuration of highs and lows during the Early Carboniferous. The dominant NW–SE

structural grain already existed in the most of the SNS, and in the area of the Dutch Cleaver Bank High this structural trend defined the boundary faults of the blocks.

Fraser & Gawthorpe (1990) noted that during the late Devonian rifting phase the north–south extensional stress regime reactivated both the NW–SE and the NE–SW trending zones of weakness, which had been inherited from the Caledonian orogeny. Strongly asymmetric grabens were formed. East–west trending basins have been well described on the British onshore (e.g. Soper et al.1987; Fraser & Gawthorpe 1990; Hollywood & Whorlow 1993). The extension is put into the context of the back-arc basin, which related to northward directed subduction, closing the Rheic Ocean. Hundreds of kilometres to the south the Variscan orogenic belt had become the dominant expression of a new collision system. The African portion of Gondwanaland collided with the European portion of Laurussia.

Middle and Late Carboniferous tectonics

From the Namurian to Westphalian C there was a long period of relative tectonic quiescence in the SNSCB. This period of roughly 20 Ma can be described as a classical sag phase according to the model of McKenzie (1978). Uniform thermal sub-

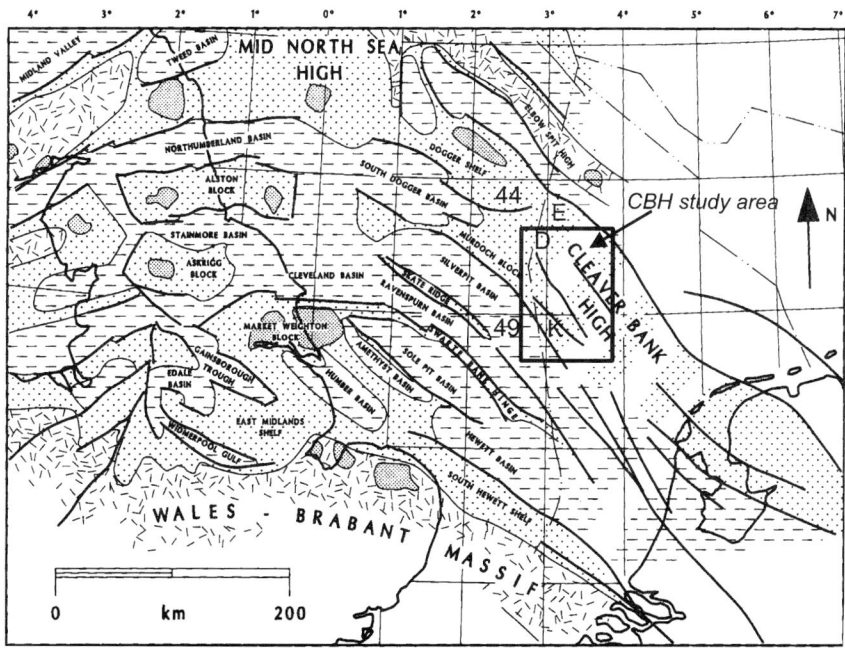

Fig. 1. Schematic structural framework during the Early Carboniferous (from Besly 1998), showing the distribution of basins (dashed) and platforms (dotted). Structural highs are often located over Late Caledonian granite masses. The CBH study area is labelled.

sidence occurred across the SNSCB, related to a gradual cooling and strengthening of the lithosphere. During Namurian, Westphalian A and B only subtle tectonic events affected the area, and Late Carboniferous post-rift sequences totalling several kilometres in thickness were deposited all over the slowly subsiding basin. This basin-fill comprises first the basal Namurian transgressive marine shales, infilling any morphological depressions remaining of the half-grabens, and the subsequent Westphalian continental deposits with a northerly source. Paralic conditions prevailed throughout most of the Westphalian, regularly interrupted by short-lived marine incursions (Ziegler 1978). It is mainly within these Late Carboniferous sequences that we can make our observations of tectonic activity. In our North Sea study area Namurian deposits are about the oldest rocks which can be identified on seismic profiles. For this reason it is also difficult to establish to what extent the morphology of the assumed Lower Carboniferous half-grabens still influenced the post-rift fill in the SNSCB.

During the evolution of the Variscan orogen the deformation front migrated gradually northwards and the back-arc seaway was closed and deformed into major nappe complexes. The loading imparted by these complexes led to the formation of a flexural foreland basin, which also migrated northwards (Besly 1998). By the Late Westphalian this flexural foreland basin had reached its northernmost position. Soon afterwards, probably during the Mid-Stephanian, Variscan inversion affected the onshore UK area, and folding and reverse faulting started there. Besly (1998) refers to an influence of west–east compression related to the Ural orogeny with respect to this compressional tectonic phase. Corfield *et al.* (1996) noted that the severity of Late Westphalian inversion depended on the orientation of the basement grain with respect to the dominantly NW–SE to NNW–SSE direction of maximum shortening. In the SNSCB with its NW–SE grain there was less strong inversion than more to the west, and the inversion which did take place had a significant component of oblique slip. They concluded that the trends of the inversion anticlines were determined by the local basement grain rather than by the direction of maximum shortening. Their estimate of the uplift in the SNSCB is 1–2 km.

Fault reactivations

Many authors have argued that it is most striking to see how much fault reactivation took place during the deformation history of the Carboniferous rocks. At least three different fault trends (east–west, NW–SE and SW–NE) have been reactivated

over and over again throughout geological history. If only the Carboniferous is considered, the question arises why features have been reactivated which were far from perpendicular to the assumed prevailing maximum stress direction associated with the Variscan events. One explanation, supported by Bartholomew *et al.* (1993), Oudmayer & de Jager (1993) and Quirk (1993), for example, would be that the existing zones of weakness were repeatedly reused because they were already there.

While there are certainly good indications to support this view, and this mechanism probably occurred at some scale, there is an interesting alternative model brought forward by Coward (1990, 1993) for at least those reactivations that occurred during the Carboniferous. He postulated a regional plate tectonic model, which is more complicated than a simple Variscan collision model with just one dominant (north–south to NW–SE oriented) stress regime. This model involves escape tectonics of a triangular plate fragment named the North Sea–Baltic block, comprising the present North Sea area, the Baltic area, and most of Britain and Scandinavia. During the Devonian and Early Carboniferous this block would have been squeezed out in an eastward direction from the colliding plates of Gondwanaland and Laurasia (Fig. 2). This lateral expulsion was also accompanied by a rotation of the block.

The implications of this model obviously are important strike-slip components in the movements. The prevailing stress regime would have been dextral transtension along the southern boundary of the block (close to the southernmost

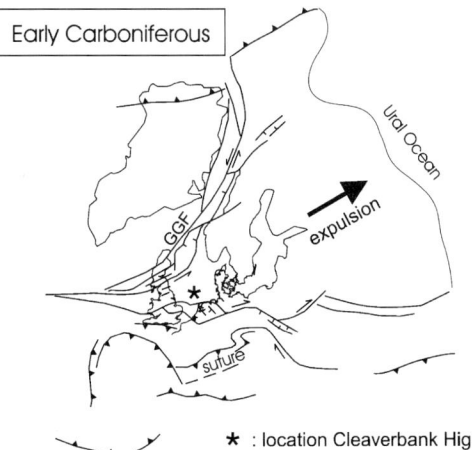

Fig. 2. Simplified Early Carboniferous tectonic map illustrating the continental expulsion of the North Sea–Baltic block, resulting in dextral transpression along the shear zones of its southern boundary, close to the CBH (after Coward 1993).

part of the North Sea area) and sinistral transtension along the northern systems such as the Great Glen fault. The release from the intendor also allowed the block to expand in the NW–SE direction. Along the shear systems pull-apart basins developed. The ones in the SNSCB would have been NW trending and under dextral transtension. Maynard *et al.* (1997) subscribe to this escape tectonics model to a large extent, but mention north–south extension within the plate (also due to the release of the confining forces as the block moved eastwards). They noted that the configuration in the Carboniferous may have been quite similar to what can be observed today in the eastern Mediterranean, where the Turkish and Greek plates are in a process of westward translation. Whatever tectonic model is accepted, it is evident that the development of the SNSCB basin was complex, and that different stress regimes occurred at the same time in the nearby region, affecting the smaller sub-basins in different ways. Maynard *et al.* (1997) noted that a simple model such as the two-phase one, involving first back-arc related north–south extension, and then thermal subsidence appears oversimplified.

In Coward's (1993) model the same wedge-shaped block was inserted back into the collision zone during the Stephanian–Autunian, because of the stress produced from the east by the Ural plate collision (Fig. 3). The sense of shear along the boundary systems reversed, and this second phase in the model fits the Late Carboniferous inversion in the SNSCB related to the approach from the south of the Variscan front, but again implies a much more complicated stress distribution than simple north–south compression. This model would imply that in the SNSCB the tectonic inversion, at the very end of the Carboniferous, must have been accompanied by left-lateral strike-slip along the existing (reactivated) east–west shear zones, such as the east–west running margins of the pull-apart basins. Leeder & Hardman (1990) postulated that some of the synrift Dinantian normal faults must have been reactivated as thrusts during the shortening phase in the Late Westphalian. Fraser & Gawthorpe (1990) report examples of such inversion from the British onshore in particular along NE–SW trending faults, which were perpendicular to the direction of maximum compressive stress. It may very well be that this thrusting had significant strike slip components along some of the fault systems.

As Besly (1998) noted, compressional tectonics started to dominate the basin from the mid-Westphalian B, but the timing of the final Variscan inversion and deformation is poorly constrained. This relates to the lack of geological information from about Westphalian D–Early Rotliegend, and therefore it is difficult to unravel exactly what happened when, during the period of 30–40 Ma (the magnitude of the hiatus depending on the position within the basin) of which little or no record was preserved. What is clear, however, is that the compression resulted in large-scale folding, uplift and erosion. The Base Permian unconformity (BPU) at the base of the Upper Rotliegend sequence is one of the most clearly defined structural unconformities within the entire stratigraphic column in the SNSCB.

Superimposed on the effects of pre-Permian deformation are the accumulated effects of all the Mesozoic and Cenozoic tectonic events. Their history can much more easily be studied by looking at the sequences younger than Carboniferous (Glennie 1998), and it is not the subject of this paper. In summary, it can be mentioned that after the Carboniferous the Cleaver Bank High underwent the same major phases of deformation known elsewhere in the SNS. It seems that during these subsequent tectonic phases the fault systems that were already present during Carboniferous were generally reactivated at least once again, and probably more than once in many cases. Most important are the Kimmerian rifting phases (during the Jurassic and Early Cretaceous), related to the opening and widening of the Atlantic Ocean, the Sub-Hercynian compression phase during the Late Cretaceous, and Miocene wrench tectonics (both related to the Alpine orogeny). Finally, in the CBH area salt tectonics (mainly during the Tertiary) also constituted a dominant element in the structural development.

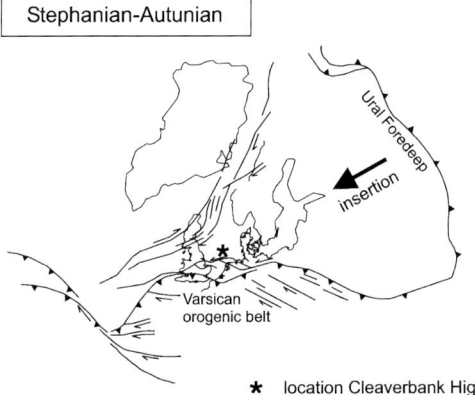

Fig. 3. The North Sea–Baltic block was pushed back into its former position during the Late Carboniferous at the time of the Variscan orogeny, resulting in sinistral transpression along the shear zones of its southern boundary, close to the CBH (after Coward 1993).

Upper Carboniferous seismic facies

Because of the depth of burial, the overlying Zechstein salt accumulations, and the specific seismic response of the Coal Measures, seismic interpretation of the intra-Carboniferous in the area was rather challenging in the days before 3D surveys. The distinct seismic response of the Westphalian Coal Measures relates to the nature of the sequences, in which basically three different lithologies (sand, shale and coal) alternate in a cyclic manner. Relevant to the seismic facies is the fact that the acoustic impedance of coal is significantly lower than that of sand and shale. As a result, strong reflections are being generated. A multiseam sequence not only generates high amplitude primary reflections, but also strong interbed multiples. The result is a seismic record which is complicated by multiples, and in which individual coal beds cannot be directly correlated to individual reflectors (Evans *et al.* 1992). Because of the interference pattern only a few high amplitude reflectors appear. The seismic package with this distinct seismic facies can be correlated to the stratigraphic units that contain most coal. In our study area of the CBH these distinct units correspond to the Coal Measures of Upper Westphalian B age (Geluk *et al.* 2002). Figure 4 shows a type-log of well E13-01, where the Coal Measures are overlain by the younger sequences of the Hospital Ground formation (Van Adrichem Boogaert & Kouwe 1993–1997).

It was observed on 2D seismic data (e.g. by Evans *et al.* 1992 and by Tantow 1993) that the Westphalian seismic sequences can be divided into two seismic facies: an upper transparent facies (*low amplitudes, low continuity* of reflectors) overlying a highly reflective seismic facies with *closely spaced, high frequency, moderate to high amplitude continuous* reflectors.

From the higher quality 3D surveys that we have studied in the CBH region, we would rather propose a three-fold division. Underlying the *high frequency, moderate to high amplitude, continuous* facies, we distinguish a third facies of *lower frequency, moderate amplitude, moderate to high continuity* reflectors. In Figure 5 this seismic facies is shown as Facies I. The boundary between Facies I and Facies II (the *high frequency, moderate to high amplitude, continuous* facies) correlates to a marked chronostratigraphic horizon, which represents a flooding surface within the Westphalian B known as the Domina marine band. In the CBH area this level is very close to the lithostratigraphically defined base of the Maurits formation, which also coincides with the top of the Caister-equivalent sandstones. In an earlier study (Geluk *et al.* 2002) we called this horizon W520. It basically

separates the Lower Westphalian B sand-prone units from the overlying Upper Westphalian B coal-rich sediments of the Maurits formation.

The boundary between the upper two seismic facies (Facies II and III) correlates to a stratigraphic level near base Westphalian C (called W600 by Geluk *et al.* 2002).

The main reason that, in spite of the distinct character of seismic Facies II (the interval between W520 and W600), it is still difficult to consistently interpret individual horizons within the Carboniferous, is the structural complexity.

Observations in the Dutch Cleaver Bank High area

The deformation of the Carboniferous sedimentary succession of the Cleaver Bank High area comprises both folding and faulting. In the 3D seismic surveys the deformation resulting from a number of different tectonic phases can be observed.

Fault systems

Figure 6 shows a map with the faults observed within the Carboniferous sequences on the 3D seismic surveys. We distinguish three types of faults according to the kind of throw that they display on the seismic data.

Type A faults are faults that show normal offset at base Permian level (most clearly seen on seismics at base Zechstein). They also show a normal offset at intra-Carboniferous levels, but these two offsets do not necessarily have to be equal. However, when the two offsets are equal, it can be concluded that the fault movements post-dated the Permian, affecting Permian and pre-Permian rocks in the same way. The majority of the type A faults have a NW–SE trend. This trend is well known in the Dutch offshore, and is usually assigned to the Kimmerian rifting phases related to the opening of the Atlantic Ocean during the Jurassic and Early Cretaceous, as well as to the later Alpine related compression and transpression. Type A faults with this known Kimmerian trend, and showing equal offsets at Permian and pre-Permian levels are abundant. Type A faults with a NE–SW trend are also observed, most frequently in the SE part of the study area. A characteristic of many of the type A faults in the CBH area, is that they have an unusually high length-to-throw ratio, something which was already observed by Oudmayer & de Jager (1993), who explain this by assuming repeated Mesozoic reactivation. Since the stress regime would only rarely have coincided with the pre-existing basement grain, they also assumed oblique slip along the faults. We confirm the observation of the high length-to-throw ratio of type A

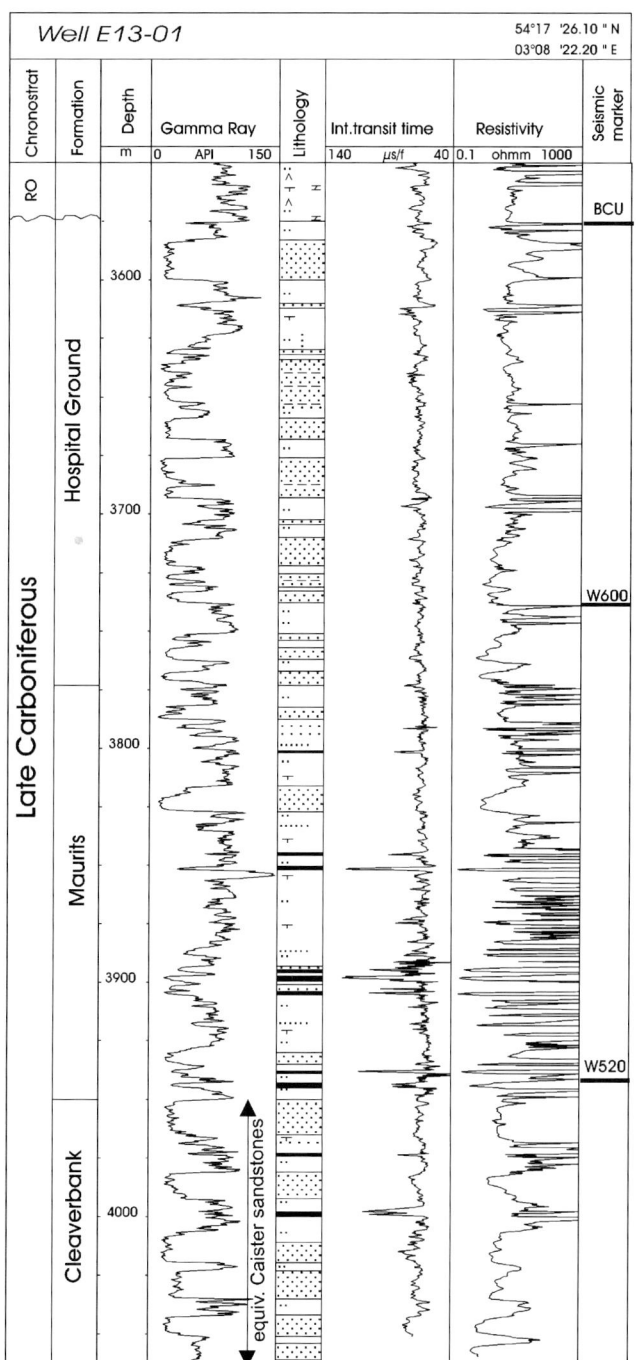

Fig. 4. Type-log for the Upper Carboniferous stratigraphy of the Dutch Cleaver Bank High from well E13-01.

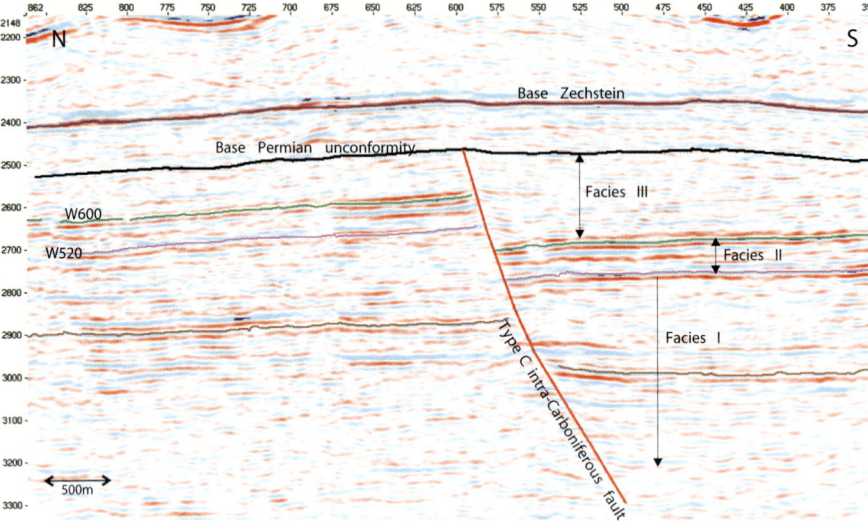

Fig. 5. Westphalian seismic facies.

faults, but do not see strong evidence for significant strike-slip components of movements along either the NW–SE or the NE–SW trending normal faults. In those cases where type A faults display different offsets at Permian and at Carboniferous levels, it must be concluded that the faults had already been active before the Late Permian, and that later (Kimmerian) tectonic phases were responsible for reactivation of these faults. In our data there are examples of both NW–SE and NE–SW normal faults which show a different (usually larger) throw at intra-Carboniferous level than at base Permian level. A third trend, only found in a limited area, for type A normal faults is NNE–SSW. In the limited area in the NW of our study area (D-quadrant), where this trend occurs, the faults seem to belong to a single complex shear zone which must have been reactivated throughout the Mesozoic and Cenozoic. In this system there are clear indications for wrenching, but there is no anomalously high length-to-throw ratio.

Type B faults are faults that show a reversed throw at base Permian level. The trend of these faults varies from east–west (mostly in the northern part of the study area) to NW–SE in the south. The throw at intra-Carboniferous level is often either smaller than at base Permian (but still reversed), or the throw is even normal. Two examples are evident in Figure 7. This means that the compressive tectonic phase which caused the reversed movements, reactivated older existing Carboniferous normal faults, but the slip was too small to compensate the larger existing Carboniferous normal throw. In view of the regional geological development of the area, it can be assumed that the com-

pression took place during the Subhercynian tectonic phase in the Late Cretaceous. Between Permian and Late Cretaceous the same fault systems may very well have been used by one or two other phases too, but this is now difficult to determine. The most important observation with regard to these type B faults is that existing Carboniferous east–west trending faults were later reactivated in an opposite sense. These faults usually dip towards the south. The hanging walls were downthrown to the south during the Carboniferous, and reversed up and to the north during the Late Cretaceous.

Type C faults are faults which have a normal throw at intra-Carboniferous level, but no throw (or significantly smaller throw) at base Permian level, proving that they were active before the Late Autunian. Sometimes these faults show a small bulge of the base Zechstein reflector, indicating a minor post Permian reactivation, but compared to the larger throw within the Carboniferous the offset at Permian level is negligible. The most frequently occurring trend for type C faults is NE–SW. In general the faults do not have the high length-to-throw ratio which is so characteristic for the NW–SE trending type A faults. However, some striking NE–SW regional lineaments, composed of an assembly of individual NE–SW faults, show up on the map. One of these is indicated in Figure 6. In addition to the NE–SW faults, the group of type C faults contains a number of NW–SE trending normal faults. Such faults only differ from the NW–SE trending type A faults that show a difference in Permian and Carboniferous throw, by the fact that the type C faults have not been reactivated by later (Kimmerian) extension phases. With respect

Fig. 6. Fault map with the faults observed on 3D seismic surveys at intra-Westphalian level. Remarkable regional NE–SW and NNE–SSW lineaments are indicated.

to the Carboniferous structural framework there is no difference. Similarly, both type A and type C groups contain NE–SW trending faults. Only type A faults have been reactivated, but the map also includes type A faults that apparently had not been active before the Permian. Figure 5 shows an example of a north–south running seismic section across one of the type C faults. In this section also a deeper event (probably of Westphalian A or Namurian age) has been interpreted in order to demonstrate a subtle increase in thickness of the Westphalian A–Lower Westphalian B from north to south. This particular fault is curved both in map view and on the vertical sections, so that it has a sort of spoon shape. The few examples of these spoon-shaped type C faults all have a rather large normal throw, and they all downthrow the hanging wall to the S or SSE.

Folding

When considering a sufficiently large area, the long wavelength (approximately 20 km) folding of the Upper Carboniferous sequences becomes obvious. The azimuth of the anticlinal axes is approximately NW–SE. On the flanks of the folds an angular unconformity is visible between the Upper Carboniferous and the Rotliegend sediments. Hence it can be concluded that the compressive tectonic phase which caused the folding, occurred after the Westphalian D, and before the deposition of the first Rotliegend sediments (some time in the Autunian). Given the hiatus in the sedimentary succession in the area, this means a period of 30–40 Ma during which the compression can have occurred. The anticlines are broad, and the areas of no (or hardly any) angular unconformity are wide. In such areas

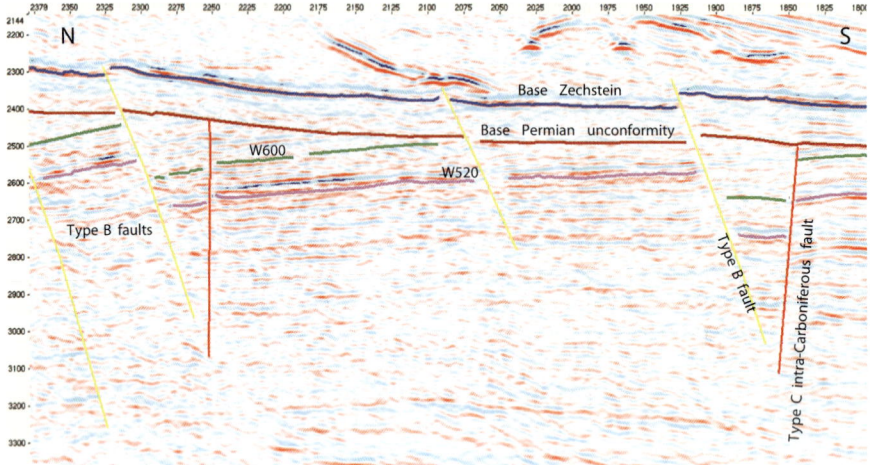

Fig. 7. North–south section showing south dipping type B reversed faults (yellow) with a normal throw at intra-Carboniferous level. Such faults generally have an east–west strike. The antithetic north dipping normal faults (red) have not been reactivated after the Early Permian.

it is not always possible to identify the base Permian unconformity on seismic profiles.

Figure 8 shows an east–west trending section, located in the northernmost blocks of the K-quadrant, across one of the anticlinal axes. On the lower part of that section (between 3.5 and 4.0 s two-way travel time, TWT) a clear expression of the anticlinal shape is present with dips which seem slightly higher than those at Westphalian B/C level (these indications are stronger when one bears in mind the higher seismic velocities at greater depth). This may be an indication of a pre-existing structural high underneath the later anticlinal axis. The trend coincides with the assumed Early Carboniferous structural framework shown in Figure 1.

Structural model for the Cleaver Bank High

In order to assist the structural analysis we constructed a so-called 'preserved thickness map' which proved to be a useful tool. This map is

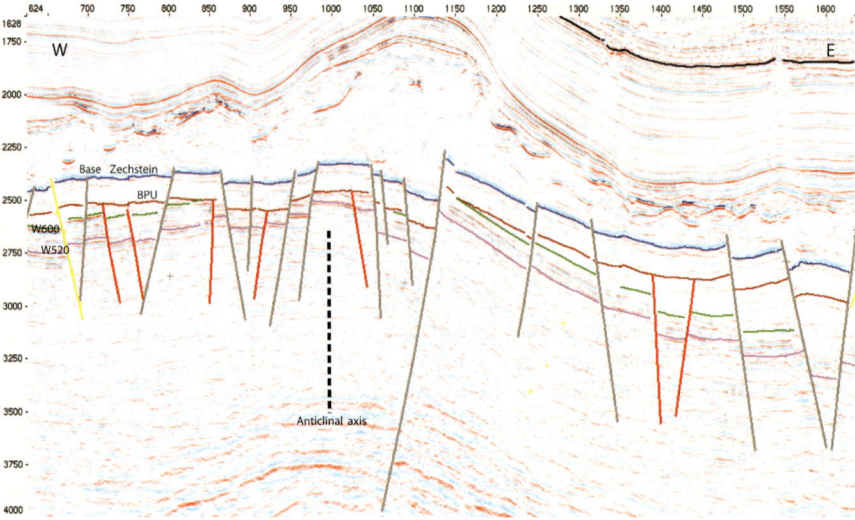

Fig. 8. East–west section through block K1 showing the position of an anticlinal axis, and the thickening of younger sequences to the east.

shown in Figure 9. It shows by colour-filled contours the thickness (in TWT) between the intra-Westphalian B marker W520, and the top of the Carboniferous. Where sudden changes in colour occur across a fault, this map indicates that the fault has been active before the Permian. In this way it even shows some faults that had not been initially interpreted. The map can also be read as a pre-Permian subcrop map. The colours are chosen in such a way that the colour classes roughly correspond to the stratigraphic units preserved immediately underneath the base Permian erosional unconformity. Pink colours represent areas where sequences of Westphalian B–Lower Westphalian C age subcrop, the yellow-orange colours relate to Upper Westphalian C, and green (approximately) to Westphalian D.

Careful analyses of this information together with the observations of the different fault systems described above, resulted in a structural model, which takes into account the timing, the sense and the relative importance of tectonic events during

Late Carboniferous–Early Permian (Autunian) times, following the period of tectonic quiescence, which ended during the Westphalian C. First of all, our observations confirm that any model for this area would include repeated reactivation of basement faults, as proposed by Bartholomew *et al.* (1993) and Oudmayer & de Jager (1993), for example.

Late Carboniferous–Early Permian tectonic phases

The NW–SE trend is mentioned by Corfield *et al.* (1996) and by Besly (1998) as being the most dominant inherited Caledonian trend, active during the Late Devonian–Early Carboniferous rifting in the SNSCB. Figure 1 suggests that the CBH, as it already existed during that time, was bounded by faults with this trend. From our observations it can be concluded that some of the faults with this NW–SE trend were indeed also reactivated during the Late Carboniferous or Early Permian, but this trend

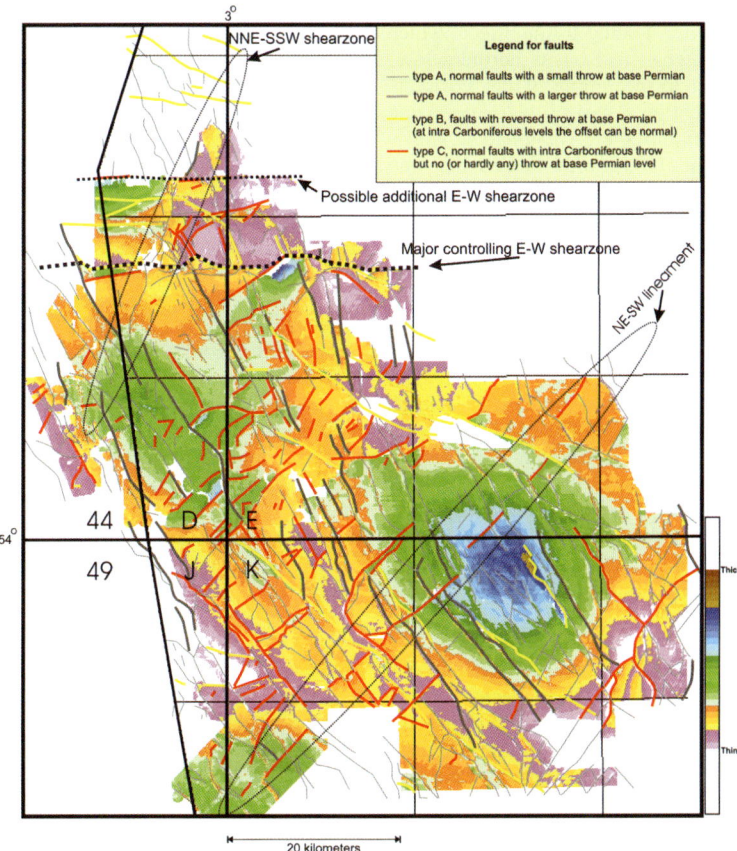

Fig. 9. Synthesis of the tectonic observations and interpretations of the dominant shear zones.

was certainly not the most important one in that period. Most of the abundant NW–SE faults now intersecting the Upper Carboniferous rocks are later Mesozoic reactivations of this Caledonian trend. From the few examples in our data it is difficult to confirm either the right lateral strike-lip component during the Stephanian–Autunian along these fault systems as postulated by Oudmayer & de Jager (1993), or the left lateral slip brought forward by Hollywood & Whorlow (1993) and by Quirk (1993). Present-day indications of wrenching along the NW–SE faults may result from Mesozoic–Cenozoic reactivation only.

If we disregard the few NW–SE faults that were active before the Mesozoic as minor, it is clear from our observations that at least two (but possibly more) major tectonic events affected the area during the Latest Carboniferous or Early Permian.

Late Carboniferous compression The large-scale regional folding probably pre-dated a Late Carboniferous or Early Permian extensional phase. The folding must be associated with a compressive phase around Westphalian D–Stephanian times. Such a timing would be in accordance with the models that relate the folding to the effects which the Variscan orogenesis had on its northern foreland. With respect to the timing of the Variscan, several authors mention a start of the compression in the wider region during the Westphalian. Hollywood & Whorlow (1993), who studied the well data of the southern part of UK quadrant 44, suggested a change to a compressional stress regime already by the end of the Westphalian B, with a culmination of the orogeny from Westphalian C–Permian. They also postulated a rifting phase at the end of the Stephanian, and refer to igneous activity. Leeder & Hardman (1990) studied 2D seismic data, and refer to the Symon unconformity (a major sequence boundary between Westphalian B and C) with respect to the approach of the Variscan compression. They suggested that the compression started in the Westphalian C.

Although indeed we have found some indications for onlap of Westphalian C sediments onto a very subtle angular unconformity, we cannot confirm that a possible equivalent of the Symon unconformity is a very dominant feature in the CBH area. In addition, it should be noted that, with the older 2D seismic data, there is a serious risk that multiples parallel to the Base Zechstein, and within the very low amplitude seismic Facies III, are mistaken for onlapping Westphalian C reflectors. Smith (1999) mentions two distinct Late Carboniferous pulses of thrusting in the Cheshire Basin in England, the first one relating to the Symon unconformity at the beginning of Westphalian C, and a second one by the end of Westphalian

C or in Westphalian D, which related to the main folding phase. It may very well be, that in a stable platform setting such as the CBH the effect of the first pulse is very minor, whereas the main folding phase affected the entire region. In that case the timing of the compression might be closer to Westphalian D.

A dominantly NW–SE strike of the folding axes can be interpreted in our study area (Fig. 9). This direction is an indication that compression from the NE as suggested by Coward (1993), and related to the Ural orogenesis (Fig. 3) may have played a larger role than the actual Variscan compression with its assumed SE–NW oriented stress. In this respect it should also be noted that the CBH area is close to the northernmost reach of Variscan influences.

Late Carboniferous and/or Early Permian extension The NE–SW trend, which was most probably also inherited from the Caledonian structural grain, was obviously active during an extensional phase sometime between Westphalian D and Autunian (Early Permian). It is difficult to be precise with regard to the timing. We assume that this extension phase post-dated the regional folding. Quirk & Aitken (1997), who studied the same area, refer to 'the rifting known as the Saalian event, during the early Permian'. However, in this area there is a hiatus in the sedimentary record representing a period of at least 30 Ma, ending with the Late Permian Rotliegend sediments, that were deposited on top of the Base Permian unconformity. The use of the term 'Saalian unconformity' in this respect may be misleading, because within a period of 30 Ma there has probably been more than one tectonic phase. The Saalian phase within the Early Permian may have been followed by later phases at the end of the Early Permian, or even the beginning of the Late Permian, or alternatively, may have been preceded by a (very) Late Carboniferous phase.

It is tempting to postulate at least two separate extensional phases, because of the different styles of the type C faults observed in the CBH area. First of all, there is a category of faults comprising small, but often deep, grabens in the very west of blocks D12 and D15 with a consistent and straight east–west strike. Given their east–west trend, it seems likely that they were reactivations of Late Devonian–Dinantian normal faults. Figure 7 shows two such small grabens. It should be noted that these both have a type B fault on their north side, but during the Carboniferous these were still simple normal faults similar to the type C faults. The type B faults were reactivated during the Late Cretaceous inversion in a reversed sense.

Secondly, we observe the so-called 'spoon-shaped' normal faults. Their fault trace in map view often bends from east–west to NE–SW, and also in the vertical section they have a listric appearance. This category is also different from the first one, because fault blocks are exclusively downthrown to the south or SE. There is clear evidence of synsedimentary tectonics at most of these faults. An example of one of these faults is shown in Figure 5. The configuration close to the assumed 'major controlling E–W shearzone' (Fig. 9) may be an indication of small pull-apart basins associated to strike-slip along the major system.

Finally, the abundant NE–SW trending straight faults of type C, present over most of the study area, except the NW corner, may either be put in a third category, or combined with the 'spoon-shaped' normal faults. We observe one larger scale regional NE–SW lineament (indicated in Figs 6 and 9), which is a grouping of many smaller NE–SW faults. This suggests that at that location there might be a link with deeper basement faults with the same trend.

Because of the observed differences between these groups of type C normal faults, all active after the Westphalian C, but before the Late Autunian, and their associated stress regimes, it seems likely that a Late Carboniferous (Westphalian D or Stephanian) extensional phase pre-dated a later Early Permian (Autunian) extensional phase.

Discussion

The one assumed major controlling east–west shear zone indicated in Figure 9 shows left-lateral offset of areas on either side with similar characteristics. Sinistral transpression was postulated by Coward (1993) along east–west boundary shear zones forming the southern margin of the North Sea–Baltic plate when it moved back to the west during the Stephanian or Autunian (Fig. 3). Such boundary shear zones would have been parallel to, and not too far from, the one interpreted by us in the D and E quadrants. According to Coward's tectonic escape model the same systems may have been used earlier in a dextral transtensional setting during the Early Carboniferous. Evidence for such movements is, however, harder to find in our data, because our interpretations are mainly restricted to the Upper Carboniferous. There are also indications for the regional importance of this major east–west shear zone, if one considers the possible extension of this zone into the UK quadrants 44 and 43. The structural style (as for example shown by O'Mara et al. 1999 for the Trent field) of some UK gas-filled structures close to this line, is very similar to the structural style in the Dutch blocks D15 and E13. The SE–NW to ESE–WNW trending structural highs in the UK sector immediately south of the extrapolated shear zone appear to end

abruptly against the zone, and a different structural style may prevail to the north of it.

Dominance of east–west trending reactivated shear zones

From the fault map (Fig. 6) it is not immediately clear how dominant the east–west systems may have been. However, from displays such as Figure 9 we interpret the occurrence of at least one, possibly a few, major controlling east–west shear-zones. At first glance they are somewhat hidden because the net throw, after the frequent reactivations in opposite senses, is often small. In addition, these systems typically show the characteristics of wrench systems (e.g. pop-ups resembling positive flower structures described by Harding (1990), laterally varying magnitude of throw, and curvature of the fault trace). Hence much of their throw is in a strike-slip sense, and therefore not easily visible.

Following the analyses of the east–west trending type C faults given above, it can be argued that there is evidence for frequent reactivation of an east–west major controlling fault system such as the one shown in Figure 9. During the Devonian–Dinantian pure extensional rifting the faults were used in a normal sense. During the Late Carboniferous compression, associated with the westward insertion of the North Sea–Baltic plate they functioned as sinistral strike-slip zones, possibly with a minor reversed dip-slip component. Locally, parts of the system were again reactivated in a normal sense during the Late Carboniferous extension, when rather deep half-grabens such as the one near well E13-01 were formed. Finally, during the Late Cretaceous Subhercynian inversion many of the east–west faults were used as dip-slip reversed faults.

Another remarkable feature in Figure 9 are the deep depressions (shown by the green and blue colours) in which the youngest Westphalian and Stephanian sequences have been preserved. The interval between the W520 (intra-Upper Westphalian B) horizon and the Base Permian unconformity alone reaches about 1200 m in thickness in the deepest depression. Therefore, the total thickness of Westphalian and Stephanian may be close to 2000 m at that point. The weak seismic signal from these youngest preserved Carboniferous sediments makes it difficult to draw any conclusions about the tectonic history, but also in view of the observed indications for repeated strike-slip movements along the dominant east–west systems, the configuration of these deep depressions resembles pull-apart basins.

Conclusions

When considering the structural history of the Cleaver Bank High area from Early Carboniferous to Early Permian, the following summary can be made. Three distinct fault trends, namely east–west, NE–SW and NW–SE, were inherited from earlier tectonic events. All three have been active during the Early Carboniferous rifting phase. After a period of relative tectonic quiescence during the Namurian and the Westphalian A and B, regional large-scale folding with dominantly NW–SE axes occurred, probably at the end of the Westphalian or beginning of the Stephanian. This compression fits the model of westward lateral movement of the North Sea–Baltic microplate, and may have led to left-lateral transpression along the existing east–west oriented zones of weakness. These east–west trending dominant fault systems parallel to the southern margin of the microplate are important major controlling fault systems, that have been reactivated during several phases throughout geological history. However, at first glance these systems are not immediately obvious from present-day fault maps. Their importance becomes clear after closer examination of the vertical sections and maps such as the 'preserved thickness map'. In the CBH area there is a lot of evidence for extensional tectonics after this phase of folding and transpression. The two different trends (east–west and NE–SW) of normal faults active before deposition of the Rotliegend sediments suggest that two separate extensional tectonic phases occurred, possibly during the Stephanian and the Autunian respectively.

The authors wish to acknowledge permission by Wintershall Noordzee b.v. and Nederlandse Aardolie Maatschappij b.v. for the publication of figures derived from their non-released 3D seismic data. We thank Mark Geluk for his support in preparing some of the data and in critically reviewing the text.

References

BARTHOLOMEW, I. D., PETERS, J. M. & POWELL, C. M. 1993. Regional structural evolution of the North Sea: oblique slip and the reactivation of basement lineaments. *In*: PARKER, J. R. (ed.) *Petroleum Geology of Northwest Europe: Proceedings of the 4th Conference*. Geological Society, London, 1109–1122.

BESLY, B. M. 1998. Carboniferous. *In*: GLENNIE, K. W. (ed.) *Petroleum Geology of the North Sea: Basic Concepts and Recent Advances* (fourth edition). Blackwell Science, Oxford, 104–136.

CORFIELD, S. M., GAWTHORPE, R. L, GAGE, M., FRASE, A. J. & BESLY B. M. 1996. Inversion tectonics of the Variscan foreland of the British Isles. *Journal of the Geological Society, London*, **153**, 17–32.

COWARD, M. P. 1990. The Precambrian, Caledonian and

Variscan framework to NW Europe. *In*: HARDMAN, R. F. P. & BROOKS, J. (eds) *Tectonic Events Responsible for Britain's Oil and Gas Reserves*. Geological Society, London, Special Publications, **55**, 1–34.

COWARD, M. P. 1993. The effect of Late Caledonian and Variscan continental escape tectonics on basement structure, Paleozoic basin kinematics and subsequent Mesozoic basin development in NW Europe. *In*: PARKER, J. R. (ed.) *Petroleum Geology of Northwest Europe: Proceedings of the 4th Conference*. Geological Society, London, 1095–1108.

EVANS, D. J., MENEILLY, A. & BROWN, G. 1992. Seismic facies analysis of Westphalian sequences of the southern North Sea. *Marine and Petroleum Geology*, **9**, 578–589

FRASER, A. J. & GAWTHORPE, R. L. 1990. Tectono-stratigraphic development and hydrocarbon habitat of the Carboniferous in northern England. *In*: HARDMAN, R. F. P. & BROOKS, J. (eds) *Tectonic Events Responsible for Britain's Oil and Gas Reserves*. Geological Society, London, Special Publications, **55**, 49–86.

GELUK, M. C., DE HAAN, H., NIO, S. D., SCHROOT, B. & WOLTERS, B. 2002. The Permo-Carboniferous gas play, southern North Sea, the Netherlands. *In*: HILLS, L. V., HENDERSON, C. M. & BAMBER, E. W. (eds) *Carboniferous and Permian of the World, XIV International Congress on the Carboniferous and Permian*. Canadian Society of Petroleum Geologists, Memoirs, **19**, 877–894.

GLENNIE, K. W. (ed.) 1998. *Petroleum Geology of the North Sea*, (fourth edition). Blackwell Science, Oxford.

HARDING, T. P. 1990. Identification of wrench faults using subsurface structural data: criteria and pitfalls. *AAPG Bulletin*, **74**(10), 1590–1609.

HARLAND, W. H. & GAYER, R. A. 1972. The Arctic Caledonides and earlier oceans. *Geological Magazine*, **109**, 289–384.

HOLLYWOOD, J. M. & WHORLOW, C. V. 1993. Structural development and hydrocarbon occurrence of the Carboniferous in the UK Southern North Sea Basin. *In*: PARKER, J. R. (ed.) *Petroleum Geology of Northwest Europe: Proceedings of the 4th Conference*. Geological Society, London, 689–696.

HORSCROFT, R., LEE, M. K., SUTTON, E. R., ROLLIN, K. E., DAVIDSON, K. & WILLIAMSON, J. P. 1992. The pre-Permian (Carboniferous) of the southern North Sea from 3D modelling and image analysis of potential field data. *61st Meeting of the European Association of Exploration Geophysicists, extended abstract*, EAGE, 328–329.

LEEDER, M. R. & HARDMAN, M. 1990. Carboniferous geology of the Southern North Sea Basin and controls on hydrocarbon prospectivity. *In*: HARDMAN, R. F. P. & BROOKS, J. (eds) *Tectonic Events Responsible for Britain's Oil and Gas Reserves*. Geological Society, London, Special Publications, **55**, 87–105.

MAYNARD, J. R., HOFMAN, W., DUNAY, R. E., BENTHAM, P. N., DEAN, K. P. & WATSON, I. 1997. The Carboniferous of western Europe: the development of a petroleum system. *Petroleum Geoscience*, **3**, 97–115.

MCKENZIE, D. P. 1978. Some remarks on the development of sedimentary basins. *Earth and Planetary Science Letters*, **40**, 25–32.

O'MARA, P. T., MERRYWEATHER, M., STOCKWELL, M. &

BOWLER, M. M. 1999. The Trent Gas Field: correlation and reservoir quality within a complex Carboniferous stratigraphy. *In*: FLEET, A. J. & BOLDY, S. A. R. (eds) *Petroleum Geology of Northwest Europe: Proceedings of the 5th Conference.* Geological Society, London, 809–821.

OUDMAYER, B. C. & DE JAGER, J. 1993. Fault reactivation and oblique-slip in the Southern North Sea. *In*: PARKER, J. R. (ed.) *Petroleum Geology of Northwest Europe: Proceedings of the 4th Conference.* Geological Society, London, 1281–1290.

QUIRK, D .G. 1993. Interpreting the Upper Carboniferous of the Dutch Cleaver Bank High. *In*: PARKER, J. R. (ed.) *Petroleum Geology of Northwest Europe: Proceedings of the 4th Conference.* Geological Society, London, 697–706.

QUIRK, D. G. & AITKEN, J. F. 1997. The structure of the Westphalian in the northern part of the southern North Sea. *In*: Ziegler, K., Turner, P. & DAINES, S.R. (eds) *Petroleum Geology of the Southern North Sea: Future Potential. Geological Society, London, Special Publications,* **123**, 143–152.

SMITH, N. T. 1999. Variscan inversion within the Cheshire basin, England: Carboniferous evolution north of the Variscan front. *Tectonophysics,* **309**, 211–225.

SOPER, N. J., WEBB, B. C. & WOODCOCK, N. H. 1987. Late Caledonian (Acadian) transpression in north-west England: timing, geometry and geotectonic significance. *Proceedings of the Yorkshire Geological Society,* **47**(3), 175–192.

TANTOW, M. S. 1993. Stratigrafie und seismisches Erscheinungsbild des Oberkarbons (Westfal, Stefan), Emsland. *Selbstverlag Fachbereich Geowissenschaften,* FU Berlin.

VAN ADRICHEM BOOGAERT, H. A. & KOUWE, W. F. P. 1993–1997. Stratigraphic nomenclature of the Netherlands; revision and update by the RGD and NOGEPA. *Mededelingen Rijks Geologische Dienst,* **50**.

ZIEGLER, P. A. 1978. North-Western Europe: Tectonics and basin development. *In*: VAN LOON, A. J. (ed.). Keynotes of the MEGS-II (Amsterdam, 1978). *Geologie en Mijnbouw,* **57**, 589–626.

Building and unfaulting fault–horizon networks

K. W. RUTTEN* & M. A. J. VERSCHUREN

Shell E&P Technology, PO Box 60, 2280AB Rijswijk, The Netherlands
Present address: Slokkert Consultancy, De Slokkert 8, 9451TD Rolde, The Netherlands (e-mail: rutten@cistron.nl)

Abstract: A fault–horizon network is a set of horizon and fault surfaces that is geologically consistent and topologically complete. It is the missing link between seismic interpretation and reservoir modelling. Building the fault–horizon network from a seismic interpretation can be a fast and automatic process. The fault slip vector field embedded in the network is used to construct tip points and fault branches, and as a powerful tool to validate network quality.

A novel, very fast method is introduced to unfault the network, i.e. to remove fault-related strain. It allows further validation of the geological consistency of the network, and it delivers an unfaulted framework for reservoir modelling. The unfaulting method approximates the three-dimensional kinematics related to faulting with the simplest possible analytical description that can be calculated from the geometry of the fault–horizon network and its fault slip vector field.

The construction of 'fault swallows' at branching faults decomposes discontinuous slip and strain at branch lines into continuous segments. With this decomposition, the unfaulting method can handle complex branching fault situations.

The methods in this paper have been developed for normal faults in deltaic arcas, but they can easily be extended to other geological settings.

In the early stages of a hydrocarbon field's life cycle, the structural complexity of faulted reservoirs is all too often underestimated, risking bypassed hydrocarbons and/or undrilled compartments. Even with 3D seismic data, diligent structural geological interpretation is still required for evaluating connectivity and sealing properties of faulted reservoirs.

The current generation of seismic interpretation workstations does not fully support the detailed and consistent representation of faulted structures: often, horizons and faults are interpreted separately and fault–horizon and fault–fault intersections are treated as separate entities. For example, a faultcut polygon in a map can be smoothed independently from the fault and horizon surfaces that intersect at the faultcut, leading to discrepancies at the fault–horizon intersection. Horizons are often shifted vertically at faults to get the unfaulted state needed for modelling porosity and permeability continuously across faults. These vertical shifts destroy fault/horizon relationships above and below the shifted horizon. Furthermore, due to software and/or hardware restrictions, structural interpretations are often not used in reservoir modelling directly but only after simplification: re-gridding at a coarser scale and removal of smaller faults. This simplification is time consuming and irreversible, making it difficult to propagate incremental

updates from seismic interpretation to the reservoir model, or conversely, to bring significant features in the reservoir model back to the seismic interpretation. In summary, the link between seismic interpretation and reservoir modelling is easily lost.

The objective of the fault–horizon network methods described in this paper is to provide this link. The network building and unfaulting methods retain full structural detail of the interpretation. The unfaulting method shifts horizons back along the faults to the unfaulted state, retaining fault–horizon relationships above and below. Because structural detail and relationships are retained, it is possible to propagate incremental updates in the seismic interpretation automatically forward to the reservoir model, and to propagate structural changes to the reservoir model automatically backward to the seismic interpretation canvas. The network building and unfaulting methods are so fast that they can be applied routinely after each interactive seismic interpretation session.

Several academic and industry groups are working towards the same goal: from the mapping and topology side the gOcad group, and from the structural geological side the Fault Analysis Group, the Princeton 3D Structure Project and several palinspastic restoration groups (see Internet sites in the list of references). The methods described in this paper distinguish themselves by concentrating on the

From: NIEUWLAND, D. A. (ed.) *New Insights into Structural Interpretation and Modelling.* Geological Society, London, Special Publications, **212**, 39–57. 0305-8719/03/$15

fault slip vector field as an essential element for building the network (construction of fault tips and branches), as a structural validation tool (following Needham *et al.* 1996) and finally as a basis for the unfaulting method. Detailed comparisons between our methods and other methods are presented towards the end of this paper.

Our fault–horizon network methods were developed for normal faults in deltaic areas. Such areas are characterized by complex, often synsedimentary linked fault systems. Hydrocarbons are sometimes present in dozens of stacked reservoirs trapped against faults (e.g. Bouvier *et al.* 1989). The methods are not necessarily limited to such tectonic settings and can be easily extended to areas characterized by, for example, contractional or strike-slip tectonic regimes.

Overview of the fault–horizon network building and unfaulting methods

The term 'fault–horizon network' is used here (and by the gOcad group) to describe a set of horizon and fault surfaces that is geologically consistent and topologically complete. Topological completeness means that fault–horizon and fault–fault intersections and terminations are explicitly defined, i.e. the network is 'watertight', as required by modelling of fluid flow in reservoirs. Geological consistency means that all aspects of the model are in full agreement with the geology

Figure 1 shows a flow diagram of our fault–horizon network methods. The seismic interpretation of faults and horizons is fed into an automated fault–horizon network building operation, which calculates intersections and truncations of faults and horizons. At the same time, the fault slip vector field is calculated and used to construct tips and branches. Consistency of the slip field is checked by visual inspection and quantitative diagnostics. If it is not consistent, the seismic interpretation has to be corrected. Once the slip field is consistent, the network is unfaulted with a simple analytical extrapolation function of the fault slip vector field. The ultimate check on validity of the network is that the strain in the network after unfaulting is geologically meaningful, and that there are no artefacts or geologically meaningless deformations in the unfaulted structure. Again, more seismic interpretation work is required as long as this is not the case.

Validating structural models

The structural validation in this paper is based on the philosophy described below, starting from the requirements of faulted reservoir modelling (see Fig. 1).

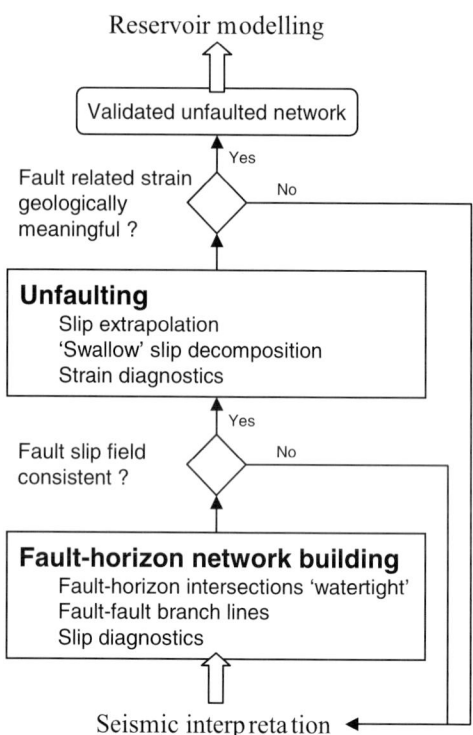

Fig. 1. Generalized flow diagram for building and unfaulting fault–horizon networks. First, a fault–horizon network is automatically built from the seismic interpretation. Then, the fault slip field in the network is checked, and in case of inconsistencies, the seismic interpretation is corrected and the network rebuilt. Once the slip field is consistent, the network is unfaulted. The strain in the network after unfaulting is again checked, returning to the seismic interpretation in case of problems. Finally, a thoroughly validated unfaulted network without loss of detail is delivered to reservoir modelling.

Any structural interpretation implies an interpretation of deformation through geological time. To validate a structural interpretation means that the deformation model implied by the interpretation is made explicit and shown to be internally consistent, geologically plausible and in agreement with regional geological history.

Geological deformation consists of faulting, folding, compaction, flow and, on a large scale, plate-tectonic translation and rotation. Most palinspastic restoration methods and software try to capture the first three components together, but often at the price of model simplification. For modelling reservoirs in deltaic areas, the strain related to faults is the most significant component and it is most sensitive to interpretation errors. To 'unfault' means to reconstruct the deformation related to faulting, or to remove the strain related to the faulting explicitly. The residual strain can then be assessed as resulting from interpretation errors, or from the other geological components (folding, compaction, flow and plate tectonic effects). As long as the strain is not geologically meaningful, the underlying seismic interpretation cannot be considered valid.

In order to be geologically meaningful, an unfaulting operation has to express the strain related to a fault in terms of its geometry and its slip field. Because mechanical rock properties at the time of faulting are poorly known, mechanical validation constraints are less useful for the seismic interpreter than geometrical or kinematical ones. The kinematical aspects are what the unfaulting method proposed here is focused on, just like other palinspastic restoration methods.

Given the pivotal role of the fault slip vector field in kinematic modelling, it has to be mapped carefully. The term 'fault-slip vector field' (or 'slip field') is used in this paper to denote the distribution of direction and magnitude of slip vectors in fault surfaces, where a slip vector connects two formerly adjacent points on opposite walls of a fault (Peacock *et al.* 2000). In order to identify such correspondence, the truncations of faults and horizons (fault–horizon cut-offs and fault–fault branches) are made explicit while building the fault–horizon network. Fault tip points are constructed by extrapolating the slip gradient. Branching faults, common in deltaic settings, will be shown to constrain the direction of slip.

In addition to visual inspection of the network in a 3D viewer, lateral and vertical slip gradients in fault plane projections are useful quantitative diagnostic tools, highlighting defects in seismic interpretation such as seismic loop skips and miscorrelations across faults. As long as the slip field is not internally consistent (Needham *et al.* 1996), the underlying seismic interpretation requires more work. Unfaulting faults with an inconsistent slip field will also highlight interpretation defects. The slip diagnostic tools are, however, closer to the source of the inconsistencies (e.g. horizon miscorrelation across a fault) than the strain diagnostics of the unfaulting operation.

Data sets

Two data sets from rollover structures in the Niger Delta are used to illustrate the methods. The first data set is a very simple one with just one fault, to illustrate the basic principles (Fig. 2). The seismic interpretation was edited for presentation purposes: other faults were removed and horizons were smoothed. The second data set (Fig. 3) is more realistic, with six faults. The area mapped here is a small collapsed crestal graben with a main fault

in the NE and several smaller faults. The main fault is synthetic to a major growth fault located outside the mapped area. A secondary fault branches off the main fault.

The vertical axis of both data sets is in seismic two-way travel time. One millisecond at the depth of the middle horizon in the first data set corresponds to about 1.37 m, in the second one to 0.86 m. Although reservoir modelling will in general be carried out *after* seismic time–depth conversion, fault–horizon networks should be built as early as possible in an interpretation project, when it may still be in seismic two-way travel time, in order to incorporate the validation steps in the interpretation routine rather than treat them as an afterthought. The fault–horizon network methods in this paper can therefore be applied before or after seismic time-to-depth conversion.

Building fault–horizon networks

Fault–horizon and fault–fault truncations and intersections are not always as sharply imaged in seismic surveys as one could wish, due to limitations of the seismic method and sometimes inappropriate seismic processing. In order to construct a 'watertight' network, such intersections have to be modelled by extrapolation from zones of good seismic data into zones of poor data (Rouby *et al.* 2000). This does not necessarily mean that the truncations and intersections are poorly defined. The definition of fault and horizon surfaces outside zones of poor seismic data leaves generally little freedom in the extrapolation. The term 'clean-up' is used here to refer to the network building operations, which are detailed below.

Fault–horizon intersections

Unreliable horizon points in the zone of poor seismic data around a fault are identified in two steps, starting with the points immediately adjacent to the fault intersection, followed by points further away that have anomalous depths relative to neighbours (Fig. 2). The latter points are replaced by extrapolations calculated with a geostatistical algorithm, which uses the faultcut mid-line (halfway between the footwall and the hanging wall) as a barrier to avoid horizon depths from the other side of the fault being used. The geostatistical depths were found to be more reliable than depths calculated by linear or curvilinear extrapolation of horizon picks on seismic sections. A straightforward algorithm then obtains the intersection lines of the fault surfaces and the extrapolated horizon surfaces. The resulting cut-off lines are as smooth and well defined as the fault surface and the horizon surface at the intersection. The horizon surface is trimmed

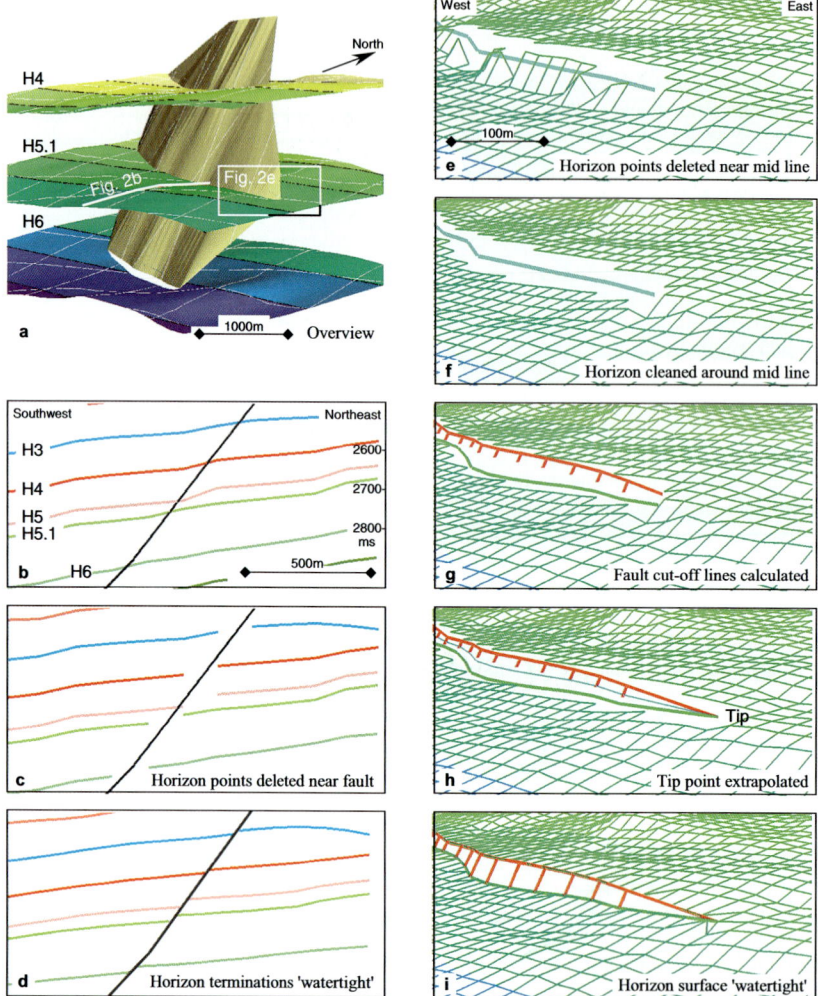

Fig. 2. Building a simple fault–horizon network, first data set. (**a**) 3D perspective overview from SE. Depth interval *c.* 2500–2900 ms seismic two-way travel time (*c.* 3425–3975 m). Depth contour interval 50 ms (*c.* 68.5 m). (**b**) Vertical section with fault picks and horizon picks as interpreted; see (a) for location. (**c**) Unreliable horizon points around fault pick deleted. (**d**) Horizons extrapolated up to the fault pick. (**e**) 3D perspective view from SE of horizon H4; grid cell size 25 × 25 m; see (a) for location. Depth colour interval is 20 ms (*c.* 27.4 m). Faultcut mid-line in blue. The grid nodes near the faultcut mid-line have been deleted. (**f**) Unreliable grid nodes around the mid-line replaced by a value calculated with geostatistics. (**g**) Cut-off lines (footwall red, hanging wall green) calculated at the intersection of the fault surface and the horizon surface. (**h**) Tip point calculated by extrapolation. (**i**) Horizon points near the cut-off lines shifted to form a 'watertight' fault–horizon intersection.

along the cut-off lines, which makes the topology locally complete or 'watertight'. In the case of a grid representation, grid nodes are shifted onto the cut-off lines. This also helps in reviewing the quality of cut-off lines without the annoying visual effect of the zigzag lines at the grid edge.

Fault tip points

Seismic data do not resolve fault tips (Walsh & Watterson 1991). With diminishing throw, the fault-related discontinuity in the seismic reflections starts to disappear when the throw is less than 4–10

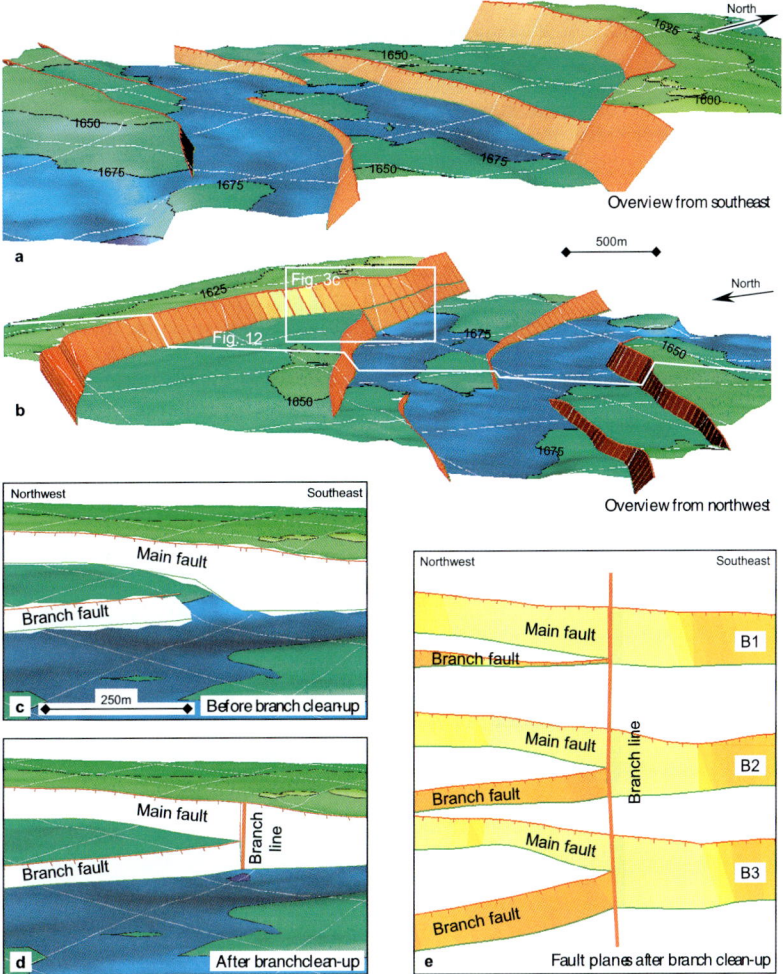

Fig. 3. Building branch fault intersections, second data set. Depth interval *c*. 1550–1750 ms (*c*. 1330–1550 m). Depth contour interval 25 ms (*c*. 21.5 m). (**a**) 3D perspective view of horizon B2 from SE. (**b**) Same horizon from NW. (**c**) Close-up view from SW; see (b) for location. Fault–horizon network building has been completed except around the branch. The branch fault ends near the main fault; the corner points were not accurately mapped. (**d**) The branch line was calculated followed by recalculation of the cut-off lines and the horizon surface around the branch. (**e**) Same 3D perspective view as (c) and (d), showing the cut-off lines and branch line in three horizons.

ms, depending on data quality, depth and seismic processing. The tip locations are therefore modelled by extrapolation (Fig. 2h), in order to avoid an unrealistically large throw gradient from the last observed throw point to the next point where the throw seems to have disappeared. Extrapolation distance is such that the lateral slip gradient at free fault tips is below 2%.

Branch lines

Faults branching along strike are a common phenomenon in deltaic settings. They are thought to develop where relay zones between parallel normal faults have breached (Davison 1994). Branch lines between linked faults can be very complex (Walsh *et al.* 1999). The implementation of the methods reported here only supports sub-vertical branch lines.

Figure 3 shows a detail of the main fault near the branch location in the second data set. There is a large slip gradient in the main fault, and the branch fault terminates abruptly near the main fault. The seismic horizon autotracker created a steep ramp in the horizon. Close inspection of the seismic data showed that the slip on the branch fault shows no sign of diminishing towards the main fault, and that the lateral slip gradient on the main fault is abrupt and high at the branch fault. In other words, there is no support for this ramp, and the branch fault should meet the main fault. Three intersecting surfaces create the bifurcation: the two faults and the horizon. The branch line is the intersection line of the two fault surfaces. The upper and lower corner points of the horizon are sited at the intersection of the branch line and the unbroken horizon surfaces; the middle corner point is sited between these two points such that the lateral slip gradient of the branch fault remains below 2%. The fault–horizon network at the branch location is now topologically complete.

The fault-slip vector field

The next step is the construction of the fault-slip vector field. This is derived from the correspondence of formerly adjacent points on opposite walls of a fault cut-off in the fault–horizon network. After the above clean-up operations the fault cut-off lines are well established. The slip vectors between them are defined relative to a local co-ordinate system defined for each fault (Fig. 4). The slip vector direction is calculated from the direction of the average fault dip and an estimate of the strike-slip component of the fault displacement. The strike-slip component can be estimated from the offset of topographical features across a fault (e.g. a river bed), from large-scale rugosity of the fault surface (Kerr & White 1996) or from the topology of branching faults (as discussed later in the context of Fig. 13). The strike-slip component may vary along strike, but only in a smooth manner.

Slip-based diagnostics

The internal consistency of the slip field is checked (see Fig. 1) by visual inspection in a 3D viewer, and by diagnostic tools using its lateral and vertical variability. The lateral and vertical variability of

the slip field of a fault is limited by rock strength. Given two adjacent points that are on the same side of a fault and that move according to corresponding fault slip vectors, the volume of rock between the points will fail if they end up too far apart. Errors and inaccuracies in seismic interpretations often lead to mapped slip gradients that exceed this failure limit.

Lateral slip variation

The 'lateral slip variation' is the variation of the slip vector, or one of its components (e.g. throw or heave), as a function of distance along the horizon cut-off lines. Barnett *et al.* (1987) proposed to use the lateral slip variation as a quality-control parameter for seismic interpretations. In coal mines, the lateral throw gradient on normal faults is never higher than 1.5% in the main part of a fault, and somewhat higher but still below 2% in transfer zones between faults (Huggins *et al.* 1995).

The lateral slip vector variation can be displayed as a function of one of its components along a fault cut-off (Fig. 5). Before clean-up of the first data set there are differences of up to 8.7 ms (11.8 m) between adjacent values. This equates to a lateral gradient of 0.397 m/m or 39.7%, which is clearly not realistic. After clean-up, the profile is smoother, because low values that were due to unreliable horizon points (see Fig. 2e) have been corrected. The maximum lateral gradient is 10.5%, which is still on the high side as compared with the above-cited literature values, but maybe not unreasonable in these unlithified delta sediments.

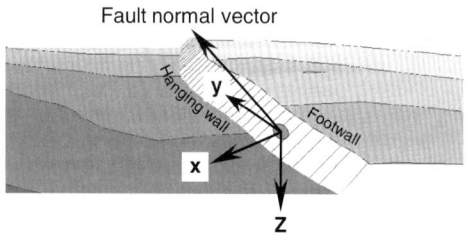

Fig. 4. Local fault co-ordinate system. The fault normal vector is orthogonal to the average fault surface. The x co-ordinate is horizontal and away from the fault, y is along strike and z is downwards.

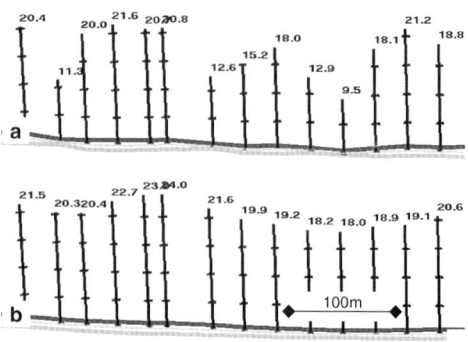

Fig. 5. Fault slip profiles along faultcut of Figure 2e–i. The line segments plotted vertically on the faultcut midline are proportional to throw. Throw values in milliseconds, tick marks spaced 5 ms (*c.* 6.8 m). (**a**) Before clean-up, showing unrealistic lateral variation. (**b**) After clean-up, with smoother lateral variation.

Vertical variation of slip

The 'vertical slip variation' is the variation in slip vectors, or in one of the slip vector components, between adjacent horizons. Large vertical slip gradients are geologically plausible in synsedimentary growth faults, but in post-sedimentary faults, such gradients imply equally large compaction gradients in layers between adjacent horizons. Such large gradients are unlikely to be geologically plausible in post-sedimentary faults. In general, the vertical slip gradients must be geologically meaningful, and between closely spaced horizons, they should be small.

A useful way to inspect slip variations is in a view perpendicular to the fault plane (Fig. 6). Values of a slip component in each horizon cut-off (or the first derivative in a given direction) are recorded on the faultcut mid-lines (half-way between footwall and hanging wall cut-off lines) and contoured on the fault surface. The contours before clean-up of the first data set show sharp angles and large gradients. After clean-up the contours are smoother, and the distance between them is larger. The distinct highs in the contour pattern suggest that the fault was formed by coalescence of two or three earlier faults.

Unfaulting method: introduction

In the unfaulting method proposed here, the three-dimensional kinematics related to faulting are approximated with the simplest possible analytical function calculated from the slip vector field: a linear extrapolation of the slip vectors away from the fault.

The main steps in the method are as follows

(Fig. 7). Each point of a horizon is projected onto the hanging wall cut-off line of a fault. At the projection point, a fault slip vector is obtained by interpolation along the fault. The displacement vector of the horizon point is taken to be parallel to the interpolated fault slip vector but in the reverse direction. Its magnitude is taken to be equal to the interpolated fault slip vector and smaller away from the fault, according to an analytic decay or other extrapolation function. The length of the extrapolation function, or the width of the zone affected by the unfaulting operation, is proportional to the slip magnitude.

The first data set is used to introduce the prin-

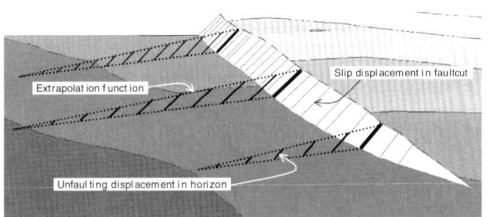

Fig. 7. The slip-extrapolation unfaulting method. The starting point of the method is the slip vector field, mapped as precisely as possible while building the fault–horizon network. The slip vectors in the faultcut are first inverted. The unfaulting displacement vectors in the horizon are then calculated by extrapolation of the slip vectors away from the fault. The orientation of the extrapolation function is orthogonal to the average fault surface defined by the fault normal vector (Fig. 4). A (variable) strike-slip component in the fault is handled by a corresponding displacement out of the plane of extrapolation (Fig. 10). The magnitude of the horizon displacement vectors is determined by an extrapolation function (Fig. 11).

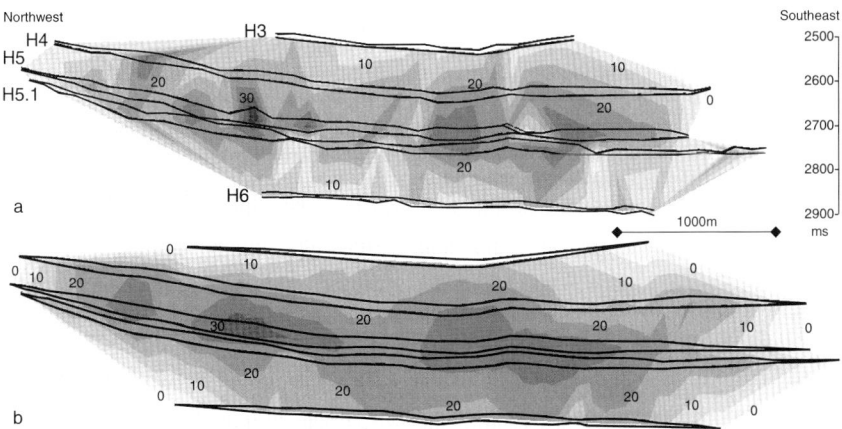

Fig. 6. Throw contours of the fault in the first data set. Cut-off lines of five horizons. Throw contour interval 5 ms (c. 6.8 m). (**a**) Contours before clean-up, with large gradients, both along horizons and between horizons. (**b**) Contours after clean-up, with smaller gradients.

ciples of the unfaulting method. Details of the method are then explained and illustrated with the second data set. Limitations of the method and a comparison with other methods are discussed at the end of this paper.

Unfaulting the first data set

Figure 8 shows the unfaulting operation on the fault–horizon network of the first data set. The fault slip vectors between the cut-off lines used in the unfaulting are shown in orange in the perspective views. In the unfaulted state, the fault gaps have disappeared and the horizons are continuous across

the fault. The horizons on the hanging wall side have been pulled up and towards the footwall cut-off to close the fault gaps. The hanging wall is hardly deformed because the strain is distributed over a large area. The three intermediate views in Figure 8 are linear interpolations between the present-day state and the unfaulted state. In a workstation display, such interpolations can be shown as a movie, which helps the interpreter to understand the structural development of complex fault systems.

Note that the section view and the 3D perspective view in Figure 8 are two views of the same unfaulting process. The unfaulting of the section

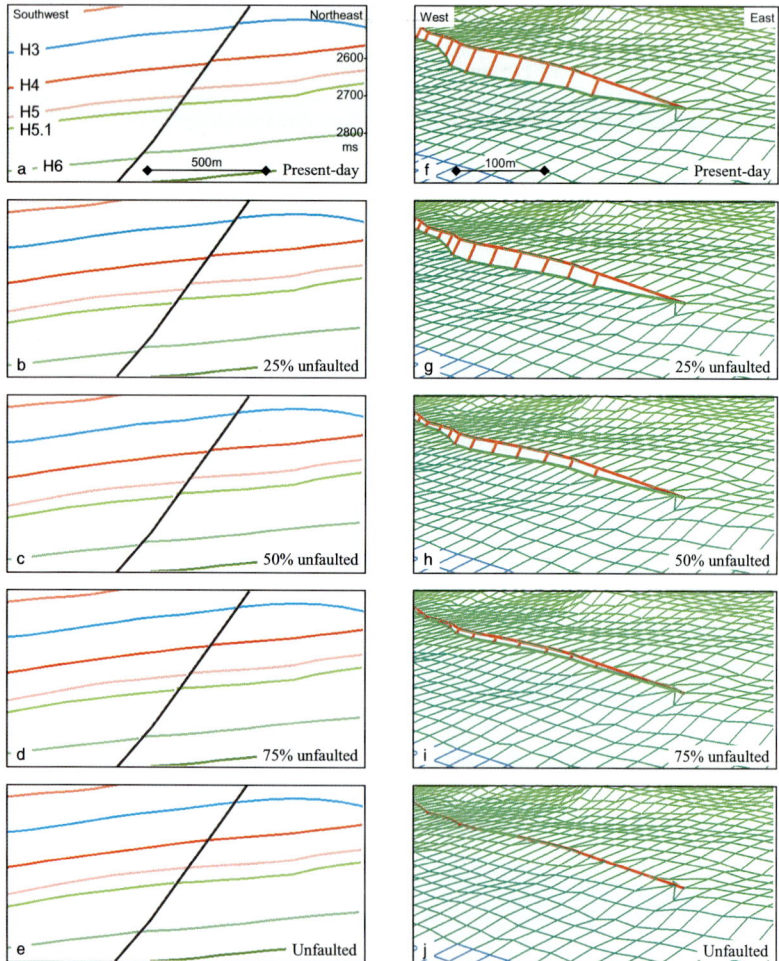

Fig. 8. Unfaulting the first data set. Unfaulting extrapolation gradient 2%. Three interpolated unfaulting steps. (**a–e**) Cross-section through the network. Same section as Figure 2b–d. Horizons unfaulted individually with the slip-extrapolation method. Because slip is consistent from one horizon to the next, unfaulting of individual horizons is consistent in 3D. (**f–j**) Perspective view of horizon H4, same view as Figure 2e–i. The horizon surface in the hanging wall is pulled up and inwards until cut-off lines coincide.

was *not* carried out in the plane of the cross-section, but for each *full* horizon surface separately. The section is just a vertical slice through the fault–horizon network, and the unfaulting operation displaces points out of the present-day section plane. Although all horizons were unfaulted separately, the unfaulting results do not show unrealistic thickness variations between the present-day and unfaulted states, because the fault–horizon network is consistent from one horizon to the next. This will be further discussed in the context of Figure 12.

Maps of horizon displacement and deformation

Maps of unfaulting results for the first data set are shown in Figure 9. In the unfaulted state, the depth contours of the horizon are continuous across the fault. The character of the depth contours away from the fault is unchanged, indicating that details in horizon structure are unchanged. Contour maps record the amount of unfaulting displacement in the North, East and Down directions. The contours

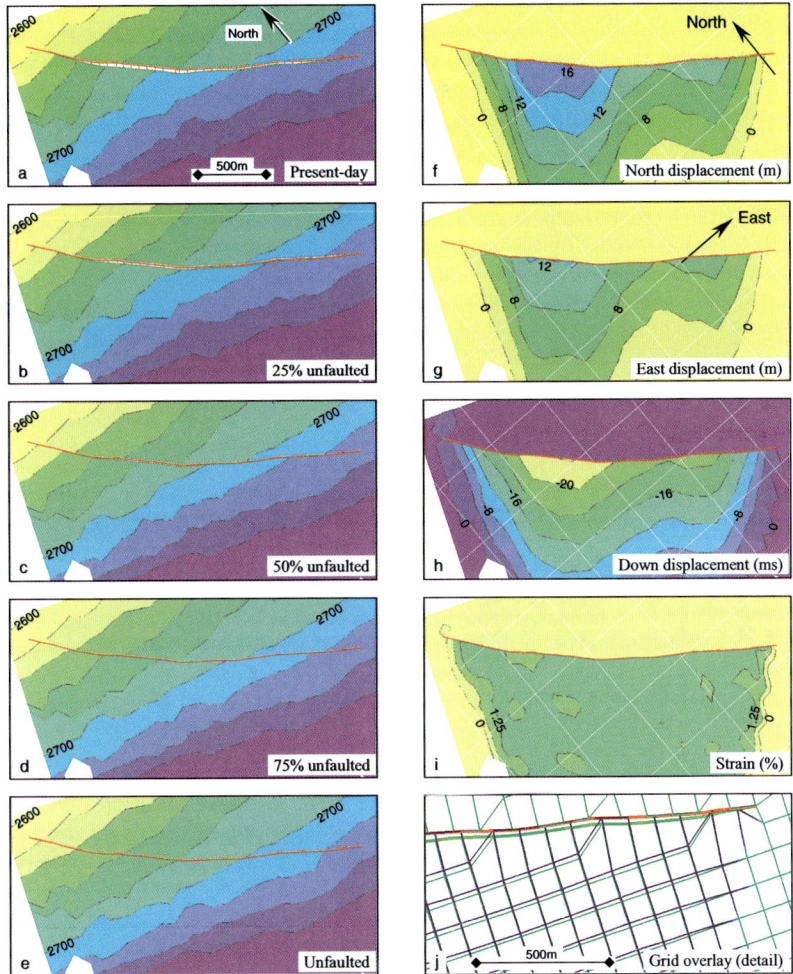

Fig. 9. Map view of horizon H5 in the first data set; no perspective, depth contour interval 20 ms (*c.* 27.4 m). Unfaulting extrapolation gradient 2%. (**a–e**) Unfaulting series with three interpolated steps. (**f–h**) Contour maps of three components of unfaulting displacement. Contour interval 4 m for horizontal, 4 ms (*c.* 5.5 m) for vertical movement. Vertical displacement is negative because depth is positive downwards. (**i**) Contour map of the horizontal strain related to the unfaulting process. Contour interval 0.625%. Deformation is approximately constant *c.* 1.25% in the middle, falling off rapidly at the edges of the displaced area. (**j**) Close-up of the horizon surface, overlaying its state before and after unfaulting. Grid cell size 100 × 100 m.

have a high value at about one-third of the fault length away from the left-hand end. This is the locus of the largest slip (see Fig. 6), and the displacement contours extend there furthest away from the fault because the length of the extrapolation is proportional to slip magnitude. Unfaulting strain is recorded in the strain contour map. Strain is defined here as the difference in area measured in the present-day state and in the unfaulted state. The strain in this data set is very low: the strain contour map shows values up to only about 1.5% and the detailed mesh overlay shows cells that are shifted laterally but that are hardly deformed.

Unfaulting method: details

The unfaulting method is an extension of section-balancing methods to three-dimensional horizon restoration. Unfaulting is applied by extrapolation of the fault slip vectors in planes oriented orthogonally to each fault (Fig. 7) rather than along an arbitrary overall cross-section. A given horizon point is unfaulted for multiple faults at the same time by summing the unfaulting displacement vectors for individual faults, each unfaulting displacement vector being calculated in the extrapolation plane through the horizon point orthogonal to that individual fault. Furthermore, the unfaulting operation can move points out of the extrapolation plane in the case of strike-slip. The orientation of the extrapolation plane is discussed first, followed by the extrapolation function itself.

Orientation of the slip extrapolation plane

The slip extrapolation plane (Fig. 7) is oriented orthogonally to the average fault surface, along the direction defined by the fault normal vector (Fig. 4). Use of a single orientation makes the unfaulting algorithm straightforward and fast. This does not mean that the direction of displacement is constant for each fault, because the direction of the unfaulting displacement vectors is given by the slip vectors (but reversed). The fault slip vectors can have a (variable) strike-slip component, which will move points out of the plane of extrapolation (Fig. 10). Thus, the direction of movement is independent of the orientation of the plane of extrapolation.

Extrapolation function

The extrapolation function relates the displacement magnitude away from the fault linearly with distance and the slip at the fault (Fig. 11). In this paper, the extrapolation function has a gradient of 2%, which corresponds with observed relations between maximum fault throw, fault length and the

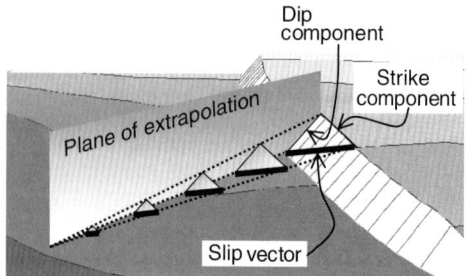

Fig. 10. Unfaulting in the presence of strike-slip. The plane of extrapolation is orthogonal to the average fault surface (Fig. 4). In the case of strike-slip, horizon points are displaced laterally out of the plane of extrapolation during unfaulting. The amount of lateral displacement depends on the strike-slip component and is proportional to the distance to the fault.

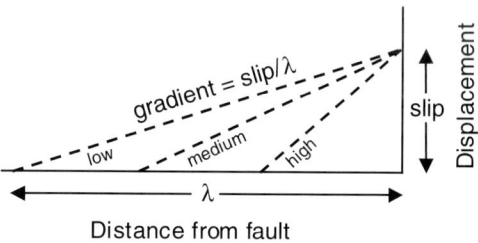

Fig. 11. The slip-extrapolation function. The gradient of the extrapolation function is a free parameter. The average horizon deformation implied by the unfaulting function is approximately equal to the gradient, e.g. a 2% gradient leads to c. 2% average strain for a single fault when the slip vector field is smooth.

extent of fault-related deformation (Gibson *et al.* 1989; review in Davison 1994).

The 2% gradient leads to average strain values of around 2% for unfaulting a single fault with a smooth slip field. In data sets with more faults, or with poorly mapped slip fields, the average strain will be higher. The minimum strain due to unfaulting is close to the gradient value, because the unfaulting displacement is distributed over the extent of the extrapolation. A sensitivity analysis on several data sets has shown that the unfaulting results are not sensitive to the type of extrapolation or the extrapolation gradient as long as the gradient is small.

Sequence of unfaulting

The unfaulting operation can be applied to all horizons simultaneously or sequentially, as well as to all faults simultaneously or sequentially, or in any combination. Sequential restoration of horizons is required when vertical throw variations correspond to unconformities or varying rates of fault growth, in order to avoid geologically meaningless thickness variations between horizons in the unfaulted state. Contemporaneous faults should in principle be unfaulted simultaneously. Even then, it may be helpful to unfault faults individually, in order to inspect their impact on horizon strain and assure the quality of interpretations. It is equally well possible to unfault the slip in parts, for instance in the case of reactivated faults or across unconformities.

Sequential unfaulting of faults is illustrated in Figure 12, which shows a section through the second data set. Unfaulting only the blue fault on the left does not change the left-hand fault block position. Faults and horizons in the graben are pulled up and leftwards to close the blue fault gap. Unfaulting only the main fault on the right leaves the right-hand fault block unchanged and pulls the graben upwards and to the right to close the main fault gap. When all faults are unfaulted together, the resulting displacement pattern is complex and difficult to predict, as it is the sum of the displacements of all faults. For a geologically consistent slip vector field, as in this example, the displacements are geologically plausible as well, that is, more or less parallel and of similar magnitude within each fault block.

Fault branches

Individual fault segments may grow and connect into linked fault systems (e.g. Davison 1994; Walsh *et al.* 1999). Where one such fault branches off another one, the slip vectors on the through-going or 'main' fault have a very large lateral gradient even after correctly calculating the intersection (see Fig. 3). On either side of the branch fault, there are two unfaulting trajectories between the hanging wall cut-off line and the footwall cut-off line (Fig. 13). On the left side of the branch fault (i.e. behind it), the trajectory passes first to the corner point on the footwall side of the branch fault, and then to the footwall of the main fault. On the right side (i.e. in front), the trajectory connects the hanging wall of the main fault directly with the footwall. The challenge for the unfaulting algorithm is to unfault both trajectories without pulling the points apart.

As a solution to this challenge, we propose to decompose the discontinuous slip on the main fault into continuous segments. To illustrate this method, a block model is used in Figure 13. The block model has three fault blocks, separated by a through-going main fault and a branching fault. Part of the slip on the main fault – towards the right of the branch line – is allocated to the branch fault. This is done by drawing a line in the main fault from the branch corner point towards the right where it converges eventually with the hanging wall cut-off line (Fig. 14). The upper part of the main fault is referred to as the 'main fault swallow', while the branch fault and the lower part of the main fault are referred to as the 'branch fault swallow'. The term 'swallow' comes from the visual aspect of the segments, with their two 'wings' that terminate with a tip point on one end and the convergence point on the other end. There is now a single unfaulting trajectory at the branch, consisting of the trajectory across the branch fault swallow and the one across the main fault swallow; the unfaulting algorithm cannot pull points apart. The two fault swallows can be unfaulted sequentially or simultaneously, both ways resulting in a smooth restoration around the branch line (Fig. 13).

Strike-slip constraint at branch

An important aspect of the 'swallow' slip decomposition method is that it provides a constraint on the strike-slip component of branching faults. Block 3 in Figure 13 is constrained to move along the branch line in order to keep in contact with the other two blocks, across the branch fault and across the main fault. Unfaulting block 3 is implemented in our method by closing the branch fault swallow gap, displacing block 3 along the branch line. The orientation of the branch line therefore determines the strike-slip component of the branch fault and of the branch fault swallow part of the main fault. The movement of blocks 2 and 3 together relative to block 1, across the main fault, is not constrained by the branch geometry. This observation is supported by finite-element work by Maerten *et al.* (1999). The lateral curvature of the main fault surface in the second data set (see Fig. 3) suggests that it is reasonable to assume pure dip-slip for it. The slip directions in Figure 13c show a slight angular discrepancy between the main fault swallow (dip-slip) and the branch fault swallow (slip along branch line). In other data sets, angular discrepancies of up to 30° have been observed.

Unfaulting the second dataset

The faults in the second data set are contemporaneous crestal collapse faults, so they should be unfaulted together (Fig. 15). The orange lines between the cut-off lines are the fault slip vectors of the fault–horizon network. The slip vectors are *not* parallel and they have an appreciable strike-slip component near the branch line. In its unfaulted state, the horizon is smooth across the closed faultcuts. The resulting topography has shal-

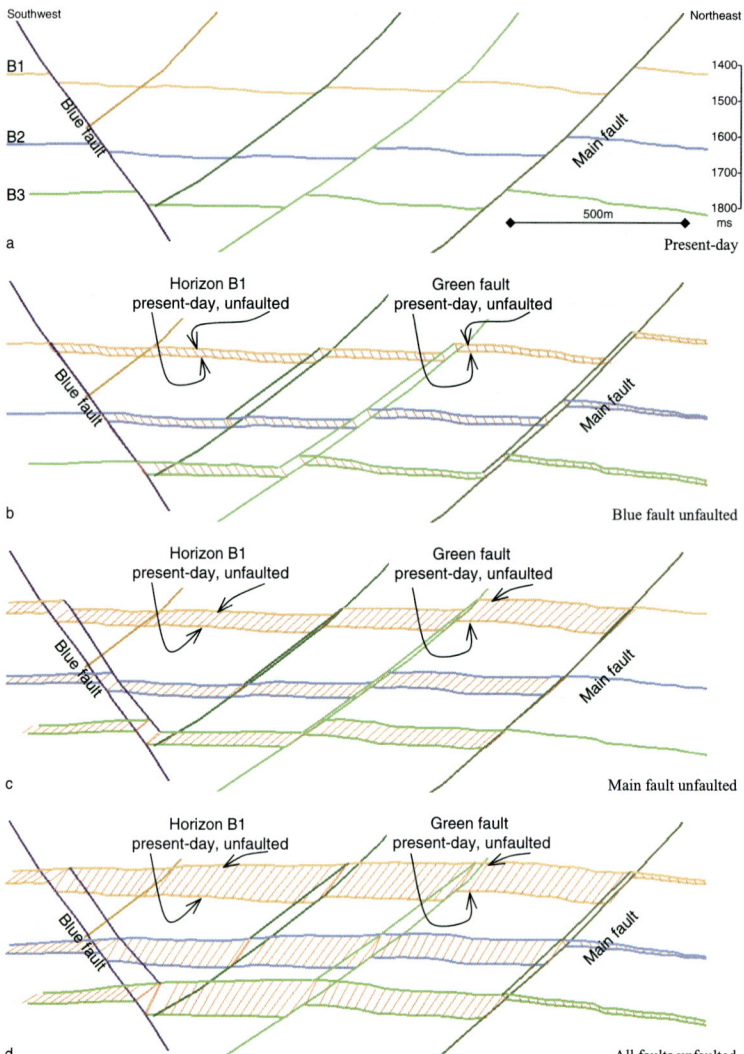

Fig. 12. Unfaulting faults sequentially in the second data set. Unfaulting displacements in orange. Horizons unfaulted individually. See Figures 3b and 16b for section location. Extrapolation gradient 2%. (**a**) Present-day state of three horizons and five faults. (**b**) Unfaulting only the blue fault. Horizon points are displaced to the left and up, parallel to the slip vectors in the blue fault. (**c**) Unfaulting only the main fault. (**d**) Unfaulting all faults simultaneously.

low valleys and a broad high in the centre. This topography is the sum of the original present-day geometry of the horizon plus the aggregate effect of the unfaulting displacements. Errors in the slip field (or, incidentally, errors in the unfaulting algorithm) would immediately be evident as sharp ridges and valleys in the unfaulted topography. Two fault scarps in the lower left corner of the area are not unfaulted because their slip field is poorly constrained.

Strain diagnostics

A map of the strain related to unfaulting the second data set is shown in Figure 16a. Its use as a diagnostic tool for validating the fault–horizon network is illustrated as follows. The strain map shows anomalously high strains, above 15%, in the lower left part. On inspection this is due to a large 12.9–22.8% lateral throw gradient near the tip point of the branch fault. This may be due to an interpret-

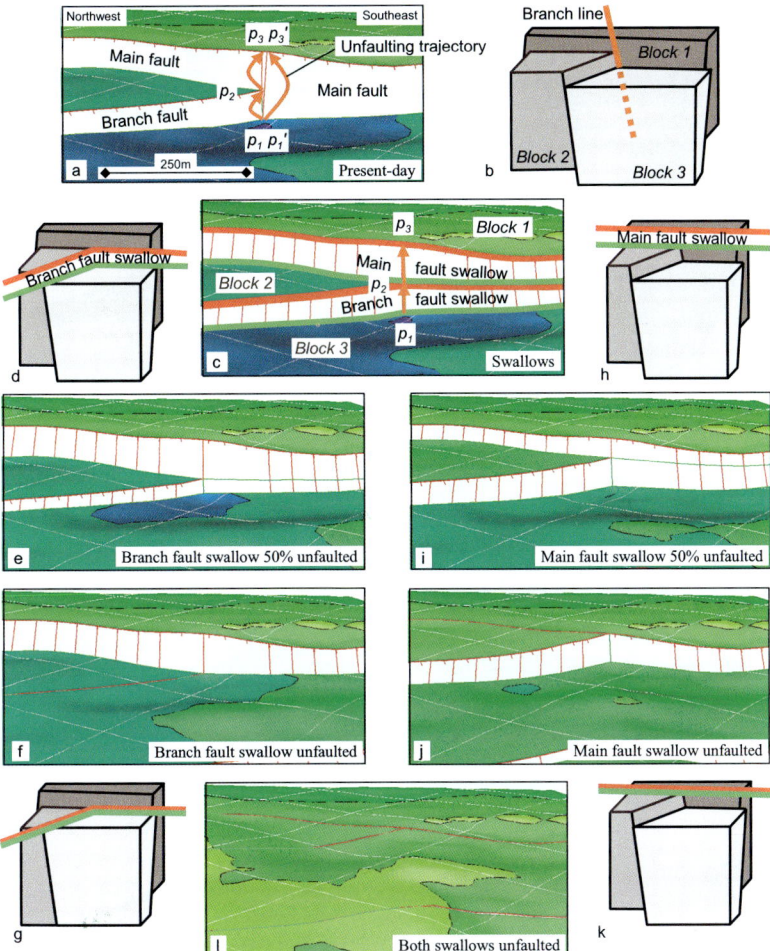

Fig. 13. Unfaulting the branching faults in the second data set, using the 'swallow' slip decomposition method. Fault slip vectors in orange. Same 3D perspective view as in Figure 3c,d. (**a**) Branch intersection as in Figure 3d. There are two different unfaulting trajectories at the corner points: $p_1p_2p_3$ behind, and p_1' p_3' in front of the branch fault. (**b**) Block diagram of the situation in Figure 13a. Block 3 must move along the branch line to maintain contact with the other two fault blocks. (**c**) The slip field on the main fault is decomposed into a main fault swallow and a branch fault swallow. Slip on the main fault swallow is pure dip slip, while on the branch fault swallow it is in the direction of the branch line. Each swallow has its own unfaulting trajectory: p_1p_2 and p_2p_3. (**d–g**) Separate unfaulting of the branch fault swallow, with one interpolated step. (**h–k**) Separate unfaulting of the main fault swallow, with one interpolated step. (**l**) Both swallows unfaulted. Same result is obtained by unfaulting the swallows sequentially or simultaneously.

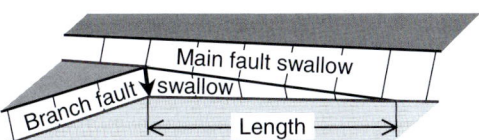

Fig. 14. 'Swallow' fault slip decomposition. Same 3D perspective as Figure 13. The length of the branch swallow 'wing' in the main fault is proportional to the magnitude of the branch slip vector at the branch location.

ation error that becomes obvious in the strain map, or it may be due to a natural increase in tip strain in the overlap zone between two en-echelon faults (Walsh *et al.* 1999). In stronger rocks, such an area would be populated by extensional fractures perpendicular to the overlapping faults.

In the rest of the map, the strain is less than 10%. When individual faults in this data set are unfaulted (see Fig. 12), the strain related to unfaulting each fault is around 2%. The aggregate unfaulting strain for a set of faults is the sum of unfaulting strains

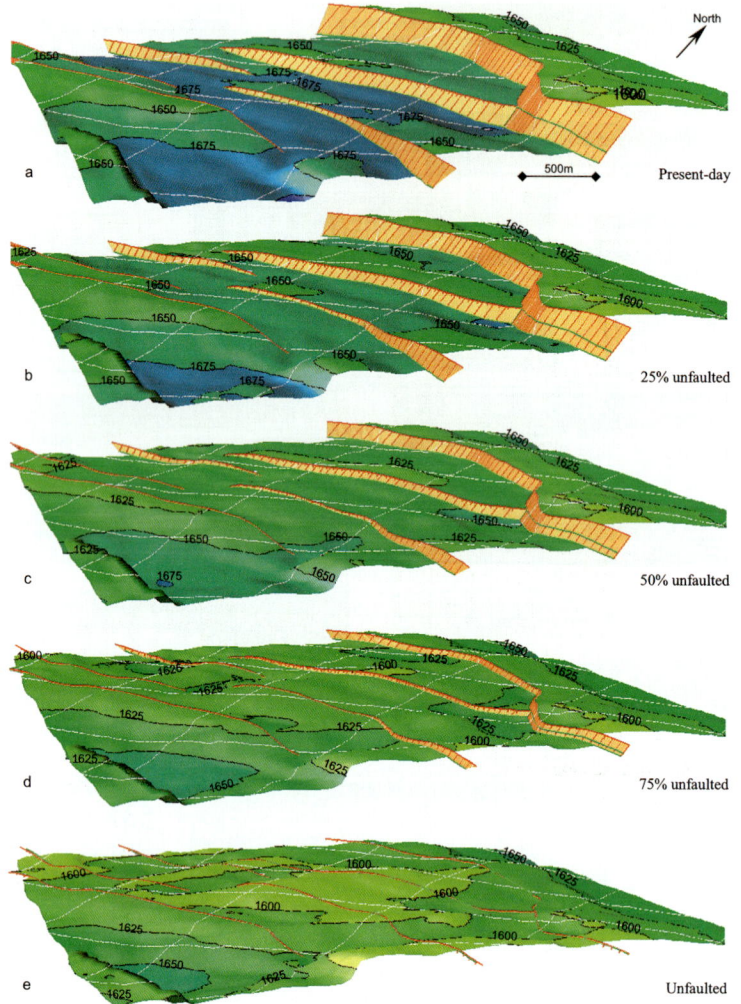

Fig. 15. Unfaulting horizon B2 in the second data set. Perspective view from the south. Depth contour interval 25 ms (*c.* 21.5 m). Unfaulting extrapolation gradient 2%. (**a–e**) Unfaulting series with three interpolated steps. Depth contours are continuous across the faults in the unfaulted state. The residual topography is determined by the initial topography of the horizon and the unfaulting displacements. In this case, the surface is acceptably smooth and flat except in the south where structure is influenced by two faults located largely outside the map area.

related to individual faults. In this case, the aggregate unfaulting strain roughly equals the constant gradient of our simple extrapolation function, times the number of faults that influence a given horizon area. The distribution and values of the strain in this example are thus geologically plausible except perhaps for the anomaly near the tip of the branch fault, which needs to be inspected in the seismic interpretation canvas.

Displacement maps

The unfaulting results can be further validated by means of maps of the aggregate displacement vectors of individual horizon points (Fig. 16b). Displacements are small in the fault block at the top because all unfaulting displacement has been allocated to the hanging wall, and there are only smaller faults facing it (see Fig. 12). The displace-

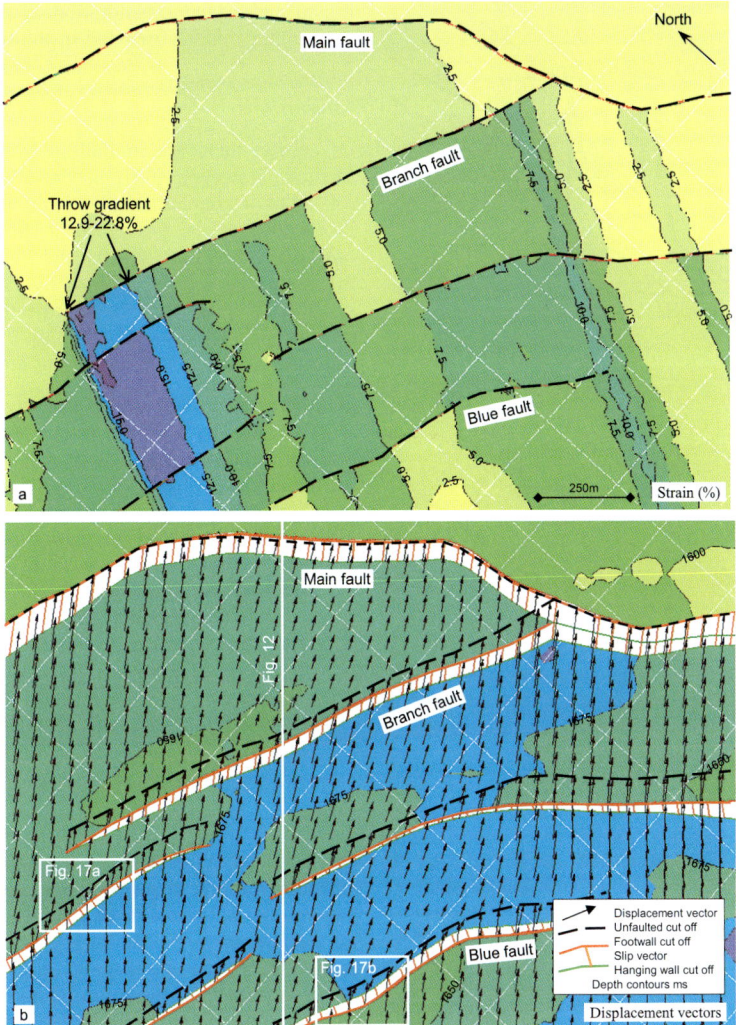

Fig. 16. Unfaulting strain and displacement in the second data set. Extrapolation gradient 2%. (**a**) Horizontal strain related to unfaulting. Contour interval 2.5%. High strains on the lower left are caused by a large lateral slip gradient near the tip of the branch fault. (**b**) Unfaulting displacement vectors of individual points on a depth contour map. Depth contour interval 25 ms. Every second vector shown.

ment patterns are not always immediately obvious, since they are the result of interference from all nearby faults. Detailed displacement patterns near two oppositely hading faults are shown in Figure 17. At first glance, the two displacement patterns look very similar, but on closer inspection the displacement arrows are complex. Both footwall and hanging wall blocks undergo the aggregate displacement of nearby faults recorded by the vectors in the footwall block. The vectors in the hanging wall block are the composite of those in the hanging wall block and the slip vectors in the fault itself.

In Figures 16 and 17, there are angular differences of up to about 30° between displacement vectors in adjacent blocks, which make a plane-strain assumption invalid for this data set. This is in agreement with other restorations of a data set in the same area (Rouby & Cobbold 1996; Kerr & White 1996).

The unfaulting results of the second data set show that our method of unfaulting by slip extrapolation can successfully cope with a typical deltaic data set with branching faults, varying fault orientations and laterally varying strike-slip.

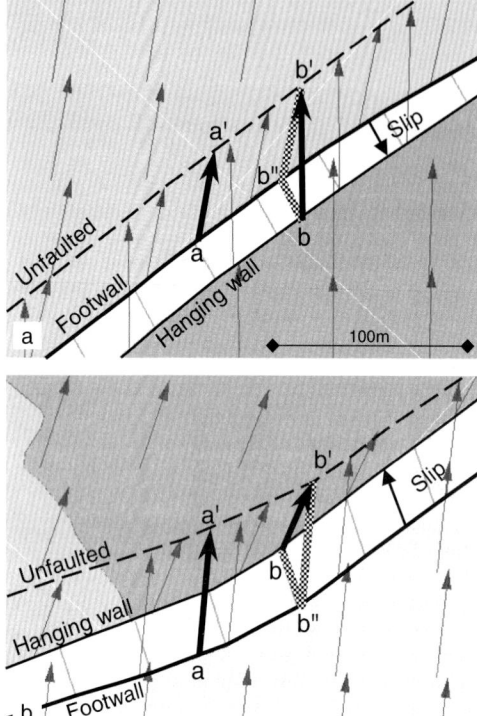

Fig. 17. Two details of unfaulting displacement vectors. See Figure 16b for locations. Vector a–a′ displaces a point on the footwall cut-off line to its unfaulted position. This vector represents the aggregate displacement due to unfaulting other faults. Vector b–b′ displaces a point on the hanging wall cut-off line to its unfaulted position. This vector has two components (in grey): b–b″ for closing the 'own' fault cut-off, and b″–b′ for the aggregate displacement due to other faults (this is a copy of a–a′). Note that the position of the footwall and hanging wall is reversed in (**a**) and (**b**). The dominant displacement direction is towards the top, for unfaulting the main fault (see Fig. 16b).

Discussion

Now that each of the steps in our unfaulting method has been described and illustrated, it is useful to put it into context, record the limitations, compare it with other methods, and discuss usability and extensibility.

The general principle of the unfaulting method is that it takes the fault slip vector field as a given, and extrapolates it with a simple analytical function that approximates fault deformation kinematically. The resulting unfaulted state is directly comparable with the slip field; there are no 'magic' steps involved. All interpretative effort can be focused on obtaining a correct slip field, rather than on try-

ing to understand a restoration of a data set that was perhaps poorly mapped in the first place.

Limitations

Of course, there are limitations to the unfaulting method. However, most of these limitations are related to our implementation in computer software rather than to the principles of the method. The only real limitations are that deformation must be moderate and mostly induced by faults: it is not possible to reconstruct large-scale thrusts or salt-related structures. Examples of current implementation-related limitations are listed below.

(1) The local fault co-ordinate system (Fig. 4) restricts the software to fault surfaces that are single-valued with respect to the fault normal.
(2) Faultcuts must have a minimum width, restricting the software to normal faulting.
(3) All fault-related deformation is attributed to the hanging wall, instead of distributing it between hanging wall and footwall in order to unfault uplifted footwalls as well. (In the two data sets used in this paper, there is no evidence of footwall uplift in the overlying sediments.)
(4) Only sub-vertical fault branch lines can be handled.

Comparison with other unfaulting methods

There are two main families of unfaulting methods and software. The first one is based on section balancing methods pioneered by Dahlstrom (1969) or on minimization of unfaulting strain in section view. The second one is based on closing the gaps between rigid fault blocks in plan or map view.

Section balancing methods move back the hanging wall along the fault trace using a choice of hanging wall deformation models (vertical shear, inclined shear, constant fault-slip, and bedding-parallel slip) reviewed by Dula (1991) and Roberts & Yielding (1994). Section balancing software is widely accepted (see Internet sites in the list of references): 3DMove (Gibbs 1983), GeoSec (Kligfield et al. 1986), Locace (Moretti & Larrére 1989), Restore (K. Duncan, described in Busbey 1991) and DepthCon2000 (J. Nicholson, pers. comm. 2001). Other section restoration methods minimize unfaulting strain of triangulated sections by rigid translation and rotation of fault blocks followed by deformation (Wickham & Moeckel 1997; Erickson et al. 2000).

In map restoration, the second family of unfaulting software, fault gaps in a map are closed by rotation and translation of rigid fault blocks (Rouby et al. 1993, 2000; Gratier et al. 1991, 1999; Williams *et al.* 1997). The accuracy of map restorations is of course dependent on the accuracy with which the fault gaps in the map are determined. In routine oil-industry workflows, fault gaps are often over-estimated leading to unrealistic displacements while unfaulting. Furthermore, the matrix solvers used in the least-squares strain minimization algorithms of map restoration software impose a practical upper limit of 1000–10 000 on the number of triangular elements in the map.

Our unfaulting method combines the principle of moving back and deforming the hanging wall that is used in the section restoration methods with the principle of closing the fault gaps that is used in the map restoration methods. Unfaulting is applied here in extrapolation planes orthogonal to each individual fault. Unfaulting displacement vectors are not constrained to these orthogonal planes but may be oriented at an angle in order to cope with strike-slip displacement. Fault gaps are closed here by reversing the fault slip vectors that are determined as accurately as possible while building the fault–horizon network. In the network, the slip vectors in one horizon are made consistent with the vectors in the next horizon so that unfaulting displacement is also consistent from one horizon to the next. The number of horizon points that can be unfaulted is in principle unlimited; the method scales linearly with the number of horizon points and with the number of faults.

Our method removes the following limitations of the two families of unfaulting methods.

(1) Most section and map restoration algorithms require fault blocks bounded by (artificial) faults on all sides. In deltaic settings with prevalent soft-domino deformation (Walsh & Watterson 1991), such fault-bounded blocks are the exception rather than the rule. In such settings, fault blocks are also a practical obstacle to unfaulting multiple sections or horizons, because fault block subdivisions are notoriously difficult to carry from one section or map to the next. Our method decomposes fault-related strain in a geologically meaningful way, avoiding artificial faults and fault blocks.

(2) In our unfaulting technique, there is no notion of a pin line. The unfaulting will not displace anything that is beyond the extent of the extrapolation function.

(3) Our unfaulting method does not require the user to specify a target surface (e.g. palaeobathymetry) towards which to reconstruct. The method only removes the fault deformation from each horizon, without prior assumption of the shape of the undeformed horizon. The depth and curvature of the horizon surface in the unfaulted state are determined by the present-day state of the surface and the displacement needed to close the fault gaps. Decompaction, uplift and unfolding are cleanly separated from the unfaulting operation, and have to be modelled separately.

Usability

The essential feature of our unfaulting method is to approximate the three-dimensional kinematics related to faulting by a simple analytical extrapolation function of the fault slip vector field. This has the following advantages with respect to ease of learning and use, as well as speed.

(1) A user has to understand and experiment with only a single degree of freedom, namely the gradient of the extrapolation function, instead of several degrees of freedom for every fault block, as is the case for the 3D inclined shear method (Buddin *et al.* 1997). This allows the user to quickly handle any number of faults.

(2) Constructing a geologically consistent fault–horizon network with a well-behaved slip vector field is a logical extension and improvement of the interpreter's daily work. Our unfaulting method reinforces this activity by describing the fault-related horizon deformation in a way that is close to the structural intuition of seismic interpreters.

(3) A complex branching fault network can be treated with the same ease as a set of isolated faults. This is thanks to the use of 'fault swallows', which are computed directly from the fault–horizon network geometry, and which decompose the discontinuous slip and strain at fault branch lines in a kinematically sound way.

(4) The analytical nature of the slip extrapolation makes it very fast. Its execution time scales linearly with the total length of fault cut-offs and with the number of points that represent the horizon. Unfaulting a horizon of the second data set with 100 000 horizon points and seven faults takes five seconds on a 433 MHz CPU. This allows unfaulting to be carried out routinely after every seismic interpretation session.

Extensibility

Our unfaulting method was developed for validation of seismic interpretations and for providing a structural framework for reservoir modelling, with emphasis on deltaic settings. For other purposes and for other geological settings, the method can be extended in small steps with more sophisticated algorithms. For example, the gradient of the linear extrapolation function can be split into separate gradients for the horizontal and vertical components of displacement, in order to allow for differences between shear and dilatory strain behaviour. A further extension can be envisaged which partitions part of the fault-related deformation to the footwall, instead of limiting it solely to the hanging wall as is illustrated in this paper. Also, the linear extrapolation function can be replaced by an impulse response function based on observations (Barnett *et al.* 1987) or elasto-dynamic theory (Ma & Kusznir 1993).

Conclusions

Tools that automate the clean-up of horizon and fault surfaces and their intersections help to build a geologically consistent and topologically complete fault–horizon network for reservoir modelling, retaining relevant detail of seismic interpretations.

The fault slip vector field of the fault–horizon network, and diagnostic tools based on it, are useful to validate the consistency of a complex fault–horizon network. The fault slip vector field is constrained by the shape of faults as well as the orientation of fault branches.

The construction of 'fault swallows' from the fault–horizon network geometry decomposes the discontinuous slip and strain at fault branches in a kinematically sound way into continuous segments.

Once the fault slip vector field of the fault–horizon network is geologically consistent and decomposed into continuous segments, unfaulting moderately deformed horizons with the slip extrapolation method is fast, intuitive and easy to use, irrespective of the number of faults or their branching complexity. The unfaulting method is not limited to plane-strain situations and handles strike-slip variation along faults. Residual strain after unfaulting provides further validation tools for the network.

The authors acknowledge the permission of Shell Technology Exploration and Production to publish this paper.

References

BARNETT, J. A. M., MORTIMER, J., RIPPON, J. H., WALSH, J. J. & WATTERSON, J. 1987. Displacement geometry in the volume containing a single normal fault. *AAPG Bulletin,* **71**, 925–937.

BOUVIER, J. D., KAARS-SIJPESTEIJN, C. H., KLUESNER, D. F., ONYEJEKWE, C. C. & VAN DER PAL, R. C. 1989. Three-dimensional seismic interpretation and fault sealing investigations, Nun River field, Nigeria. *AAPG Bulletin,* **73**, 1397–1414.

BUDDIN, T. S., KANE, S. J., WILLIAMS, G. D. & EGAN, S. S. 1997. A sensitivity analysis of 3-dimensional restoration techniques using vertical and inclined shear constructions. *Tectonophysics,* **269**, 33–50.

BUSBEY, A. B. 1991. Restore. *Geobyte,* August, 51–53.

DAHLSTROM, C. D. A. 69. Balanced cross sections. *Canadian Journal of Earth Sciences,* **6**, 743–757.

DAVISON, I. 1994. Linked fault systems; extensional, strike-slip and contractional. *In:* HANCOCK, P. L. (ed.) *Continental Deformation.* Pergamon Press, Oxford, 121–142.

DULA, W. F. 1991. Geometric models of listric normal faults and rollover folds. *AAPG Bulletin,* **75**, 1609–1625.

ERICKSON, G. S., HARDY, S. & SUPPE, J. 2000. Sequential restoration and unstraining of structural cross sections: applications to extensional terranes. *AAPG Bulletin,* **84**, 234–249.

GIBBS, A. 1983. Balanced cross-section construction from seismic sections in areas of extensional tectonics. *Journal of Structural Geology,* **5**, 153–160.

GIBSON, J. R., WALSH, J. J. & WATTERSON, J. 1989. Modelling of bed contours and cross-sections adjacent to planar normal faults. *Journal of Structural Geology,* **11**, 317–328.

GRATIER, J. -P., GUILLIER, B., DELORME, A. & ODONNE, F. 1991. Restoration and balance of a folded and faulted surface by best-fitting of finite elements: principle and applications. *Journal of Structural Geology,* **13**, 111–115.

GRATIER J. P., HOPPS, T., SORLIEN, C. & WRIGHT, T. 1999. Recent crustal deformation in southern California deduced from the restoration of folded and faulted strata. *Journal of Geophysical Research,* **104**, 4887–4899.

HUGGINS, P., WATTERSON, J., WALSH, J. J. & CHILDS, C. 1995. Relay zone geometry and displacement transfer between normal faults recorded in coal-mine plans. *Journal of Structural Geology,* **17**, 1741–1755.

KERR, H. G. & WHITE, N. 1996. Application of an inverse method for calculating three-dimensional fault geometries and slip vectors, Nun River field, Nigeria. *AAPG Bulletin,* **80**, 432–444.

KLIGFIELD, R., GEISER, P. & GEISER, J. 1986. Construction of geologic cross sections using microcomputer systems. *Geobyte,* **1**, 60–66.

MA, X. Q. & KUSZNIR, N. J. 1993. Modelling of near-field subsurface displacements for generalized faults and fault arrays. *Journal of Structural Geology,* **15**, 1471–1484.

MAERTEN, L., WILLEMSE, E. J. M., POLLARD, D. D. & RAWNSLEY, K. 1999. Slip distributions on intersecting normal faults. *Journal of Structural Geology,* **21**, 259–271.

MORETTI, I. & LARRÉRE, M. 1989. LOCACE: Computer-aided construction of balanced geological cross sections. *Geobyte,* **4**, 17–24.

NEEDHAM, D. T., YIELDING, G. & FREEMAN, B. 1996. Analysis of fault geometry and displacement patterns. *In:* BUCHANAN, P. G. & NIEUWLAND, D. A. (eds) *Modern Development In Structural Interpretation, Validation And Modelling.* Geological Society, London, Special Publications, **99**, 189–199.

PEACOCK, D. C. P., KNIPE, R. J. & SANDERSON, D. J. 2000. Glossary of normal faults. *Journal of Structural Geology,* **22**, 291–305.

ROBERTS, A. & YIELDING, G. 1994. Continental Extensional Tectonics. *In*: HANCOCK, P. L. (ed.) *Continental Deformation*. Pergamon Press, Oxford, 223–250.

ROUBY, D. & COBBOLD, P. R. 1996. Kinematic analysis of a growth fault system in the Niger Delta from restoration in map view. *Marine and Petroleum Geology,* **13**, 565–580.

ROUBY, D., COBBOLD, P. R., SZATMARI, P., DEMERCIAN, S., COELHO, D. & RICI, J. A. 1993. Least-squares palinspastic restoration of regions of normal faulting – application to the Campos basin (Brazil). *Tectonophysics,* **221**, 439–452.

ROUBY, D., XIAO, H. & SUPPE, J. 2000. 3-D restoration of complexly folded and faulted surfaces using multiple unfolding mechanisms. *AAPG Bulletin,* **84**, 805–829.

WALSH, J. J. & WATTERSON, J. 1991. Geometric and kinematic coherence and scale effects in normal fault systems. *In*: ROBERTS, A. M., YIELDING, G. & FREEMAN, B. (eds) *The Geometry of Normal Faults*. Geological Society, London, Special Publications, **56**, 193–203.

WALSH, J. J., WATTERSON, J., BAILEY, W. R. & CHILDS, C. 1999. Fault relays, bends and branch-lines. *Journal of Structural Geology,* **21**, 1019–1026.

WICKHAM, J. & MOECKEL, G. 1997. Restoration of structural cross-sections. *Journal of Structural Geology,* **19**, 975–986.

WILLIAMS, G. D., KANE, S. J., BUDDIN, T. S. & RICHARDS, A. J. 1997. Restoration and balance of complex folded and faulted rock volumes: flexural flattening, jigsaw fitting and decompaction in three dimensions. *Tectonophysics,* **273**, 203–218.

Internet sites for fault-horizon network building software

FAPS: http://www.badleys.co.uk
gOcad: http://www.earthdecision.com

Internet sites for palinspastic restoration software

DepthCon2000: http://www.digitalgeology.co.uk
GeoSec: http://www.paradigmgeo.com
Locace: http://www.beicip.com
3Dmove: http://www.mve.com
Restore: ftp://element.ess.ucla.edu/restore

Finite strain variations along strike in mountain belts

D. HINDLE

GeoForschungsZentrum Potsdam, Projektbereich 3.1, Telegrafenberg C223, D-14473 Potsdam, Germany (e-mail: hindle@gfz-potsdam.de)

Abstract: In order to quantify finite strain from displacement fields derived from structural restorations in mountain belts, a simple technique using triangular elements (based on the finite element technique) is presented. This adequately describes the discontinuous displacement of a fold-thrust belt, whilst overcoming the difficulties posed by discontinuities which are integrated across appropriate elements for the purpose of strain calculation. The technique has been applied to two fold-thrust belts at contrasting scales, the Central Andes and the Neuchâtel region of the Jura mountains (Switzerland). In both cases, structural restorations have produced the finite displacement fields thought to be involved in their generation. Subsequent calculation of strain variation along strike shows that there is a strong relationship between first-order structure (the scale of major fold axes) and the trends of axis orientations. In spite of the contractional setting of both belts, strong along-strike extension is predicted but this is always in the context of a generally parallel to slightly convergent displacement field, and such extensions are generally related to differential shear accommodated on strike-slip or tear faults. Most elements showing along-strike extension are nevertheless predicted to undergo tectonic thickening. Second-order structures (faults necessary to accommodate local predicted deformation) are also calculated and the enormous difference between slip directions on small-scale faults and the regional 'far field' displacement direction is demonstrated. This is thought to be important when trying to infer large-scale tectonic information from very small-scale structures found in the field.

Along strike, mountain belts often exhibit a high order (i.e. small scale) spatial, structural complexity, and low order (i.e. large scale) simplicity. Geologists characterize regions of orogens in map view as cylindrical or non-cylindrical, and sometimes a general pattern described as an arc or orocline can be deduced. Many geologists have speculated in this way in the past, inferring displacement patterns intuitively from the structural trends seen in mountain belts such as the Alps or the Andes (e.g. Argand 1992; Carey 1955). Hindle *et al.* (2000) and Hindle & Burkhard (1999) have calculated the large-scale relationships between along-strike variation of shortening/displacement in thrust belts and finite strain and shown that some of the earlier intuitive models were physically unlikely. A broad correspondence between predicted finite strains based on relatively simple displacement models and the structural grain of the Jura mountains was demonstrated using a general model similar to that of Ferrill & Groshong (1993) for arcuate shaped mountain belts. The possibility of making a more detailed assessment of finite strain remains by using a slightly modified mathematical technique from that of Hindle *et al.* (2000) and combin-

ing it with plan view restorations based on structural cross-sections. This paper sets out to demonstrate a method based on part of the finite element technique for quantifying finite strain in as much detail as one knows the finite displacement field. The aim of such a calculation is to validate a restoration scheme for a region, to give a true picture of the strain due to differential displacement along strike and predicted local convergence or divergence of displacement vectors, and also to predict some structural trends such as fold or fault orientations that might be found at a smaller scale.

Plan view restoration techniques

Many techniques have been used for plan view restoration and these can be subdivided according to the amount of detail necessary in an initial data set to successfully use a particular method of restoration. The first group are all examples of serial 2D cross-section balancing, with data integrated along strike between sections to give a compatible plan view restoration. This general approach has sometimes been named 'block mosaic'. One of the earlier examples is the work of Laubscher (1965) in

From: NIEUWLAND, D. A. (ed.) *New Insights into Structural Interpretation and Modelling.* Geological Society, London, Special Publications, **212**, 59–74. 0305-8719/03/$15

the Eastern Jura mountains of Switzerland. More recently, Philippe (1995) restored the entire Jura mountain chain. In the Andes, Kley (1999) produced a similar restoration based on the principles outlined by Laubscher. This is a first-order technique requiring a relatively simple data set, i.e. geological maps, some seismic data and interpretative work to draw structural cross-sections. This approach is highly applicable at the large (i.e. mountain belt) scale where the fine details of additional displacement due to folding as opposed to faulting are of relatively little consequence.

Recent work has also spread into 3D unfolding, namely restoring the folding part of the displacement field (McCoss 1988; Gratier *et al.* 1991; Gratier & Guiller 1993). This requires more detailed structural knowledge of a region, usually in the form of a structure contour map of a particular target horizon for restoration. Gratier *et al.*'s (1991) method uses data very closely by genuinely unfolding a folded surface by effectively 'ironing' it flat. In its most basic form, the method represents a whole, folded section of an horizon as a series of rigid triangular elements in 3D co-ordinate space, and transforms them back to 2D space whilst preserving their area. The result is overlapping rigid elements which are then iteratively refitted to minimize voids and overlaps. Any region of a horizon will then have a new, flattened boundary which could be compared to the projection of its original boundary into the horizontal plane. Usually, a surface will not only be folded but also

cut by faults. Faults allow the surface to be decomposed into discrete regions which are completely surrounded by faults and can be unfolded individually before the new element boundaries are iteratively refitted (e.g. Rouby *et al.* 1993, 2000). Again, refitting may be done by hand or may be automated, depending largely on the quality of data available.

The Central Andes and The Jura Mountains

The Central Andes and the region of the Jura mountains in the canton of Neuchâtel, Switzerland are regions of mountain belts at contrasting scales (Fig. 1), and consequently provide examples where the different approaches to horizon or regional restoration outlined above are applicable. Both are classic examples of arcuate belts, and have been interpreted as resulting from several possible mechanisms.

The Andes mountains extend an enormous distance along strike (6000 km) and preserve a remarkable amount of topographic elevation in the form of the gigantic Altiplano-Puna plateau. They result from deformation along the active margin between the South American plate and the subducting Nazca oceanic plate. The majority of the deformation forming the present-day structure of the chain is probably Late Tertiary (*c.* post-25 Ma), with likely pulses of tectonic activity within this period. Many ideas relate relative plate motions to the overall generation of the belt (e.g. Pardo-

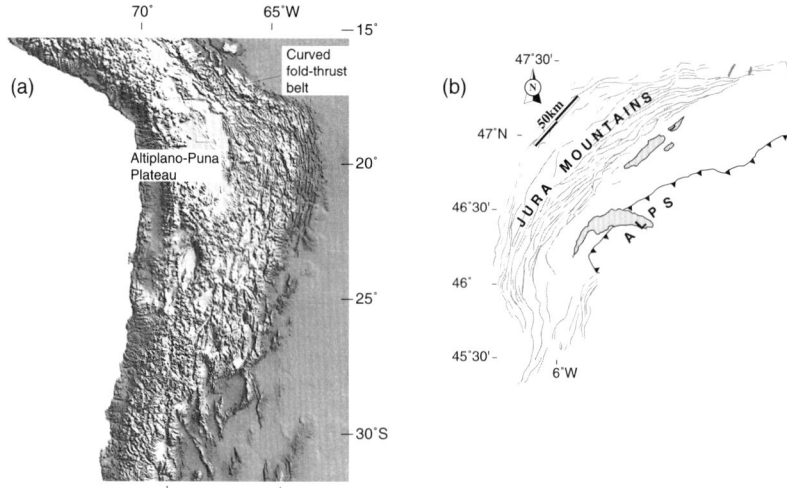

Fig. 1. Images of the Andes mountains and the Jura mountains. (**a**) Digital elevation model (USGS 30 arc second DEM of the Andes, high res. GIF from Cornell University website, http://www.geo.cornell.edu/geology/cap/CAP_gen/CAP_topo.html) of the Andes. Highlighted are the Altiplano-Puna high plateau region and the curved fold and thrust belt. (**b**) Trace of Jura fold axes and again the curvature of the belt along strike.

Casas & Molnar 1987; Gephart 1994; Norabuena *et al.* 1998, 1999; Silver *et al.* 1998). Evidence for some relationship between tectonic activity and convergence direction and rate appears compelling, but the precise mechanisms involved remain controversial. The Central Andean region shows a strong curvature in structural trend along strike, which is well defined in the fold-thrust belts of the Eastern Cordillera and the Subandean Belt. The Andean foreland is strongly segmented along strike between regions of thin- and thick-skinned deformation (Kley *et al.* 1999) but the region dealt with here is essentially consistently thin-skinned in the western most part and thick-skinned with a mid-crustal detachment in the east. The Andes provided some of the inspiration for the orocline model of (Carey 1955), but were later reinterpreted in terms of a primary arc by Sheffels (1995). Serial 2D cross-section data and application of the block mosaic technique to make a plan view restoration by Kley (1999) has given a picture of the overall displacement field involved in generating the chain, and shows a relatively parallel field with a strong maximum of displacement and shortening occurring in the region of Bolivia between 15° and 20°S, supporting further a primary arc style of formation (Fig. 2).

The Jura mountains have a long documented history of investigation (Schardt 1906, 1908; Buxtorf 1916; Aubert 1945; Laubscher 1961, 1965; Pavoni 1961; Burkhard & Sommaruga 1998). Many theories of their evolution have been proposed by these authors, but the most recent arguments (e.g. Burkhard & Sommaruga 1998; Sommaruga 1999)

demonstrate that they are a consequence of the generally parallel, differential displacement and shortening along strike of a Mesozoic cover of material which has slid freely over a weak, Triassic evaporite detachment, in a classic, thin-skinned style. The Jura is also now interpreted as a primary arc (Hindle & Burkhard 1999; Hindle *et al.* 2000). What follows describes the estimation of finite strain variation across the Central Andes after Kley (1999) and across the Neuchâtel Jura (Switzerland) after Hindle (1997).

Strain calculation – past and present

Past, a review

Howard (1968*a*, *b*, *c*) produced some of the earliest work directly quantifying strain in the horizontal plane from finite displacements. The approach he outlined was one of graphically determining displacement gradients across a region based on the known displacements of certain points. Once local gradients of displacement are known on a grid of some form, finite strain can be calculated at each grid point. It is important to note that this technique treats strain as a continuous function of position, which contrasts with the approach to be used here. Cobbold (1979) and Gratier *et al.* (1989) also approached the question of strain calculation. However, neither determines strain estimates so directly from displacements as both Howard's (1968*a*, *b*, *c*) and this present work.

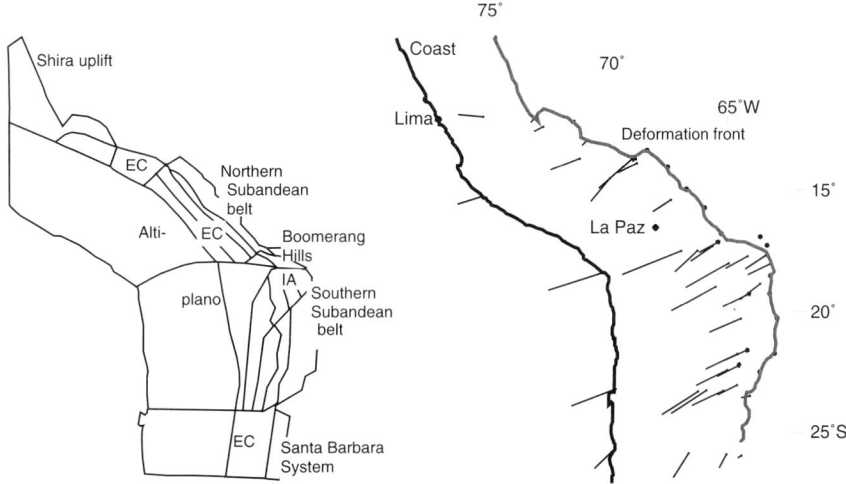

Fig. 2. Structural units used for the block mosaic restoration of the Central Andes and the finite displacement field of the Central Andes from Kley (1999), showing the broadly parallel trend of displacement for the past 25 Ma.

Triangular decomposition method

Hindle (1997) presents the mathematical basis for the direct determination of strain from displacement presented here. This is described more fully in the appendix to this paper. In contrast to Howard's 'smooth' functions of strain, this method relies on decomposing a displacement field's components piecewise into 'linear' (triangular) elements. The term 'linear' refers to the mathematical description of strain over each triangular element. The strain tensor is composed of constants, and the displacement functions associated with it are linear. A finite strain tensor is always defined as some combination of products of the displacement/deformation gradient matrix and its transpose (Malvern 1969; Means 1976) (see appendix).

A triangular decomposition of the Andean displacement field is shown in Figure 3. The choice of triangles has to take account of the distribution of faults across the region and a number of ways to do this exist. For instance, consider the case where a region is restored into a series of fault bounded blocks (Fig. 4a). Between the blocks lie regions of footwall which at the present day lie buried beneath the hanging wall of thrusts (Fig.

4b). The result is that two points lying either side of a fault will map to the same point in the deformed state. Consequently, we need some way of choosing which particular vector of the two possible ones we wish to use. This will depend strongly on the situation being examined, but the physical consequences of one choice or another are relatively simple (Fig. 4c). When vectors related to a fault are defined along the fault's 'hanging wall margin' then the associated deformation in the fault will be represented in elements lying to the 'footwall' side of the fault. Alternatively, if the vectors defined are based on the 'footwall margin' displacement, then fault-related strain will be found in elements lying to the fault's hanging wall side. Whichever choice is made, the same convention, namely vectors defined on footwall or hanging wall margins should be applied across an entire region. Strike-slip or tear faults which relay displacement between different thrust faults also need careful attention (Fig. 4d). Any such strike-slip zone when sampled locally will show differential displacement accumulated on different thrust faults relative to some remote pin, and this will be sampled differently according to how elements are defined (Fig. 4d).

One final situation is possible, namely, an

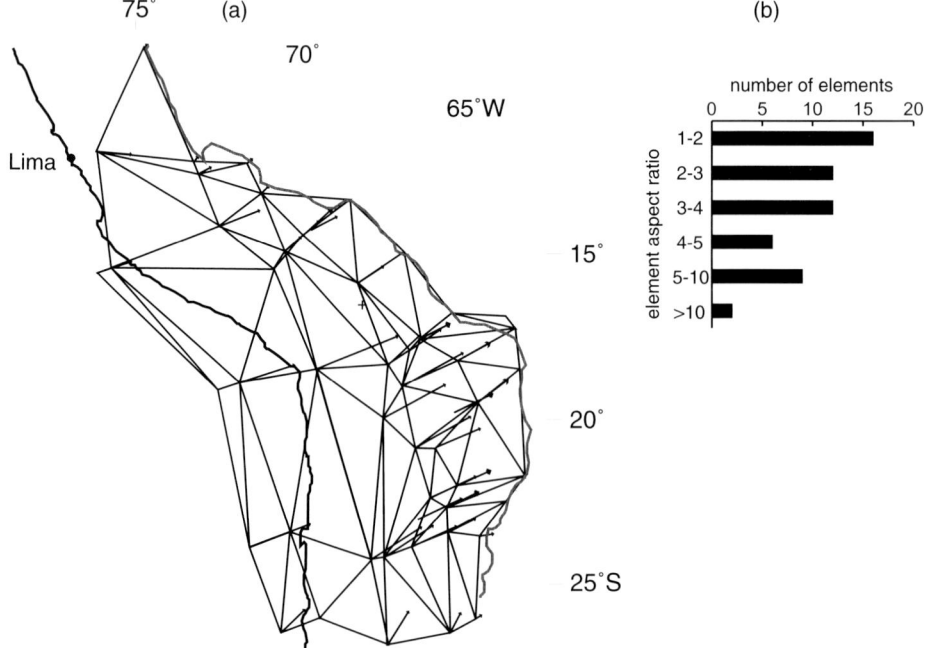

Fig. 3. (**a**) Triangular decomposition of the Andean displacement field. Elements are shown in their retrodeformed state with displacement vectors at their vertices pointing to the present-day position of a node. (**b**) Histogram of the 'shape' of elements (aspect ratio) – a measure of their 'stability' relative to small perturbations of input parameters (in this case, small variations of displacement vectors).

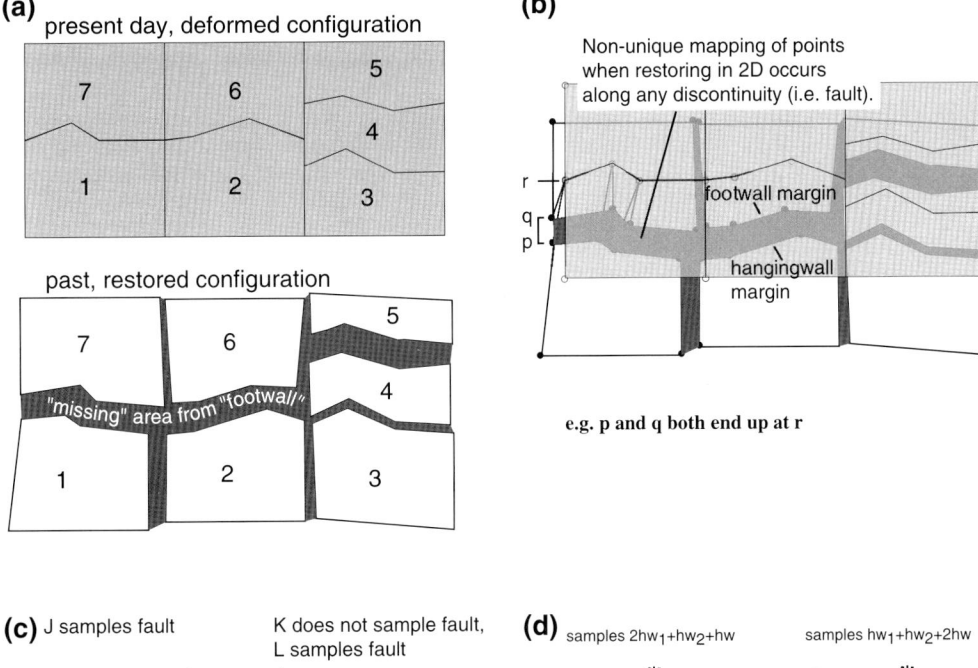

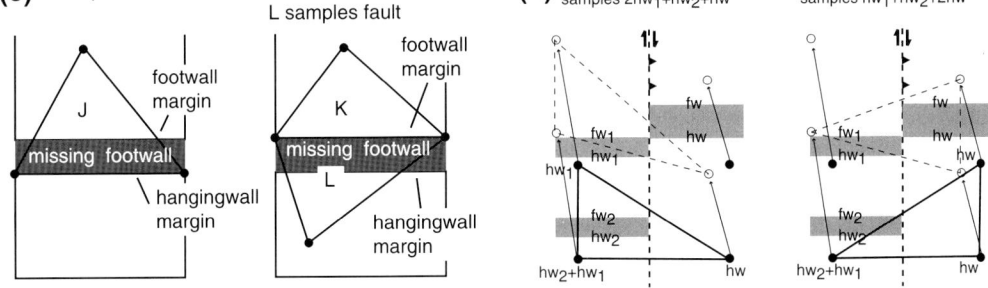

Fig. 4. The relationship between block mosaic restoration and 'continuous' displacement as represented by triangular (linear) gridding. (**a**) The block mosaic restoration aims to restore material to its original position by assuming that a region can be broken up into fault-bounded blocks which can be repositioned so that they are separated by amounts corresponding to fault offsets (missing footwall) known from cross-section balancing or seismic data. (**b**) When drawing displacement vectors between known points in the deformed and undeformed states, the problem arises that two points map onto one point (due to the presence of discontinuities). (**c**) We can choose which one of two possible vectors we wish to use according to how we wish to sample fault strain in a subsequent triangular decomposition of a displacement field. Here, J relies on drawing vectors from hanging wall margins of blocks whereas K works from the footwall vectors, pushing the strain from the same fault into the preceding element L. (**d**) An element can be thought to sample the linear sum of fault displacements affecting its three nodes as shown for the contrasting elements straddling a tear fault. Note that as nodes are more hinterlandward, they accumulate the sum displacement of more faults. Note also the contrast in what is sampled by each element according to which three of four possible displacement vectors are connected to form an element.

element can be defined joining points purely in one block, and not straddling any fault, and such elements will sample internal strain (e.g. possible folding) in that block. Whilst elements are defined based on consistently sampling fault patterns across a region, the shape of the resulting elements nevertheless cannot be ignored. The shape of a triangle is often characterized by its aspect ratio (longest edge/smallest altitude) and this parameter relates directly to the numerical stability of an element, namely how much a solution (e.g. strain tensor) varies as a function of an input parameter (e.g. small changes to displacement vectors). The element aspect ratios for the Andes, are in *c.* 75% of cases less than 5 (Fig. 3, histogram) which ensures reasonable quality of analysis.

Finite strain estimation

The Central Andes

The Central Andean chain with its arcuate form might be expected to yield interesting strain patterns on the basis of displacements. The displacement field itself between 10°S and 25°S shows very strong differential displacement along strike. Figure 2 shows a slight modification of Kley's (1999) displacement field for the last 25 Ma in the Central Andes. The modification concerns the Interandean zone (the nose of the bend) where what was originally considered to be one broad region with a single, main, basal detachment fault has been split into two. There is reasonable evidence from outcrop patterns in this area that such an assumption holds (e.g. Pareja *et al.* 1978). The original map view balancing was carried out following the principles of the block mosaic technique (Laubscher 1965, 1981; Baby *et al.* 1993). The Central Andes have been broken up into respective units (Subandean, Interandean, Eastern Cordillera, Altiplano), which are in turn cut into a number of blocks along strike. Deformation occurring on thrusts and folds in these structural units is then attributed to the 'gaps' between these blocks, thereby distributing shortening approximately correctly in map view. Block restoration then involves keeping the 'fault separations' between blocks consistent with known data whilst minimizing overlaps of blocks. Kley (1999) allowed some shear of the rigid blocks to minimize overlaps. This shear is assumed to represent distributed deformation in the blocks not represented in the block separations.

The resultant displacement field shows convergence of vectors at the northern and southern extremities of the model, with the vectors becoming progressively more parallel and orogen normal as the bend axis is approached, where we also reach maximum displacement values. A triangular grid has been produced from the displacement field as described above (Fig. 3). It should be noted that the model is modified from Kley's restoration to include a rigid coastal margin of South America. The total displacement of the system, represented by the westernmost set of vectors, remains the same, but is transferred inland so that a deformation gradient begins only on the western flank of the Altiplano-Puna (Fig. 3, westernmost elements).

The calculated strain values (Table 1, Fig. 5) show strong variation across the chain. The first-order feature is the strong along-strike swing in the orientation of the strain axes. However, this is not smoothly distributed along the chains' length. The northern and southern Subandean ranges (see Kley 1999) are relatively cylindrical, and have a principal orientation difference of *c.* 40° to 50°. The northern Subandean ranges have an approximate, general short axis orientation of *c.* −40° to −50° (positive anticlockwise from an approximate north-trending line) i.e. roughly NW–SE oriented, and a relatively homogeneous direction field. The southern ranges lie *c.* −90° to +80°, approximately east–west oriented, and again relatively homogeneous in direction. However, the central part of the bend in the orocline (named the Interandean zone) shows a very sharp change in orientation of principal strain axes over a relatively short distance. There are two trends of orientation, one on the northern flank (around 17°S) of *c.* −15° to −30° (NNW–SSE) and one on the southern flank, *c.* −35° to −75° (between NW and SW). The Altiplano–Puna elements show a marked change in orientation around the point of inflection of the west coast of South America (at 18°S). North of this line, the short axes trend NW through to NNE. South of it they trend SW through to SSW. Consequently, both the Altiplano–Puna and Interandean ranges have strongly heterogeneous strain fields. One of the most tempting comparisons to make is between the predicted strain axis orientations and the topography as defined by digital elevation models. There is for instance a very strong visual correlation between a portion of the long axis trends and the major trends of topography in the eastern parts of the fold-thrust belt (Fig. 5b). This is very strong in the southern Subandean zone. The conclusion that the finite strain pattern and fold axis orientations correlate in fold-thrust belts was reached by Hindle *et al.* (2000), in the Jura mountains and seems to be the case in the Andes.

The extremes of the values of strain axes merit examination. They represent the local, bulk finite strain, reflecting the contribution of faults over the triangular element, and are given in terms of principal stretches, s1 and s2. Consequently, shortening of 80% (s1 = 0.20, element 52) is found, occurring in an elongated element along the deformation front, and is partly a consequence of the gridding. Most of the shortening values in the range of 70–80% occur in elements with two nodes which remain fixed. This means that they lie on the deformation front, so any movement of the third, free node will lead to large deformations in an elongated element. Care is needed when interpreting the percentage shortening and elongation over an element as a result. However, the strain values still reflect the displacement field in general, and consequently, the large strains (generally 50%+ shortening i.e. s1 < 0.5) are concentrated in a narrow band along the Subandean chains and within the Interandean zone. These values reflect the distribution of displacement and deformation over the chain. The total displacements that are measured at the coast are not uniformly distributed across the width of

Table 1. *Strain parameters for the Central Andes*

Element	Short axis s_1	Long axis s_2	Axis angle $\alpha°$ from N	Dilatation $s_1 \times s_2$
1	0.87	1.05	15.94	0.91
2	0.66	1.06	29.77	0.71
3	0.87	1.19	−44.13	**1.03**
4	0.92	1.32	−31.17	**1.21**
5	0.92	1.14	47.71	**1.05**
6	0.83	1.11	58.17	0.93
7	0.76	1.33	42.28	**1.00**
8	0.79	1.33	37.94	**1.05**
9	0.67	1.05	−80.98	0.70
10	0.46	1.34	−40.71	0.62
11	0.49	1.17	−41.23	0.58
12	0.46	1.18	−37.63	0.54
13	0.47	1.02	−29.36	0.48
14	0.64	1.14	26.46	0.74
15	0.90	1.80	−10.17	**1.62**
16	0.26	1.14	−55.62	0.30
17	0.27	1.09	−57.02	0.29
18	0.78	1.02	75.33	0.80
19	0.85	1.02	76.57	0.86
20	0.75	1.09	53.7	0.81
21	0.64	1.39	−30.75	0.89
22	0.83	0.95	−48.5	0.79
23	0.65	1.21	−8.81	0.79
24	0.76	1.08	−60.37	0.82
25	0.57	1.00	−41.65	0.57
26	0.70	1.10	−25.64	0.78
27	0.48	1.94	−15.78	0.92
28	0.91	**2.14**	−4.15	**1.95**
29	0.84	1.03	85.35	0.65
30	0.87	1.01	78.55	0.89
31	0.76	1.16	−80.54	**1.15**
32	0.77	1.31	−75.57	**1.00**
33	0.79	0.91	89.93	0.72
34	0.78	1.03	67.06	0.80
35	0.24	1.16	−18.64	0.28
36	0.35	1.07	−50.56	0.37
37	0.45	1.01	−48.11	0.45
38	0.44	1.06	−45.44	0.47
39	0.60	1.04	−36.64	0.62
40	0.48	2.01	−15.05	0.97
41	0.37	1.10	−33.13	0.41
42	0.45	0.87	−63.24	0.39
43	0.56	1.02	−82.93	0.57
44	0.49	1.65	−88.05	0.81
45	0.56	0.98	82.73	0.55
46	0.44	1.19	−79.61	0.52
47	0.49	1.04	−84.5	0.51
48	0.40	1.22	72.3	0.49
49	0.32	1.19	77.99	0.38
50	0.47	1.04	87.28	0.49
51	0.46	0.95	82.98	0.44
52	**0.20**	1.01	−53.56	0.21
53	0.26	1.28	−23.28	0.34
54	0.00	1.00	0	1
55	0.41	1.10	−33.84	0.45
56	0.44	1.03	−77.02	0.45
57	0.32	0.88	−61.58	0.28
58	0.28	1.12	−79.9	0.31
59	0.51	1.18	−88.59	0.60
60	0.51	1.15	73.43	0.58
61	0.49	1.31	71.18	0.64
62	0.47	1.05	89.18	0.50
63	0.51	1.27	−76.75	0.65
64	0.60	1.07	−82.51	0.64
65	0.69	1.16	−67.8	0.80

Extremes of s_1 and s_2 are in bold typeface as are all dilatations greater than one.

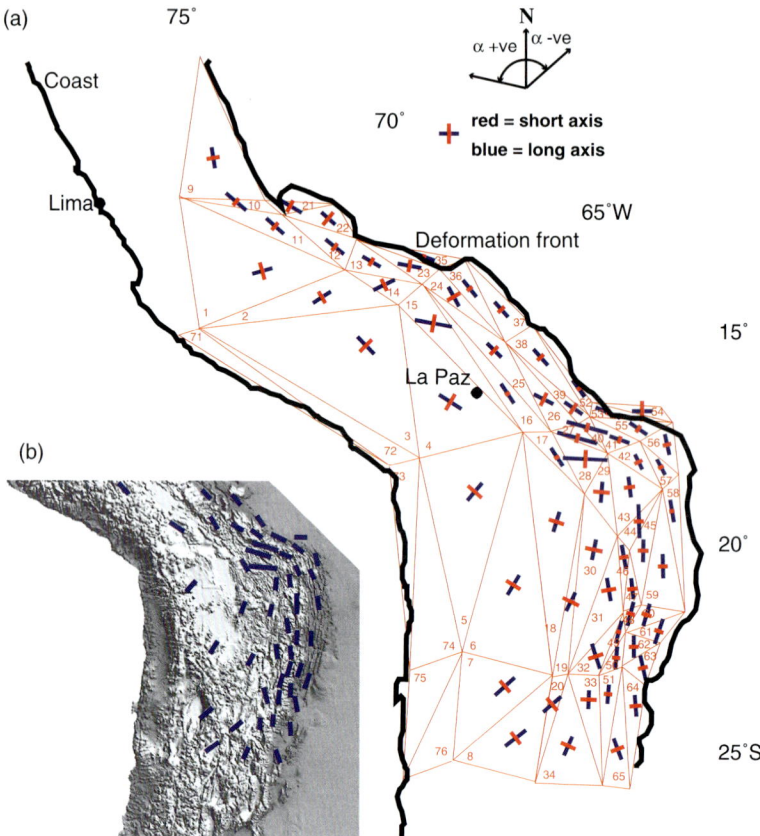

Fig. 5. The finite strain field of the Central Andes. (**a**) Calculated strains based on the gridding described of Kley's (1999) displacement field (see also Table 1 for numerical values. (**b**) Close comparison between the structure highlighted by the digital elevation model and the long axes of the strain ellipses.

the belt, but instead diminish rapidly to zero only east of the eastern margin of the Altiplano–Puna region. The finite extension values (actually stretches are given) also show strong elongation in some places (e.g. element 28: s2 = 2.14). The extensions are mostly localized in a region of the Interandean zone. These values of extension seem contradictory since they are found in a fold-thrust belt where the dominant process is 'contraction', i.e. there is tectonic-induced vertical thickening. Multiplying both strain axes together gives a much better picture of what happens to the ground. If the normalized product of these two axes, which lie in the horizontal plane, is greater than one, this implies that the whole element will have thinned, to conserve volume in some imaginary strain ellipsoid. We see that only eight of 65 elements show such stretch and thinning, and some with s2 > 2.0 nevertheless 'thicken' tectonically (e.g. element 40).

The Neuchâtel Jura

The Neuchâtel Jura has been restored using 3D unfolding (Gratier *et al.* 1991). The region dealt with is approximately 50 km by 50 km and sits in the central part of the arcuate chain of the Jura mountains (Fig. 6). The central part of the Jura is a region with relatively low differential displacement along strike (Philippe 1995). This has an important influence on heterogeneity in the strain field we predict. However, lying across the region is an almost north trending 'tear' fault (La Tourne–La Ferrière fault; Figs 6 and 7) which must be a consequence of relaying displacement between different thrusts along strike. The 3D geometry of a structural surface (top Argovien) was available from a pre-existing, highly detailed, structure contour map (Kiraly 1969) which was subsequently digitized. This serves as a very robust model for restoration since most major faults were already drawn onto

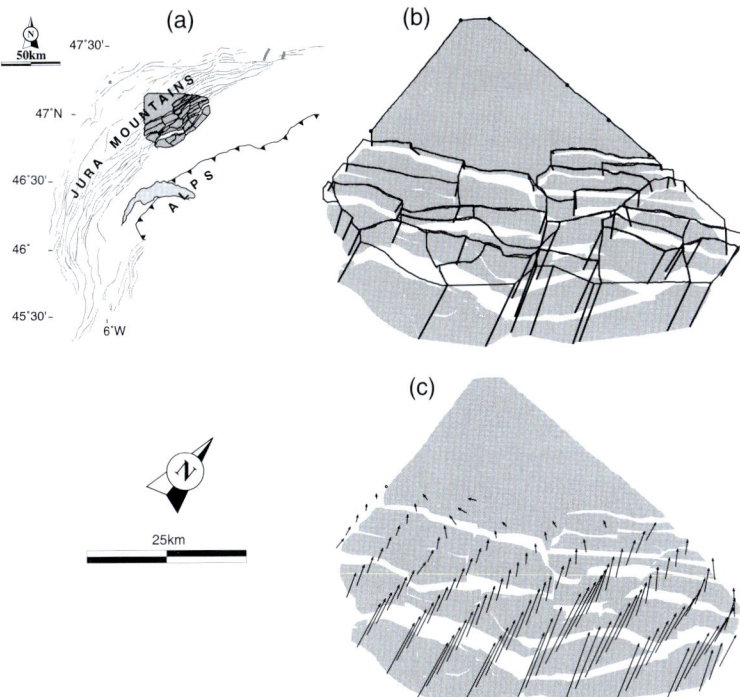

Fig. 6. Displacement field for the Neuchâtel Jura from Hindle (1997). (**a**) shows the location of the region restored in the Jura. (**b**) shows the original vector field connecting known points in the deformed and undeformed state which have been derived from a combination of unfolding the individual, fault bounded blocks with Gratier's (1991) technique and refitting the blocks whilst respecting fault offsets calculated from balanced cross sections. (**c**) shows the result of interpolating these vectors to produce a smoothed displacement field, locally consistent with the original field in (**d**).

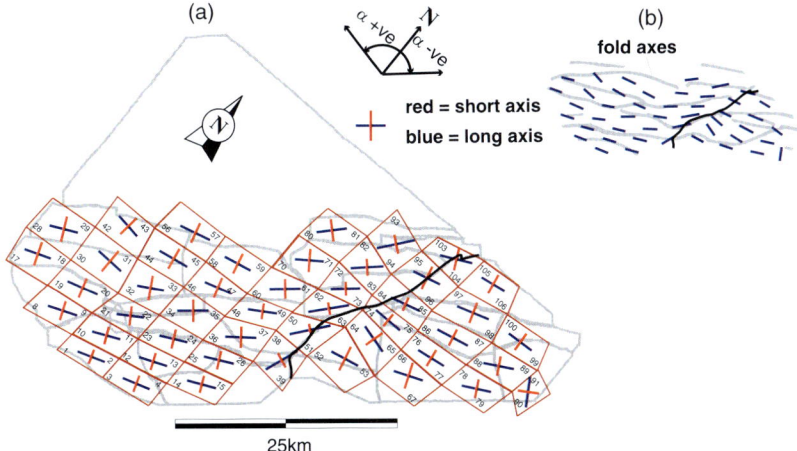

Fig. 7. The strain axis pattern produced from the restoration of the Neuchâtel Jura. (**a**) Detail of strain axes for each (combined) element as described in the text (see also Table 2 for numerical values). (**b**) A more detailed comparison of long axes and fold axes.

it, making the task of dividing the map into separate, fault-bounded units relatively simple. Individual blocks were unfolded using the computer software UNFOLD (Gratier & Guiller 1993), and new block outlines were pieced together, whilst still accounting for considerable amounts of 'missing' footwall (Fig. 6), data coming from cross-sections constructed across the region (Hindle 1997).

The restoration shows that total displacement varies little along strike. The first row of displacement vectors are all of very similar magnitude. This contrasts with the large-scale Andean displacement field where strong variation along strike is found. For the Jura example, a different approach for generating the displacement field has been used. The same rules for linking points consistently between retro-deformed and present-day positions have been respected to produce an initial set of displacement vectors (Fig. 6b). Subsequently, these have been interpolated over the Jura region onto a regular grid as an alternative to using the raw displacement data (Fig. 6c). The displacement vectors are quite parallel across the region although there is a noticeable anticlockwise twist of the shorter vectors towards the Jura foreland. A different approach for evaluating strain has also been used, namely to combine pairs of triangles with two common vertices in an area-weighted average strain tensor over what is effectively an initially regular, square element. The initial triangular elements are also regular and identical because the displacement field has been interpolated onto a regular grid with aspect ratio c. 1.41, removing any shape-induced problems from the analysis.

Results of the analysis are presented in Table 2 in the same way as the Andean data. However, each homogeneous strain measure is applicable over the whole area of two triangles. The orientations of principal axes has an almost sigmoidal pattern along strike (long axes), with the lateral extremities of the direction field trending close to north (short axes), whilst the central part of the region, close to the tear fault, trends abruptly NE (Fig. 7). There is a general correspondence between fold axis trends and long axis orientations. The maximum shortening calculated is 59% ($s1 = 0.41$, element 62/63; Table 2). All of the highest percentage shortenings are found adjacent to the tear fault. Maximum extension found is 40% ($s2 = 1.4$), again adjacent to the tear fault. The tear fault appears to be a large influence on the finite strain field, but it should also be noted from the displacement field that there is no actual offset recorded at the ends of the tear fault after restoration. The feature appears to be purely a result of local, differential transfer of displacement between thrusts with the global sum of displacement of all the thrusts either side of the tear fault identical.

Discussion

The data presented from both the Andes and the Jura suggest a strong link between finite strain and various structural features. We can investigate this at various levels of detail and try to offer explanations as to why any such links exist. Various scales of investigation and explanation are necessary, starting at the largest scale, namely that of morphological features visible on digital terrain models. However, there are possibilities to make smaller scale predictions of structural trends which should relate directly to field data geologists can collect.

Relationship between bulk strain and structural features

In both the Andes and Jura mountains there is a strong correlation between first-order features (fold axes, topography) and finite strain parameters (long and short axis orientations) calculated from displacement and heavily influenced by large displacement localized on faults (Fig. 4c, d). A physical explanation of this correlation is lacking. In thin-skinned thrust belts, many geometric models of fault-related folding (Suppe 1983; Wilkerson et al. 1991; Shaw et al. 1994) show that fault geometry should control folding. In the brittle regime, geometric techniques provide a robust tool for examining fold-thrust relationships. Their extension into pseudo-3D balanced models is particularly interesting, since examples from seismic data (e.g. Shaw et al. 1994) support a very strong fault ramp control on fold geometry, fold hinge orientation and plunge, which together are the first-order structural feature described and related to finite strain axes in this paper. Shaw et al.'s (1994) models allow differential displacement along strike of a single thrust by creating more structural relief (geometrically balanced) to compensate for greater displacement and predict plunging structures along strike and non-cylindrical fold axes. Hence, the method predicts essentially the same features as the geometric-mechanical arguments in this paper but by a completely independent and non-mechanical method.

Shaw et al. (1994) described the 3D evolution of a thrust sheet which undergoes differential displacement over a ramp and has a 'simple shear' type of deformation, i.e. there is no deformation gradient parallel to the direction of transport (Fig. 8a). To directly compare the structural trends of this model to those from a geometric-mechanical model we need appropriate boundary conditions for strain modelling. A simple shear-only geometry is unrealistic because it neglects the deformation gradient parallel to transport (i.e. finite shortening

Table 2. *Strain parameters for the Jura mountains*

Element	Short axis s_1	Long axis s_2	Axis angle $\alpha°$ from N	Dilatation $s_1 \times s_2$
1/2	0.50	0.99	18.97	0.50
3/4	0.58	1.02	18.08	0.59
8/9	0.66	1.05	21.84	0.69
10/11	0.60	1.00	24.86	0.60
12/13	0.59	1.00	22.39	0.59
14/15	0.59	1.02	25.03	0.60
17/18	0.82	0.95	20.51	0.78
19/20	0.70	1.12	16.66	0.78
21/22	0.64	0.97	32.85	0.63
23/24	0.66	0.97	26.37	0.64
25/26	0.63	1.05	26.60	0.66
28/29	0.86	1.00	25.98	0.86
30/31	0.89	1.03	−1.43	0.92
32/33	0.84	0.94	29.73	0.78
34/35	0.75	1.07	40.06	0.81
36/37	0.72	1.07	37.78	0.77
38/39	0.54	1.24	69.31	0.67
42/43	0.78	0.92	−7.32	0.72
44/45	0.76	1.08	9.89	0.82
46/47	0.72	0.96	16.74	0.69
48/49	0.53	1.00	31.73	0.53
50/51	0.42	**1.40**	51.41	0.59
52/53	0.88	1.03	12.49	0.91
56/57	0.72	1.12	14.39	0.81
58/59	0.73	1.00	12.45	0.73
60/61	0.66	1.06	40.67	0.70
62/63	**0.41**	1.40	49.02	0.57
64/65	0.68	1.03	−14.98	0.71
66/67	0.89	1.04	22.05	0.93
70/71	0.77	0.98	35.53	0.76
72/73	0.47	1.01	40.97	0.48
74/75	0.52	1.06	−0.07	0.55
76/77	0.63	1.02	8.47	0.64
78/79	0.64	1.03	28.09	0.66
80/81	0.74	1.03	46.60	0.76
82/83	0.56	1.11	42.81	0.63
84/85	0.58	1.05	0.75	0.61
86/87	0.61	1.17	12.33	0.71
88/89	0.66	1.06	28.85	0.69
90/91	0.72	1.03	−58.22	0.74
93/94	0.65	1.31	51.53	0.85
95/96	0.83	1.01	14.12	0.84
97/98	0.73	1.18	18.19	0.86
99/100	0.68	0.96	3.50	0.65
103/104	0.76	1.14	26.40	0.86
105/106	0.72	0.91	9.67	0.65

Extremes of s_1 and s_2 are in bold typeface.

in a fold-thrust belt). Consequently, an extra amount of shortening should be added parallel to the transport direction as part of the input to any strain model. Strain axes produced in this way are of very comparable orientation to predicted fold hinges from a geometric balanced model (Fig. 8b and c), which goes some way to explaining the correlation found in the Andes and Jura.

Validating structural restorations by using strain values

The preceding arguments and the data from the Andes and the region of the Neuchâtel Jura show that first-order structures and strain axes should have systematically similar orientations. It is reasonable to ask whether this allows us to 'invert'

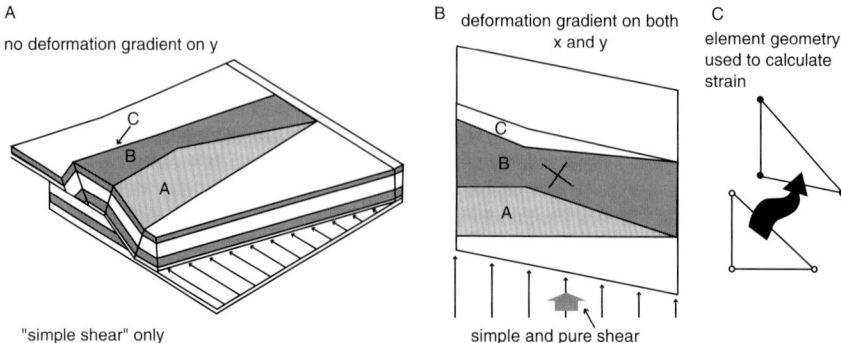

Fig. 8. (**a**) A 3D, balanced model of a fold developing by differential displacement over a thrust-ramp (adapted from Shaw *et al.* 1994). (**b**) A comparison between predicted fold hinge traces by the vertical projection method, and a possible finite strain axis for similar deformation geometry – simple shear (differential arrows) plus pure shear applied everywhere (solid arrow) (**c**) The mapping of an initial triangle onto a deformed triangle used for the strain calculation.

the procedure and use the correspondence between strain axes and structural grain to validate a structural restoration. This is especially the case in areas of relatively short periods and likely constant direction of shortening (Andes *c.* 30 Ma, Jura *c.* 10 Ma). In such cases, if a restoration produced strain patterns opposed to the structural trends found in the field, there would be good grounds to doubt the validity of the predicted displacement field. An intriguing question regards the local values of the extension and shortening values, and the element stretches. Any region where actual stretch (equivalent to tectonic thinning) has occurred in a fold-thrust belt might be questionable. However, it is feasible to have such thinning (thinking for instance of predictions made for Coulomb wedges undergoing changing boundary conditions). It seems reasonable therefore to propose these parameters as a check that can be used in validating any plan view restoration of a thrust belt.

Predicting lower order structures

The numerical technique employed here produces a series of homogeneously strained elements. The homogeneous strain values raise interesting possibilities for predicting 'lower order' structures (e.g. small faults for accommodating internal deformation of large thrust sheets). The strain tensor describes the change of shape of any part of a single element, a homogeneous quantity. An idea of Reches (1978) was that any 3D, homogeneous strain of the crust could be accommodated on a four-fault pattern with a bi-conjugate symmetry around the principal strain axes (Nieto-Samaniego 1999). The faults are oriented in space such that they minimize the energy dissipated by shearing deformation on all faults in a volume. This opens up the possibility to calculate such fault distri-

butions and the senses of shear upon them for the element strain values, and compare them to field structures.

As an example, fault sets and slip senses are shown (Fig. 9) for elements north and south of the Arica bend in the Andes. We see four fault orientations in each case quite oblique to the strain axes, and slip vectors giving a strong oblique slip component on the fault: faults are in all cases contractional. It is unknown whether such orientations of faults would be found in nature, and if so, at what scale they would exist. There are arguments against such fault sets existing; for instance, pre-existing heterogeneities may well dominate a system and absorb substantial amounts of deformation even though their orientation would not be ideal as predicted by this theory. Normal stresses acting on

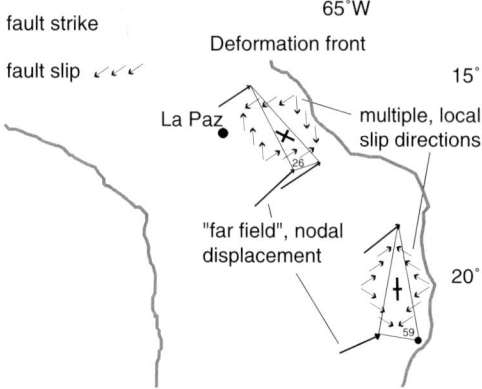

Fig. 9. The relationship between boundary displacements (nodal displacements) and predicted slip vectors on ideal fault sets after Reches (1978). Note the enormous difference between the two sets of vectors, and the orientations of the finite strain axes (centre of elements).

fault planes are also neglected in calculating the energy dissipation (Reches 1978). However, as a first approximation it raises interesting questions, and could be used as an alternative for explaining fault kinematic data, so often interpreted in terms of local stresses and plate motion. This faulting theory would suggest that such data are much more intricately related to local displacement gradients, in turn controlled by regional displacement patterns. There is a counter-intuitive contrast between regional 'far field' or boundary displacements (nodal displacements of elements) which are part of the relatively well ordered and parallel, global displacement field and the individual slip directions that are predicted on the fault planes. The slip directions it must be remembered are slip systems used to accommodate boundary displacements and their net effect is to allow the change of shape the boundary displacement imposes. This theoretical prediction highlights the danger of making regional-scale extrapolations about displacement based on local tectonic indicators such as fault slickenslides.

Conclusions

Block mosaic restoration can be used to draw a logical, and consistent displacement field for the purposes of calculating finite strain. The block mosaic displacement field is discontinuous, but by respecting the displacements locally (i.e. at all known displacement vectors) the net deformation due to the field can be simulated by a continuous, but piecewise displacement function represented by a triangular grid.

The finite strains calculated by this technique are for the case of both the Andes mountains and the Neuchâtel Jura, consistent with the structural trends at a large scale to be found in the field (namely those trends defined by fold axes).

A logical relationship can be demonstrated between fold hinge trends predicted by geometric models of fault-ramp folding in three dimensions and the likely predicted finite strain axes for such a situation.

There are grounds for using finite strain calculations as a check on the validity of block mosaic restorations, since real structural trends and finite strain axes from such a restoration would be expected to be similar.

The theoretical relationship between local structures and 'far field' boundary displacements has been demonstrated and shows that even in a quite parallel displacement field, local indicators of thrust displacement sense may well be oblique to the true regional sense of displacement. This has major consequences for field geology and extrapolations based on very small-scale observations.

I would like to thank J. Kley for work on describing strain in the Andes, J. P. Gratier and M. Burkhard for work with the Neuchâtel Jura, and O. Besson for the mathematics. K. Rutten and B. Windley are thanked for helpful reviews. The work in the Jura mountains was supported by Swiss National Science Foundation Grants No. 20–43055.95 and 20–50535.97, and the present work in the Andes by Leibnizförderprogram grant no. ON70/10–1.

Appendix

First, a simple example of strain calculation is worked through for non-mathematicians. The mathematics used for calculating strain from the relative displacement of three points can be split into stages.

1. Work out a displacement function

If we use a simple initial triangle with its vertices at (0,0), (1,0), (0,1), the displacement equations to transform this triangle onto any other are:

$$x_1 = A_{11}s + A_{12}t + A_{13}, \quad x_2 = A_{21}s + A_{22}t + A_{23} \tag{A1}$$

where s and t are the co-ordinates of the initial triangle. The values of A_{ij} are functions of the co-ordinates of the three vertices of the new triangle, (a,b) (c,d) and (e,f) (Fig. A1a).

$$A_{11} = (c-a), A_{12} = (e-a), A_{13} = a, \tag{A2}$$

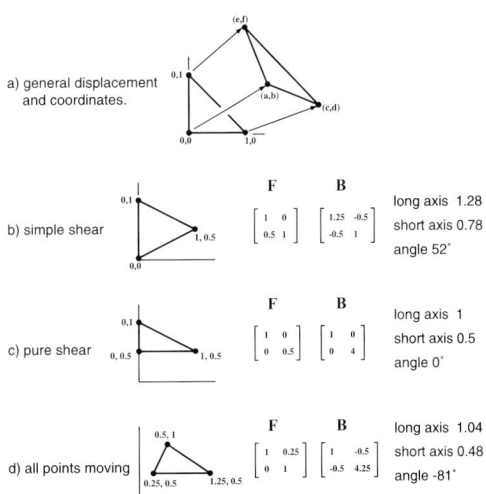

Fig. A1. (a) An initial triangle ((0,0) (0,1) (1,0)) transformed onto a new triangle ((a,b) (c,d) (e,f)). (b) Graphic showing the deformed triangle described in the first section of the appendix. **F** is the corresponding deformation gradient or Jacobian matrix, **B** the strain tensor. (c, d) Further examples of displacement strain relationships.

$A_{21} = (d–b)$, $A_{22} = (f–b)$, $A_{23} = b$

Suppose for instance we have just one non-zero displacement, hence $d = 0.5$, and all other vertices are the same as for the initial triangle. (Fig. A1b). Then

$A_{11} = 1$, $A_{12} = 0$, $A_{13} = 0$, $A_{21} = 0.5$, $A_{22} = 1$, $A_{23} = 0$

and the displacement equations are

$$x_1 = 1s + 0t + 0, \quad x_2 = 0.5s + 1t + 0 \qquad (A3)$$

This is a very simple example (of *simple* shear), but any shape of triangle could be produced this way.

2. Find this function's Jacobian matrix

This is just a matrix containing all the partial derivatives of the displacement function. In our case it has a general form

$$\mathbf{F} = \begin{bmatrix} \dfrac{\partial x_1}{\partial s} & \dfrac{\partial x_1}{\partial t} \\ \dfrac{\partial x_2}{\partial s} & \dfrac{\partial x_2}{\partial t} \end{bmatrix}, \qquad (A4)$$

and our example works out as

$$\begin{bmatrix} 1 & 0 \\ 0.5 & 1 \end{bmatrix}$$

This matrix is also known as a 'deformation gradient matrix' and its derivation is thoroughly explained in Means (1976). Notice the matrix contains only constants, which is always the case, no matter what displacements are applied to a triangle.

3. Multiply the transpose of the inverse of the Jacobian by its inverse

A matrix $\begin{bmatrix} a_{11} & a_{12} \\ a_{21} & a_{22} \end{bmatrix}$ has inverse $\dfrac{1}{a_{11}a_{22}-a_{12}a_{21}} \bullet$

$\begin{bmatrix} a_{22} & -a_{12} \\ -a_{21} & a_{11} \end{bmatrix}$ and transpose $\begin{bmatrix} a_{11} & a_{21} \\ a_{12} & a_{22} \end{bmatrix}$. Hence,

for our example, the inverse is $\begin{bmatrix} 1 & 0 \\ -0.5 & 1 \end{bmatrix}$ and

the transpose of the inverse is $\begin{bmatrix} 1 & -0.5 \\ 0 & 1 \end{bmatrix}$. We

multiply these in the order

$$\begin{bmatrix} 1 & -0.5 \\ 0 & 1 \end{bmatrix} \bullet \begin{bmatrix} 1 & 0 \\ -0.5 & 1 \end{bmatrix} = \begin{bmatrix} 1.25 & -0.5 \\ -0.5 & 1 \end{bmatrix}$$

which gives us the strain tensor. Notice this tensor is symmetric, i.e. $a_{12} = a_{21}$ (in fact this strain tensor is always symmetric).

4. Find the eigenvalues and eigenvectors of this new matrix

Eigenvalues l_1, l_2, of the matrix $\begin{bmatrix} a_{11} & a_{12} \\ a_{21} & a_{22} \end{bmatrix}$ are given by the roots of the equation $(l - a_{11})(l - a_{22}) - a_{12}a_{21} = 0$. Hence, for our case $l_1 = 1.64$, $l_2 = 0.61$ (to 3sf), and the principal axes are given by $1/\sqrt{l}$, giving 0.78 and 1.28 respectively for this example. Notice how in a case of simple shear, where there is only motion in one direction (parallel to the t axis), there is still a strong finite extension (1.28–28% extension). The orientation of the strain axes comes from the equation arctan $\left[\dfrac{l - a_{11}}{a_{12}}\right]$. If we take the value determined from the smaller eigenvalue, we get an angle +52.0°, and this is the angle of the short axis positive, anticlockwise from the t (or x_2) axis. One final point to note: if

$$a_{12} = a_{21} = 0$$

then the matrix is called 'diagonal', and has the special (and very convenient) property that a_{11} and a_{22} are the eigenvalues, and the eigenvectors are oriented along the co-ordinate axes. This type of strain tensor represents a constant shortening in a particular co-ordinate direction, for instance (a,b) = (0,0.5) (c,d) = (1,0.5) (e,f) = (0,1) (see Fig. A1).

This shows in a simple, worked example how to 'map' a simple triangle onto any other triangle. However, the paper has described a technique which maps any triangle onto any other: the initial grid in the Andes is full of irregular triangles. The maths to solve this problem is a bit more specialized (what follows is not really for the non-mathematical), but works approximately as follows. Our 'easy' triangle with vertices (0,0), (1,0) and (0,1) (general s,t co-ordinate system) is now used as a reference element. Hence, we can define two separate functions transforming this reference element onto two different (initial and final) triangles using the procedure described above. Thus

$$x = f(s,t), \quad X = g(s,t) \qquad (A5)$$

where f transforms the reference element onto the final deformed (x) region and g transforms it onto the undeformed, initial (X) region (Fig. A2). But we need to find the function passing from X to x. This is the composite function $f \cdot g^{-1}$. So, we define

$$h = f g^{-1}$$
$$h: X \in \mathbf{R}^2 \rightarrow x \in \mathbf{R}^2 \qquad (A6)$$

which we do not know explicitly, but we can calculate it in terms of f and g from Equation A6. Now, we define the general deformation gradient matrix as follows

$$\frac{\partial x_i}{\partial X_j} = \mathbf{F}_{ij}, \; i = 1, 2, j = 1, 2 \qquad (A7)$$

and as before, the strain tensor we will use is

$$[\mathbf{F}^{-1}]^T \bullet \mathbf{F}^{-1} \qquad (A8)$$

In order to find an equivalent expression to (A8) in terms of h, we define a global derivative operator $\mathbf{D}$, and then find an expression for $\mathbf{D}h$ in terms of f and g. So,

$$\mathbf{D}h = \mathbf{D}[f_\circ g^{-1}] = \mathbf{D}f \cdot \mathbf{D}g^{-1} \qquad (A9)$$

then, we find an expression for $\mathbf{D}h^{-1}$. So, from Equation A9,

$$\mathbf{D}h^{-1} = \mathbf{D}[f_\circ g^{-1}]^{-1} = \mathbf{D}g \cdot \mathbf{D}f^{-1} \qquad (A10)$$

Then, from Equation A8, a general solution for the strain tensor in terms of f and g is of the form,

$$[\mathbf{D}h^{-1}]^T \cdot \mathbf{D}h^{-1} = [\mathbf{D}f^{-1}]^T \cdot \mathbf{D}g^T \cdot \mathbf{D}g \cdot \mathbf{D}f^{-1} \qquad (A11)$$

As before, we require the eigenvalues and eigenvectors of the matrix produced by Equation A11.

References

ARGAND, E. 1922. La tectonique de l'Asie. In: *Congrès Géologique International Extrait du compte-rendu du XIIIe congrès géologique internationale*, Brussels, 171–372.

AUBERT, D. 1945. Le Jura et la tectonique d'écoulement. *Memoire de la Société Vaudoise des Sciences Naturelles*, **12**, 93–152.

BABY, P., GUILLIER, B., OLLER, J. & MONTEMURRO, G. 1993. Modèle cinématique de la Zone Subandine du coude de Santa Cruz (entre 16°S et 19°S, Bolivie) déduit de la construction de cartes équilibrées. *Comptes Rendus de l'Académie des Sciences de Paris*, **317(II)**, 1477–1483.

BURKHARD, M. & SOMMARUGA, A. 1998. Evolution of the Western Swiss Molasse Basin: structural relations with the Alps and the Jura belt. *In*: MASCLE, A., PUIGDEFABREGAS, C., LUTERBACHER, H. P. & FERNANDEZ, M. (eds) *Cenozoic Foreland Basins of Western Europe*. Geological Society, London, Special Publications, **134**, 279–298.

BUXTORF, A. 1916. Prognosen und Befunde beim Hauensteinbasis und Grenchenberg tunnel und die Bedeutung der letzteren für die Geologie der Juragebirges. *Verhandlungen Naturforschender Gesellschaft, Basel*, **27**, 185–254.

CAREY, S. W. 1955. The orocline concept in geotectonics. *Proceedings of the Royal Society of Tasmania*, **89**, 255–288.

COBBOLD, P. R. 1979. Removal of finite deformation using strain trajectories. *Journal of Structural Geology*, **5**, 299–305.

FERRILL, D. A. & GROSHONG, R. H. 1993. Kinematic model for the curvature of the northern Subalpine Chain, France. *Journal of Structural Geology*, **15(3–5)**, 523–541.

GEPHART, J. W. 1994. Topography and subduction geometry in the Central Andes: Clues to the mechanics of a noncollisional orogen. *Journal of Geophysical Research*, **99(B6)**, 12279–12288.

GRATIER, J. -P. & GUILLER, B. 1993. Compatibility constraints on folded and faulted strata and calculation of total displacement using computational restoration (UNFOLD program). *Journal of Structural Geology*, **15(3–5)**, 391–402.

GRATIER, J. P., MÉNARD, G. & ARPIN, A. 1989. Strain-displacement compatibility and restoration of the ChaTnes Subalpines of the Western Alps. *In*: COWARD, M. P., DIETRICH, D. & PARK, R. G. (eds) *Alpine Tectonics*. Geological Society, London, Special Publications, **45**, 65–81.

GRATIER, J. -P., GUILLIER, B., DELORME, A. & ODONNE, F. 1991. Restoration and balance of a folded and faulted surface by best-fitting of finite elements: principle and applications. *Journal of Structural Geology*, **13(1)**, 111–115.

HINDLE, D. 1997. *Quantifying stresses and strains from the Jura arc and their usefulness in choosing a deformation model for the region*. Thesis, Université de Neuchâtel.

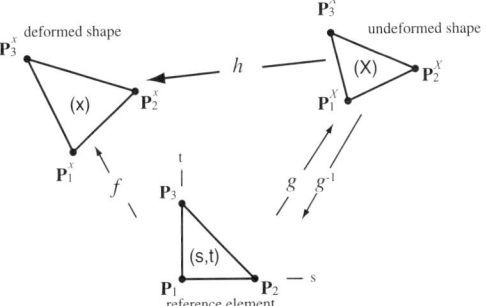

Fig. A2. Functions f, g and h as defined by the nodal displacements of a reference triangular element (s, t) onto an initial $(\mathbf{X})$ element and a final $(\mathbf{x})$ element. As can be seen, the path of h is equivalent to the composite of $f \cdot g^{-1}$.

HINDLE, D. & BURKHARD, M. 1999. Strain, displacement and rotation associated with the formation of curvature in fold belts; the example of the Jura arc. *Journal of Structural Geology*, **21**, 1089–1101.

HINDLE, D., BESSON, O. & BURKHARD, M. 2000. A model of displacement and strain for arc-shaped applied to the Jura arc. *Journal of Structural Geology*, **22**, 1285–1296.

HOWARD, J. H. 1968a. Recent deformation at Buena Vista Hills, California. *American Journal of Science*, **266**, 737–757.

HOWARD, J. H. 1968b. The role of displacements in analytical structural geology. *Geological Society of America Bulletin*, **79**, 1846–1852.

HOWARD, J. H. 1968c. The use of transformation constants in finite homogeneous strain analysis. *American Journal of Science*, **266**, 497–506.

KIRALY, L. 1969. Bref commentaire œ la carte structurale de la surface Argovien-Sequaniendans le canton de Neuchâtel. *Bulletin de la Société Neuchâteloise des Sciences Naturelles*, **92**, 71–73.

KLEY, J. 1999. Geologic and geometric constraints on a kinematic model of the Bolivian orocline. *Journal of South American Earth Sciences*, **12**, 221–235.

KLEY, J., MONALDI, C. R. & SALFITY, J. A. 1999. Along-strike segmentation of the Andean foreland; causes and consequences. *Tectonophysics*, **301**, 75–94.

LAUBSCHER, H. 1965. Ein kinematisches Modell der Jurafaltung. *Eclogae Geologicae Helvetiae*, **58(1)**, 232–316.

LAUBSCHER, H. P. 1961. Die Fernschubhypothese der Jurafaltung. *Eclogae Geologicae Helvetiae*, **54(1)**, 221–282.

LAUBSCHER, H. P. 1981. The 3D propagation of décollement in the Jura. *In*: McClay, K. R. & PRICE, N. J. (EDS) *Thrust and Nappe Tectonics*. Geological Society, London, Special Publications, **9**, 311–318.

MALVERN, L. E. 1969. *Introduction to the Mechanics of a Continuous Medium*. Prentice-Hall, New York.

McCOSS, A. M. 1988. Restoration of transpression/transtension by generating the three-dimensional, segmented helial loci of deformed lines across structure contour maps. *Journal of Structural Geology*, **10**, 109–120.

MEANS, W. D. 1976. *Stress and Strain; Basic Concepts of Continuum Mechanics for Geologists*. Springer-Verlag, New York.

NIETO-SAMANIEGO, A. 1999. Stress, strain and fault patterns. *Journal of Structural Geology*, **21**, 1065–1070.

NORABUENA, E., LEFFLER-GRIFFIN, L., MAO, A., DIXON, T., STEIN, S., SACKS, S., OCOLA, L. & ELLIS, M. 1998. Space Geodetic observations of Nazca-South America Convergence across the Central Andes. *Science*, **279**, 358–362.

NORABUENA, E., DIXON, T., STEIN, S. & HARRISON, C. 1999. Decelerating Nazca-South America and Nazca-Pacific Plate motions. *Geophysical Research Letters*, **26(22)**, 3405–3408.

PARDO-CASAS, F. & MOLNAR, P. 1987. Relative motion of the Nazca (Farallon) and South American plates since Late Cretaceous time. *Tectonics*, **6(3)**, 233–248.

PAREJA, J., VARGAS, C., SUAREZ, R., BALLON, R., CARRASCO, R. & VILLAROEL, C. 1978. *Mapa geologico de Bolivia, scale 1:1000000*. GEOBOL/YPFB, La Paz.

PAVONI, N. 1961. Faltung durch Horizontalvershiebung. *Eclogae Geologicae Helvetiae*, **54(2)**, 515–534.

PHILIPPE, Y. 1995. *Rampes latérales et zones de transfert dans les chaines plissées*. PhD thesis, Université de Savoie.

RECHES, Z. 1978. Analysis of faulting in three-dimensional strain field. *Tectonophysics*, **47**, 109–129.

ROUBY, D., COBBOLD, P. R., SZATMARI, P., DEMERCIAN, S., COELHO, D. & RICI, J. A. 1993. Least-squares palinspastic restoration of regions of normal faulting-application to the Campos Basin, Brazil. *Tectonophysics*, **221**, 439–452.

ROUBY, D., XIAO, H. & SUPPE, J. 2000. 3D restoration of complexly folded and faulted surfaces using multiple unfolding mechanisms. *AAPG Bulletin*, **84(6)**, 805–829.

SCHARDT, H. 1906. deux coupes générales à travers la chaîne du Jura. *Archives des Sciences Physiques et Naturelles, Genève*, **XXIII**.

SCHARDT, H. 1908. Les causes du plissement et des chevauchements dans le Jura. *Eclogae Geologicae Helvetiae*, **X**, 484–488.

SHAW, J. H., HOOK, S. C. & SUPPE, J. 1994. Structural trend analysis by axial surface mapping. *American Association of Petroleum Geologists Bulletin*, **78(5)**, 700–721.

SHEFFELS, B. M. 1995. Is the bend in the Bolivian Andes an orocline? *In*: TANKARD, A. J., SUAREZ, S. & WELSINK, H. J. (eds) *Petroleum Basins of South America*. AAPG Memoirs, **62**, 511–522.

SILVER, P. G., RUSSO, R. M. & LITHGOW-BERTELLONI, C. 1998. Coupling of South American and African plate motion and plate deformation. *Science*, **279**, 60–63.

SOMMARUGA, A. 1999. Décollement tectonics in the Jura foreland fold-and-thrust belt. *Marine and Petroleum Geology*, **16**, 111–134.

SUPPE, J. 1983. Geometry and kinematics of fault-bend folding. *American Journal of Science*, **283**, 684–721.

WILKERSON, M. S., MEDWEDEFF, D. A. & MARSHAK, S. 1991. Geometrical modeling of fault-related folds: a pseudo-three-dimensional approach. *Journal of Structural Geology*, **13(7)**, 801–812.

New aspects of tectonic stress inversion with reference to the TENSOR program

D. DELVAUX[1,2] & B. SPERNER[3]

[1]*Royal Museum for Central Africa, Department of Geology-Mineralogy, B-3080 Tervuren,*
Belgium (e-mail: ddelvaux@africamuseum.be and damien.delvaux@skynet.be)
[2]*Present address: Vrije University, Amsterdam, The Netherlands*
[3]*Geophysical Institute, Karlsruhe University, Hertzstrasse 16, 76187 Karlsruhe, Germany*

Abstract: Analysis of tectonic stress from the inversion of fault kinematic and earthquake focal mechanism data is routinely done using a wide variety of direct inversion, iterative and grid search methods. This paper discusses important aspects and new developments of the stress inversion methodology as the critical evaluation and interpretation of the results. The problems of data selection and separation into subsets, choice of optimization function, and the use of non-fault structural elements in stress inversion (tension, shear and compression fractures) are examined. The classical Right Dihedron method is developed in order to estimate the stress ratio R, widen its applicability to compression and tension fractures, and provide a compatibility test for data selection and separation. A new Rotational Optimization procedure for interactive kinematic data separation of fault-slip and focal mechanism data and progressive stress tensor optimization is presented. The quality assessment procedure defined for the World Stress Map project is extended in order to take into account the diversity of orientations of structural data used in the inversion. The range of stress regimes is expressed by a stress regime index R', useful for regional comparisons and mapping. All these aspects have been implemented in a computer program TENSOR, which is introduced briefly. The procedures for determination of stress tensor using these new aspects are described using natural sets of fault-slip and focal mechanism data from the Baikal Rift Zone.

Analysis of fault kinematic and earthquake focal mechanism data for the reconstruction of past and present tectonic stresses are now routinely done in neotectonic and seismotectonic investigations. Geological stress data for the Quaternary period are increasingly incorporated in the World Stress Map (WSM) (Müller & Sperner 2000; Müller *et al.* 2000; Sperner *et al.* 2003).

Standard procedures for brittle fault-slip data analysis and stress tensor determination are now well established (Angelier 1994; Dunne & Hancock 1994). They commonly use fault-slip data to infer the orientations and relative magnitude of the principal stresses.

A wide variety of methods and computer programs exist for stress tensor reconstruction. They are either direct inversion methods using least square minimization (Carey-Gailhardis & Mercier 1987; Angelier 1991; Sperner *et al.* 1993) or iterative algorithms that test a wide range of possible tensors (Etchecopar *et al.* 1981) or grid search methods (Gephart 1990*b*; Hardcastle & Hills 1991; Unruh *et al.* 1996). The direct inversion methods are faster but necessitate more complex mathemat-

ical developments and do not allow the use of complex minimization functions. The iterative methods are more robust, use simple algorithms and are also more computer time intensive, but the increasing computer power reduces this inconvenience.

This paper presents a discussion on the methodology of stress inversion with, in particular, the use of different types of brittle fractures in addition to the commonly used fault-slip data, the problem of data selection and the optimization functions. Two methodologies for stress inversion are presented: new developments of the classical Right Dihedron method and the new iterative Rotational Optimization method. Both methods use of the full range of brittle data available and have been adapted for the inversion of earthquake focal mechanisms. The interpretation of the results is also discussed for two important aspects: the quality assessment in view of the World Stress Map standards and the expression of the stress regime numerically as a Stress Regime Index for regional comparisons and mapping.

All aspects discussed have been implemented in the TENSOR program (Delvaux 1993*a*), which can

From: NIEUWLAND, D. A. (ed.) *New Insights into Structural Interpretation and Modelling.* Geological Society, London, Special Publications, **212**, 75–100. 0305-8719/03/$15

be obtained by contacting the first author. A program guideline is also provided with the program package.

Stress inversion methodologies

Stress analysis considers a certain volume of rocks, large enough to sample a sufficiently large data set of slips along a variety of different shear surfaces. The size of the volume sampled should be much larger than the dimensions of the individual brittle structures. For geological indicators, relatively small volumes or rock ($100–1000$ m^3) are necessary to sample enough fault-slip data, while for earthquake focal mechanisms, volumes in the order of $1000–10\,000$ km^3 are needed.

Stress inversion procedures rely on Bott's (1959) assumption that slip on a plane occurs in the direction of the maximum resolved shear stress. Inversely, the stress state that produced the brittle microstructures can be partly reconstructed knowing the direction and sense of slip on variably oriented fault planes. The slip direction on the fault plane is inferred from frictional grooves or slickenlines. The data used for the inversion are the strike and dip of the fault plane, the orientation of the slip line and the shear sense on the fault plane. They are collectively referred to as fault-slip data. Focal mechanisms of earthquakes are also used in stress inversion. The inversion of fault-slip data gives the four parameters of the reduced stress tensor: the principal stress axes $\sigma 1$ (maximum compression), $\sigma 2$ (intermediate compression) and $\sigma 3$ (minimum compression) and the Stress Ratio $R = (\sigma 2 - \sigma 3)/(\sigma 1 - \sigma 3)$. The two additional parameters of the full stress tensor are the ratio of extreme principal stress magnitudes ($\sigma 3/\sigma 1$) and the lithostatic load, but these two cannot be determined from fault data only. We refer to Angelier (1989, 1991, 1994) for a detailed description of the principles and procedures of fault-slip analysis and palaeostress reconstruction.

We are aware of the inherent limitations of any stress inversion procedures that apply also to the discussion proposed in this paper (Dupin *et al.* 1993; Pollard *et al.* 1993; Nieto-Samaniego & Alaniz-Alvarez 1996; Maerten 2000; Roberts & Ganas 2000).

The question was raised as to whether fault-slip inversion solutions constrain the principal stresses or the principal strain rates (Gephart 1990*a*). We will not discuss this question here, and leave readers to form their own opinions on how to interpret the inversion results. The brittle microstructures (faults and fractures) are used in palaeostress reconstructions as kinematic indicators. The stress inversion scepticals (e.g. Twiss & Unruh 1998) argue that kinematic indicators are strain markers

and consequently they cannot give access to stress. Without entering in such debate, we consider here that the stress tensor obtained by the inversion of kinematic indicators is a function that models the distribution of slip on every fault plane. For this, there is one ideal stress tensor, but this one is only certainly active during fault initiation. After faults have been initiated, a large variety of stress tensors can induce fault-slip by reactivation.

Stress and strain relations

In fault-slip analysis and palaeostress inversion, we consider generally the activation of pre-existing weakness planes as faults. Weakness planes can be inherited from a sedimentary fabric such as bedding planes, or from a previous tectonic event. A weakness plane can be produced also during the same tectonic event, just before accumulating slip on it, as when a fault is neoformed in a previously intact rock mass. The activated weakness plane F can be described by a unit vector **n** normal to F (bold is used to indicate vectors). The stress vector σ acting on the weakness plane F has two components: the normal stress ν in the direction of **n** and the shear stress τ, parallel to F. These two stress components are perpendicular to each other and related by the vectorial relation $\sigma = \nu + \tau$.

The stress vector σ represents the state of stress in the rock and has $\sigma 1$, $\sigma 2$ and $\sigma 3$ as principal stress axes, defining a stress ellipsoid. The normal stress ν induces a component of shortening or opening on the weakness plane in function of this sign. The slip direction **d** on a plane is generally assumed to be parallel to the shear stress component τ of the stress vector σ acting on the plane. It is possible to demonstrate that the direction of slip **d** on F depends on the orientations of the three principal stress axes, the stress ratio $R = (\sigma 2 - \sigma 3)/(\sigma 1 - \sigma 3)$ and the orientation of the weakness plane **n** (Angelier 1989, 1994).

The ability of a plane to be (re)activated depends on the relation between the normal stress and shear stress components on the plane, expressed by the friction coefficient:

$$\phi = a\,\tan\frac{|\tau|}{|\nu|}$$

If the characteristic friction angle ϕ of the weakness plane F with the stress vector σ acting on it overcomes the line of initial friction, the weakness plane will be activated as a fault. Otherwise, no movement will occur on it. This line is defined by the *cohesion factor* and the *initial friction angle* ϕ_o.

Data types and their meaning in stress inversion

Determination of palaeostress can be done using two basic types of brittle structures: (1) faults with slip lines (slickensides) and (2) other fracture planes (Angelier 1994; Dunne & Hancock 1994). In the following discussion as in program TENSOR, under the category 'fault', we understand only fault planes with measurable slip lines (slickensides), while we refer to all other types of brittle planes that did not show explicitly traces of slip on them as 'fractures'.

Brittle structures other than the commonly used slickensides can also be used as stress indicators (Dunne & Hancock 1994). They are generally known as 'joints' or 'fractures'. The term 'joint' formally refers to planes with no detectable movements on them, both parallel and perpendicular to the plane (Hancock 1985). Others use the term joint to refer to large tensional fractures and in this case, tension fractures are mechanically similar to joints (Pollard & Aydin 1988). To avoid confusion, we use here the general term 'fracture' for all planar surfaces of mechanical origin, which do not bear slip lines on the fracture surface. Brittle fractures as understood here bear some information on the stress stage from which they are derived (Dunne & Hancock 1994). In order to assess the relationship between fracture planes and stress axes, it is important to determine their genetic classes. They may form as *tension fractures* (tension joints) or *shear fractures* that may appear as conjugate pairs. *Hybrid tension* fractures with a shear component cannot be considered for stress analysis, as the mechanical conditions for their formation are intermediate between those responsible for tension fractures and for shear fractures.

Compression fractures form a particular type that can also be used in palaeostress analysis. The pressure-solution type of cleavage plane can be considered as a *compression fracture* if it can be demonstrated that no shear occurred along the plane. *Stylolites seams* are considered separately from *compression fractures*, for which the stylolite columns tend to form parallel to the direction of $\sigma1$.

For faults, one can measure directly the orientations of the fault plane and the slip line, and determine the slip sense (normal, inverse, dextral or sinistral) using the morphology of the fault surface and secondary structures associated as in Petit (1987). In vectorial notation, this corresponds to **n** (unit vector normal to the fault plane) and **d** (slip direction, defined by the orientation of the slip line and the sense of slip on the plane). These form the so-called fault-slip data.

Data on slip surface and slip direction can also be obtained indirectly by combining several brittle structures, as suggested by Ragan (1973). Conjugate sets of shear fractures $\mathbf{n}_1$ and $\mathbf{n}_2$ can be used for reconstructing the potential slip directions on the fracture planes, and to infer the orientations of principal stress axes. In conjugate fracture systems, $\sigma1$ bisects the acute angle $\mathbf{n}_1{}^{\wedge}\mathbf{n}_2$, $\sigma2$ is determined by the intersection between $\mathbf{n}_1$ and $\mathbf{n}_2$, and $\sigma3$ bisects the obtuse angle. The slip directions $\mathbf{d}_1$ and $\mathbf{d}_2$, respectively on shear planes defined by $\mathbf{n}_1$ and $\mathbf{n}_2$, are perpendicular to $\sigma2$ in a way that the two acute wedges tend to converge.

Similarly, the slip direction on a shear plane without observable slip line can be inferred if tension fractures are associated at an acute angle to the shear plane (Ragan 1973). For shear plane $\mathbf{n}_s$ and associated tension fracture $\mathbf{n}_t$, $\sigma1$ is parallel to the tension fracture, $\sigma2$ is determined by the intersection between $\mathbf{n}_s$ and $\mathbf{n}_t$, and $\sigma3$ is perpendicular to the tension fracture (parallel to $\mathbf{n}_t$). The slip direction $\mathbf{d}_s$ on the shear plane defined by $\mathbf{n}_s$ is perpendicular to $\sigma2$ so the block defined by the acute angle $\mathbf{n}_s{}^{\wedge}\mathbf{n}_t$ tend to move towards the block defined by the obtuse angle.

In these two cases the slip direction **d** on the shear planes **n** is reconstructed and these can be used in the inversion as additional fault-slip data.

For the stress inversion purpose, we distinguish between three types of brittle fractures.

- Tension fractures (plume joints without fringe zone, tension gashes, mineralized veins, magmatic dykes), which tend to develop perpendicular $\sigma3$ and parallel to $\sigma1$. The unit normal vector **nt** represents an input of the direction $\sigma1$.
- Shear fractures (conjugate sets of shear fractures, slip planes displacing a marker), which form when the shear stress on the plane overcomes the fault friction. It corresponds to the input of a fault plane $\mathbf{n}_s$, but without the slip direction **d**.
- Compression fractures (cleavage planes), which tend to develop perpendicular to $\sigma1$ and parallel to $\sigma3$. The vector $\mathbf{n}_c$ represents an input of the direction $\sigma1$.

For the stylolites, it is more accurate to use the orientation of the stylolite columns as a kinematic indicator for the direction of maximum compression ($\sigma1$) instead of the plane tangent to the stylolite seam (i.e. input of direction $\sigma1$).

Earthquake focal mechanisms are determined geometrically by the orientations of the **p**- and **t**-kinematic axes bisecting the angles between the fault plane and the auxiliary plane. They can be determined also by the orientation of one of the two nodal plane ($\mathbf{n}_1$ or $\mathbf{n}_2$) and the associated slip vector ($\mathbf{d}_1$ or $\mathbf{d}_2$), or by the orientation of the two nodal planes ($\mathbf{n_1}$ and $\mathbf{n_2}$) and the determination of the regions of compression and tension (i.e. input of either $\mathbf{n_1}$ and $\mathbf{n_2}$ or $\mathbf{p}$ and $\mathbf{t}$).

In addition to the orientation data, it is also important to record qualitative information for all fault-slip data: the accuracy of slip sense or fracture type determination (slip sense confidence level), a weighting factor (between 1 and 9, as a function of the surface of the exposed plane), whether the fault is neoformed or has been reactivated, the type and intensity of slip striae, the morphology and composition of the fault or fracture surface, estimations of the relative timing of faulting for individual faults, based on cross-cutting relationships or fault type.

Data selection and separation into subsets

After measuring the fault and fracture data in the field or compiling a catalogue of earthquake focal mechanisms, the data on brittle structures are introduced in a database. This raw assemblage of geometrical data forms a *data set* (or *pattern* if we follow the terminology of Angelier 1994). The raw data set is used as a starting point in stress inversion. For rock masses that have been affected by multiple tectonic events, the raw data set consists of several *subsets* (or *systems*) of brittle data. A subset is defined as a group of faults and fractures (in fault-slip analysis) or of focal mechanisms (in seismotectonic analysis) that moved during or have been generated by a distinct tectonic event. Moreover, the movement on all the weakness planes of the subset can be fully described in a mechanical point of view by the stress tensor characteristic of the tectonic event.

For an appropriate constraint on a stress tensor, data subsets should be composed of more than two families of data. Using the definition of Angelier (1994), a family is a group of brittle data of the same type and with common geometrical characteristics. In stress inversion, movement on a system of weakness planes is modelled by adjusting the four unknowns of the reduced stress tensor. Therefore, the stress tensor will be better constrained for data subsets with the largest amount of families of data of different type and orientation. This concept is used in the diversity criteria for the quality ranking procedure, described later.

In raw data sets, it is frequently observed that all the brittle data do not belong to a single subset as defined above. This is often due to the action of several tectonic events during the geological history of a rock mass. But the observed misfits can also be the consequence of other factors such as measurement errors, the presence of reactivated inherited faults, fault interaction, non-uniform stress field and non-coaxial deformation with internal block rotation (Dupin *et al.* 1993; Pollard *et al.* 1993; Angelier 1994; Nieto-Samaniego & Alaniz-Alvarez 1996; Twiss & Unruh 1998; Maerten 2000; Robert & Ganas 2000).

The frequent occurrence of multiple-event data sets and the numerous possible sources of misfits have important implications in fault-slip analysis and palaeostress reconstruction. It necessitates the separation of the raw data sets into subsets, each characterized by a different stress tensor. This check is necessary even in the case of a single-event data set, as there are several possibilities to create outliers. Errors might occur during field work (e.g. uneven or bent fault planes, incorrect reading), during data input (incorrect transmission), or during data interpretation (e.g. not all events are detected due to the lack of data). A certain percentage of misfitting data (*c.* 10–15%) is normal, but they have to be eliminated from the data set for better accuracy of the calculated results.

In the iterative approach for stress tensor determination, data are excluded on the basis of a misfit parameter that is calculated for each fault or fracture as a function of the model parameters ($\sigma 1$, $\sigma 2$, $\sigma 3$ and the stress ratio R) that best fit the entire set of data. A first stress model is determined on the raw data set, and then the data with the largest misfit are separated from the raw data set. After a first separation, this process is repeated and the original data set is progressively separated (split) into a subset containing data more or less compatible with the stress model calculated, and non-compatible data which remain in the raw data set. After the separation of a first subset from the raw data set, this process is repeated again on the remaining data of the raw data set, to eventually separate a second subset. We will discuss this procedure in more detail later when presenting the Right Dihedron and the Rotational Optimization methods.

In summary, data separation is performed during stress inversion as a function of misfits determined with reference to the stress model calculated on the data set. This is done in an interactive and iterative way and the two processes (data separation into subset and stress tensor optimization for that subset) are intimately related.

The first step of the selection procedure starts in the field. Faults or fractures of the same type, with the same morphology of fault surface, the same type of surface coating or fault gauge are likely to have been formed under the same geological and tectonic conditions. Already in the field, they can be tentatively classified into different families, and families associated into subsets.

Cross-cutting relations might help to differentiate between different families of faults, but it is not always easy to interpret these relations in terms of successive deformation stages. If relations between two types of structures are consistent and systematic at the outcrop scale (i.e. a family of normal faults systematically younger than a family of strike-slip faults), they can be used to differentiate fault families in the field and to establish their relative chronology. But this is not enough to differentiate brittle systems of different generation, as data subsets should be composed of several families of structures in order to provide good constraints during stress inversion. As much qualitative field information and as many observations as possible of the relations between pairs of fault or fractures are needed. But they are often insufficient alone to identify and differentiate homogeneous families of fault and fractures related to a single stress event.

The starting point for fault separation and stress inversion can be the first separation performed in the field. If this is not possible, a rapid analysis of the p and t axes associated with the faults and fractures can help differentiate between different families of faults and fractures, based on their kinematic style and orientation (e.g. normal, strike-slip and reverse faulting, tension and compressional joints). This might be helpful if the measured faults and fractures belong to deformation events of markedly different kinematic styles. The separation done in the field has to be checked and refined during the inversion. In most cases, however, the selection of

data relies mostly on an interactive separation during the inversion procedure.

However, we want to warn of pure automated data separation, because this might result in completely useless subsets. For better clarity we illustrate this risk with a 2D data set (depending on two parameters) instead of a 4D example (depending on four parameters), as it would be necessary for fault-slip data (three principal stress axes plus stress ratio). Figure 1 shows how the automated separation of a data set with two clusters results in three subsets, all of them representing only parts of the two clusters of the original data set. The same happens if multi-event fault-slip data sets are separated automatically. This problem can be handled by first making a rough 'manual' separation and then doing the first calculation. In the 2D example of Figure 1 this can easily be done by separating the data of the two clusters into different subsets. The manual separation of fault-slip data is not so straightforward because it is a 4D problem. The first requirement for a successful separation are field observations which indicate the existence of more than one event and can be used to discriminate the character of the different events. The best indicators for a multi-event deformation are differently oriented slip lines on one and the same fault plane which, in the best case, show different min-

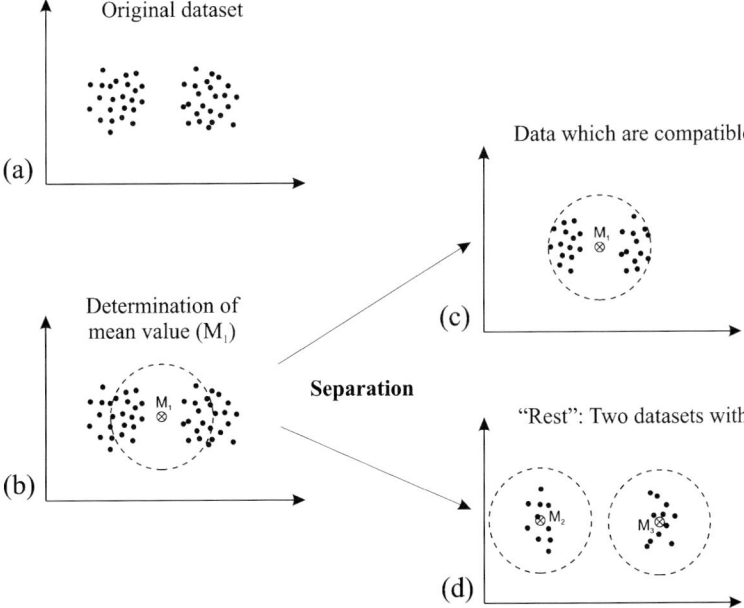

Fig. 1. Separation of an artificial data set into subsets using an automated procedure. The original data set shows two clusters (**a**), but the calculated mean M_1 lies in between them (**b**). Separation according to the deviation from M_1 results in a data set compatible to M_1 (**c**) and two other data sets with 'own' means M_2 and M_3 (**d**). Thus, the automated separation leads to three subsets (instead of two) with three different means M_1-M_3. The mean of the largest subset M_1 (c) has nothing in common with the means of the two clusters in the original data set (a).

eralization, so that the mineralization can then be used as a separation criterion. Additionally to field observations, obviously incompatible data can be separated through careful inspection of the data plots. As shown in Figure 2 some of the fault planes might have similar orientations, but different slip lines (e.g. down-dip versus strike-slip movements; see encircled data in Fig. 2). Separating these data into different subsets can serve as a starting point for the assignment of the other data, so that consistent subsets emerge (Fig. 2). These subsets are then checked for compatibility by feeding them into a computer method like the Right Dihedron and the Rotational Optimization implemented in the program TENSOR.

In this paper, we strongly advise making an initial separation in function of field criteria, a careful observation of the data plots and p-t analysis. The raw data set or the preliminary subsets should then be further separated into subsets while optimizing the stress tensor using successively the improved Right Dihedron method and the Rotational Optimization method in an interactive way.

Improved Right Dihedron method

The well-known Right Dihedron method was originally developed by Angelier & Mechler (1977) as a graphical method for the determination of the range of possible orientations of $\sigma 1$ and $\sigma 3$ stress axes in fault analysis. The original method was translated in a numeric form and implemented in different computer programs. We discuss here a series of improvements that we developed to widen the applicability of this method in palaeostress analysis. These new developments concern (1) the estimation of the stress ratio R, (2) the complementary use of tension and compression fractures, and (3) the application of a compatibility test for data selection and subset determination using a Counting Deviation.

Although the Improved Right Dihedron method will still remain downstream in the process of palaeostress reconstruction, it now provides a preliminary estimation of the stress ratio R and data selection. It is typically designed for building initial data subsets from the raw data set, and for making a first estimation of the four parameters of the reduced stress tensor. The Improved Right Dihedron method forms a separate module in the TENSOR program. The original method is described first briefly, before focusing on the improvements.

General principle The Right Dihedron method is based on a reference grid of orientations (384 here) pre-determined in such a way that they appear as a rectangular grid on the stereonet in lower hemisphere Schmidt projection. For all fault-slip data, compressional and extensional quadrants are determined according to the orientation of the fault plane and the slip line (Fig. 3a–c), and the sense of movement. These quadrants are plotted on the reference grid and all orientations of the grid falling in the extensional quadrants are given a *counting value* of 100% while those falling in the compressional quadrants are assigned 0%. This procedure is repeated for all fault-slip data (Fig. 3f). The counting values are summed up and divided by the number of faults analysed. The grid of counting values for a single fault defines its characteristic *counting net*. The resulting grid of average counting values for a data subset forms the *average*

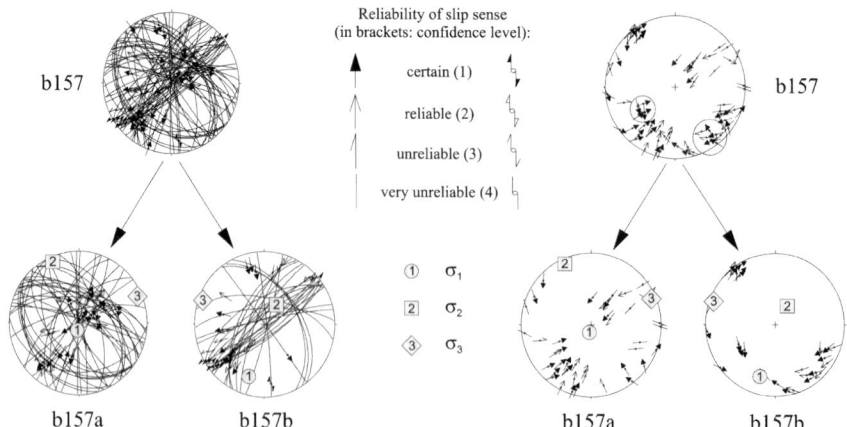

Fig. 2. Separation of a fault-slip data set (b157) into two subsets (b157a and b157b). Obviously incompatible slickensides that have similar fault orientations, but different slip directions (for examples see encircled data) are separated into different data sets. The calculated palaeostress axes for the subsets are plotted.

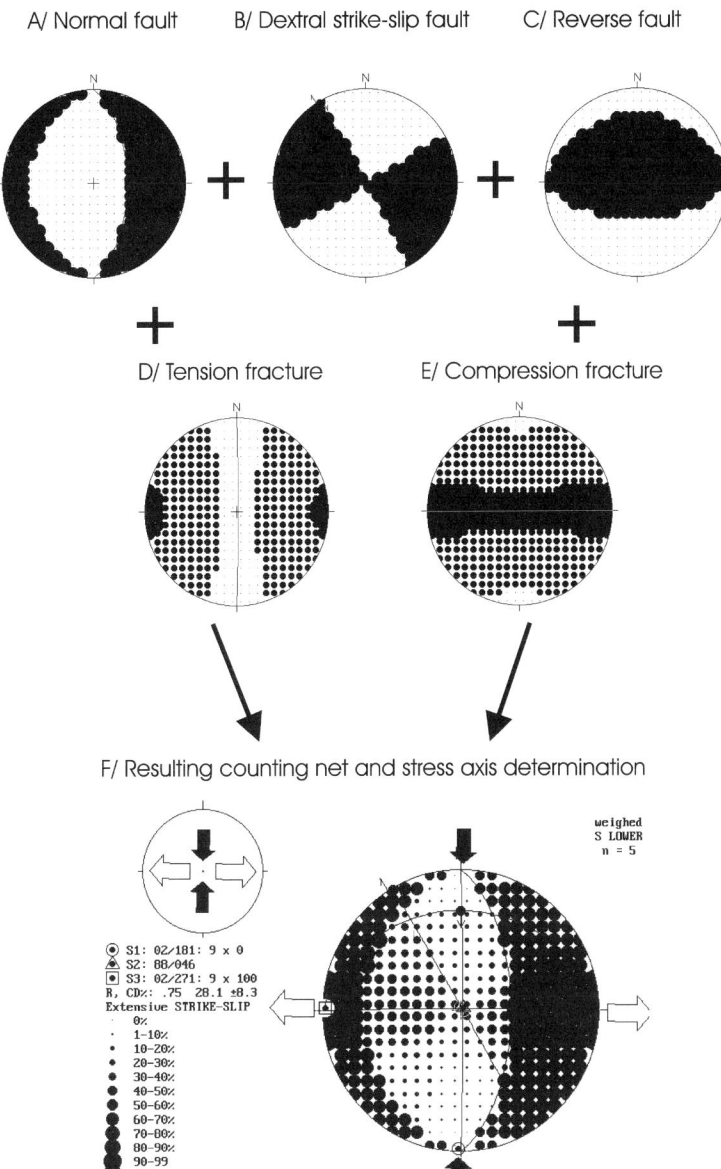

Fig. 3. Principle of the Right Dihedron method (Schmidt projections, lower hemisphere). (**a–e**) different kinds of simple fault-slip data used with their characteristic counting net and projection of the fault or fracture plane, and their combination to produce the resulting average counting net (**f**). The orientation of $\sigma 1$ and $\sigma 3$ axes are computed as the mean orientation of the points on the counting grid that has respectively values of 0 and 100. The two axes are set perpendicular to each other and the orientation of the intermediate $\sigma 2$ axis is obtained, orthogonal to both $\sigma 1$ and $\sigma 3$. The counting value of the point on the counting net which is the closest to $\sigma 2$ serves for the estimation of the R ratio: $R \cong (100 - S2val)/100$, where S2val is the counting value of the nearest point on the counting grid.

counting net for this subset. The possible orientations of $\sigma 1$ and $\sigma 3$ are defined by the orientations in the average counting net that have values of 0% and 100%, respectively.

This method is particularly suitable for the stress analysis of earthquake focal mechanisms. For fault-slip data, it gives only a preliminary result, as it does not verify the Coulomb criteria. Problems occur for data sets with only one orientation of fault planes (and with identical slip direction). In this case, σ1 and σ3 deviate by 15° from the 'correct' position, if they are placed in the middle of the compressional/extensional quadrants. For conjugate faults the position in the middle of the 0%/100% area corresponds with σ1 and σ3 (Fig. 4). Moreover, it can be applied only when the sense of movement is given; otherwise the compressional and extensional quadrants are undetermined.

The graphical method gives a range of possible orientations of σ1 and σ3. These correspond to the grid orientations on the resulting counting net with respectively values of 0% and 100%. The mean orientations of these reference points give respectively the most probable orientations for σ1 and σ3. Because σ1 and σ3 are determined independently, they are not always perpendicular to each other.

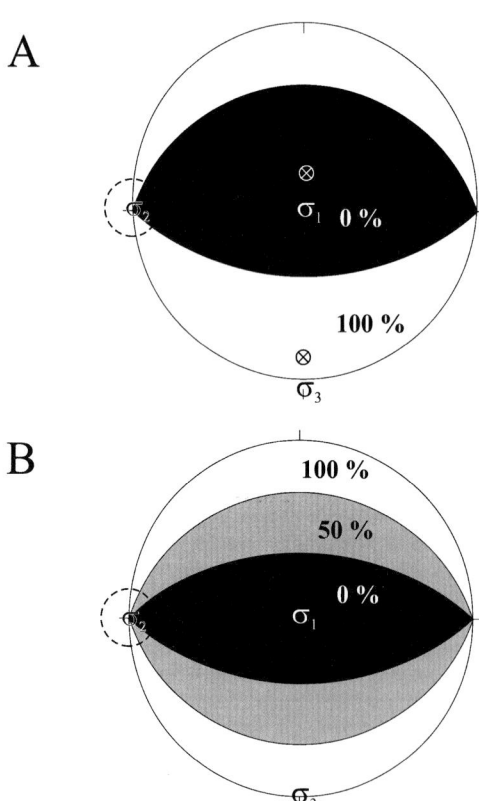

Fig. 4. Uncertainty in determination of stress axes with the Right Dihedron method. (**a**) For a single normal fault the middle of the compression and of the extension quadrant (crosses) is not identical with σ1 and σ3 (15° off). (**b**) Two conjugate normal faults and their compression and extension quadrants: if σ1 and σ3 are placed in the middle of the 0% and 100% area respectively, they are correctly located.

They can be set orthogonally by choosing either σ1 or σ3 fixed and rotating the other axis around a rotation axis defined as the normal to the plane containing σ1 and σ3. The orientation of the intermediate σ2 axis can then easily be deduced.

One problem with this method is that it does not determine the stress ratio R ($R = (\sigma2 - \sigma3)/(\sigma1 - \sigma3)$) and that σ1 and σ3 are undefined when the extreme values on the counting net do not reach 0% and 100%.

Estimation of stress ratio R The stress ratio R, defined as equivalent to $(\sigma2 - \sigma3)/(\sigma1 - \sigma3)$ is one of the four parameters determined in the stress inversion, with the three principal stress axes σ1, σ2 and σ3. Until now, the Right Dihedron method has given only an estimate of the orientations of σ1, σ2 and σ3, but not of the stress ratio R. A careful observation of the Right Dihedron counting nets, however, shows that their patterns differ as a function of the type of stress tensor (extensional, strike-slip or compressional). We therefore investigated a way to express this pattern as a function of a parameter that would be an estimation of the stress ratio R.

A way to do this is to compare the orientation of the previously determined σ2 axis with the distribution of counting values on the average counting net. Using the position of the σ2 axis on the counting grid, we found that a good estimation of the stress ratio R can be obtained with the empiric relation:

$$R \cong (100 - S2val)/100$$

where S2val is the counting value of the point on the reference grid nearest to the orientation of σ2. This formula is only valid for large fault populations with a wide variety of fault plane orientations.

The accuracy of this method for the estimation of the R ratio has been validated using models with synthetic sets of faults obtained by applying different stress tensors on a set of pre-existing weakness planes of different orientation and computing the shear stress component τ on the plane and the friction angle φ. The differents sets are then submitted to stress analysis using the Improved Right Dihedron method. The values of the stress ratio R obtained are in general close to the ones used to produce the models, within a range of R ± 0.1 (Fig. 5). Similarly, the orientation of the stress axes generally match within a few degrees the ones used to generate the synthetic sets. Experience gained using this method in conjunction with other methods of direct inversion (like the Rotational Optimization method described later) on a large number of sites shows that the stress ratio R esti-

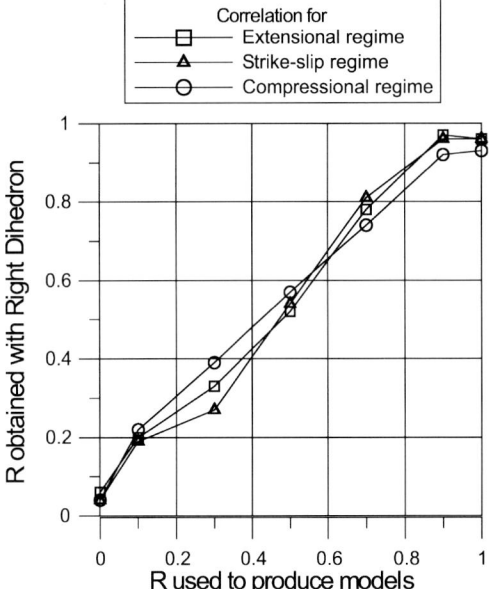

Fig. 5. Relation between stress ratios R used to produce models of synthetic fault sets by applying different stress tensors on a set of 152 pre-existing weakness planes of different orientation, and R values obtained by analysing these sets with the Improved Right Dihedron method. Models were produced with $R = 0, 0.1, 0.3, 0.5, 0.7, 0.9$ and 1.0. for extensional, strike-slip and compressional stress regimes.

mated using the Improved Right Dihedron method is generally close to the one obtained by the Rotational Optimization method (see hereafter).

We conclude that for all types of stress tensors (extensional, strike-slip and compressional) with $\sigma 1 > \sigma 2 > \sigma 3$ and $0.25 < R < 0.75$, the Improved Right Dihedron method successfully estimates the four parameters of the stress tensor. In the extreme case of flattening ($\sigma 2 = \sigma 3$, $R = 0$) or constriction ($\sigma 1 = \sigma 2$, $R = 1$) only one of the extreme values will be well defined (0% for flattening, 100% for constriction); the other one will have medium values and a circular distribution. Therefore only $\sigma 1$ is well defined when $R \cong 0$ and $\sigma 3$, when $R \cong 1$.

Use of compression and tension fractures as palaeostress indicator The Right Dihedron method also allows the use other types of brittle data such as compression and tension fractures as defined above for estimating the four parameters of the stress tensors (Fig. 3d, e). For tension fractures, $\sigma 3$ is considered as oriented within a cone angle of β degrees around the normal to the plane and $\sigma 1$ is located at an acute angle ($\leq \beta$) of the tension plane. The opposite is true for compression fractures, with

$\sigma 1$ located within the cone angle β and $\sigma 3$ at an acute angle ($\leq \beta$) to the tension plane. The orientations between the cone angle β and the surface generated by the revolution of a line inclined at an angle β from the fracture plane are considered as intermediate.

It is possible to define on the counting nets, areas in compression (value 0), in extension (value 100), and intermediate areas (value 50). The individual counting nets are summed up and averaged as in the case of faults, to obtain the average counting net. This allows the combined analysis of the three different types of data (slickensides with known sense of movement, tension and compression fractures, Fig. 3).

For the value of the angle β we use the common initial friction angle of 16.7° given by Byerlee (1978). The use of this value is justified by the shear stress/normal stress relations in the initial friction law (Jaeger 1969). If the angle between the tension fracture or the normal to the compression fracture and $\sigma 1$ is larger than 16.7°, the resolved shear stress is theoretically high enough to cause slip on that plane, which is in contradiction to the postulated nature of that plane. Other values for the angle β can be used, without modifying the general principle.

When working with a database composed only of fracture data, this method provides a way to estimate the stress ratio R, while this could not be determined by formal inversion.

Counting Deviation Another important improvement of the original Right Dihedron method is the development of a parameter for estimating the degree of compatibility of the individual counting nets with the average counting net of the subset. It relies on the calculation of a Counting Deviation CD (expressed in %) for each datum by comparing its counting net with the average counting net. Fault or fracture data with a low CD value contribute in a positive way to the average counting net (reinforce the extreme counting values) and the data with higher CD values contribute in a negative way (weaken the extrema).

The principle of Counting Deviation and its use in compatibility testing are presented hereafter with reference to Figure 6. The first (and forward) step is to compute the average counting net by summing the values of each point on the reference grid for all the individual counting nets (Fig. 6a–c), taking into account the weighting factor associated with each datum (Fig. 6d). After this operation, the second step is performed in a reverse way. Each individual counting net is compared with the average counting net. As a result, a differential counting net is obtained for each datum by subtracting for each orientation of the reference grid the coun-

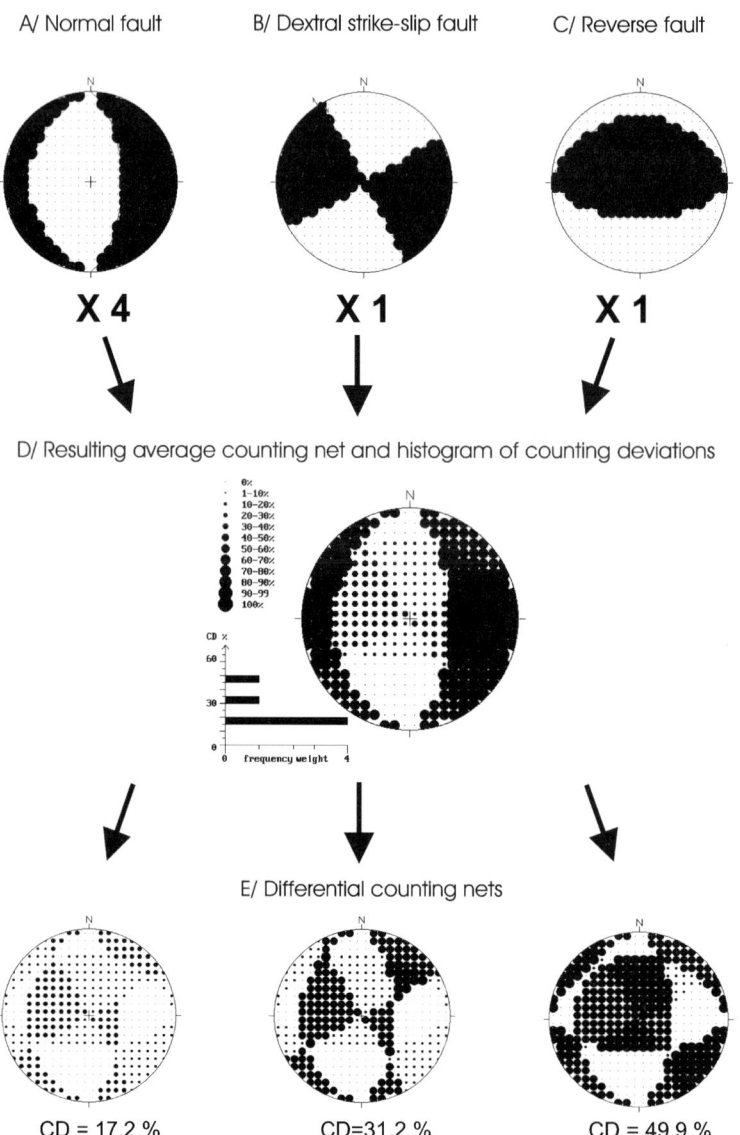

Fig. 6. Use of the Counting Deviation CD to filter the data. As in Figure 3, data from three different types of brittle structure (**a–c**) are used to produce the average counting net (**d**), with consideration of the weighting factor. For all data, each counting value of the individual counting nets is multiplied by the weighting factor (4 for the normal fault and 1 for the other data). The stronger influence of the normal fault on the resulting average counting net is clearly visible (d). The differential counting nets (**e**) are obtained by subtracting respectively the counting values of the individual counting nets (a, b or c) from the ones of the average counting net (d), without application of the weighting factor. The Counting Deviation (CD) is the average counting value for the 384 points on the differential counting nets, expressed as a percentage (e). On the average counting net diagram (d), the CD values are displayed in a histogram, weighted according to the weighting factor associated with the individual data. The average counting deviation for the whole data set is 25 ± 12.3% (1σ). In this case, the reverse fault has a CD value above the average CD + 2σ and can be eliminated from the data set. If the process is repeated on the two remaining data, the dextral strike-slip fault will in turn be eliminated.

ting values of the individual nets from the those of the average counting net (Fig. 6e). The Counting Deviation (CD) is the average of all the 384 values on the differential counting net. As the counting values in the counting net are expressed as percentages, the unit for the Counting Deviation is also the per cent.

The compatibility of individual data with the entire data subset is estimated by the dispersion of the CD values away from the arithmetic mean of all the CD values. The homogeneity of the data set is expressed by the standard deviation of the CD values: smaller standard deviations suggest that most data of the subset are compatible with the final counting net and hence possibly belong to the same kinematic event. In the TENSOR program, the fault planes can be rejected if their CD values are larger than the arithmetic mean $+1\sigma$ or $+2\sigma$, depending on the choice of the user. This is used to control the compatibility of each data with the final result, and to reject it from the subset if necessary.

Using this system, a first solution is computed on the initial data set (single-event deformation) or on the manually separated subsets (multi-event deformation), and the incompatible faults are rejected. The procedure is repeated several times until well-defined areas of 0 and 100 values are obtained in the counting net. (An example of progressive separation is given in Fig. 8a and Table 1 for fault-slip data set and in Fig. 9a and Table 3 for a focal mechanism data set.) The final result and the corresponding subset will serve as a starting point for the rotational optimization.

In summary, the Improved Right Dihedron method allows a first estimation of the orientations of the principal stress axes and of the stress ratio R, and a first filtering of compatible fault-slip data. The selected fault-slip population and the preliminary tensor can be used as a starting point in the iterative inversion procedures like the Rotational Optimization method described hereafter.

Rotational Optimization method

The Rotational Optimization method presented here is a new iterative inversion procedure. As for all iterative inversion methods, it is based on the testing of a great number of different stress tensors, with the aim of minimizing a misfit function. In principle the whole range of orientations for the three stress axes and the stress ratio R has to be tested to find the minimum value of the misfit function (grid search). Considering these four parameters, the problem is close to a four-dimensional one, with an additional constraint that the three principal stress axes have to be orthogonal. This

leads to a large number of different configurations of the stress tensor to be tested with the data set.

To find the solution in a more direct way, we developed the Rotational Optimization method and propose to initiate the search procedure using the stress tensor estimated with the Right Dihedron method. It allows restriction of the search area during the inversion, so that the whole grid does not have to be searched.

In the following, we introduce a few classic theoretical notions, discuss in detail the misfit functions to minimize, and present the Rotational Optimization procedure with the help of a natural example from the Baikal Rift Zone in Siberia.

Shear stress construction and misfit parameters The basic idea in the direct inversion is to test a number of stress tensor on all faults (and fractures) of the data set by computing the direction, sense and magnitude of the shear stress (τ) acting on the plane, the magnitude of the related normal stress (σ_n), the characteristic dihedral angle 2θ and the friction angle ϕ.

In TENSOR, this is done using the method of Means (1989). In the inversion of fault-slip data, the isotropic part of the stress tensor is missing. Thus the reduced stress tensor is identical with the deviatoric part of the stress tensor. The magnitudes are expressed in a relative way because the absolute values cannot be determined using geological data only (see Angelier 1989). By convention, the magnitude of $\sigma 1$ is fixed at 100 and the magnitude of $\sigma 3$ at 0 (in abitrary units). The relative magnitudes are therefore in the range $0 = \sigma 3 \leq \sigma 2 \leq \sigma 1 = 100$. The magnitude of the $\sigma 2$ axis is fixed by the stress ratio R, as a function of the magnitudes $\sigma 1$ and $\sigma 3$.

The occurrence of slip on a pre-existing rock discontinuity is governed by friction laws. On a Mohr diagram, the corresponding point is enclosed between the initial friction curve and the maximum friction line (failure curve). The relation between the minimum shear stress (τ) and the normal stress (ν) is approximately linear (Angelier 1989). Hence, the sliding criteria can be simplified by assuming that the friction angle ϕ must be greater than the initial friction angle and smaller than the maximum friction angle. The friction criteria of Byerlee (1978) are used here as default values (initial friction angle ϕ_o of 16.7° and maximum angle ϕ_{max} of 40.4°). If $\phi < \phi_o$, no reactivation of pre-existing discontinuities will occur; if $\phi > \phi_{max}$ failure will occur with the development of a new fracture.

Minimization functions A great advantage of the iterative approach for stress tensor inversion is that the complexity of the function to minimize is not a limiting factor and the function can be easily changed without changing the algorithm.

In general, the minimization function has the following form:

$$F_j = \frac{\sum (f_j(i) \times w(i))}{n \times \sum w(i)} \qquad (1)$$

where $w(i)$ is the weight of the individual data and $f_j(i)$ is the function that has to be minimized.

In the following, by fault-slip data we understand fault planes (defined by the unit vector $\mathbf{n}$ normal to the slip plane) and associated slip vectors $\mathbf{d}$ on the slip planes, either observed or reconstructed as explained above. Faults with associated slip striae are also known as *slickenside*. In this category, we include also fault planes with slip direction only (direction of movement unknown). For those fault-slip data, the most classic misfit parameter is the angular deviation between the slip vector $\mathbf{d}$ and the shear stress $\mathbf{t}$ (slip deviation α), computed in function of the stress tensor and the orientation of the slip plane $\mathbf{n}$:

$$f_1(i) = \alpha(i) \ (= \text{function F1 in TENSOR program})$$

For slickensides with known sense of movement, the slip vector is uniquely defined and α ranges between 0 and 180°. When the sense of movement is not known, only the orientation of the slip vector is defined, and two opposite directions are possible. In this case, the maximum slip deviation is at 90° from the slip direction. It is generally assumed that for a slip deviation of $\alpha \leq 30°$ the observed fault movement is compatible with the theoretical shear vector.

It is also common to use a least-square function for the minimization. This implies a Gaussian-type distribution of individual misfits. A good example is function S4 Angelier (1991), already proposed by the same author in 1975:

$$f_2(i) = \sin^2(\alpha(i)/2)$$
$$(= \text{function F2 in TENSOR program})$$

These two functions are still insufficient in the case of newly formed conjugate fault systems. In this case $\sigma 1$ and $\sigma 3$ might lie anywhere in the plane perpendicular to the two conjugate fault sets, because the slip deviation will be minimum for any configuration of stress axes. As long as $\sigma 2$ is parallel to the intersection line of the two fault systems, the orientation of $\sigma 1$ and $\sigma 3$ will not influence the slip deviation. Thus additional constraints are necessary, taking into account the ability of the fault to slip (Angelier 1991). This can be either the

friction angle ϕ or the shear stress magnitude $|\tau|$ on the fault plane, which both should be maximized. As the mechanical properties of the faulted rocks are generally unknown in standard palaeostress investigation, it is more appropriate to use $|\tau|$ to express the tendency of the fault to slip (see also Morris *et al.* 1996). Similarly, the normal stress component on the slip surface (ν) also influences the ability of the fault to slip. It should be minimum in order to lower the fault friction and hence to favour slip. These three parameters can be taken into account simultaneously by combining in the same function the slip deviation α and the normal stress magnitude $|\nu|$ that should be minimized, with the shear stress magnitude $|\tau|$ that should be maximized. This is done in function $f_3(i)$:

$$f_3(i) = (f_2(i) \times 360) + ((\text{Tinv}(i) + |\nu(i)| - 29.7)/p)$$

The first term is for the minimization of the deviation angle α and corresponds to the $f_2(i)$ defined above, multiplied by 360 to ensure that this term remains dominant with regards to the second term. The second term allows increasing the tendency of the fault to slip, by maximizing $|\tau|$ while minimizing $|\nu|$. $\text{Tinv}(i)$ maximizes $|\tau|$ when minimizing $f_3(i)$: $\text{Tinv}(i) = \sigma_1/2 - |\tau(i)|$ with $\sigma_1/2$ corresponding to the maximum possible value of $|\tau|$. $|\nu(i)|$ minimizes the normal stress component on the slip surface.

With magnitudes of σ_1 and σ_3 fixed respectively at 0 and 100, the minimum possible value for the expression $(\text{Tinv}(i) + \text{Nnorm}(i))$ is 29.7 on a Mohr circle construction. To allow this term to converge to zero in the most favourable cases, 29.7 is substracted from the result of this expression. The proportionality factor p also allows the first term to be kept dominant with regard to the second term. In TENSOR, it is set by default to 5, but can be modified.

The TENSOR program can also use fracture data to constrain the stress tensors, and is not restricted to the analysis of slickenside data only. To implement this, other minimization functions have been developed.

For shear fractures (slip surface with observed movement but without slip lines), the second term of function $f_3(i)$ can be used as defined above:

$$f4(i) = ((\text{Tinv}(i) + |\nu(i)| - 29.7)/p)$$

For tensional fractures (plume joints, mineralized veins, magmatic dykes), the normal stress $|\nu(i)|$ applied to the fracture surface should be minimal to favour fracture opening. Simultaneously the shear stress magnitude $|\tau(i)|$ should also be minimal to prevent slip on the plane:

$f5(i) = (|\nu(i)| + |\tau(i)|)/p)$

For compressional fractures, the normal stress applied to the fracture should be maximal ($\sigma1 - |\nu(i)|$), while the shear stress $|\tau(i)|$ should be minimal:

$f6(i) = (\sigma1 - |\nu(i)| + |\tau(i)|)/p)$

It is important to be aware that the stress ratio R can only be calculated with the functions that minimize the slip deviation α ($f1$–$3(i)$). This restriction has been implemented in the Rotational Optimization procedure in TENSOR. When working with compression or tension fractures only, a rough estimation of the stress ratio R can be obtained by the Right Dihedron method. This value is then maintained during the following procedures in the Rotational Optimization.

In program TENSOR, all functions described above should used for data sets containing only the corresponding data type. For mixed data sets (containing different types of data together) a composite function has been implemented. It allows the use of mixed data sets, by combining the optimization procedure for slickenside data and tension, shear or compression fractures. The contribution of each type of data is adapted in a way that they all have to be minimized to improve the quality of the tensor. In the general optimization function F_j, the individual functions $f(i)$ are adapted to the type of data as follows:

- for fault planes with slip lines: $f3(i)$
- for shear fractures: $f4(i)$
- for tension fractures: $f5(i)$
- for compression fractures and stylolites: $f6(i)$

This composite function ($F5$ in TENSOR) has been proved very efficient in palaeostress inversion of mixed data sets.

Interactive Rotational Optimization and kinematic separation of fault-slip data Stress tensor determination using the Rotational Optimization procedure consists in a controlled 4D grid search involving successive rotations of the tensor around the three principal stress axes ($\sigma1$, $\sigma2$ and $\sigma3$) and equivalent testing of the stress ratio R. For each stress axis, the rotation angle is determined for which the misfit function has its minimum value. The minimum value of the misfit function for each run is determined by taking the minimum of the polynomial regression curve adjusted to the results of each test, in a graphic representation with the rotation angle as abscissa and the value of the misfit function as ordinate (Fig. 7). The tensor is then rotated accordingly. The same procedure is repeated successively for the following stress axis. After rotation around

the three stress axes, the value of R is determined in a similar way by testing a range of possible values of R. Each run involves first the adjustment of the stress axes, then the R ratio. The process is repeated several times until the tensor is stabilized, so that further rotations of the stress axes or modifications of the stress ratio do not improve the results.

The starting point for the first run is the results from the Improved Right Dihedron analysis (Fig. 8a). The tensor is rotated successively around each stress axis within a range of $\pm45°$ in steps of 22 to 5° and the full range of R values is checked in steps of 0.25 to 0.12 between 0 and 1. The configuration ($\sigma1$–3, R) which gives the lowest value for the optimization function is used as the starting point for the next run with smaller rotation angles and smaller steps for R. During the following runs, these values are progressively narrowed to $\pm5°$ for the stress axes rotation and ±0.1 for the stress ratio check.

The palaeostress tensor obtained is defined for the population of fault-slip data for which it was computed. However, the raw population of fault-slip data measured in the field is usually not homogeneous and not all fault-slip data can be attributed to a single stress tensor. This results from the fact that fault-slip data do not always fulfil Bott's (1959) basic assumption, which suffers from a series of limitations. The main limitations are: inhomogeneous stress field, pre-existing anisotropies, interaction between different faults or segments of a fault zone, asymmetrical stress tensor and thus rotation of the entire rock body or of internal blocks relative to the stress field (Dupin *et al.* 1993; Pollard *et al.* 1993; Twiss & Unruh 1998). Additionally, the fault pattern can be complicated by the existence of two or more subsequent deformation events. Therefore, palaeostress analysis involves the separation of fault-slip data into populations that can each be characterized by a unique stress tensor.

A maximum slip deviation α of 30° is set as upper limit for defining whether a fault-slip datum is compatible to a stress tensor. When using the composite function (function $F5$ in TENSOR), the value of the function is also used to check if fractures defined as shear, tensional or compressional fractures are compatible with the tensor. This function is defined in such a way that it reaches the value of 0 for best-fit situations and may reach a maximum of 20 for compressional and tensional fractures and 22 for shear fractures.

In the rotational optimization procedure, the separation is performed progressively during the inversion. This can be done after each optimization run rather than after the final determination of the tensor. To ensure an efficient fault separation, the

Rotation around σ1 axis
Rotation angle : +34°
Function F5 value: 20.515

Rotation around σ2 axis
Rotation angle : 0°
Function F5 value : 20.5

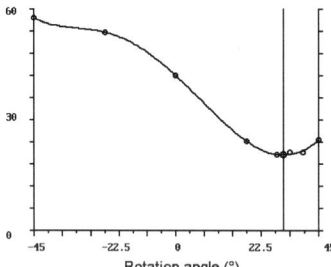

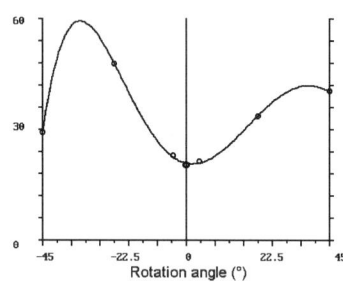

Rotation around σ3 axis
Rotation angle : -16°
Function F5 value : 18.23

Optimisation of R Ratio
R value : 0.75
Function F5 value : 13.49

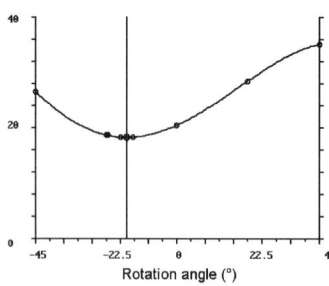

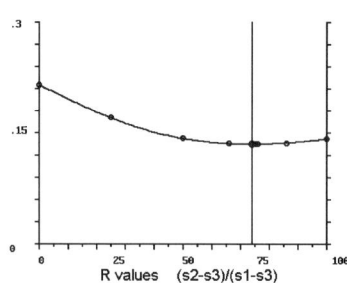

Fig. 7. Principle of the 4D Rotational Optimization. The stress tensor is rotated successively around the σ1, σ2 and σ3 axes by ±12.5 and 45°, a polynomial regression curve is computed by least square minimization using values of the optimization function (F5 here). The minimum of the function is found then additional rotations of ±5° from the angle corresponding to the minimum value of the function are again performed. A new regression curved is computed and the minimum is taken to define the rotation angle to apply to the initial tensor (+34° for the rotation around σ1). The equivalent rotation is performed and the procedure is repeated for the next two axes. A similar procedure is done for optimizing the stress ratio R, by checking a range of possible values between 0 and 1. The range of rotation angle sand values of R is progressively decreased to provide finer constraints during the next runs.

procedure of tensor optimization and fault separation is executed successively with a decreasing value of the maximum slip deviation (e.g. from 4° to 30°; Fig. 8 and Table 1). Because a stress tensor is adjusted on a particular fault population, a new stress tensor has to be recalculated after each modification of the fault population. Using this procedure, the first tensor obtained corresponds to the subset, which is represented in the original population by the greatest number of faults. The rejected faults are then submitted to the same procedure, to extract the next subset.

Figure 8 and Table 1 present an example of progressive stress tensor optimization and fault separation using successively the Right Dihedron and the Rotational Optimization methods on a natural fault-slip data set measured along a border fault of the Central Baikal basin in the area of Zama (Delvaux *et al.* 1997b, 1999). This fault data set contains a minority of reverse or thrust faults related to an older brittle event, and a majority of normal or oblique-slip faults related to Late Cenozoic extension.

We want to point out again that an uncontrolled automated separation presents a great risk because it might lead to useless results (see remarks above). Thus, we would recommend first doing a rough separation using field observations, careful inspec-

(a)

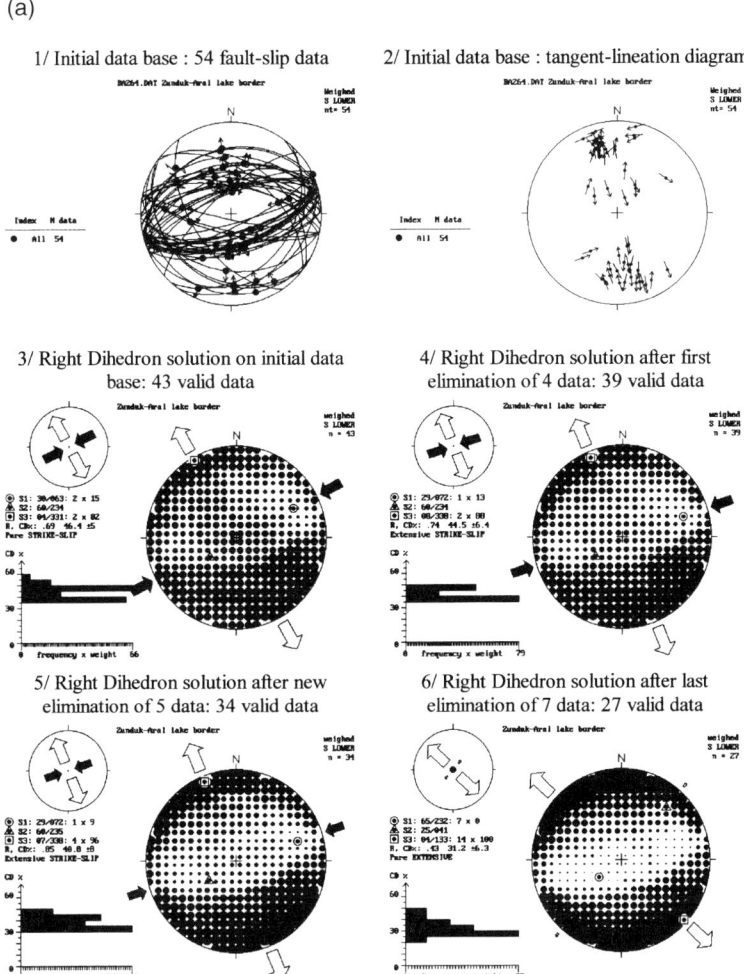

1/ Initial data base : 54 fault-slip data

2/ Initial data base : tangent-lineation diagram

3/ Right Dihedron solution on initial data base: 43 valid data

4/ Right Dihedron solution after first elimination of 4 data: 39 valid data

5/ Right Dihedron solution after new elimination of 5 data: 34 valid data

6/ Right Dihedron solution after last elimination of 7 data: 27 valid data

Fig. 8. Example of progressive kinematic separation and stress tensor optimization on a natural polyphase fault-slip data set measured along a border fault of the Central Baikal basin in the area of Zama (Delvaux *et al.* 1997b, 1999). The initial data base contains 54 fault-slip data, some related to an older compression stage but the majority of them related to Late Cenozoic extension. All stereograms have Schmidt lower hemisphere projections. The different parameters used for estimating the quality results are reported in Table 1. (**a**, 1 and 2) Initial data base; (3–6) progressive fault separation using the Right Dihedron, leading to a first subset and starting tensor for subsequent rotational optimization. (**b**, 7) After successive optimizations and elimination of incompatible data (8–12), the final solution is obtained for the first separated set, represented by the greatest number of fault-data (13–14). For the remaining 25 fault-slip data that were excluded from the initial data base (**c**, 15 and 16), a new series of separation was done using Right Dihedron (17), leading to a second subset which was again progressively optimized and separated (18 and 19) until the final solution is reached for the second set (20). Cross-cutting relations and fault plane morphology observed in the field show that the second set corresponds to an older stress state, compatible with the Early or Mid-Palaeozoic local stress field (Delvaux *et al.* 1995b) and that the first set is compatible with the Late Cenozoic stress field (Delvaux *et al.* 1997b).

tion of data plots, and then only starting the optimization procedure.

At the end of the procedure, when two or more subsets are separated from the original fault population, it is necessary to test the stress tensors of each subset on the total fault population, to check if the fault separation has been done in an optimal way. Effectively, the faults-slip data are considered compatible with a stress tensor as soon as the deviation angle α is less than 30°. In the process of

(b)

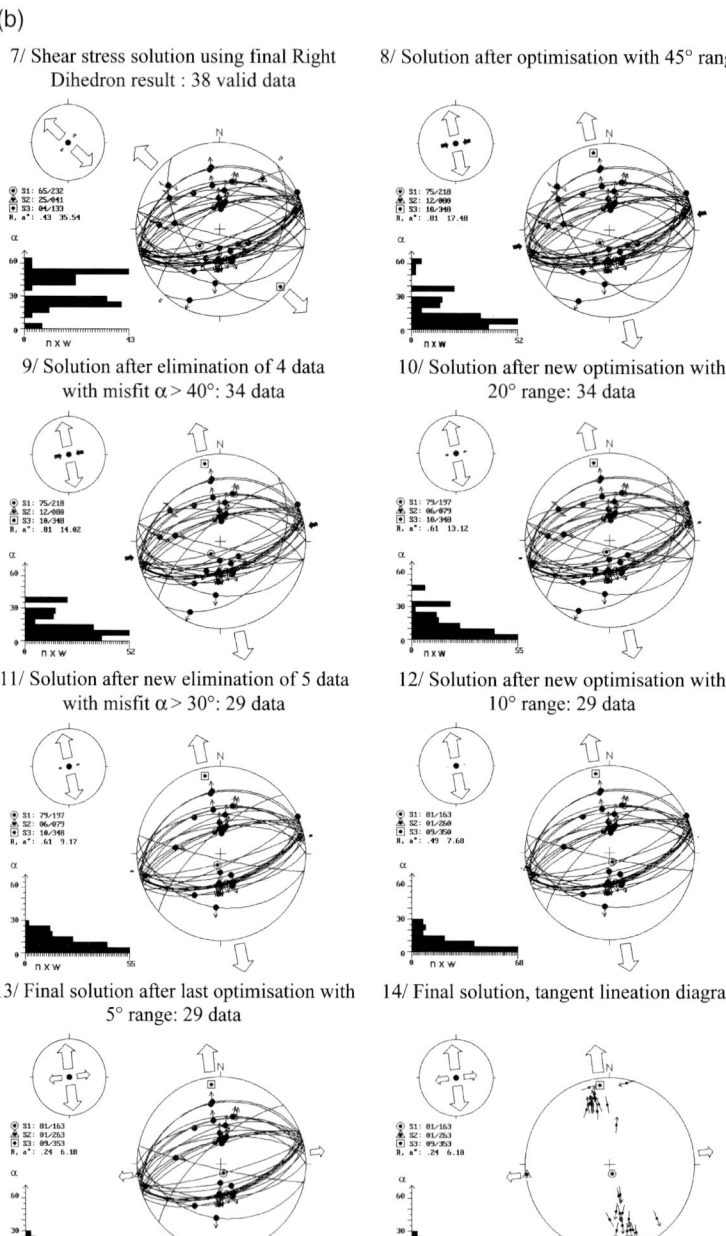

7/ Shear stress solution using final Right Dihedron result : 38 valid data

8/ Solution after optimisation with 45° range

9/ Solution after elimination of 4 data with misfit α > 40°: 34 data

10/ Solution after new optimisation with 20° range: 34 data

11/ Solution after new elimination of 5 data with misfit α > 30°: 29 data

12/ Solution after new optimisation with 10° range: 29 data

13/ Final solution after last optimisation with 5° range: 29 data

14/ Final solution, tangent lineation diagram

Fig. 8. Continued.

stress tensor inversion and fault separation, a given fault-slip datum is attributed to the first population for which it is compatible with the characteristic stress tensor. But it can also be compatible with the stress tensors of the other subsets, sometimes with an even smaller deviation angle α. During the final cross-check (not shown in the example of Fig.

8), each fault datum is attributed to the population for which the composite function $F5$ is minimum. Because each subset was modified, a new optimization run has to be performed on each subset.

As explained above, the deviation angle α alone is not enough to unambiguously discriminate between faults of different data subsets or between

(c)

15/ Remaining 25 fault-slip data

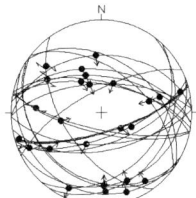

16/ Remaining, tangent lineation diagram

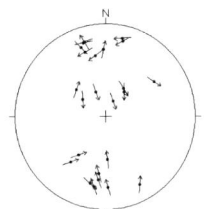

17/ Final Right Dihedron solution after elimination of 5 data

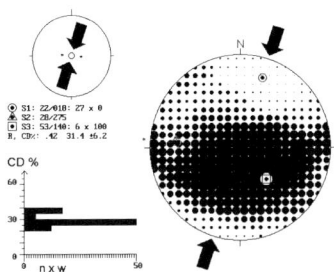

18/ Solution after successive optimisations and data eliminations with respectively 45° and 20° range and α > 40° and 30°

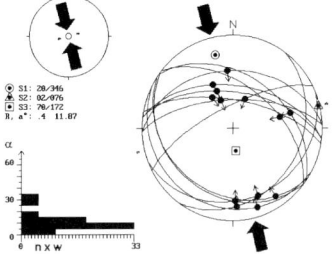

19/ Final solution after additional optimisations with 10° and 5° range

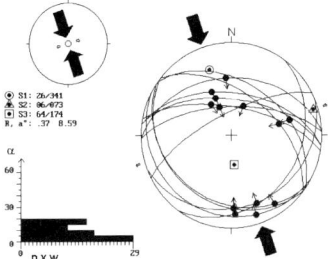

20/ Final solution, tangent lineation diagram

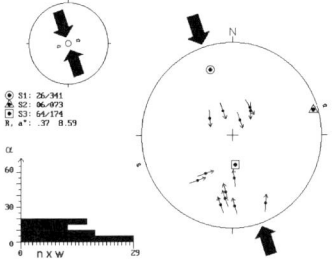

Fig. 8. Continued.

focal planes of the same focal mechanism, when they all fit the basic requirement that $\alpha < 30°$ and $\theta > \theta_0$. The criteria on the slip deviation α and the friction angle θ are used to eliminate fault-slip data or focal planes from the data set, while the composite function is generally used as a minimization function during the inversion and for the finer discrimination between data subsets or focal planes.

Interpretation of results and quality assessment

The quality of the results obtained by stress inversion is dependent on a number of factors such as the number of data per subset, the slip deviation, the type of data, and even on the experience of the user. Thus, for a rating of the quality and a better comparability of the results it is necessary to introduce a quality ranking. A first quality ranking scheme was proposed by Delvaux *et al.* (1995*b*). It was later improved by Delvaux *et al.* (1997*b*). In the first release of the World Stress Map (WSM; Zoback 1992), the quality ranking proposed for geological indicators was defined in a way that all stress tensors obtained from the inversion of Quaternary-age fault-slip data obtained the best quality. In co-operation with the WSM project (Sperner *et al.* 2003) and on the basis of these earlier approaches, the quality ranking scheme included in the TENSOR program was further

Table 1. *Value of the parameters used for estimating the quality of the results of progressive stress tensor optimization and fault-slip separation using successively the Right Dihedron and Rotational Optimization for the example in Figure 8*

First subset, Right Dihedron

Step	Difference between extreme counting values	Average counting deviation (%, $\pm 1\sigma$)
3	67	46.4 $\pm$5
4	75	44.5 $\pm$ 6.4
5	87	40.8 $\pm$ 8.0
6	100	31.2 $\pm$ 6.3

First subset, Rotational Optimization

Step	n	n/nt	CL_w	a	F5	DT_w	Plen	Slen	QRw	QRt	QI
7	38	0.7	0.73	35.5	42.0	1.0	0.82	0.68	E	E	6.1
8	38	0.7	0.73	17.5	19.2	1.0	0.82	0.68	D	D	12.5
9	34	0.63	0.75	14.0	11.6	1.0	0.83	0.72	C	C	12.2
10	34	0.63	0.75	13.1	11.3	1.0	0.83	0.72	C	C	13.0
11	29	0.54	0.76	9.2	6.4	1.0	0.83	0.75	B	C	13.2
12	29	0.54	0.76	7.7	5.7	1.0	0.83	0.75	B	C	15.7
13	28	0.52	0.76	6.2	5.0	1.0	0.83	0.75	B	C	17.5

Second subset, Right Dihedron

Step	Difference between extreme counting values	Average counting deviation (%, $\pm 1\sigma$)
17a	86	39.5 $\pm$ 7.6
17	100	31.4 $\pm$ 6.2

Second subset, Rotational Optimization

Step	n	n/nt	CL_w	a_w	F5	DT_w	Plen	Slen	QRw	QRt	QI
18a	20	0.37	0.9	28.8	30.1	1.0	0.60	0.54	E	E	13.1
18	13	0.24	1.0	11.9	7.4	1.0	0.77	0.59	D	D	9.9
19a	13	0.24	1.0	9.0	5.4	1.0	0.77	0.59	D	D	12.9
19b	13	0.24	1.0	8.2	5.1	1.0	0.77	0.59	D	D	14.2
19	13	0.24	1.0	8.6	5.0	1.0	0.77	0.59	D	D	13.6

This data set contains fault-slip data belonging to two different stress stages. The subset represented by the largest number of data will be extracted first. The CLw parameter progressively increases due to the elimination of data with less constrained slip sense determination. For the first subset, the values of *Slen* increases and limit the final quality of the tensor QRt to C quality, while the WSM quality QRw remains to B. This is justified by the fact that the retained fault population consists mainly in two conjugated sets of fault-slip data. As a result, the stress ratio *R* is not well constrained.

developed to meet the more strict requirements of the new release of the WSM project.

In accordance with the new ranking scheme for the WSM, the quality ranges from A (best) to E (worst), and is determined as a function of threshold values of a series of criteria. The threshold values have been chosen more or less arbitrarily and then tested with field data and checked to find whether they gave a good range of quality according to our 'feeling' of the results. A given quality is assigned if the threshold values corresponding to that particular rank are met for all the criteria.

For the WSM quality ranking scheme the following parameters are used, with threshold values as defined in Table 2:

n, number of fault-slip data used in the inversion
n/nt, ratio of fault-slip data used relative to the total number measured
α_w, deviation between observed and theoretical slip directions
Cl_w, slip sense confidence level for individual faults, applied for the TENSOR program:

Certain	1.00
Probable	0.75
Supposed	0.50
Unknown	0.25

In the case of fractures, Cl_w corresponds to the confidence level in type determination (shear, tensional, compressional) and for focal mechanisms, to the quality of focal mechanism determination

DT_w, fault-slip data type:

Slickenside	1.00
Focal mechanism	0.75
Tension/compression fracture	0.50
Two conjugate planes	0.50
Stylolite column	0.50
Movement plane + tension fracture	0.50
Shear fracture	0.25

As discussed in Sperner et al. (2003), the diversity of orientation of fault planes and of slip lineations is also an important parameter to consider. It is clear that the stress tensors are better constrained if the fault planes and the slip lines have a large variety of orientations rather than if they are all parallel to each other. The distribution of orientation data can be expressed by the normalized length of a vector obtained by addition of unit vectors representing the poles of the fault planes $\mathbf{p}$ or the slip direction $\mathbf{s}$:

$$Plen = \frac{\sqrt{p_x{}^2 + p_y{}^2 + p_z{}^2}}{n}$$

$$Slen = \frac{\sqrt{s_x{}^2 + s_y{}^2 + s_z{}^2}}{n}$$

where p_x, p_y, p_z and s_x, s_y, s_z are respectively the Cartesian co-ordinates of the unit vector $\mathbf{p}$ and $\mathbf{s}$ (expressed as direction cosines). The normalized lengths range between 1 for unimodal populations (all poles/slip lines parallel) to ±0.6 for homogeneously distributed populations.

Unfortunately the computation of such distribution has to be done on the original fault-slip data, which are generally not given in published results. For this reason, it was decided not to use these two criteria in the definition of the WSM quality rank for geological indicators (here named QRw, Table 2).

In the TENSOR program, the diversity criteria using *Plen* and *Slen* were also implemented and combined with the five first parameters, to determine the TENSOR quality rank QRt (Table 2). In this way, *Plen* and *Slen* are additional criteria which lower the quality of the result if the diversity of orientation of data is insufficient regarding to the WSM quality rank QRw.

A quality index (QI) expressed numerically has been previously proposed by Delvaux et al. (1995b, 1997b), based on the following formula:

$$QI = n^2 \times (n/nt)/\alpha_w$$

with threshold values of $QI \geq 1.5$ for A quality, $1.5 > QI \geq 0.5$ for B quality, $0.5 > QI \geq 0.3$ for C quality and $QI < 0.3$ for D quality. However, this system proved to be unsatisfactory for a reliable quality assessment and we prefer to abandon it.

Table 2. *Threshold values as defined in Sperner et al. (2003) for the individual criteria used in the quality ranking scheme for the WSM Quality Rank QRw (n to DTw)*

WSM quality rank QRw	n	n/nt	CL_w	α_w	DT_w	Plen	Slen	Tensor quality rank QRt
A	≥25	≥0.60	≥0.70	≤9	≥0.90	≤0.80	≤0.80	A
B	≥15	≥0.40	≥0.55	≤12	≥0.75	≤0.85	≤0.85	B
C	≥10	≥0.30	≥0.40	≤15	≥0.50	≤0.92	≤0.92	C
D	≥6	≥0.15	≥0.25	≤18	≥0.25	≤0.95	≤0.95	D
E	<6	<0.15	<0.25	>18	<0.25	>0.95	>0.95	E

Proposed threshold values for the *Plen* and *Slen* are added for the Tensor quality rank QRt.

Stress regime index

The stress regime can be expressed numerically by the stress regime index R' defined in Delvaux *et al.* (1997*b*). The main stress regime is a function of the orientation of the principal stress axes and the shape of the stress ellipsoid: extensional when $\sigma1$ is vertical, strike-slip when $\sigma2$ is vertical and compressional when $\sigma3$ is vertical. For each of these three regimes, the value of the stress ratio R is fluctuating between 0 and 1. When the value is close to 0.5 (plane stress), the stress regimes are said 'pure' extensional/strike-slip/compressional. The transition between the three regimes is expressed by opposed values of R. An extensional regimes with $R = 1$ is equivalent to a strike-slip regime with $R = 1$. Similarly, a strike-slip regime with $R = 0$ is equivalent to a compressional regime with $R = 0$.

To facilitate the representation of the range of stress regimes, Delvaux *et al.* (1997*b*) defined a stress regime index R' which expresses numerically the stress regime as follows:

- $R' = R$ when $\sigma1$ is vertical (extensional stress regime)
- $R' = 2 - R$ when $\sigma2$ is vertical (strike-slip stress regime)
- $R' = 2 + R$ when $\sigma3$ is vertical (compressional stress regime).

In the WSM, the naming of the stress regimes is correspondingly: normal faulting/strike-slip faulting / thrust faulting regime. The R' index forms a continuous progression from 0 (radial extension = flattening) to 3 (radial compression = constriction), while R is successively evolving from 0 to 1 in the extensional field, 1 to 0 in the strike-slip field and again 0 to 1 in the compressional field. R' has values of 0.5 for pure extension, 1.0 for extensional strike-slip, 1.5 for pure strike-slip, 2.0 for strike-slip compressional and 2.5 for pure compression. The intermediate stress regimes are sometimes also named 'transtension' for the transition between extension and strike-slip and 'transpression' for the transition between strike-slip and compression.

The stress tensors are displayed in map view by symbols representing the orientation and relative magnitude of the horizontal stress axes, as suggested by Guiraud *et al.* (1989) and further developed by Delvaux *et al.* (1997*b*).

In regional studies it is sometimes necessary to obtain the mean regional stress directions of a series of closely related sites. In Delvaux *et al.* (1997*b*), we introduced the concept of a 'weighed mean tensor', without defining it clearly. This was based on the calculation of mean SHmax, Shmin and σv stress axis by vectorial addition, taking into account the number of fault-slip data used for the

stress inversion of each tensor. Similarly, the mean stress ratio was computed as the average stress ratio index R' defined above. In this procedure, the orientations of $\sigma1$, $\sigma2$ and $\sigma3$ axes are assessed to SHmax, Shmin and Sv as a function of the stress regime (extensional, strike-slip or compressional), expressed as azimuth and plunge. Now, we prefer to use simply the average SHmax azimuth (as defined in the World Stress Map) and the average stress regime index R' as defined above. These two parameters describe fully the orientations of the stress axes (in terms of SHmax and Shmin, assuming Sv vertical), and the stress regime ($R' =$ 0–1 for extensional regime, 1–2 for strike-slip regimes and 2–3 for compressional regimes).

Applications

Inversion of earthquake focal mechanism data

Earthquake focal mechanisms are defined by two orthogonal nodal planes, one of them being the plane that accommodated the slip during seismic activation (fault plane) and the other being the auxiliary plane. In the absence of seismological or geological criteria, both nodal planes are potential slip planes and they cannot be discriminated. The Right Dihedron method is particularly well adapted for the stress inversion of focal mechanisms, as it also uses two orthogonal planes to define compressional and extensional quadrants. The Improved Right Dihedron method is useful for a first estimation of the stress tensor, and also for an initial separation of mechanisms from the database. The preliminary stress tensor and filtered focal mechanism data set are used as a starting point in the Rotational Optimization procedure (Fig. 9 and Table 3).

In the Right Dihedron method, both nodal planes of an incompatible focal mechanism are eliminated simultaneously as they both have the same Counting Deviation values. In the Rotational Optimization procedure, all the nodal planes, which have slip deviations greater than the threshold value of, for example, $40°$ at the beginning of the analysis procedure or $30°$ at the end, are eliminated. The two nodal planes have generally different values for the slip deviation α. If one of the planes has α greater than the threshold value, it is considered as an auxiliary plane and is eliminated. If both α values are higher than the threshold value, the entire focal mechanism is eliminated. If both α values are lower than the threshold value, the two nodal planes are kept for further processing. This results in the progressive selection of the probable fault plane for each focal mechanism.

The final choice of the presumed fault plane for

(a)

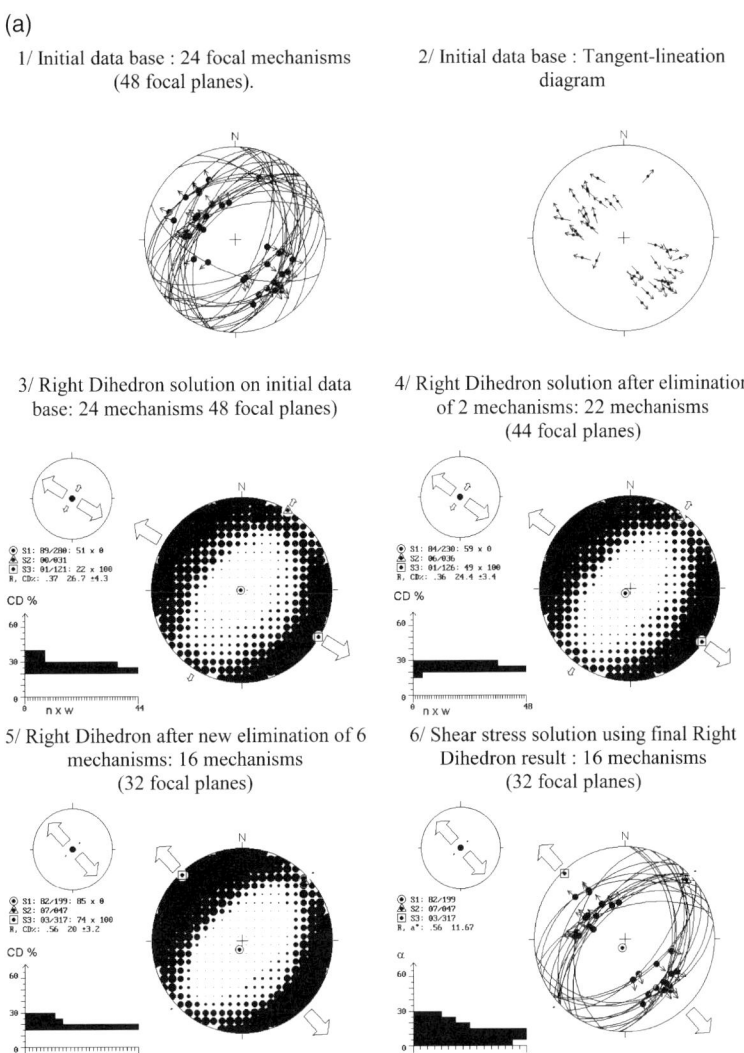

1/ Initial data base : 24 focal mechanisms (48 focal planes).

2/ Initial data base : Tangent-lineation diagram

3/ Right Dihedron solution on initial data base: 24 mechanisms 48 focal planes)

4/ Right Dihedron solution after elimination of 2 mechanisms: 22 mechanisms (44 focal planes)

5/ Right Dihedron after new elimination of 6 mechanisms: 16 mechanisms (32 focal planes)

6/ Shear stress solution using final Right Dihedron result : 16 mechanisms (32 focal planes)

Fig. 9. Example of progressive kinematic separation and stress tensor optimization for earthquake focal mechanism data from the Central Baikal region (taken from Delvaux *et al.* 1999). All stereograms with Schmidt lower hemisphere projections. The different parmeters used for estimating the quality results are reported in Table 3. The initial data base is composed of 24 focal mechanisms, each expressed by two focal planes (**a**, 1 and 2). The Right Dihedron procedure first eliminates incompatible focal mechanisms (3–5) and produces a starting solution for the Rotational Optimization (5 and 6). The Rotational Optimization progressively improves the tensor and selects one focal plane for each mechanism on the basis of the value of the composite function, and eventually eliminates focal planes whose slip deviation α is $>30°$ (**b**, 9–12).

each mechanism is done as a function of the value of the composite function during the rotational optimization. As explained above, this function combines the minimization of the misfit angle α and the maximization of the shear stress $|\tau|$ on every plane. The composite function is therefore also convenient for the determination of the fault and auxiliary planes of focal mechanism if both

have a slip deviation α lower than $30°$. The fault plane will be the one with the smallest value of the composite function.

In this selection, any additional constraint (geological or seismological) on the fault plane has to be considered as dominant over the selection which minimizes the composite function. It should be kept in mind that the orientation of the activated

(b)

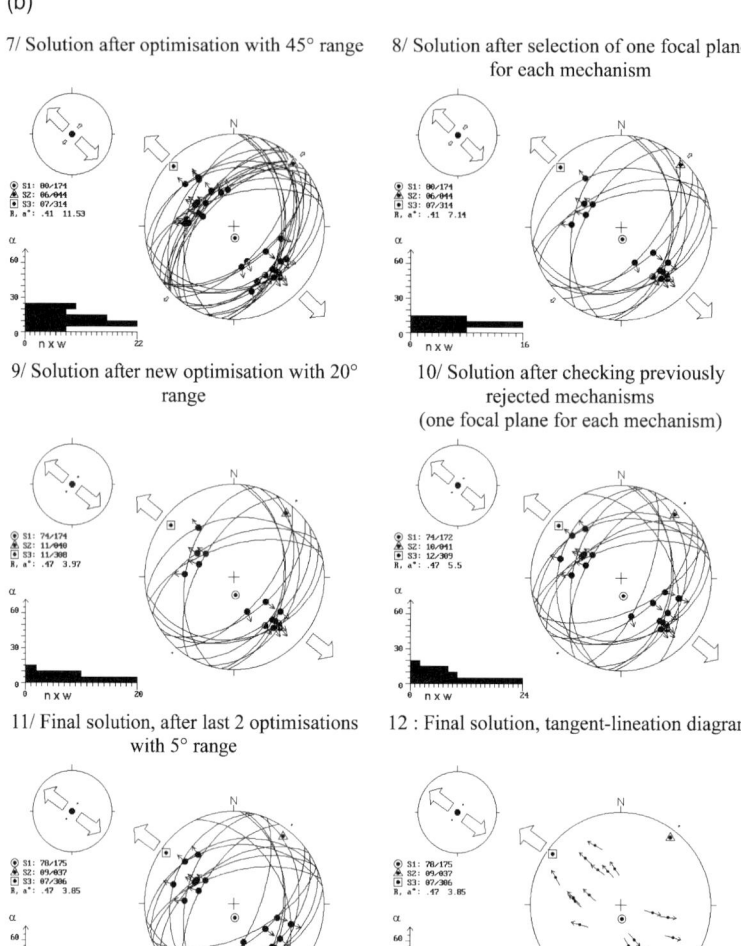

7/ Solution after optimisation with 45° range

8/ Solution after selection of one focal plane for each mechanism

9/ Solution after new optimisation with 20° range

10/ Solution after checking previously rejected mechanisms (one focal plane for each mechanism)

11/ Final solution, after last 2 optimisations with 5° range

12 : Final solution, tangent-lineation diagram

Fig. 9. Continued.

fault plane can be influenced by pre-existing planes and might be more unfavourably oriented than a new plane.

A special option has been implemented in the Rotational Optimization module of TENSOR that rejects automatically the non-compatible mechanisms and discriminates between the fault plane and the auxiliary plane of a focal mechanism, for any given stress tensor.

A similar (but less automated) procedure was used in Petit *et al.* (1996) for stress inversion of focal mechanisms from the Baikal Rift Zone. They compared their results with those obtained using the Carey-Gailardis & Mercier (1987) method on the same data set. It proved that, in most cases, the

two methods give similar results. The differences result mainly from the different selections of fault and auxiliary planes, not from the inversion itself.

Tests with synthetic fault sets

To validate the results obtained with the Right Dihedron and the Rotational Optimization methods, we performed a series of tests using a synthetic set of fault and fracture data representing a strike-slip regime with north–south compression, vertical intermediate axis and $R = 0.4$. This was produced by applying the corresponding tensor on a set of 152 pre-defined weakness planes of various orientations distributed homogeneously on a ster-

Table 3. *Value of the parameters used for estimating the quality of the results of progressive stress tensor optimisation and focal mechanism data separation using successively the Right Dihedron and Rotational Optimisation for the example in Figure 8*
Right Dihedron

Step	Difference between extreme counting values	Average counting deviation (%, $\pm 1\sigma$)
3	100	26.7 ± 4.3
4	100	24.4 ± 3.4
5	100	20.0 ± 3.2

Rotational Optimization

Step	nt	N	n/nt	CL_w	a	F5	DT_w	Plen	Slen	QRw	QRt	QI
6	48	32	0.67	1.0	11.7	6.2	0.75	0.68	0.68	B	B	31.3
7	48	32	0.67	1.0	11.5	5.5	0.75	0.68	0.68	B	B	31.7
8	24	16	0.67	1.0	7.1	2.7	0.75	0.70	0.71	B	B	34.3
9	24	16	0.67	1.0	4.0	1.9	0.75	0.70	0.71	B	B	61.6
10	24	22	0.92	1.0	4.9	3.0	0.75	0.69	0.71	B	B	82
11	24	22	0.92	1.0	3.9	2.3	0.75	0.69	0.71	B	B	104.9

For the Rotational Optimization, the procedure started on all focal planes ($nt = 48$) then one focal plane is selected for each mechanism at step 8 ($nt = 24$). In step 10, mechanisms previously excluded by the Right Dihedron procedure are reincorporated ($n = 16$–22). The highest possible quality for focal mechanisms is quality B because of the uncertainty in the differentiation between actual focal plane and auxiliary plane for each mechanism. The values of *Plen* and *Slen* increase with increasing separation, reflecting a decreasing in diversity of focal planes and associated slip directions. In this example the quality index (QI) regularly increases

eonet (Fig. 10–1). To simulate the different types of brittle structures, we used an initial friction angle of 10°, a cohesion factor of 10 (arbitrary units) and a ratio $\sigma 3/\sigma 1$ of 0.2. On a Mohr diagram (Fig. 10–2), the individual data are plotted in different areas, corresponding to the types of structures generated: 88 faults, 9 tension fractures, 24 compression fractures and 31 shear fractures. When working separately with all the faults, tension fractures and compression fractures, the stress tensors obtained with both methods are always close to that used for generating the data set (Fig. 10–3 to 10–6). The orientations of the stress axes always fit closely to those used for generating the data set. The R value obtained for the set of 88 faults is 0.51 with the Right Dihedron method and 0.4 for the Rotational Optimization method (exactly the original value). For the tension and compression fracture sets, the R value is estimated only with the Right Dihedron method and the same value is used in the Rotational Optimization. The R values obtained (0.75 for the set of 9 tension fractures and 0.35 for the set of 24 compression fractures) reflect the weaker constraint on R using fracture instead of fault data. The relatively small number of data in the tension fracture set might also influence the accuracy of the result.

Regional applications

The program TENSOR is now in use in more than 30 different laboratories worldwide and has been used in a large spectrum of geotectonic settings, for the late Palaeozoic to the Neotectonic period. These include the East African rift (Delvaux *et al.* 1992, 1997*a*; Delvaux 1993*b*,), the Baikal Rift Zone (Delvaux *et al.* 1995*b*, 1997*b*; Petit *et al.* 1996; San'kov *et al.* 1997), the Dead Sea Rift (Zain Eldeen *et al.* 2002), the Oslo graben (Heeremans *et al.* 1996), the Apennines (Cello *et al.* 1997; Ottria & Molli 2000), the Neogene evolution of Spain (Stapel *et al.* 1996; Huibergtse *et al.* 1998) and the Altai belt in Siberia (Delvaux *et al.* 1995*a*, 1995*c*; Novikov *et al.* 1998). Inversion of focal mechanisms has also been performed using this program (Petit *et al.* 1996).

Conclusions

New aspects of the tectonic stress inversion have been discussed. The separation of data into subsets during the inversion is an integral part of the stress analysis process. It has to be performed with much care and has to rely first on an initial 'manual' separation on the basis of careful field observations.

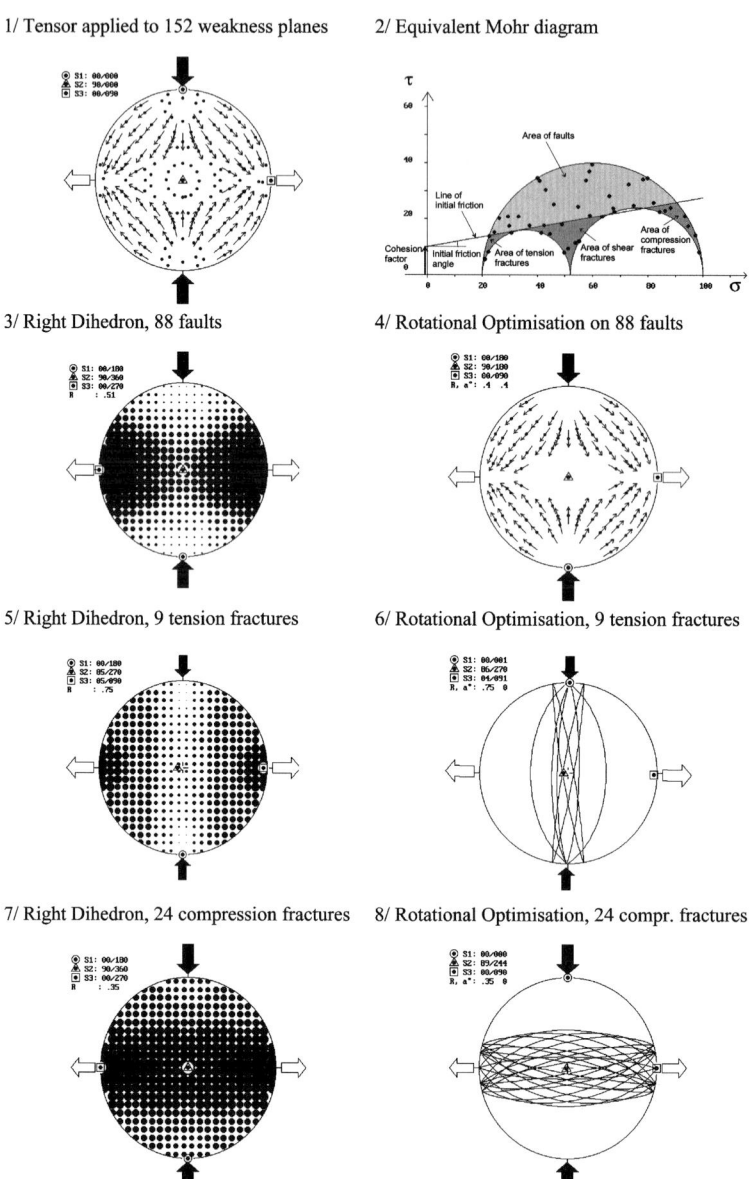

Fig. 10. Example of stress analysis of a synthetic data set produced by applying a strike-slip stress tensor (north-south, $\sigma1$, vertical $\sigma2$, $R = 0.40$) on a reference set of 152 pre-existing weakness planes. (1) Tangent-lineation diagram, lower hemisphere projection of all 152 planes, with arrow showing the directions of slip for activated faults. (2) Mohr diagram illustrating the areas of faults and different types of fracture. (3 and 4) Right Dihedron and Rotational Optimization (tangent-lineation diagram) solutions for the set of 88 faults. (5 and 6) Right Dihedron and Rotational Optimization (great circle projection) solutions for the set of nine tension fractures. (7 and 8) Right Dihedron and Rotational Optimization (great circle projection) solutions for the set of 24 compression fractures.

Not only faults with slip lines (slickensides) can be used in the stress inversion, but also fractures (tension, shear and compressional) and focal mechanisms.

The improved Right Dihedron method allows (1)

estimatation of the stress ratio R, (2) the use of tension and compression fractures in combination with slickenside data, and (3) an initial separation of the data to be performed on the basis of the Counting Deviation. The stress tensor and preliminary data set obtained after the use of the Right Dihedron are used as a starting point for the Rotational Optimization. This method is based on a controlled grid search with rotational optimization of a range of misfit functions. Of the different misfit functions proposed, the composite function allows simultaneous minimization of the slip deviation α for slickensides, maximization of shear stress τ for fault planes and shear fractures, minimization of normal stress ν for tension fractures and maximization of normal stress ν for compression fractures and stylolites.

A quality ranking scheme was developed in collaboration with the World Stress Map to estimate the quality of the results obtained. It is based on a series of parameters, for which threshold values are pre-defined. The type of stress regime is also expressed numerically by a stress regime index R'. The application of the Right Dihedron and Rotational Optimization is illustrated by a synthetic data set and natural data sets of fault-slip data and focal mechanism data from the Baikal Rift Zone.

All the aspects discussed here were implemented in the TENSOR program developed by the first author (available on request). This program is a tool for controlled interactive separation of fault-slip or focal mechanism data and progressive stress tensor optimization using successively the Improved Right Dihedron method and the iterative Rotational Optimization method.

The development of this method for stress tensor inversion and the related Quick-Basic computer program was performed by D. Delvaux during a series of projects related to the East African and Baikal rift systems, funded by the Belgian federal state (former FRFC-IM , now SSTC) and hosted at the Royal Museum for Central Africa. The Quality ranking procedure was set up jointly with the World Stress Map Project of Karlsruhe University. This paper was written during an ISES Visiting Research Fellowship of D. Delvaux at the Vrije Universiteit Amsterdam. I thank S. Cloetingh for his invitation and support and D. Nieuwland for his editorial efforts.

References

ANGELIER, J. 1989. From orientation to magnitudes in paleostress determinations using fault slip data. *Journal of Structural Geology*, **11**, 37–50.

ANGELIER, J. 1991. Inversion directe et recherche 4-D: comparaison physique et mathématique de deux méthodes de détermination des tenseurs des paléocontraintes en tectonique de failles. *Comptes Rendus de l'Académie des Sciences de Paris*, **312(II)**, 1213–1218.

ANGELIER, J. 1994. Fault slip analysis and paleostress reconstruction. *In*: HANCOCK, P. L. (ed.) *Continental Deformation*. Pergamon, Oxford, 101–120.

ANGELIER, J. & MECHLER, P. 1977. Sur une méthode graphique de recherche des contraintes principales également utilisable en tectonique et en seismologie : la méthode des dièdres droits. *Bulletin de la Société Géologique de France*, **7(19)**, 1309–1318.

BOTT, M. H. P. 1959. The mechanism of oblique-slip faulting. *Geological Magazine*, **96**, 109–117.

BYERLEE, J. D. 1978. Friction of rocks. *Pure & Applied Geophysics*, **116**, 615–626.

CAREY-GAILHARDIS, E. & MERCIER, J. -L. 1987. A numerical method for determining the state of stress using focal mechanisms of earthquake populations: application to Tibetan teleseisms and microseismicity of southern Peru. *Earth and Planetary Sciences Letters*, **82**, 165–179.

CELLO, G., MAZZOLI, S., TONDI, E. & TURCO, E. 1997. Active tectonics in the central Apennines and possible implications for seismic hazard analysis in Peninsular Italy. *Tectonophysics*, **272**, 43–68.

DELVAUX, D. 1993a. The TENSOR program for paleostress reconstruction: examples from the east African and the Baikal rift zones. *In: Terra Abstracts*. Abstract supplement No. 1 to *Terra Nova*, **5**, 216.

DELVAUX, D. 1993b. Quaternary stress evolution in East Africa from data of the western branch of the East African rift. *In*: THORWEIHE, U. & SCHANDELMEIER, H (eds) *Geoscientific Research in Northern Africa*. Balkema, Rotterdam, 315–318.

DELVAUX, D., LEVI, K., KAJARA, R. & SAROTA, J. 1992. Cenozoic paleostress and kinematic evolution of the Rukwa - North Malawi rift valley (East African rift system). *Bulletin des Centres de Recherches Exploration-Production elf aquitaine*, **16(2)**, 383–406.

DELVAUX, D., FERNANDEZ-ALONSO, M., KLERKX, J. *et al.* 1995a. Evidences for active tectonics in Lake Teletskoye (Gorny-Altai, South-Siberia). *Russian Geology and Geophysics*, **36(10)**, 100–112.

DELVAUX, D., MOEYS, R., STAPEL, G., MELNIKOV, A. & ERMIKOV, V. 1995b. Paleostress reconstructions and geodynamics of the Baikal region, Central Asia. Part I: Paleozoic and Mesozoic pre-rift evolution. *Tectonophysics*, **252**, 61–101.

DELVAUX, D., THEUNISSEN, K., VAN DER MEER, R. & BERZIN, N. 1995c. Formation dynamics of the Gorno-Altaian Chuya-Kurai depression in Southern Siberia: paleostress, tectonic and climatic control. *Russian Geology and Geophysics*, **36(10)**, 26–45.

DELVAUX, D., KERVYN, R., VITTORI, E., KAJARA, R. S. A. & KILEMBE, E. 1997a. Late Quaternary tectonic activity and lake level fluctuation in the Rukwa rift basin, East Africa. *Journal of African Earth Sciences*, **26(3)**, 397–421.

DELVAUX, D., MOEYS, R., STAPEL, G. *et al.* 1997b. Paleostress reconstructions and geodynamics of the Baikal region, Central Asia. Part II: Cenozoic tectonic stress and fault kinematics. *Tectonophysics*, **282(1–4)**, 1–38.

DELVAUX, D., FRONHOFFS, A., HUS, R. & POORT, J. 1999. Normal fault splays, relay ramps and transfer zones in the central part of the Baikal Rift basin: insight from digital topography and bathymetry. *Bulletin des Centres de Recherches Exploration–Production Elf Aquitaine*, **22(2)**, 341–358 (2000 issue).

DUNNE, W. M. & HANCOCK, P. L. 1994. Paleostress analysis of small-scale brittle structures. *In*: HANCOCK, P. L. (ed.) *Continental Deformation*. Pergamon, Oxford, 101–120.

DUPIN, J. -M., SASSI, W. & ANGELIER, J. 1993. Homogeneous stress hypothesis and actual fault slip: a distinct element analysis. *Journal of Structural Geology,* **15**, 1033–1043.

ETCHECOPAR, A., VASSEUR, G. & DAIGNIERES, M. 1981. An inverse problem in microtectonics for the determination of stress tensors from fault striation analysis. *Journal of Structural Geology,* **3**, 51–56.

GEPHART, J. W. 1990a. Stress and the direction of slip on fault planes. *Tectonics,* **9**(4), 845–858.

GEPHART, J. W. 1990b. FMSI: A Fortran program for inverting fault/slickenside and earthquake focal mechanism data to obtain the regional stress tensor. *Computer & Geosciences,* **16**(7), 953–989.

GUIRAUD, M., LABORDE, O. & PHILIP, H. 1989. Characterization of various types of deformation and their corresponding deviatoric stress tensor using microfault analysis. *Tectonophysics,* **170**, 289–316.

HANCOCK, P. L. 1985. Brittle microtectonics: pinciples and practice. *Journal of Structural Geology,* **7**, 437–457.

HARDCASTLE, C. H. & HILLS, L. S. 1991. Brute3 and Select: QuickBasic 4 programs for determination of stress tenros configurations and separation of heterogeneous populations of fault-slip data. *Computer & Geosciences,* **17**(1), 23–43.

HEEREMANS, M., LARSEN, B. T. & STEL, H. 1996. Paleostress reconstructions from kinematic indicators in the Oslo Graben, southern Norway: new constraints on the mode of rifting. *In*: CLOETINGH, S., BEN-AVRAHAM, Z., SASSI, W., HORVATH, F. (eds) *Dynamics of Basin Formation and Inversion, Tectonophysics,* **266**, 55–79 (Corrections: *Tectonophysics,* 1997, **277**, 339–344).

HUIBERGTSE, P, ALEBEEK, H., ZALL, M. & BIERMANN, C. 1998. Paleostress analysis of the northern Nijar and southern Vera basin: contraints for the Neogene displacement history of major strike-slip faults in the Betic Cordilleras, SE Spain. *Tectonophysics,* **300**, 79–101.

JAEGER, J. C. 1969. *Elasticity, Fracture and Flow with Engineering and Geological Applications.* Chapman & Hall, London.

MAERTEN, L. 2000. Variation in slip on intersecting normal faults: Implications for paleostress inversion. *Journal of Geophysical Research,* **105**(B11), 25553–25565.

MEANS, W. D. 1989. A construction for shear stress on a generally oriented plane. *Journal of Structural Geology,* **11**, 625–627.

MORRIS, A. P., FERRIL, D. A. & HENDERSON, D. B. 1996. Slip tendency analysis and fault reactivation. *Geology,* **24**, 275–278.

MÜLLER, B. & SPERNER, B. 2000. Contemporary Tectonic Stress: Advances in Research and Industry. *Eos, Transactions, American Geophysical Union,* **81**(13), 137–140.

MÜLLER, B., REINECKER, J. & FUCHS, K. 2000. The 2000 release of the World Stress Map. World Wide Web address: http://www.world-stress-map.org

NIETO-SAMANIEGO, A. F. & ALANIZ-ALVAREZ, S. A. 1996. Origin and tectonic interpretation of multiple fault patterns. *Tectonophysics,* **270**, 197–206.

NOVIKOV, I. S., DELVAUX D. & AGATOVA A. R. 1998. Neotectonics of the Kurai Ridge (Gorny-Altai). *Russian Geology and Geophysics,* **39**(7), 970–977.

OTTRIA, G. & MOLLI, G. 2000. Superimposed brittle structures in the late-orogenic extension of the Northern Apennine: results from the Garrara area (Alpi Apuane, NW Tuscany). *Terra Nova,* **12**(2), 52–59.

PETIT, C., DÉVERCHÈRE, J., HOUDRY-LÉMONT, F., SANKOV, V. A., MELNIKOVA, V. I. & DELVAUX, D. 1996. Present-day stress field changes along the Baikal rift and tectonic implications. *Tectonics,* **15**(6), 1171–1191.

PETIT, J. P. 1987. Criteria for the sense of movement on fault surfaces in brittle rocks. *Journal of Structural Geology,* **9**, 597–608.

POLLARD, D. D. & AYDIN, A. 1988. Progress in understanding jointing over the past century. *Bulletin of the Geological Society of America,* **100**, 1181–1204.

POLLARD, D. D., SALTZER, S. D. & RUBIN, A. 1993. Stress inversion methods: are they based on faulty assumptions? *Journal of Structural Geology,* **15**(8), 1045–1054.

RAGAN, D. M. 1973. *Structural Geology. An Introduction to Geometrical Techniques* (second edition). John Wiley & Sons, Chichester.

ROBERTS, G. P. & GANAS, A. 2000. Fault-slip directions in central and southern Greece measured from striated and corrugated fault planes: Comparison with focal mechanism and geodetic data. *Journal of Geophysical Research,* **105**(B10), 23443–23462.

SAN'KOV, V. A., MIROSHNICHENKO, A. I., LEVI, K. G., LUKHNEV, A., MELNIKOV, A. I. & DELVAUX, D. 1997. Cenozoic tectonic stress field evolution in the Baikal Rift Zone. *Bulletin des Centres de Recherches Exploration-Production Elf Aquitaine,* **21**(2), 435–455.

SPERNER, B., RATSCHBACHER, L. & OTT, R. 1993. Fault-Striae analysis: A Turbo-Pascal program package for graphical presentation and reduced stress tensor calculation. *Computer & Geosciences,* **19**(9), 1361–1388.

SPERNER, B., MÜLLER, B., HEIDBACH, O., DELVAUX, D., REINECKER, J. & FUCHS, K. 2003. Tectonic stress in the Earth's crust: advances in the World Stress Map project. *In*: NIEUWLAND, D. A. (ed.) *New Insights into Structural Interpretation and Modelling.* Geological Society, London, Special Publications, **212**, 101–116.

STAPEL, G., MOEYS, R. & BIERMANN, C. 1996. Neogene evolution of the Sobras basin (SE Spain) determined by paleostress analysis. *Tectonophysics,* **255**, 291–305.

TWISS, R. J. & UNRUH, J. R. 1998. Analysis of fault slip inversions: Do they constrain stress or strain rate? *Journal of Geophysical Research,* **103**, 12205–12222.

UNRUH, J. R., TWISS, R. J. & HAUKSSON, E. 1996. Seismogenic deformation field in the Mojave block and implications for tectonics of the eastern California shear zone. *Journal of Geophysical Research,* **101**, 8335–8361.

ZAIN ELDEEN, U., DELVAUX, D. & JACOBS, P. 2002. Tectonic evolution in the Wadi Araba segment of the Dead Sea Rift, South-West Jordan. European Geophysical Society, Special Publication Series, **2**, 1–29.

ZOBACK, M. L. 1992. First- and second-order patterns of stress in the lithosphere: The World Stress Map Project. *Journal of Geophysical Research,* **97**, 11703–11728.

Tectonic stress in the Earth's crust: advances in the World Stress Map project

B. SPERNER[1], B. MÜLLER[1], O. HEIDBACH[1], D. DELVAUX[2,3], J. REINECKER[1]
& K. FUCHS[1]

[1]*Geophysical Institute, Karlsruhe University, Hertzstrasse 16, 76187 Karlsruhe, Germany*
(e-mail: blanka.sperner@gpi.uni-karlsruhe.de)
[2]*Royal Museum for Central Africa, Leuvensesteeweg 13, 3080 Tervuren, Belgium*
[3]*Present address: Vrije University, Amsterdam, The Netherlands*

Abstract: Tectonic stress is one of the fundamental data sets in Earth sciences comparable with topography, gravity, heat flow and others. The importance of stress observations for both academic research (e.g. geodynamics, plate tectonics) and applied sciences (e.g. hydrocarbon production, civil engineering) proves the necessity of a project like the World Stress Map for compiling and making available stress data on a global scale. The World Stress Map project offers not only free access to this global database via the Internet, but also continues in its effort to expand and improve the database, to develop new quality criteria, and to initiate topical research projects. In this paper we present (a) the new release of the World Stress Map, (b) expanded quality ranking schemes for borehole breakouts and geological indicators, (c) new stress indicators (drilling-induced fractures, borehole slotter data) and their quality ranking schemes, and (d) examples for the application of tectonic stress data.

Tectonic stress is felt most extremely during its release in catastrophic earthquakes, although less spectacular tectonic stress is a key safety factor for underground constructions such as tunnels, caverns for gas storage or deposits of nuclear waste (Fuchs & Müller 2001). Furthermore, the economic aspect is of increasing importance in hydrocarbon recovery. Knowledge of the stress field is used in the hydrocarbon industry to foresee stability problems of boreholes and to optimize reservoir management through tectonic modelling in combination with correlation with other data sets such as 3D-structural information, e.g. location and orientation of faults.

Geodetic measurements (e.g. Global Positioning System, GPS; Very Long Baseline Interferometry, VLBI; Satellite Laser Ranging, SLR) have become available in increasing numbers and enhanced quality. They provide surface displacement vectors from which the strain rate field can be deduced. The combined interpretation of stress and strain rate data provides unique challenges for Earth scientists in a number of fields. In tectonics, plate boundary forces confine the kinematics of plate motion and the dynamics of plate deformation resulting, for example, in major differences between intraplate and plate boundary deformation zones. The width of the latter even varies for the different types of plate boundaries (Gordon & Stein 1992). These differences express the laterally and vertically heterogeneous deformation pattern of the crust due to its composition, mechanical properties and tectonic setting. The comparison of the strain rate field at the surface with earthquake data from the seismogenic part of the crust will give information on the depth variation of strain. The relationship of strain and stress is of special importance at active fault zones, where accumulated stress is abruptly released in earthquakes.

The scale of stress impact ranges from a continent-wide scale in order to explain tectonic processes, to a metre scale at reservoir and construction sites. The global database World Stress Map (WSM) provides information on the contemporary tectonic stress in the Earth's crust in a compact and comprehensive way. The WSM project was initiated as a task force of the International Lithosphere Program under the leadership of M. L. Zoback. The database is now maintained and expanded at the Geophysical Institute of Karlsruhe University as a research project of the Heidelberg Academy of Sciences and Humanities. The WSM team regards itself as 'brokers' for these fundamental data: data of different types from all over the world are integrated into a compact database following standardized procedures for quality assign-

From: NIEUWLAND, D.A. (ed.) *New Insights into Structural Interpretation and Modelling.* Geological Society, London, Special Publications, **212**, 101–116. 0305-8719/03/$15

ment and for data format conversion. The resulting database is available to all via the Internet (http://www.world-stress-map.org) for further investigations.

We are indebted to numerous individual researchers and working groups all over the world for providing data for the database and we are looking forward to receiving new data for upgrading the database in the near future. For its successful continuation the WSM project is dependent on further data release from industry and academia.

Since the presentation of the WSM project in a special volume of the *Journal of Geophysical Research* in 1992 (Vol. 97, no. B8), relevant changes took place, not only concerning the amount of data available in the WSM database, but also regarding new methods for stress investigations (e.g. drilling-induced fractures, borehole slotter) as well as improved quality criteria for geological indicators and for borehole breakouts. Here, we give an overview about the state-of-the-art of the database complemented by an outlook on the capability of stress data for practical applications in industry and academia. Thereby, we concentrate on the new features of the WSM; for basic information about the project we refer readers to the *Journal of Geophysical Research* special volume (as mentioned above).

Database and Internet access

Information about the tectonic stress field in the Earth's crust can be obtained from different types of stress indicators, namely earthquake focal mechanisms, well bore breakouts, hydraulic fracturing and overcoring measurements, and young (Quaternary) geological indicators like fault-slip data and volcanic alignments. The reliability and comparability of the data are indicated by a quality ranking from A to E, with A being the highest quality and E the lowest (for further investigations only the most reliable data with a quality of A, B or C should be used). An overview about the quality ranking scheme is given in Zoback (1992) as well as on the WSM website (http://www.world-stress-map.org).

The WSM website gives a detailed description of the database comprising not only data on the orientation of the maximum horizontal compression (S_H), which are usually plotted on the stress maps, but also information such as the depth of the measurement, the magnitude, the lithology and so on. The complete database can be downloaded from the website either as dBase IV or as an ASCII file. In addition, the website provides instructions for stress regime characterization, abstracts from the two WSM Euroconferences, interpretations of the stress field in selected

regions, guidelines for data analysis (which until now are available for borehole breakouts), and much more.

Numerous stress maps for different regions of the world are available on the Internet as postscript files. As an additional service the program CASMO (**C**reate **A** **S**tress **M**ap **O**nline) offers the possibility to create user-defined stress maps, for example by adding topography, plotting only data of a specific type (e.g. only earthquakes), or plotting data from a certain depth range. CASMO is available on request and takes less than two hours for a return of the stress map via e-mail. All stress maps are plotted with GMT provided by Wessel & Smith (1991, 1998; http://www.soest.hawaii.edu/gmt/); plate boundaries come from the PLATES project (http://www.ig.utexas.edu/research/projects/plates/plates.html).

The new release of the WSM

The new release of the WSM (Müller *et al.* 2000; Fig. 1) encompasses 10 920 data records, each with up to 56 detail entries. As in the earlier releases (e.g. Zoback 1992; Müller *et al.* 1997) most data come from earthquakes (63%; with Harvard University being the major 'data contributor'; CMT solutions; http://www.seismology.harvard.edu/projects/CMT/) and borehole breakouts (22%; Table 1, Fig. 2). About two-third of the data (68%) have a quality of A, B or C and thus can be used for further investigations (D and E quality data are too unreliable, but are kept for book-keeping purposes in order to inform future researchers that these data have already been analysed). New data mainly came from Europe, Australia and America.

The WSM database is designed as a tool to work with stress data. Due to the structure of the database the data can be selected according to a number of criteria such as type, location, regime, depth, and so on. Additionally, data from mid-ocean ridges which may be directly related to plate boundary processes and which had so far been excluded from the WSM database are now included. These data with less than 2° distance to the next plate boundary are marked by 'PBE' (**P**ossible Plate **B**oundary **E**vent) in the last field of the database (field PBE), so that they can easily be filtered if necessary.

New database structure

A three-letter country code is used for the numbering of the data; it is based on ISO 3166 provided by the United Nations (http://www.un.org/Depts/unsd/methods/m49alpha.htm). We opted for a new

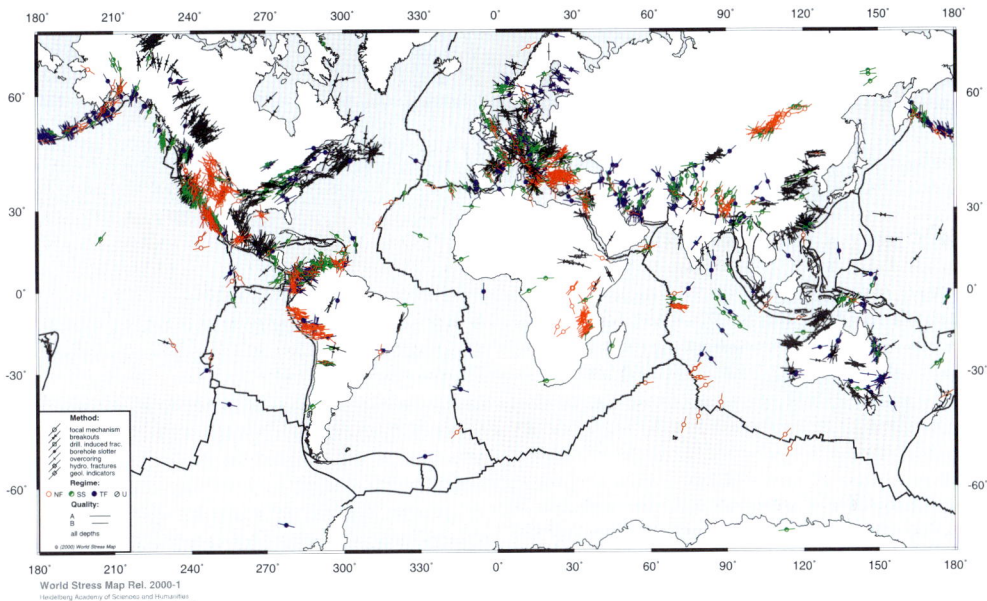

Fig. 1. The new release of the World Stress Map (WSM 2000). Lines show the orientation of the maximum horizontal compression (S_H); different symbols represent different data types, different symbol sizes characterize different data qualities; different colours and fillings indicate different stress regimes (red/unfilled, normal faulting (NF); green/half-filled, strike-slip faulting; (SS); blue/filled, thrust faulting (TF); black, unknown (U)).

Table 1. *Distribution of data types in the WSM database*

Data type (abbreviations)	Number (A–E)	Percentage	Number (A–C)	Percentage
Focal mechanism (FMS, FMA, FMC)	6893	63.12	4894	66.30
Borehole breakouts (BO, BOC, BOT)	2414	22.11	1683	22.80
Geological: fault-slip (GFI, GFM, GFS)	373	3.42	321	4.35
HydroFrac (HF, HFG, HFM, HFP)	314	2.88	205	2.77
Overcoring (OC)	615	5.63	104	1.41
Geological: volcanic alignments (GVA)	223	2.04	101	1.37
Drilling-induced fractures (DIF)	44	0.40	36	0.49
Borehole slotter data (BS)	33	0.30	29	0.39
Petal centreline fracture (PC)	9	0.08	9	0.12
Shear wave splitting (SW)	2	0.02	0	0
Sum:	10920	100.00	7382	100.00

For details about the data types see http://www.world-stress-map.org

country code to eliminate confusion occurring with the old country code. The latter subdivided some countries into several (tectonic) units. However, this subdivision was not done consistently (some countries like Mexico were subdivided into numerous subunits, while other large countries were not subdivided) we decided for the option one country = one code according to international standards. Information about the tectonic unit can be given/found in the fields 'LOCALITY' or 'COMMENT'. For the same reasons, the oceans are no longer subdivided into different regions, but all get the code 'SE'. The three-letter country code is noted in the new field 'ISO' at the very beginning of each data record; the field 'SITE' with the old numbering remains part of the database to guarantee the comparability with older releases. To avoid the millennium bug (Y2K problem) all fields containing a date, i.e. the fields 'DATE', 'REF1' to 'REF4', and 'LAST—MOD', were extended by two characters and the year is now shown as four-digit number.

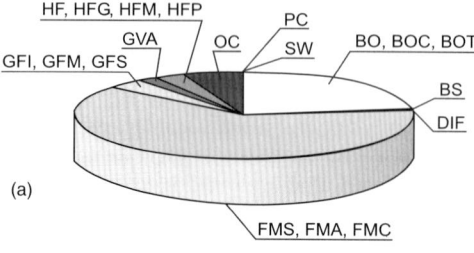

(a)

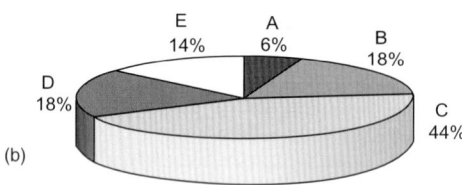

(b)

Fig. 2. Distribution of the WSM 2000 data with respect to (**a**) data type (see Table 1 for abbreviations) and (**b**) quality (A, best, E, worst quality). Most data come from earthquakes (63%; FMS, FMA, FMC) and borehole breakouts (22%; BO, BOC, BOT). The most frequent quality is C due to the fact that most earthquake data are Harvard CMT solutions which are assigned quality C (for discussion see Zoback 1992).

New quality ranking schemes

During the operation of the WSM database and from discussion with other researchers, the original quality ranking schemes have been modified and improved. This is sometimes related to clarity of the quality assignment instructions and sometimes to the use of different tools for recording of the data. Up to now, the modification is complete for three types of stress indicators, geological indicators (especially fault-slip data), borehole breakouts, and overcoring data. Others will follow.

Expanded quality ranking scheme for geological indicators (GFI, GFM, GFS)

Many methods for analysing fault-slip data (e.g. Allmendinger *et al.* 1989; Angelier 1984) and numerous computer programs have been developed

(e.g. Sperner *et al.* 1993; Sperner 1996; Delvaux *et al.* 1997). Nevertheless, some disagreement prevails concerning the meaning of the determined orientations, whether they represent strain rate or stress orientations; for a review see Twiss & Unruh (1998). We do not intend to interfere with that discussion or to describe all these methods and the related software. Instead we want to provide a useful tool for quality ranking.

Quality ranking for fault-slip data has so far been defined through the following criteria:

A quality: inversion of fault-slip data for best fitting mean deviatoric stress tensor using Quaternary age faults (GFI data).
B quality: slip direction on fault plane, based on mean fault attitude and multiple observations of the slip vector; maximum stress at 30° to fault (GFM data).
C quality: attitude of fault and primary sense of slip known, no actual slip vector (GFS data).
D quality: (a) offset core holes, (b) quarry pop-ups, or (c) postglacial surface fault offsets.

Due to this quality ranking GFI data (inversion of fault-slip data) are *always* assigned quality A. However, the quality of these data is dependent on several different criteria such as the number of data and the average misfit between calculated and theoretical slip direction. We use these criteria to introduce a new quality ranking for GFI data. For each of these criteria 'subqualities' ranging from 'a' to 'e' are defined (Table 2). The overall quality of the analysis is given by its *lowest* subquality.

Quality criteria for GFI data

(a) *Number of data.* The number of data used for the calculation of the stress axes.
(b) *Percentage of data.* The percentage of data relative to the total number of data measured in an outcrop. With more than one event recorded in an outcrop, the quality will (in most cases) decrease reflecting the difficulty to correctly separate the data into subsets.
(c) *Confidence number.* The reliability of the results of fault-slip analysis is very sensitive to the

Table 2. *Quality ranking scheme for fault-slip data (GFI)*

Subquality	Number of data	Percentage of data	Confidence number (slip sense determ.)	Fluctuation (°) (average misfit)	Data type
a	≥25	≥60	≥0.70	≤9	≥0.90
b	≥15	≥45	≥0.55	≤12	≥0.75
c	≥10	≥30	≥0.40	≤15	≥0.50
d	≥6	≥15	≥0.25	≤18	≥0.25
e	<6	<15	<0.25	>18	<0.25

The overall quality is defined by the *lowest* subquality of any of these criteria.

reliability of slip sense determination. Hardcastle (1989) introduced four confidence levels (1–4) for slip sense determination which are assigned to the single faults during field measurements. We gave these confidence levels different weights in the following way:

1, absolutely certain:	1.00
2, certain:	0.75
3, uncertain:	0.50
4, very uncertain or unknown:	0.25

The 'confidence number' is defined as the average of these values for all data used in the final analysis. If no confidence levels were assigned during field measurements the overall quality assigned due to the 'rest' of the criteria is taken and downgraded by one class (e.g. from quality B to C).

(d) *Fluctuation.* The fluctuation is defined as the arithmetic mean between the measured and the theoretical slip directions (the theoretical slip direction is assumed to be parallel to the direction of maximum shear stress along the fault plane).

(e) *Type of data.* Information from other structures like tension joints can give additional constraints on the calculated stress tensor. The computer program TENSOR written by Delvaux (Delvaux *et al.* 1997; Delvaux & Sperner 2003) includes these data into the stress analysis by using different weights for the different types of structures:

Slickenside	1.00
Tension/compression joint	0.50
Two conjugated planes	0.50
Movement plane + tension joint	0.50
Shear joint	0.25

The average value for all structures used in the stress analysis gives the 'data-type number' for Table 2. For pure fault-slip data sets this number (1.00) has no relevance.

(f) *Diversity of orientation of fault planes and of lineations.* Fault-slip analysis is based on an equation system with four unknowns (Angelier & Goguel 1979) due to the search for the orientation of the three stress axes and for the stress ratio. Thus at least four differently oriented fault planes are necessary to reveal an unambiguous result. Unbalanced data sets with fault planes and lineations with similar orientations will not give clear results for both the orientation of the stress axes and the stress ratio. Data sets including only two fault plane orientations (conjugated fault sets) give better, but still not optimum restrictions concerning the stress axes, while the stress ratio is still undefined. Thus, the diversity of the measured data ensures the non-ambiguity of the calculated result. A measure of the diversity is given by the normalized length of the mean vector of the fault plane poles and of the mean slip line vectors. This normalized length is close to 1 for parallel planes or parallel lineations and could be used as an additional quality criterion. But in contrast to the other quality criteria (which are all standard measures in fault-slip analysis) the diversity cannot be easily obtained from published data. It can only be estimated from the data plots, but for an accurate calculation the original data must be available in digital form. Thus, we gave up our initial intention to include the diversity into the quality ranking, but we plead for a careful handling of fault-slip data concerning this problem. Data sets which show a clearly unbalanced fault plane orientation might be downgraded by the WSM authorities.

In this context the quality ranking of two other geological indicators, i.e. GFM and GFS data (see definition above), has to be critically reviewed. Their quality assignment is justified for recently active, large-scaled structures as described by Zoback (1992), i.e. for earthquake fault scarps. But the fact that these data build one data class together with GFI data (fault-slip inversion; see above), which are derived from microstructures, brings the danger that small-scaled structures are used in the same way. In this case, the GFM and GFS data are downgraded to quality C and D, respectively (Table 3); the motivation for this is given in the following paragraphs.

GFM data are derived from single fault planes under the assumption of an angle of 30° between maximum compression axes and fault plane (with the fault plane pole, the lineation and the compression axis lying in one plane). For newly formed faults, e.g. in Quaternary rocks, this assumption might be reasonable, but for reactivated faults other angles are as probable as the usually taken 30° (e.g. San Andreas Fault with *c.* 80°; Zoback *et al.* 1987). This uncertainty makes it necessary to downgrade data of this type to quality C, instead of B as in the old quality ranking scheme. This is additionally justified when taking into account that data from fault-slip inversion (GFI data) with quality B are based on at least 15 fault-slip data with an average misfit angle less than 12°. GFM data do not reach this level of accuracy and thus represent a lower quality (i.e. quality C).

The next lower level of quality is represented by GFS data which are derived from fault planes with

Table 3. *Additional constraints for the quality ranking of GFM and GFS data*

Data type	GFM	GFS
Large scale (earthquake related)	B	C
Small scale (microstructures)	C	D

unknown slip vector, but with 'primary sense of slip known' (due to the old quality ranking scheme). Usually this 'primary sense of slip' is derived from the offset of planar structures (e.g. bedding planes). As long as no other planar structure, oblique to the first one, gives additional information on the slip direction, no statement about the slip direction can be made. The offset of one-dimensional features (e.g. longish fossils) gives an unambiguous indication for the slip direction, but they are rarely found offset in nature. Thus, the offset along a fault usually does not clearly show the slip direction from which to derive the orientation of the principal stress axes. Consequently we downgraded the quality assignment for this type of data (GFS data) to quality D.

Expanded quality ranking scheme for borehole breakouts (BO, BOC, BOT)

Breakouts develop when the stress concentration at the borehole wall exceeds the rock strength. They occur in the direction of the minimum horizontal stress S_h, 90° off the orientation of the maximum horizontal stress S_H. In the last two decades of breakout investigation logging tools have improved and the horizontal as well as the vertical resolution of the recorded data have increased. A variety of breakout interpretation methods based on these high resolution tools have been developed. To assist the individual researcher and to keep data comparable the WSM team set up guidelines of data analysis for borehole breakout investigation.

(a) *Criteria for breakout identification.* Numerous authors have treated the problem of breakout identification (e.g. Fordjor *et al.* 1983; Plumb & Hickman 1985; Peska & Zoback 1995; Zajac & Stock 1997). For the WSM database we suggest the following criteria for the identification of breakouts from borehole geometry tools such as BGT, SHDT, FMS, FMI. These criteria are modified after Plumb & Hickman (1985) and Zajac & Stock (1997).

(1) Tool rotates freely below and above the breakout interval. Rotation has to stop in the breakout interval.
(2) Caliper difference has to exceed 10% of bit size.
(3) The smaller caliper has to range between bit size and 110% of bit size.
(4) The length of the breakout zone has to be at least 1 m.
(5) The direction of elongation must not consistently coincide with the azimuth of the high side of the borehole. Thus, for boreholes which deviate more than 1° from vertical, the angular difference between the direction of

the hole azimuth and the azimuth of the greater caliper has to exceed 15°.

(b) *Recommendations for breakout analysis*
(1) Use breakout identification criteria (see above).
(2) Use visual control (for caliper data): if the analysis is purely based on the statistics of Pad-1 azimuth (Fig. 3a), direction A would

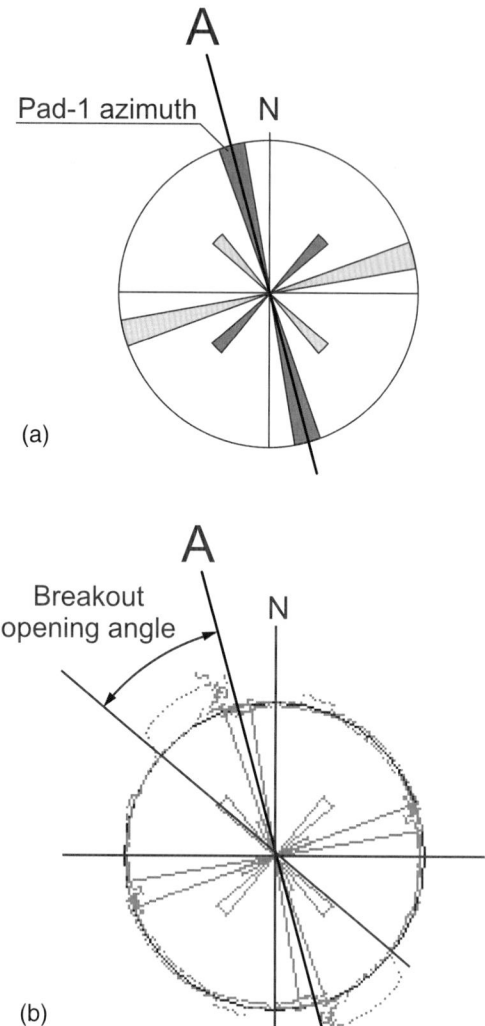

Fig. 3. (a) Distribution of four-arm caliper data (Pad-1,3 azimuth: dark grey; Pad-2,4 azimuth: light grey). Due to the Pad-1,3 maximum orientation 'A' seems to represent the breakout orientation. (**b**) Contour plot of caliper data. The distribution of the data indicates that the tool stuck to one side of the breakout (due to cable torque), so that orientation 'A' does not represent the correct breakout azimuth.

be obtained as breakout azimuth. But, due to cable torque, the tool may stick to one side of the breakout and A may be off the true breakout azimuth. Contour plots help to avoid this source of error (Fig. 3b).

(3) Use circular statistics (e.g. Mardia 1972).

(4) Correct the S_H azimuth according to magnetic declination at the time of logging.

(5) Check for key seats: a means to check for key seat effects is to plot breakout orientation with depth including the direction of hole azimuth (Fig. 4).

(c) *Quality ranking scheme for borehole breakouts.* For borehole breakouts the quality is ultimately linked to the standard deviation, the number

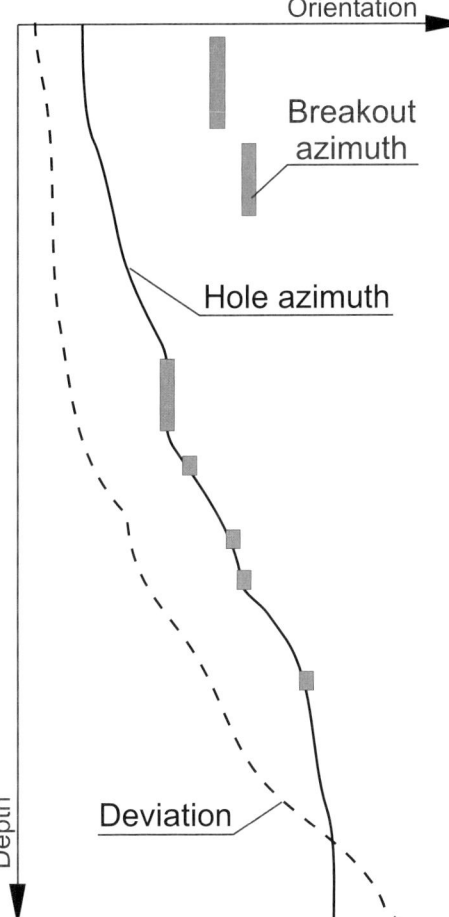

Fig. 4. Identification of key seats. In the deeper section the breakouts follow the direction of the hole azimuth indicating possible key seats. Deviation: angle between borehole and vertical direction; hole azimuth: dip azimuth of non-vertical borehole.

of breakouts and the total length of breakouts (Zoback 1992) (Table 4). The overall quality of the analysis is given by its *lowest* subquality.

Expanded quality ranking scheme for overcoring data (OC)

The previous quality ranking for overcoring measurements contained some unclear formulations (e.g. 'multiple consistent measurements') which will be removed. For the sake of consistency, overcoring data will be treated like borehole slotter data because they face similar problems in quality ranking. (For details see section on Borehole slotter data and Table 6.)

New methods

Two new types of stress indicators are now included in the WSM database: drilling-induced fractures (DIF) and borehole slotter data (BS). Their quality rankings are described in detail below.

Drilling-induced fractures (DIF)

The availability of high resolution logging tools such as **F**ormation**M**icro**S**canner (FMS), **F**ormation**M**icro**I**mager (FMI) or **B**orehole **T**ele**v**iewer (BHTV) enables the analysis of tensile failure phenomena at the borehole wall. In the WSM database tensile failure from induced hydraulic testing (hydrofrac) is already included. In addition to the stimulated hydrofracs, unintentional tensile failure created during drilling may occur (Fig. 5a). These drilling-induced fractures (DIF) can be used for stress determination (Wiprut *et al.* 1997; Fig. 5b). Identification criteria have been proposed, for example by Brudy & Zoback (1999) and Ask (1998). We basically follow their criteria with some minor modifications.

Table 4. *Quality ranking scheme for borehole breakouts (BO, BOC, BOT)*

Subquality	Number of data	Total length (m)	Standard deviation (°)
a	≥10	≥300	≤12
b	≥6	≥100	≤20
c	≥4	≥30	≤25
d	<4	<30	≤40
e	–	–	>40

The overall quality is defined by the *lowest* subquality of any of these criteria.

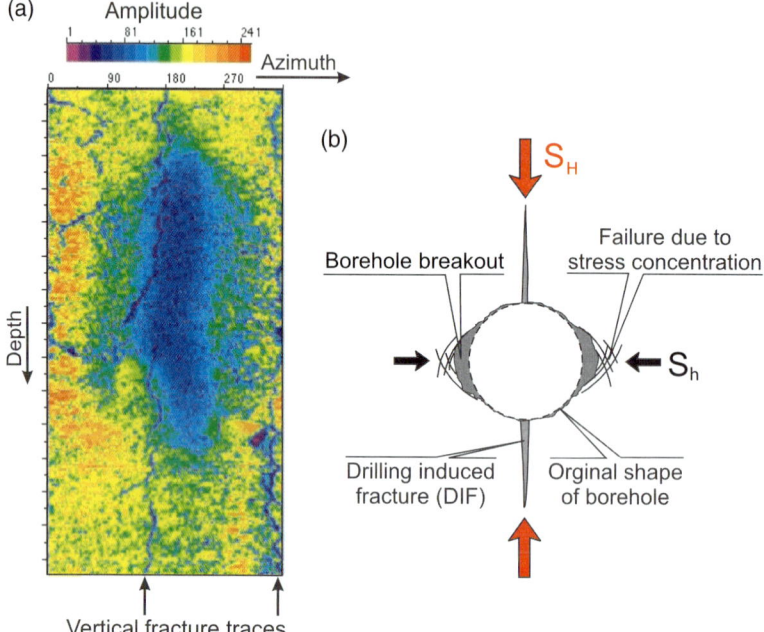

Fig. 5. (**a**) Borehole televiewer image with vertical fracture traces which are interpreted as drilling-induced fractures. Colours correspond to the amplitude of the reflected acoustic (ultrasonic) pulse and thus to the reflection coefficient of the borehole wall (low amplitude = rough, fractured borehole wall such as in breakout intervals). (**b**) Sketch of a borehole with drilling-induced fractures parallel to the maximum horizontal compression (S_H) and borehole breakouts perpendicular to S_H.

(a) *Criteria for DIF identification*

(1) Fractures that are subparallel to the borehole axis, occur in pairs on opposite sides of the borehole wall and are not connected to each other (not to be misinterpreted as sinusoidal steep fractures).

(2) Short en-echelon fracture traces (also occurring in pairs) but inclined with respect to the borehole axis. They result from stress distribution at the borehole wall when the borehole axis does not coincide with one of the principal stress axes.

(3) The length of DIFs ranges between 0.1 and 2 m.

(4) If breakout observations are available DIF and breakout orientations should differ by 90°.

(5) The shape of DIFs should be irregular since perfectly straight features could result from mechanical wear of the borehole.

(b) *Quality ranking scheme for DIFs.* The quality ranking scheme for DIFs is based on the number of fractures which had been identified, the standard deviation of their average and the total fracture length (Table 5). The quality ranking scheme is similar in structure to the breakout quality ranking.

Borehole slotter data (BS)

Borehole slotter data are obtained by a stress relief method (no overcoring) which is described in detail by Bock & Foruria (1983), Bock (1993), and Becker & Werner (1994). Results of borehole slotter measurements agree with data from other methods (e.g. borehole breakouts; Becker & Werner 1994). Becker & Paladini (1990) developed a quality ranking scheme for *in situ* stress data which we used as a basis for the (new)

Table 5. *Quality ranking scheme for drilling induced fractures (DIF)*

Subquality	Number of data	Total length (m)	Standard deviation (°)
a	≥10	≥300	≤12
b	≥6	≥100	≤20
c	≥4	≥30	≤25
d	<4	<30	≥40
e	<4	<30	>40

The overall quality is defined by the *lowest* subquality of any of these criteria.

Table 6. *Quality ranking scheme for borehole slotter (BS) and overcoring (OC) data*

Subquality	Number of data	Depth of measurement (m)	Standard deviation (°)
a	≥11	≥300	≤12
b	≥8	≥100	≤20
c	≥5	≥30	≤25
d	≥2	≥10	≤40
e	<2	<10	>40

The overall quality is defined by the *lowest* subquality of any of these criteria.

WSM quality ranking. We adapted it to the already existing criteria for similar stress indicators like overcoring or breakout data. Important criteria are the depth of measurement, the number of data and the standard deviation of the principal stress axes orientation (Table 6). Borehole slotter measurements normally are within the first 30 m of a borehole, so that topographic effects or stress deflections around underground openings might influence the result. Thus the distance to topographic features (e.g. quarry walls) and excavation walls in the case of underground openings is important: the distance to topographic features must be larger than three times the height of the topographic feature and the distance to excavation walls must be larger than two times the excavation radius (Fig. 6). All measurements at smaller distances are assigned quality E.

Quality assignment and depth of measurement

We are aware that the quality assignment cannot be transferred from one method to another. Researchers may argue that the assignment of qualities A or B to borehole slotter (BS) or overcoring (OC) data is restricted to measurements at depths greater than 100 m, which is in most cases equival-

ent to measurements in tunnels or mines since the drillholes for BS and OC data usually are less deep. In contrast to borehole slotter and overcoring data, the quality ranking for geological fault-slip data (GFI), drilling-induced fractures (DIF), and borehole breakouts (BO) does not contain such a depth criterion. For the latter two it is considered to be unnecessary because these data result from measurements in industrial wells at several hundred metres depth. In addition, the depth condition for BOs and DIFs is substituted by the conditions of total length of BO or DIF data, respectively.

GFI data are surface data in terms of measurements. However, their origin may be from greater, but unknown, depth. To account for the missing depth condition in the quality ranking of the GFI data a larger number of data is required. For example, quality A can be obtained with a minimum of 25 data for GFI measurements, while only 11 are needed for BS and OC measurements.

Application of tectonic stress data

Numerous fields of applications exist for tectonic stress data. They are not restricted to pure 'academic' research projects, but also include investigations with large economic impact, e.g. in hydrocarbon or geothermal energy production. In the following we give examples for both types of application. For the 'academic' example we selected the eastern Mediterranean region where numerous stress data are available and where geodetic measurements (GPS, SLR) provide the possibility to compare both data sets. The 'economic' examples deal with the stability of underground openings and with hydrocarbon reservoir management.

Stress and velocity in the eastern Mediterranean region

Tectonic overview The Neogene to recent tectonic evolution of the eastern Mediterranean region is driven by the north–south convergence between the African and Arabian plates, and the Eurasian plate.

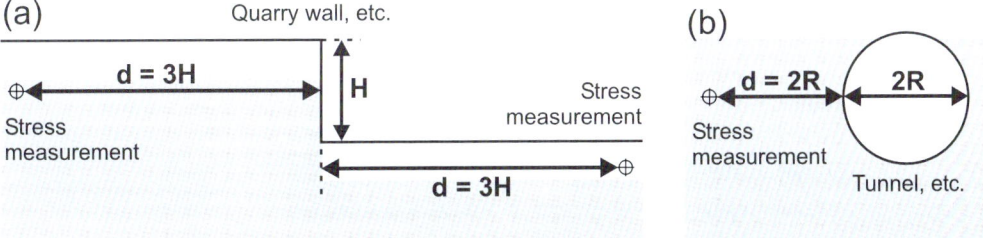

Fig. 6. Minimum distance *d* of stress measurements from (**a**) topographic features (three times the height of the topographic feature) and (**b**) underground openings (two times the excavation radius).

The African plate moves at a velocity of 9 mm a⁻¹ and the Arabian at 25 mm a⁻¹ relative to the Eurasian plate (DeMets *et al.* 1994). An assemblage of lithospheric blocks trapped in this convergence zone moves independently between Africa, Arabia and Eurasia resulting in oblique continental collisions and thus in the close vicinity of collision and subduction zones (e.g. McKenzie 1972). Subduction occurs beneath the Hellenic Arc, while continental collision already started in its lateral continuations towards the NW (Dinarides) and towards the east (Cyprean Arc) (Fig. 7). The southwestward retreat of the Hellenic Arc subduction zone causes backarc extension in the Aegean and enables the westward extrusion of the Anatolian block out of a zone of continental collision at the eastern end of the Anatolian block (e.g. McKenzie 1970; Le Pichon & Angelier 1979; Royden 1993). This collision is the result of the north-northwestward movement of the Arabian plate which is separated from the African plate by the Dead Sea fault and the Red Sea rift.

Stress data The majority of the stress data comes from fault-slip analysis (for a compilation see Mercier *et al.* 1987) and from earthquake data (e.g. Jackson & McKenzie 1988). In the Aegean and western Anatolia normal faulting is the dominant stress regime with S_H orientations trending NW to NE indicating an approximately north–south extension (Fig. 8). In the northern part of central Anatolia strike-slip faulting is the prevailing stress regime.

North–south extension in the northern Aegean Sea and western Anatolia had been interpreted to be the expression of backarc extension due to subduction zone retreat of the Hellenic Arc (e.g. Müller *et al.* 1992). The fan-shaped arrangement of S_h (minimum horizontal stress; $S_h \perp S_H$) might result from the suction force acting radially outward (i.e. southwest- to southeastward) caused by slab rollback as shown in the numerical models of Meijer & Wortel (1996). Strike-slip movements in this region and in northern central Anatolia are most probably related to the westward directed lateral extrusion of the Anatolian block with dextral movements along the North Anatolian fault (Fig. 7).

Velocity data (GPS, SLR) Geodetic observations were initiated in the mid-1980s, so that today accurate data about the velocity vectors are available for most of the observation sites (e.g. Noomen *et al.* 1996; Reilinger *et al.* 1997; Kaniuth *et al.*

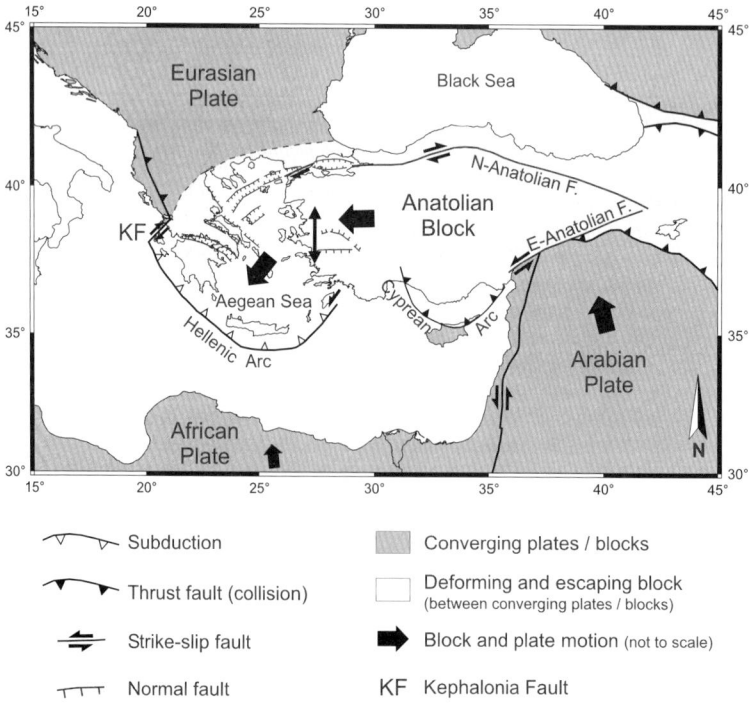

Fig. 7. Tectonic map of the eastern Mediterranean region. Lateral extrusion of the Anatolian block towards the west is triggered by continental collision of the Arabian plate with the Eurasian plate in combination with subduction zone retreat at the Hellenic arc.

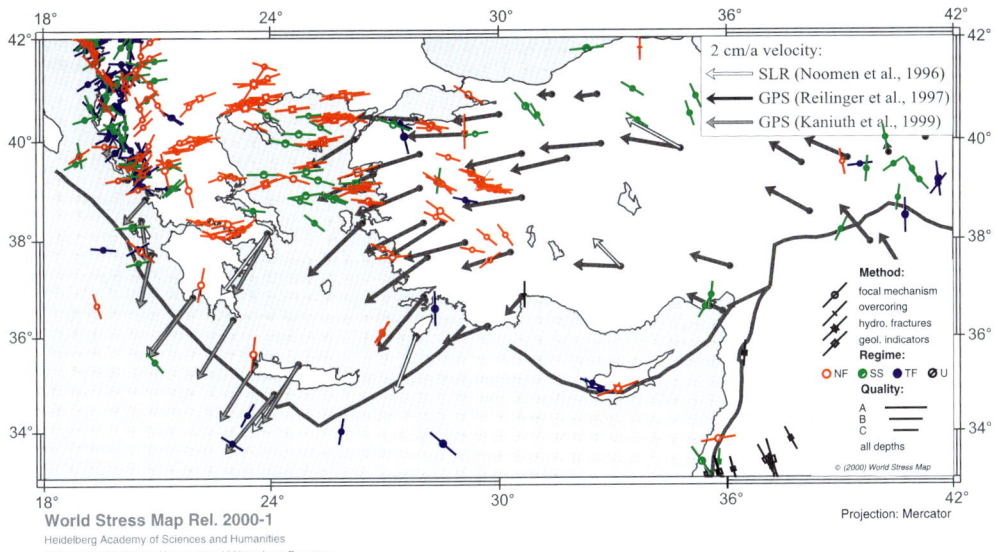

Fig. 8. Velocity and stress data for the eastern Mediterranean region. GPS data come from Reilinger *et al.* (1997) and Kaniuth *et al.* (1999), SLR data from Noomen *et al.* (1996), and stress data from the World Stress Map database (Müller *et al.* 2000). For further explanations of the stress symbols see Figure 1.

1999; Fig. 8). They indicate west-directed movements of Anatolia and southwestward movements of the Aegean. The transition zone between both blocks (western Turkey) is characterized by west-southwestward movements. The influence of plate boundary forces and different material properties on these movements have been investigated by finite element modelling. For details and references see Heidbach & Drewes (2003). Other models reconstructing the velocity vectors in the eastern Mediterranean assume two individually moving rigid lithospheric blocks: (1) the clockwise rotating Aegean block encompassing the southern and central parts of the Aegean and the southern part of Greece (south of the Gulf of Korinth); and (2) the counterclockwise rotating Anatolian block which is bordered by the North and East Anatolian faults (Fig. 7; e.g. Drewes 1993; Le Pichon *et al.* 1995). This approach fits with palaeomagnetic data which also show opposite rotation senses for both blocks (e.g. Kissel *et al.* 1987; Kissel & Laj 1988). In contrast Reilinger *et al.* (1997) modelled their set of velocity vectors with a single Aegean-Anatolian block. In general a good agreement exists, but significant discrepancies occur in western Anatolia and the southern Aegean.

Comparison of stress and velocity data For the eastern Mediterranean both types of data are explained by identical processes (see above). Nevertheless, there is no straightforward relation between them and a number of aspects have to be taken into account when comparing stress and velocity data.

(1) Velocity fields reflect not only strain (i.e. internal deformation), but also translation and rotation (some people use the expression 'strain' in a wider sense including translation and rotation, but in this paper we use it in the more rigorous way which excludes these two processes and is restricted to distance changes between the points of a body). Thus, first rigid body movements have to be removed. For large-scale velocity fields, which have to take into account the curvature of the Earth, all rigid plate or block movements are rotations around a Euler pole. Accordingly, Reilinger *et al.* (1997) calculated the internal deformation of the eastern Mediterranean from geodetic data and found that it is small in central Anatolia. In contrast, the plate and block boundaries are characterized by broad deformation zones.

(2) Axes of principal strain rate and stress are only in some cases parallel to each other (coaxial deformation, e.g. pure shear); in most cases they are oriented obliquely to each other (non-coaxial deformation, e.g. simple shear). Additionally, the relationship between stress and strain rate might be non-linear (e.g. strain hardening or softening).

(3) Stress data are compiled throughout the whole crust while geodetic data exclusively reflect surface movements. Decoupling between different layers might result in different orientations of stress axes at different depths (e.g. Brereton & Müller 1991; Jarosinski 1998). In the eastern Mediter-

ranean S_H orientations from the seismogenic upper 15 km of the crust are in good agreement with S_H orientations from the numerous surface data (geological indicators; Fig. 8). In contrast, the stress regime shows a depth dependence in the central and northern Aegean: surface data indicate a normal faulting regime, while data from depths of more than 5 km show strike-slip faulting. The change from normal faulting to strike-slip faulting regime is in contradiction to the regime changes observed elsewhere with depth which are caused by increasing overburden and thus are characterized by a change from strike-slip or thrust faulting to normal faulting. One explanation could be a gravitational collapse of the thicker crust with its higher topography in the northernmost part of the Aegean towards the central Aegean with its thinner crust and its lower topography. In this case a decoupling of the southward sliding uppermost crust from the underlying rocks would be necessary.

(4) Both data types represent different time periods. While geodetic data reflect movements during the last decade or even shorter periods, stress data come from several decades (earthquake data) or even from several tens to hundreds of thousands of years (geological indicators which come from Quaternary rocks). Deformation of the Earth's crust takes place in a non-continuous way, implied, for example, by the locally observed clustering of earthquakes through time (a seismically active period is followed by an inactive period) or by progressive failure along a fault (Stein *et al.* 1997). The seismic cycle of strong earthquakes can be as long as several hundreds of years (Jackson *et al.* 1988). For the eastern Mediterranean region the majority of the stress data comes from earthquakes and geological indicators thus reflecting a much longer time period than the geodetic measurements in Figure 8 which cover the period 1988–1994 (Reilinger *et al.* 1997). Nevertheless, McClusky *et al.* (2000) found that the geodetic data for the North and East Anatolian faults are in agreement with slip rates for the last 4–5 Ma revealed from geological data. However, the data of McClusky *et al.* (2000) do not include observations *after* the Izmit earthquake in August 1999. Earthquakes have a strong effect on the velocity field even at distances of hundreds of kilometres or more depending on the magnitude of the event. For the $M_w = 8.0$ earthquake near Antofagasta (north Chile) Klotz *et al.* (1999) observed a coseismic slip of 10 cm at a distance of 300 km.

(5) The processing of geodetic data has not yet been standardized. Thus, the results depend on the choice of the reference system and of the models included into the processing. Other critical factors are the number of observation sites and the obser-

vation time. This makes a comparison of the results of different research groups difficult or even impossible. McClusky *et al.* (2000) did a compilation of GPS data for the eastern Mediterranean which encompasses more observation sites and a longer observation period (1988–1997) compared to earlier compilations by Reilinger *et al.* (1997). The differing results of these two compilations led to different interpretations of the block movements in this region. Reilinger *et al.* (1997) 'saw' only one block, the Anatolian-Aegean block, while McClusky *et al.* (2000) 'needed' two blocks, the Anatolian and the Aegean blocks, to explain the velocity vectors (see above).

Conclusions This list of 'complications' as well as the data from the eastern Mediterranean region indicate that much more extra work is necessary to understand the relationship between stress, strain and velocity, and how these parameters depend on changes in plate tectonic processes and material properties (e.g. change of viscosity due to enhanced heat flow) as well as on local influences by seismic activity. On a global scale the ILP project 'Global Strain Rate Map' under the guidance of W. Holt (SUNY Stony Brook) and J. Haines (Cambridge University) and the WSM are prerequisites for this kind of study.

Stability of underground openings

Any underground construction causes severe modifications of the initial *in situ* rock stress at the location of the site: artificial openings create fresh surfaces which modify the boundary conditions at the location of the walls (mostly zero surface stress). This leads to a spatial distortion of the stress pattern with stress accumulations around the opening. The stress redistribution for circular holes in an anisotropically stressed rock is classically described by the Kirsch equations (Kirsch 1898). Stress redistribution around tunnels and shafts has been treated analytically as well as numerically for more complicated designs (numerous examples are summarized in Amadei & Stephansson 1997).

Since rock is less stable in tension, tensile stress concentrations may open pre-existing fractures or even fracture the rock, thus increase the weathering and leading to further instability. Compressive stresses in the vicinity of underground openings can result in different kinds of rock failure, e.g. rock bursts or side-wall spalling. In boreholes the stress-induced compressive failure is called borehole breakout (Fig. 5b).

The stability of construction sites does not depend only on the rock type but also on the orientation and magnitude of the tectonic stress in combination with the design and orientation of the

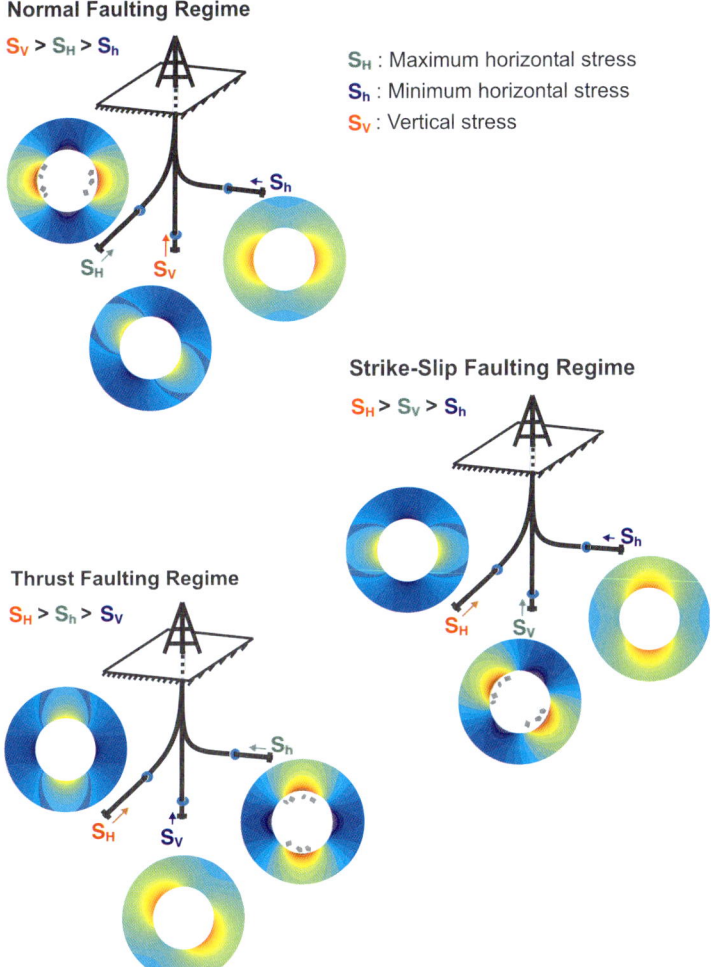

Fig. 9. Stability of underground openings due to their orientation relative to the principal stress axes. In all three stress regimes, openings parallel to the intermediate stress axis (green arrows; S_H in normal faulting, S_V in strike-slip faulting, and S_h in thrust faulting regime) are most endangered for rock failure with the failure occurring in the direction of the minimum stress axes (blue arrows; S_h in normal and strike-slip faulting, S_V in thrust faulting regime). Discs show stress concentrations (yellow to red colours) around the borehole.

underground opening. In normal faulting regimes, boreholes (shafts and mines) are most likely to fail under compression sideways when they are drilled parallel to the S_H direction (Fig. 9). This is in agreement with the observation that in places where vertical stresses are high, problems in the sidewalls of mines occur. Especially in areas with high topography and steep valley slope (for example Norwegian fjords) the tunnels that parallel the valley (parallel to S_H) show rockbursts in the tunnel wall (Amadei & Stephansson 1997). In thrust faulting regimes boreholes fail at the top and bottom if drilled parallel to S_h. Again this is in agreement with mining observations where prob-

lems in roofs and floors occur for areas of high horizontal stress. Under strike-slip conditions vertical boreholes and shafts are most endangered for failure because they experience high stress concentrations in the direction parallel to S_h.

Hydrocarbon reservoir management

In the future, the discovery of new gigantic hydrocarbon reservoirs will be rare, thus companies will focus on the optimization of the production of existing fields. This covers exploitation of smaller reservoirs or marginal fields and activities in the deeper offshore areas. Tectonic stress in combination with the actual fracture network controls the fluid flow anisotropy and thus the drainage pattern within the reservoir. Numerous examples of a strong correlation between the preferential direction of fluid flow in a reservoir and the orientation of S_H are described by Heffer & Dowokpor (1990), Heffer & Koutsabeloulis (1995), and Heffer *et al.* (1995). Thus, knowledge about the orientation of S_H can be used for an optimized arrangement of the production wells (Fig. 10).

One distinguishes the fracture-induced permeability from the intrinsic permeability of the rock which is connected with the available pore space. Both together form the effective permeability (e.g. Connolly & Cosgrove 1999). In particular, oil or gas recoveries from reservoirs with low intrinsic permeabilities require widespread networks of interconnected fractures (Lorenz *et al.* 1988; Hickman & Dunham 1992).

Oil recovery is enhanced by the creation of new (artificial) fractures (pathways). A fracture develops away from the borehole by pumping water or gels into the target area. It propagates perpendicularly to the least principal stress. Thus, advance knowledge of the stress orientations enables prediction of the orientations of stimulation fractures and thus enhances the economic success of field activities.

Outlook

The WSM is designed as a secure repository of stress data from academic research and economic investigations, and as a tool to work with the data. It has a long-term perspective (continuation until at least 2008) and thus enables oil companies, for example, to come back to hydrocarbon fields which had been considered as uneconomic earlier. The data are readily available via the Internet so that time-consuming searching in archives can be avoided. Additionally, with CASMO we provide a user-friendly tool for the creation of stress maps. The growth of the database by *c.* 20% compared with the 1997 release (1997: 9147 data sets; 2000: 10 920 data sets) justifies the hope that the data coverage on a global scale will further improve during the next few years, so that some of the large data gaps can be closed.

Stress observations are fundamental Earth science data which are used to relate global dynamic modelling of lithospheric stresses with plate movements. Other scientific challenges are the correlation between crustal dynamics and velocity anisotropies in the upper mantle, regional investigations of tectonic motions, and the comparison of stress data with modern geodetic observations as described for the Mediterranean. The increase of stress observations will promote further research projects leading to a better understanding of the sources and the effects of tectonic stresses.

Investigation of tectonic stress is not only of scientific interest, but also has an economic impact by providing information on the stability of underground openings like boreholes, tunnels, or radioactive waste depositories and thus enhances the value of the data gained by the industry. Moreover, preventing rock failure by choosing an appropriate orientation of underground openings and precautions for stabilization can be considered as a crucial security aspect for people working in mines and tunnels.

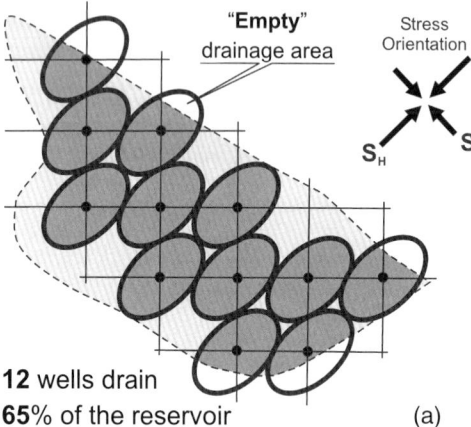

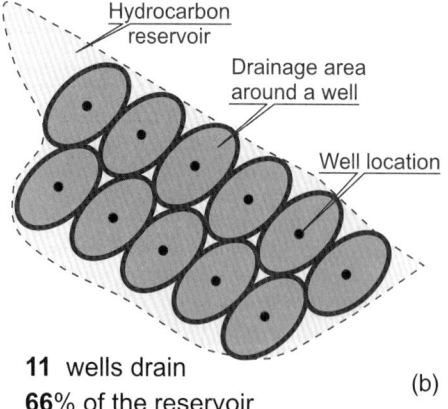

Fig. 10. Stress-controlled flow anisotropy (modified after Bell 1996). Due to the preferred orientation of fractures parallel to S_H, the drainage area of each well is elliptical with the long axis parallel to S_H. (**a**) The arrangement of wells due to geometric aspects (north–south and east–west grid) requires 12 wells for draining 65% of the reservoir. (**b**) The arrangement due to reservoir flow characteristic requires only 11 wells for 66% drainage and is thus more economic.

The World Stress Map project is financed by the Heidelberg Academy of Sciences and Humanities. We thank A. Becker and D. Nieuwland for their helpful and constructive reviews.

References

ALLMENDINGER, R. W., GEPHARDT, J. W. & MARRETT, R. A. 1989. *Notes on Fault Slip Analysis.* Notes to the Geological Society of America short course 'Quantitative interpretation of joints and faults'. Cornell University, Ithaca.

AMADEI, B. & STEPHANSSON, O. 1997. *Rock Stress and its Measurement.* Chapman & Hall, London.

ANGELIER, J. 1984. Tectonic analysis of fault-slip data sets. *Journal of Geophysical Research,* **89**, 5835–5848.

ANGELIER, J. & GOGUEL, J. 1979. Sur une méthode simple de détermination des axes princepaux des contraintes pour une population de failles (A simple method for determining the principal axes of stress for a fault population). *Comptes Rendus de l'Académie des Sciences, Paris, Série D,* **288**, 307–310.

ASK, M. V. S. 1998. *In-situ and laboratory stress investigations using borehole data from the North Atlantic Ocean.* PhD thesis, Department of Civil and Environmental Engineering, Royal Institute of Technology, Stockholm.

BECKER, A. & PALADINI, S. 1990. In situ Spannungen in Nord- und Mitteleuropa. *Schriftenreihe Angewandte Geologie Karlsruhe,* **10**, 1–63.

BECKER, A. & WERNER, D. 1994. Strain measurements with the borehole slotter. *Terra Nova,* **6**, 608–617.

BELL, J. S. 1996. In situ stresses in sedimentary rocks (part 2): Applications of stress measurements. *Geoscience Canada,* **23**, 135–153.

BOCK, H. 1993. Measuring in situ rock stress by borehole slotting. *Comprehensive Rock Engineering – Principles, Practice & Projects,* **3**, 433–443.

BOCK, H. & FORURIA, V. 1983. A recoverable borehole slotting instrument for in situ stress measurements, not requiring overcoring. *Proceedings of the International Symposium on Field Measurements in Geomechanics, Zürich,* 15–29.

BRERETON, R. & MÜLLER, B. 1991. European stress: contributions from borehole breakouts. *Philosophical Transactions of the Royal Society of London,* Series A, **337**, 165–179.

BRUDY, M. & ZOBACK, M. D. 1999. Drilling-induced tensile wall-fractures: implications for determination of in-situ stress orientation and magnitude. *International Journal of Rock Mechanics and Mining Sciences and Geomechanics Abstracts,* **36**, 191–215.

CONNOLLY, P. & COSGROVE, J. 1999. Prediction of fracture-induced permeability and fluid flow in the crust using experimental stress data. *AAPG Bulletin,* **83**, 757–777.

DELVAUX, D. & SPERNER, B. 2003. New aspects of tectonic stress inversion with reference to the TENSOR program. *In:* NIEUWLAND, D. A. (ed.) *New Insights into Structural Interpretation and Modelling.* Geological Society, London, Special Publications.

DELVAUX, D., MOEYS, R., STAPEL, G. *et al.* 1997. Paleostress reconstruction and geodynamics of the Baikal region, Central Asia, Part 2, Cenozoic rifting. *Tectonophysics,* **282**, 1–38.

DEMETS, C., GORDON, R. G., ARGUS, D. F. & STEIN, S. 1994. Effect of recent revisions to the geomagnetic reversal time scale on estimates of current plate motions. *Geophysical Research Letters,* **21**, 2191–2194.

DREWES, H. 1993. A deformation model of the Mediterranean from space geodetic observations and geophysical predictions. *International Association of Geodesy, Symposium,* **112**, 373–378.

FORDJOR, C. K., BELL, J. S. & GOUGH, D. I. 1983. Breakouts in Alberta and stress in the North American plate. *Canadian Journal of Earth Sciences,* **20**, 1445–1455.

FUCHS, K. & MÜLLER, B. 2000. World Stress Map of the Earth: a key to tectonic processes and technological applications. *Naturwissenschaften,* **88**, 357–371.

GORDON, R. G. & STEIN, S. 1992. Global tectonics and space geodesy. *Science,* **256**, 333–342.

HARDCASTLE, K. C. 1989. Possible paleostress tensor configurations derived from strike-slip data in eastern Vermont and western New Hampshire. *Tectonics,* **8**, 265–284.

HEFFER, K. J. & DOWOKPOR, A. B. 1990. Relationship between azimuths of flood anisotropy and local earth stresses in oil reservoirs. *Proceedings of the North Sea oil and gas reservoir conference,* 251–260.

HEFFER, K. J. & KOUTSABELOULIS, N. C. 1995. Stress effects on reservoir flow; numerical modelling used to reproduce field data. *In:* DE HAAN, H. J. (ed.) *New Developments in Improved Oil Recovery.* Geological Society, London, Special Publications, **84**, 81–88.

HEFFER, K. J., FOX, R. J., McGILL, C. A. & KOUTSABELOULIS, N. C. 1995. Novel techniques show links between reservoir flow directionality, Earth stress, fault structure and geomechanical changes in mature waterfloods. *Society of Petroleum Engineers, Annual Technical Conference & Exhibition, Dallas,* SPE 30711.

HEIDBACH, O. & DREWES, H. 2003. 3D Finite element model of major tectonic processes in the Eastern Mediterranean. *In:* NIEUWLAND, D. A. (ed.) *New Insights into Structural Interpretation and Modelling.* Geological Society, London, Special Publications.

HICKMAN, R. G. & DUNHAM, J. B. 1992. Controls on the development of fractured reservoirs in the Monterey Formation of central California. *In:* LARSEN, R. M., BREKKE, H., LARSEN, B. T. & TELLERAAS, E. (eds) *Structural and Tectonic Modelling and its Application to Petroleum Geology.* Norwegian Petroleum Society, Special Publications, **1**, 343–353.

JACKSON, J. & McKENZIE, D. 1988. The relationship between plate motions and seismic tremors, and the rates of active deformation in the Mediterranean and Middle East. *Geophysical Journal of the Royal Astronomical Society,* **93**, 45–73.

JACKSON, J., HAINES, J. & HOLT, W. 1988. The horizontal velocity field in the deforming Aegean Sea region determined from the moment tensors of earthquakes. *Journal of Geophysical Research,* **97**, 17 657–17 684.

JAROSINSKI. M. 1998. Contemporary stress field distortion in the Polish part of the Western Outer Carpathians and their basement. *Tectonophysics,* **297**, 91–119.

KANIUTH, K., DREWES, H., STUBER, K. *et al.* 1999. Crustal deformations in the central Mediterranean derived from the WHAT A CAT GPS project. *Proceedings of the 13th Working Meeting on European VLBI for Geodesy and Astronomy*, Wettzell, 192–197.

KIRSCH, G. 1898. Die Theorie der Elastizität und die Bedürfnisse der Festigkeitslehre. *Zeitschrift des Vereines Deutscher Ingenieure*, **42**, 797.

KISSEL, C. & LAJ, C. 1988. The Tertiary geodynamical evolution of the Aegean arc: a paleomagnetic reconstruction. *Tectonophysics*, **146**, 183–201.

KISSEL, C., LAJ, C., SENGÖR, A. M. C. & POISSON, A. 1987. Paleomagnetic evidence for rotation in opposite senses of adjacent blocks in the northeastern Aegea and Western Anatolia. *Geophysical Research Letters*, **14**, 907–910.

KLOTZ, J., ANGERMANN, D., MICHEL, G. E. *et al.* 1999. GPS-derived deformation of the central Andes including the 1995 Antofagasta $M_w = 8.0$ earthquake. *Pure Applied Geophysics*, **154**, 709–730.

LE PICHON, X. & ANGELIER, J. 1979. The Hellenic arc and trench system: a key to the neotectonic evolution of the eastern Mediterranean area. *Tectonophysics*, **60**, 1–42.

LE PICHON, X., CHAMOT-ROOKE, N., LALLEMANT, S., NOOMEN, R. & VEIS, G. 1995. Geodetic determination of the kinematics of central Greece with respect to Europe: Implications for eastern Mediterranean tectonics. *Journal of Geophysical Research*, **100**, 12675–12690.

LORENZ, J. C., WARPINSKI, N. R., TEUFEL, L. W., BRANAGAN, P. T., SATTLER, A. R. & NORTHROP, D. A. 1988. Results of the Multiwell Experiment: in situ stresses, natural fractures, and other geological controls on reservoirs. *EOS Transactions, American Geophysical Union*, **69**, 817, 825–826.

McCLUSKY, S., BALASSANIAN, S., BARKA, A. *et al.* 2000. Global Positioning System constraints on plate kinematics and dynamics in the eastern Mediterranean and Caucasus. *Journal of Geophysical Research*, **105**, 5695–5719.

McKENZIE, D. P. 1970. Plate tectonics of the Mediterranean region. *Nature*, **226**, 239–243.

McKENZIE, D. P. 1972. Active tectonics of the Mediterranean region. *Geophysical Journal of the Royal Astronomical Society*, **30**, 109–185.

MARDIA, K. V. 1972. *Statistics of Directional Data*. Academic Press, New York.

MEIJER, P. TH. & WORTEL, M. J. R. 1996. Temporal variation in the stress field of the Aegean region. *Geophysical Research Letters*, **23**, 439–442.

MERCIER, J. L., SOREL, D. & SIMEAKIS, K. 1987. Changes in the state of stress in the overriding plate of a subduction zone: the Aegean Arc from the Pliocene to the Present. *Annales Tectonicae*, **1**, 20–39.

MÜLLER, B., ZOBACK, M. L., FUCHS, K., MASTIN, L., GREGERSEN, S., PAVONI, N., STEPHANSSON, O. & LJUNGGREN, C. 1992. Regional patterns of tectonic stress in Europe. *Journal of Geophysical Research*, **97**, 11783–11803.

MÜLLER, B., WEHRLE, V. & FUCHS, K. 1997. The 1997 release of the World Stress Map. WWW address: http://www.world-stress-map.org.

MÜLLER, B., REINECKER, J., HEIDBACH, O. & FUCHS, K. 2000. The 2000 release of the World Stress Map. WWW address: http://www.world-stress-map.org.

NOOMEN, R., SPRINGER, T. A., AMBROSIUS, B. A. C. *et al.* 1996. Crustal deformations in the Mediterranean area computed from SLR and GPS observations. *Journal of Geodynamics*, **21**, 73–96.

PESKA, P. & ZOBACK, M. D. 1995. Compressive and tensile failure of inclined wellbores and determination of in situ stress and rock strength. *Journal of Geophysical Research*, **100**, 12791–12811.

PLUMB, R. A. & HICKMAN, S. H. 1985. Stress-induced borehole elongation: A comparison between the four-arm dipmeter and the borehole televiewer in the Auburn Geothermal well. *Journal of Geophysical Research*, **90**, 5513–5522.

REILINGER, R. E., McCLUSKY, S. C., ORAL, M. B. *et al.* 1997. Global Positioning System measurements of present-day crustal movements in the Arabia-Africa-Eurasia plate collision zone. *Journal of Geophysical Research*, **102**, 9983–9999.

ROYDEN, L. 1993. The tectonic expression of slab pull at continental convergent boundaries. *Tectonics*, **12**, 303–325.

SPERNER, B. 1996. Computer programs for the kinematic analysis of brittle deformation structures and the Tertiary tectonic evolution of the Western Carpathians (Slovakia). *Tübinger Geowissenschaftliche Arbeiten, Reihe A*, **27**, 1–120.

SPERNER, B., RATSCHBACHER, L. & OTT, R. 1993. Fault-striae analysis: a Turbo Pascal program package for graphical presentation and reduced stress tensor calculation. *Computers & Geosciences*, **19**, 1361–1388.

STEIN, R. S., BARKA A. A. & DIETRICH, J. H. 1997. Progressive failure on the North Anatolian fault since 1939 by earthquake stress triggering. *Geophysical Journal International*, **128**, 594–604.

TWISS, R. J. & UNRUH, J. R. 1998. Analysis of fault slip inversions: Do they constrain stress or strain rate? *Journal of Geophysical Research*, **103**, 12205–12222.

WESSEL, P. & SMITH, W. H. F. 1991. Free software helps map and display data. *EOS Transactions, American Geophysical Union*, **72**, 441.

WESSEL, P. & SMITH, W. H. F. 1998. New, improved version of the Generic Mapping Tools released. *EOS Transactions, American Geophysical Union*, **79**, 579.

WIPRUT, D., ZOBACK, M. D., HANSSEN, T. H. & PESKA, P. 1997. Constraining the full stress tensor from observations of drilling-induced tensile fractures and leak-off test: Applications to borehole stability and sand production on the Norwegian margin. *International Journal of Rock Mechanics and Mining Sciences*, **34**, 417.

ZAJAC, B. J. & STOCK, J. M. 1997. Using borehole breakouts to constrain the complete stress tensor: Results from the Siljan Deep Drilling Project and offshore Santa Maria Basin, California. *Journal of Geophysical Research*, **102**, 10083–10100.

ZOBACK, M. D., ZOBACK, M. L., MOUNT, V. S. *et al.* 1987. New evidence on the state of stress on the San Andreas fault system. *Science*, **238**, 1105–1111.

ZOBACK, M. L. 1992. First- and second-order patterns of stress in the lithosphere: The World Stress Map Project. *Journal of Geophysical Research*, **97**, 11703–11728.

A four-year study of shear-wave splitting in Iceland: 1. Background and preliminary analysis

T. VOLTI[1,2] & S. CRAMPIN[1,3]

[1]*School of Geosciences, University of Edinburgh, Grant Institute, West Mains Road, Edinburgh, EH9 3JW, UK (e-mail: tvolti@glg.ed.ac.uk; scrampin@ed.ac.uk)*
[2]*Japan Marine Science and Technology Centre, Yokohama, 237–0061, Japan*
[3]*Edinburgh Anisotropy Project, British Geological Survey, Edinburgh, EH9 3LA, UK*

Abstract: A four-year study of seismic shear-wave splitting in Iceland was designed to seek temporal variations before earthquakes. Shear-wave splitting is observed routinely in Iceland whenever shear-waves arrive within the shear-wave window of seismic stations, and whenever adequate data are available, temporal and spatial variations in shear-wave splitting are observed before both earthquakes and volcanic eruptions. Shear-wave splitting is caused principally by the stress-aligned fluid-saturated microcracks and pore throats in almost all *in situ* rocks. Fluid-saturated microcracks are the most compliant elements of the rock mass, and changes in splitting can be directly interpreted and modelled as the effects of changing stress on the microcrack geometry in the rock mass often at considerable distances from the immediate earthquake source zone. Such changes were found and are reported in Part 2 of this study. This chapter presents the background, preliminary observations, and analysis of shear-wave splitting in Iceland.

Parts 1 and 2 (Volti & Crampin 2003) of this study summarize the results of a four-year study of seismic shear-wave splitting data in Iceland. The European Commission funded PRENLAB Projects, 1996 to 2000, used Iceland as a natural laboratory for earthquake prediction research. PRENLAB extended the SIL seismic network (South Iceland Lowland network of seismic stations; Stefánsson *et al.* 1993; Böðvarsson *et al.* 1999), and made various geophysical and geological studies throughout Iceland, including studies of shear-wave splitting above small earthquakes. Iceland is on an offset of the Mid-Atlantic Ridge, and this study concentrates on the major concentration of seismicity over the transform zone which is onshore in SW Iceland (described by Menke *et al.* 1994) and is sometimes known as the South of Iceland Seismic Zone (SISZ). Shear-wave splitting is observed whenever shear-waves were recorded within the shear-wave window below seismic stations.

We suggest that the observed changes in time-delays are not earthquake precursors in the usual sense of the word. Shear-wave splitting is monitoring the build-up of stress before larger earthquakes and hence is monitoring the primary driving force of all earthquakes. Without the stress build-up, there would be no earthquakes. Consequently, with appropriate observations, changes in shear-wave splitting can always be observed before earthquakes, and inferred changes in crack geometry can be directly interpreted in terms of changes in stress and the approach of fracture criticality and earthquakes.

In this chapter, we describe the background of shear-wave splitting and measurement techniques, and the preliminary observations in Iceland. Changes in the time-delays between split shear-waves were identified before both earthquakes and eruptions at the few stations where there were adequate temporal and spatial data to monitor temporal variations. The time and magnitude of an *M*5 earthquake in Iceland was successfully stress-forecast (Crampin *et al.* 1999*a*). This and other temporal variations are reported in Part 2 (Volti & Crampin 2003).

Forewarnings

The major problem with this study is that we are unable to resolve the exceptionally large (±80%) scatter in the measured time-delays between the split shear-waves (see for example fig. 1 of Volti & Crampin 2003). The five possible sources of scatter identified in conventional (non-critical) geophysics

From: NIEUWLAND, D. A. (ed.) *New Insights into Structural Interpretation and Modelling.* Geological Society, London, Special Publications, **212**, 117–133. 0305-8719/03/$15

cannot explain the large scatter, and we have to appeal to the geophysics of critical systems. This we suggest is plausible. The 1.5% to 4.5% shear-wave velocity anisotropy observed in most rocks in the uppermost 15 km of the crust (Crampin 1994) imply distributions of very closely spaced fluid-saturated microcracks which are near to levels of fracture criticality at *c.* 5.5% shear-wave velocity anisotropy (Crampin & Zatsepin 1997). This suggests that the microcrack distributions in most rocks are close to a critical state. Critical states imply heterogeneities and clustering at all scales in both time and space (Bruce & Wallace 1989; Crampin & Chastin 2001). Consequently, seismic waves travelling through rocks on the verge of criticality will be scattered and in particular the scattering will vary with time. Since it can be shown that shear-wave splitting is controlled by and is extremely sensitive to changes in low-level deformation (Crampin & Zatsepin 1997), the effects will be most severe on the scattering of shear-wave splitting.

Background

In anisotropic elastic solids, seismic shear-waves travelling at the group velocity split into two nearly perpendicular polarizations, which travel at different velocities and have polarizations and velocities that are fixed for each direction of travel in each anisotropic symmetry system (Crampin 1981). Shear-waves with approximately orthogonal polarizations separate in time and write characteristic signatures into the particle motion that can be identified and analysed either in polarization diagrams (hodograms) or by rotating horizontal seismograms parallel and perpendicular to the preferred polarizations. The distinctive signature of azimuthally varying shear-wave splitting, where the faster polarization is approximately parallel to the direction of maximum horizontal stress, is widely observed in almost all *in situ* rocks below a critical depth, usually between 500 m and 1 km below the surface (Crampin 1994). At shallower depths, shear-wave splitting may be disturbed by near-surface stress-release anomalies and the effects are often dominated by lithological and petrological phenomena.

Such stress-aligned shear-wave splitting was first positively identified in the shear-wave window above small earthquakes in the Turkish Dilatancy Projects (Crampin *et al.* 1980, 1985). Since then, stress-aligned shear-wave splitting has been observed in a great variety of sedimentary, igneous and metamorphic crustal rocks below the critical depth in controlled-source exploration seismology, as well as above small earthquakes (reviewed by Crampin 1994; see also Crampin 1996; Winterstein

1996). The splitting is caused by propagation through distributions of fluid-saturated grain-boundary cracks and low aspect-ratio pores that become aligned in the stress field (Crampin 1994; Crampin & Zatsepin 1997). These distributions of aligned 'cracks' are known as *extensive-dilatancy anisotropy* (EDA) and the individual inclusions as EDA-cracks (Crampin *et al.* 1984). There are only a few well-understood exceptions where *in situ* rocks do not contain EDA-cracks (Crampin 1994, 1999).

Evolution of fluid-saturated cracks

Fluid-saturated EDA-cracks are stress-sensitive and highly compliant. This means that cracks respond immediately if the stress in the rockmass changes, however marginally. The mechanism for the response (the mechanism for deformation) is fluid migration by flow or diffusion along pressure-gradients (Brodie & Rutter 1985; Rutter & Brodie 1991) between neighbouring grain-boundary cracks, and low aspect-ratio pores at different orientations to the stress field (Zatsepin & Crampin 1997; Crampin & Zatsepin 1997). Since shear-wave splitting is sensitive to details of microcrack geometry, changes in shear-wave splitting can be used to monitor stress-induced changes to microcrack geometry. The evolution of such stressed fluid-saturated microcracks under changing conditions has been modelled by *anisotropic poro-elasticity* (APE) (Zatsepin & Crampin 1997; Crampin & Zatsepin 1997). APE-modelling of fluid-saturated crack evolution in typical stress fields in the Earth leads to crack distributions of approximately parallel, approximately vertical cracks below the critical depth (Crampin & Zatsepin 1997). Such cracks have hexagonal anisotropic symmetry (transverse isotropy) with a horizontal axis of symmetry (TIH-anisotropy), or minor perturbations thereof.

APE modelling matches a very wide range of some 15 different static and dynamic phenomena associated with shear-waves, cracks, and the occurrence of earthquakes (Crampin 1999). APE shows that the dominant effect of increasing stress, in the low-level pre-fracturing stage before earthquakes or eruptions occur, is to increase the aspect-ratios of cracks perpendicular to the minimum compressional stress (Crampin & Zatsepin 1995, 1997), confirming the empirical hypotheses of Peacock *et al.* (1988). The good match of APE-modelling to observations of a variety of phenomena (Crampin 1999) is strong confirmation that the major cause of shear-wave splitting is microcracks rather than macrocracks, which would be much less compliant to small perturbations.

Previous observations of temporal changes

Changes in shear-wave polarizations, which would indicate changes in stress direction, have not been reliably observed. Changes in time-delays vary with azimuth and incidence angle within the shear-wave window. Consequently, the shear-wave window has been divided into two segments (Fig. 1) sensitive to crack aspect-ratios and crack densities, respectively. Changes in crack aspect-ratio of near-vertical parallel cracks cause changes in time-delays along ray-paths within Band-1 of the shear-wave window (Crampin 1999). Band-1 is the double-leafed solid angle of ray-paths with angles 15° to 45° either side of the average plane of the crack distributions (between 45° and 75° to the crack normal). Time-delays between split shear-waves in Band-1 are sensitive to crack aspect-ratio and hence sensitive to increasing stress, where increases of aspect-ratio increase the *average* time-delays in Band-1 (Crampin & Zatsepin 1997; Crampin 1999). Arrivals in Band-2 (ray-paths ±15° to the crack plane) are sensitive principally to crack density.

The effects of increases of stress on crack aspect-ratios can continue until the cracking is so pronounced that shear strength is lost and fractures

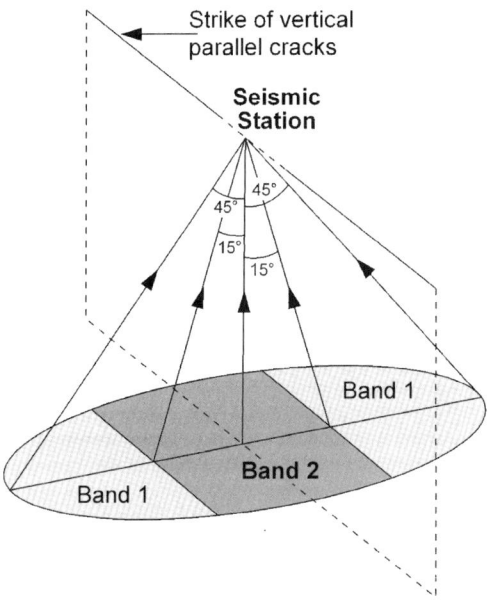

Fig. 1. Geometry of Band-1 and Band-2 in the shear-wave window in a medium with distributions of parallel vertical cracks. The circle is a horizontal slice of the vertical cone within which source events are in the effective shear-wave window of a seismic station on a horizontal free-surface, where there are typical low-velocity surface layers.

and earthquakes can occur. This happens at a level of cracking known as fracture criticality. First identified in observations by Crampin (1994), Crampin & Zatsepin (1997) show that fracture criticality is equivalent to the percolation threshold of stress-aligned fluid-saturated cracks (at about 5.5% shear-wave velocity anisotropy). At fracture criticality, the cracking is so extensive that through-going cracks necessarily exist, and there is fracturing, faulting, earthquakes, and in some cases volcanic eruptions. Following fracturing and faulting, pore fluids are released, the stress is relaxed and, without fluids to support aspect-ratios, cracks close, and the level of cracking relaxes to subcritical.

Note that the observed levels of fracture criticality, when earthquakes occur at time-delays of 11 to 14 ms km^{-1}, is higher in Iceland (Volti & Crampin 2003) than typically found elsewhere (Crampin 1999). Levels of observed fracture criticality vary strongly (Crampin 1993b) with properties of the matrix rock such as Poisson's ratio, and with properties of the pore-fluid (pressure, acoustic velocity, viscosity, etc.). The expected cause is the high heat-flow in Iceland.

Variations in the time-delays between split shear-waves in Band-1 have been observed before several earthquakes worldwide. These are: the Ms 6, North Palm Springs Earthquake of 8 July 1986, in southern California (Peacock et al. 1988; Crampin et al. 1990, 1991); an Ms 4, Parkfield earthquake (Liu et al. 1997); and before larger earthquakes in two isolated swarms (Crampin 1993a) at Enola, Arkansas (Booth et al. 1990), and Hainan Island, South China Sea (Gao et al. 1998). Time-delays (normalized by path length to ms km^{-1}) indicate aspect-ratios increasing until the (normalized) level of fracture criticality is reached when the earthquake or eruption occurs. Both rates of increase of time-delays, implying increasing aspect-ratios, and levels of fracture criticality, just before larger earthquakes occur, can be estimated from the variation in time-delays before larger earthquakes.

Continuing studies at Parkfield following Liu et al. (1997) would be possible. However, despite extensive investigations, such as the series of five papers concluding with Nadeau & McEvilly (1997), to our knowledge, no further studies of temporal variations of shear-wave splitting have been made at Parkfield.

The high seismicity, the extensive seismic network, and the rapid analysis and location procedures developed by the Icelandic Meteorological Office during the PRENLAB Project provided both the restrictive recording geometry and excellent facilities for monitoring changes in shear-wave splitting before earthquakes. This chapter reports and interprets the changes in shear-wave splitting seen before earthquakes and volcanic eruptions, and discusses their implications for monitoring pre-fracturing deformation in crustal rocks and stress-forecasting earthquakes elsewhere, away from the high seismicity of SW Iceland.

It is worth noting that one of the aims of the SIL network was to record and locate small magnitude earthquakes (Stefánsson *et al.* 1993). The earthquakes in Table 1 show that this study would not have been possible without data from small magnitude earthquakes.

Preliminary notes

(1) Observations of temporal changes in shear-wave splitting above small earthquakes due to the build-up of stress before larger earthquakes or volcanic eruptions require a restrictive source-receiver geometry with seismic activity adequately distributed, both temporally and spatially, within the shear-wave window at each station. Earthquakes are typically clustered in space and time and the need for well-distributed activity within about 8 km (the average focal depth) of the station for adequate data imposes severe constraints, even in areas of comparatively high seismicity such as SW Iceland.

(2) The shear-wave window is the vertical cone of directions bounded by $\sin^{-1}(Vs/Vp)$ which is equal to 35.26° for Poisson's ratio of 0.25, where shear-wave seismograms are not disturbed by *S*-to-*P* conversions at the free surface (Booth & Crampin 1985). However, because of ray curvature due to low-velocity near-surface layers, the effective straight-line cone of epicentral distance over focal depth can usually be extended to ± *c.* 45°. Note that because of ray curvature through the low-velocity near-surface layers, wider angle arrivals will also be in the shear-wave window (Menke *et al.* 1994). These will give consistent polarization data, which are controlled primarily by near-recorder structure, but the more complicated ray-path precludes interpretation of time-delays between split shear-waves. Since analysis of time-delays is the key to monitoring the

Table 1. *Earthquakes in figures*
(i) Earthquakes for seismograms in Figure 2

Station	Date*	Time (hr:min:s)	Lat. (°N)	Lat. (°W)	Mag. (*M*)	Depth (km)	Epi. Dist. (km)
a) KRI	1999 06 28	20:26:07.69	63.897	22.091	1.44	4.5	2.2
b) BJA	1998 11 17	15:35:58.11	63.934	21.384	1.06	5.7	4.2
c) SAU	1999 03 16	13:02:26.80	64.006	20.385	0.24	6.0	2.3
d) GRI	1998 09 05	01:00:09.64	66.635	17.932	0.55	12.2	11.0

(ii) Earthquakes for seismograms in Figure 5a

Station	Date*	Time (hr:min:s)	Lat. (°N)	Lat. (°W)	Mag. (*M*)	Depth (km)	Epi. Dist. (km)
a) BJA	1998 07 12	01:33:13.06	63.935	21.387	0.07	6.5	4.3
b) BJA	1998 07 12	08:06:31.87	63.938	21.392	0.65	6.6	4.5
c) BJA	1998 07 12	09:35:52.18	63.936	21.390	0.23	6.2	4.4
d) BJA	1998 08 23	15:18:00.05	63.962	21.321	0.72	5.4	2.0
e) BJA	1998 09 05	04:59:02.49	63.945	21.442	0.63	8.4	6.8
f) BJA	1998 09 07	01:52:48.28	63.950	21.348	0.15	6.5	2.2
g) BJA	1998 09 15	13:04:01.76	63.938	21.404	0.21	6.8	5.0
h) BJA	1998 09 30	21:24:16.82	63.940	21.373	0.13	7.3	3.5
i) BJA	1998 10 18	18:00:01.70	63.918	21.270	0.90	7.1	3.5

(iii) Earthquakes for seismograms in Figure 5b

Station	Date*	Time (hr:min:s)	Lat. (°N)	Lat. (°W)	Mag. (*M*)	Depth (km)	Epi. Dist. (km)
a) SAU	1999 03 15	05:26:17.14	63.981	20.401	0.06	5.9	1.3
b) SAU	1999 03 16	06:54:41.06	64.007	20.380	0.12	5.7	2.5
c) SAU	1999 03 16	09:06:56.82	64.005	20.384	0.22	5.8	2.3
d) SAU	1999 03 16	13:02:26.80	64.006	20.385	0.24	6.0	2.3
e) SAU	1999 03 16	15:29:42.95	64.004	20.378	0.16	5.4	2.4
f) SAU	1999 03 16	19:37:52.66	64.006	20.379	0.25	6.3	2.5
g) SAU	1999 03 17	07:26:30.38	64.006	20.386	0.37	5.9	2.3
h) SAU	1999 03 18	07:54:32.37	64.006	20.381	0.04	6.6	2.5
i) SAU	1999 03 19	09:29:54.34	64.004	20.389	0.11	5.7	2.0

*Year month day.

effects of changing stress (Volti & Crampin 2003), we restrict observations to shear-wave arrivals within the straight-line cone $\pm c. 45°$.

(3) Observations of normalized time-delays between split shear-waves, the key parameter in this chapter, show a very large scatter with the amplitude of the variations seen before earthquakes (a progressive increase ending in an abrupt decrease) less than the amplitude of the scatter. Unfortunately, schemes for statistical analysis for data including sporadic but meaningful abrupt changes in level do not yet exist, and would be difficult and time-consuming to develop. Consequently, without established techniques, we prefer to keep statistics to a minimum in order to present the data without possibly incorrect or incomplete statistical specifications.

(4) Earthquake magnitudes referred to by 'M' are magnitudes reported by the SIL seismic network in Iceland, which are approximately equivalent to the body-wave magnitude mb.

Techniques for measuring shear-wave splitting

The most important parameters characterizing shear-wave splitting are the polarization direction of the first (faster) split shear-wave (projected onto the horizontal plane), and the time-delay between the two split shear-waves. In the Earth's crust, where azimuthally aligned shear-wave splitting is typically caused by nearly vertical, nearly parallel EDA-cracks, the polarizations give some estimate of the strike of the cracks and hence the direction of maximum horizontal stress. The (normalized) time-delays give some measure of the magnitude of the stress, or rather the effects of the stress on the geometry of the crack distributions, along the specific ray-path directions.

These parameters can be measured visually. Attempts have been made to make measurements automatically by computer algorithms but these are successful only in particular conditions, which are not found in Iceland.

Visual measurements

The procedures used for visual measuring of shear-wave splitting are described by Chen et al. (1987) and Liu et al. (1997) (see also Peacock et al. 1988). Figure 2 shows the three plots (i, ii and iii), required to accurately measure polarizations and time-delays. These are typical examples of shear-wave splitting from each of the four stations KRI, BJA, SAU and GRI. Figure 2(i) shows three-component seismograms: vertical, and two horizontal components, rotated into radial and trans-

verse orientations. Within the shear-wave window, the horizontal component seismograms contain most of the shear-wave energy and most of the information (Crampin 1985). The vertical component seismograms are used to distinguish between the shear-wave motion and the possible P-to-S or S-to-P converted waves, which are polarized vertically (SV-waves), and nearly radial horizontally, respectively (Crampin 1990a). Note that care must be taken to avoid misinterpreting phases converted at local topographic irregularities, which can produce anomalous signals and anomalous measurements (Crampin 1990a). Clear and impulsive onsets, typical of shear-waves recorded within the shear-wave window in Iceland, usually make the identification of the first shear-wave arrival relatively unambiguous. Occasionally (one case in ten, say), signal-to-noise ratios are low, so that no clear shear-wave onset can be identified. In such cases, the record is discarded.

The polarization direction of the first arrival is determined from the corresponding polarization diagram in Figure 2(ii). The first motion of the shear-wave is usually sufficiently linear for the polarization direction to be identified on the horizontal 'LRTA' polarization diagram (Left, Right, Towards, and Away from the source), where the initial polarization vector is marked by an arrow. When narrowly elliptical motion is observed, the average polarization is chosen. The comparatively few cases without abrupt onsets, usually because of poor signal-to-noise ratios, are discarded. Weights or qualities (1, 2, 3) are assigned to the measurements according to their perceived reliability (1, good; 2, reasonable; 3, poor).

The time-delay is determined from the onset of the second arrival. This can be identified as an abrupt change of direction in the polarization diagrams, and the time-delay measured from the samples (ticks) in the diagram. However, the preferred technique is the *Ando rotation* (Ando et al. 1980), where horizontal seismograms are rotated into components which are parallel and perpendicular to the direction of polarization of the first (faster) shear-wave polarization (Fig. 2(iii)). This allows time-delays to be measured directly from the difference in arrival times of the two orthogonal shear-wave traces. Identification of the second arrival is not always easy as second split shear-waves may be obscured by both signal-generated coda following the P-wave and by background noise. When in any doubt, a weight 3 is assigned to ambiguous seismograms and the measurement is discarded.

Shear-wave splitting is a robust and pervasive phenomenon but the detailed waveforms are subtle and complicated. The *exact* appearance of split shear-wave arrivals depends among others on fre-

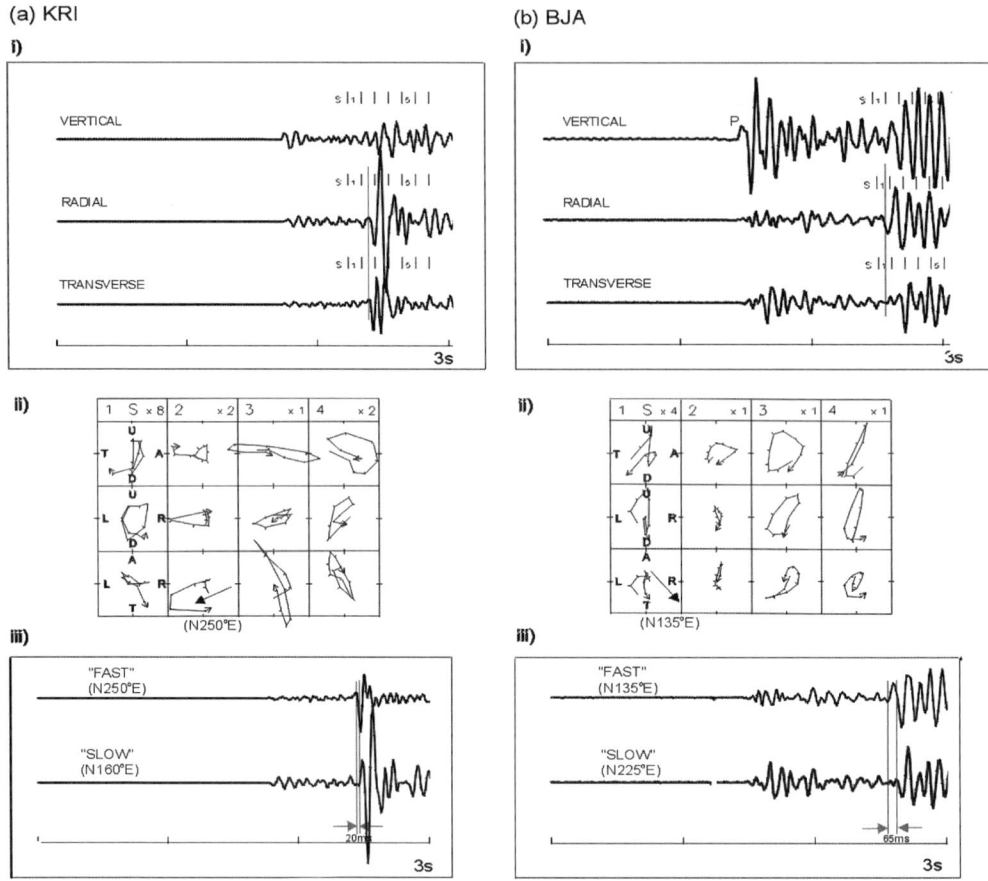

Fig. 2. Typical examples of seismograms of local earthquakes recorded within the shear-wave window for the earth-quakes listed in Table 1i. The sampling rate is 100 samples per second. At each station (i) seismograms are rotated into vertical, horizontal radial, and horizontal transverse directions, with numbered 0.1 s intervals for polarization diagrams of the shear-wave motion. (ii) Mutually orthogonal polarization diagrams, where U, D, L, R, T and A refer respectively to Up, Down, Left, Right, Towards and Away directions from the source. The number top left is the polarization interval in (i); top right is the amplitude factor (number of relative multiplications of the traces); the arrowhead marks the arrival, and the arrow marks the horizontal vector polarization, of the leading split shear-wave. (iii) Horizontal seismograms rotated into polarization directions from (ii) of the faster and slower split shear-waves, where the time-delay is the difference in arrival times in milliseconds.

quency and amplitude of the arrival, source polariz-ation, angle of emergence, azimuth of arrival, attenuation, structure along the ray-path, P-wave coda, proximity to shear-wave singularities, as well as the polarizations and time-delays of the split shear-waves themselves. As Figure 2 indicates, it is not always simple to measure shear-wave split-ting, and subjectivity is difficult to avoid.

Some 12 years ago, a test of repeatability of sub-jective visual techniques was made at the British Geological Survey. Five seismologists, some with-out familiarity with shear-wave splitting, indepen-dently measured azimuthal directions and time-delays on 20 plots of waveforms and polarization

diagrams from a variety of earthquakes, following the prescription of Chen et al. (1987). Some 80% of the readings were identical (within $\pm 10°$ of azi-muth, and within one or two samples equivalent to ± 0.02 s). The remaining 20% were ambiguous, and would have been discarded in monitoring exer-cises. The conclusion was that visual techniques did provide sufficient objectivity and sufficient repeatability.

Although the above guidelines for visual measurement confer consistency so that different observers obtain closely similar measurements, it would clearly be beneficial if automatic-reading techniques could be used. The main problem of

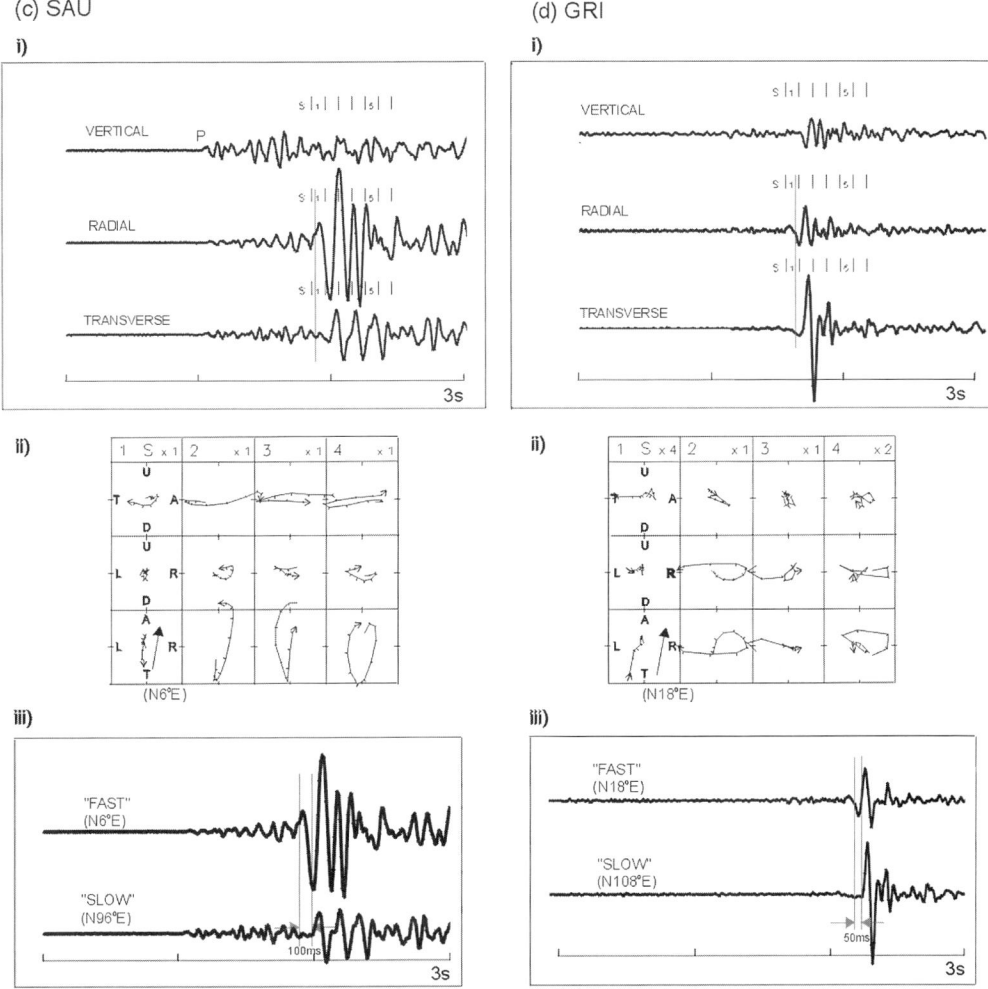

Fig. 2. Continued.

automatic techniques is that although they aim to be objective, none to date give reliable measurements for shear-wave splitting along crustal raypaths (even for the comparatively classic examples of shear-wave splitting in Iceland). In addition, when they fail for crustal shear-waves, they typically fail spectacularly and yield gross errors which are not always identified (see Crampin *et al.* 1991), whereas visual techniques are usually free of gross errors. Consequently, automatic techniques require stringent visual checking, to confirm they are a true image of the seismograms. Without reliable automatic techniques, careful visual checking is still required, and we consider it more direct to avoid having to assess the reliability of automatic techniques and only use visual techniques.

Automatic measurements

The waveforms of shear-wave splitting in the upper mantle are much simpler than those in the crust but because of longer wavelengths and lower-frequency signals, upper mantle shear-wave splitting frequently has elliptical polarizations which are difficult to measure visually. Consequently, over the years a number of automatic techniques have been developed for successfully measuring shear-waves in the mantle (e.g. Bowman & Ando 1987; Silver & Chan 1991). Splitting in the crust, however, has waveforms complicated by multiple arrivals and surface interactions and, although a number of automatic techniques have been suggested over the years (Shih *et al.* 1989; Aster *et al.* 1990, 1991), none have been wholly successful. Classic examples of shear-wave splitting with impulsive shear-wave arrivals, with good signal-to-noise ratios, and with significant time-delays are comparatively easy to measure both visually and automatically. Such classic arrivals are usually rare but are comparatively common in Iceland, which generally has good recordings of shear-wave splitting. The principal difficulty in both visual and automatic techniques is that the behaviour of shear-wave splitting in the crust varies widely from region to region, station to station, and between different ray-paths within the shear-wave window at the same station (see causes of differences in previous section). (There may also be short-term temporal instabilities, discussed below, so that observations from identical source and receivers may show varying time-delays.) The visual observer can make appropriate adjustments but, to our knowledge, no automatic technique has sufficient flexibility to have wide application.

It is comparatively straightforward to design automatic techniques that will reliably read polarizations and time-delays of classic shear-wave splitting at a particular station with a particular source-receiver geometry. The temptation is to apply these techniques outside their region of competence, when they may be seriously inadequate.

To our knowledge, the most successful automatic measurements have been made by the cross-correlation function (CCF) technique of Gao when applied to an isolated swarm of small earthquakes on Hainan Island, Southern China (Gao *et al.* 1998; this technique is similar to that of Silver & Chan 1991). The CCF algorithm of Gao *et al.* calculates the cross-correlation function between two rotated horizontal shear-wave components. For each angle of rotation a cross-correlation coefficient (CCC) is evaluated, with values between 0 and 1. The polarizations of faster shear-wave and the time-delays between fast and slow waves can be estimated at the largest values of the CCC. The technique obtained reliable variations of time-delays with a much reduced scatter at two stations within the shear-wave window which showed the changes in behaviour expected in Band-1 before a M_L 3.7 earthquake (Gao *et al.* 1998).

Since it produced the most consistent results yet obtained with automatic techniques for shear-waves with ray-paths in the crust, we applied the CCF technique of Gao *et al.* (1998) to a subset of the Icelandic data at Station BJA. Figure 3 shows time-delays processed visually contrasted with those processed with the automatic CCF. Data are from 1 January to 10 November 1998 during which time there were two earthquakes with *M* of *c.* 5 near the station. The CCF technique was applied to all earthquakes in this interval that had good *visual*

estimates of shear-wave splitting. Only readings with values of CCC larger than (an arbitrary) 0.6 were retained. Only 21 of the 37 visually processed events had CCC greater than 0.6 (including two with zero time-delay erroneously indicating no splitting). Such automatic techniques are, not surprisingly, much less flexible and less adaptable than visual techniques. The time and magnitude of the November 1998 event were successfully stress-forecast from visual data (Volti & Crampin 2003; Crampin *et al.* 1999*a*). This forecast was made before the foreshock and aftershock data had been processed and made available over the Internet. Stress-forecasting would not have been possible with the omission of so many events by the automatic technique.

The success of the CCF technique applied to the Hainan Island data (Gao *et al.* 1998) in contrast to the much less successful application to Icelandic data in Figure 2, is almost certainly because the Hainan Island data were from an *isolated* swarm (Crampin 1993*a*) with a small hypocentral diameter. This meant that there was only a limited range of ray-paths to each of the two stations within the shear-wave window, so that the CCF technique was repeatedly accessing similar waveforms. This is close to the ideal time-lapse configuration where comparatively similar source and receiver geometry (Crampin 1990*b*) produce comparatively consistent time-delays.

Clearly automatic techniques for estimating shear-wave splitting parameters from earthquake arrivals in the crust are desirable but difficult to design. Comparatively simple deterministic techniques are unlikely to be sufficient, except in exceptional circumstances, as in the isolated swarms on Hainan Island (Gao *et al.* 1998). Combinations of visual and automated techniques, such as those used by Menke *et al.* (1994), are still largely subjective. Widely applicable wholly automated programs would require considerable sophistication, and wholly successful techniques have not yet been developed. As a consequence, shear-wave splitting in Iceland was measured using the visual techniques of the previous section.

Probably the most effective automated analysis would use artificial neural network (ANN) technology. Such techniques have been applied to exploration data (Dai & MacBeth 1994) with some success. However, the difficulty of selecting appropriate ANN training sets would still make application to shear-waves from earthquake signals complicated and would preserve whatever subjectivity was in the training sets.

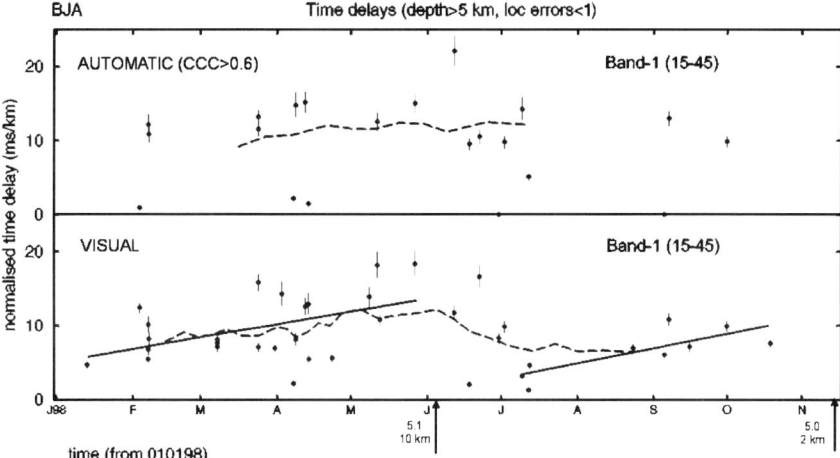

Fig. 3. A comparison of visual and automatic techniques for measuring time-delays. Time-delays, normalized by path length, from 1 January to 13 November 1998 in Band-1 of the shear-wave window at station BJA. Time-delays are obtained by, at top, the automatic CCF technique for cross-correlation coefficients (CCC) greater than 0.6 and, at bottom, the visual techniques outlined in the text. Vertical lines through the normalized time-delay points are error bars (observed time-delays in milleseconds divided by the ends of the range of location errors). Dashed lines are nine-point moving averages. Straight lines through the visual measurements are least-square estimates, beginning just before a minimum of nine-point average and ending at a large earthquake. Arrows with magnitudes and epicentral distances from BJA indicated mark the times of *M*5 earthquakes.

Shear-wave splitting in Iceland (1996–1999)

The persistent seismicity of much of Iceland, and the seismic network and processing system developed for the SIL seismic network in Iceland, particularly in SW Iceland, satisfy the restrictive geometrical conditions required for monitoring shear-wave splitting before earthquakes (Crampin 1991*a*). The northern leg of the transform zone in SW Iceland, the Tjörnes fracture zone, is mostly offshore but is also of interest, as a number of destructive earthquakes (*Ms* > 6.5) have occurred there in the past. Most of the seismic stations are located in the vicinity of these two transform zones, with the major activity in SW Iceland. The map in Figure 4 shows the SIL network during 1996 to 1999 with rose diagrams of shear-wave polarizations, read by visual techniques, superimposed on roundels of equal-area polar maps of the polarizations of the faster split shear-waves.

Many of the rose diagrams in Figure 4, particularly BJA, HAU, SAN, SOL and SAU, are exceptionally linear, reflecting the classic examples of shear-wave splitting common in Iceland (Fig. 2). Other rose diagrams show some scatter (discussed in the next sections) but there is a predominantly NE–SW alignment in SW Iceland and a NNE–SSW alignment in north-central Iceland, indicating marginally different directions of the dominant compressional stress in southern and northern

transform zones of the Mid-Atlantic Ridge in Iceland. The NE–SW alignment in SW Iceland was previously identified in SW Iceland from observations of shear-wave splitting by Menke *et al.* (1994), who showed that, as is typically the case, the polarizations of the faster shear-waves were parallel to numerous dykes, fissures, tensional cracks, and micro-earthquake focal mechanisms.

Note that the polarization directions of stress-aligned shear-wave splitting are exceptionally sensitive to the shear-wave window (Booth & Crampin 1985). Generally the stations showing largest scatter of polarizations in Figure 4 are those with severe local surface or subsurface topography where the azimuth and angle of incidence of arrivals distort the effective shear-wave window.

Examples of shear-wave splitting

In order to demonstrate some of the range of variations in shear-wave splitting time-delays, we plot in Figure 5 horizontal seismograms rotated into the preferred fast and slow polarization directions for two suites of earthquakes. We select two sequences of typical seismograms illustrating particular phenomena.

Figure 5a shows the sequence of *increasing (normalized) time-delays* from 12 July 1998 to 18 October 1998 in Band-1 of the shear-wave window at Station BJA. These are plotted in Figure 1*a* of Volti & Crampin (2003), and were the sequence

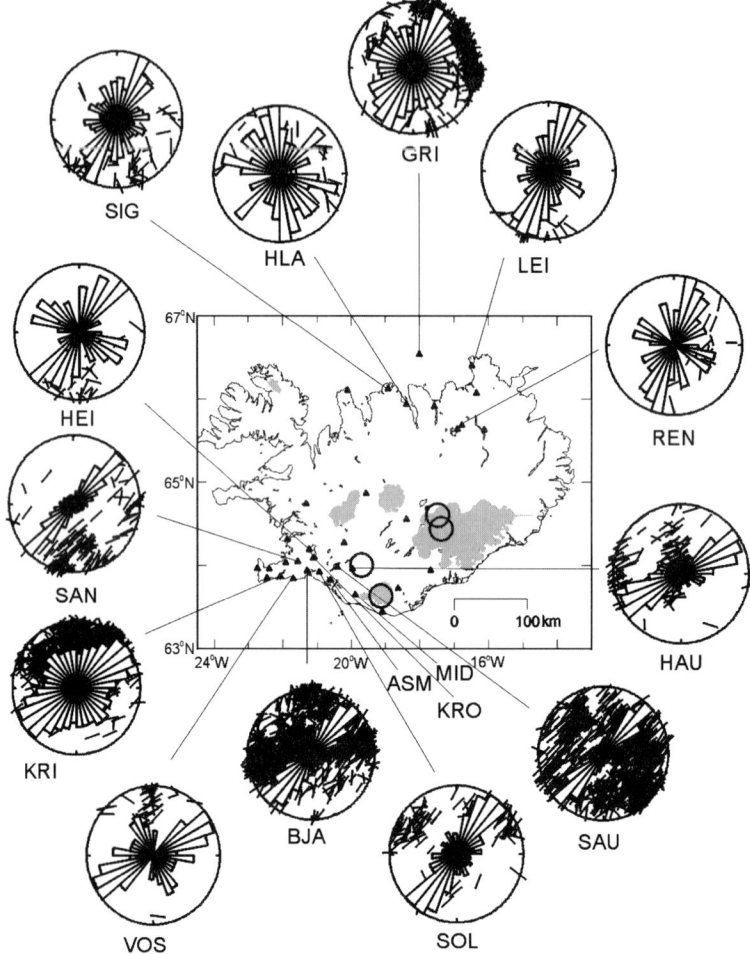

Fig. 4. Map of Iceland, showing the seismic network and analyses of shear-wave splitting. Triangles are SIL seismic stations, several of which were installed during the four years 1996 to 1999. Shaded areas are ice fields of which the largest is Vatnajökull. Open circles mark the active volcanoes of, from north to south, Bárðarbunga, Grímsvötn, Hekla and Katla. The roundels are equal-area polar plots of polarizations of the faster split shear-wave arrivals in the shear-wave window (out to 45°) during the four years, containing rose diagrams (with 10° petals) of these polarizations. Roundels are shown only for those named stations where there are more than ten arrivals within the shear-wave window. The named stations without roundels (ASM, KRO, MID) have sufficient arrivals but the waveforms and polarizations are severely disturbed by local rifting and/or local surface or subsurface topography.

which allowed the $M5$ earthquake to be successfully stress-forecast (Crampin *et al.* 1999*a*). The time-delays for this increase show less scatter about the mean value than is typical for other increases.

Vertical 'stripes' of time-delays, where a wide range of time-delays occurs in a very short space of time, are common in plots of time-delay variations (see Fig. 1 of Volti & Crampin 2003). They are particularly common in aftershock sequences. Typically such arrivals are from earthquakes in tight source volumes of limited dimensions. Figure

5b shows time-delays from such a sequence of earthquakes from 15 to 17 March 1999, observed at Station SAU; these are plotted in Band-2 of the shear-wave window in Figure 1b of Volti & Crampin (2003). Only the first earthquake in the stripe, (a) in Table 1iii, is from outside a 1 km diameter source volume.

The time-delays in Figure 5b (both non-normalized in seismograms and normalized in equal-area plots) show an exceptionally large scatter from 2 to 16 ms km^{-1} ($\pm 80\%$). It is difficult to explain this by any conventional geophysical mechanism

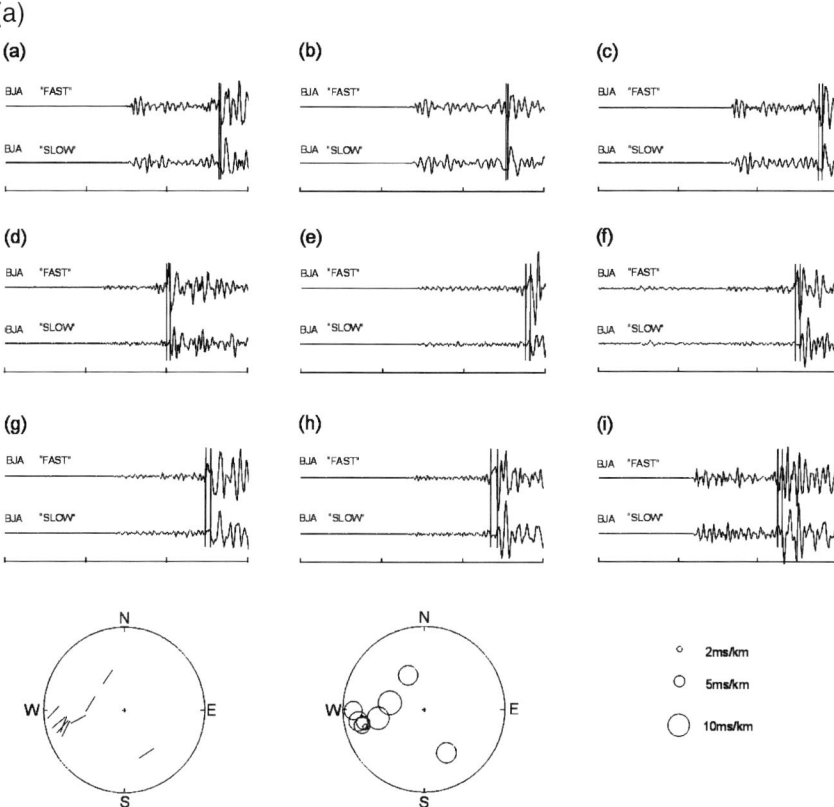

Fig. 5. Examples of shear-wave splitting in sequences of seismograms. Upper diagrams are horizontal seismograms rotated into fast and slow polarization directions with arrivals marked with vertical bar. Roundels are equal-area polar plots out to incident angles of 45°: (left) bars of fast shear-wave polarizations; and (right) circles of shear-wave time-delays with diameters proportional to delay. (**a**) Data from nine earthquakes listed in Table 1ii showing increasing time-delays. (**b**) Data from nine earthquakes listed in Table 1iii showing variation in normalized time-delays for earthquakes in a 'stripe' of time-delays, eight of which have nearly identical foci.

(see section on scatter, below). Note that some seismograms from the localized source volume in Figure 5b show strong similarities but there are no exactly similar seismograms. In general, pairs of similar seismograms are rare in seismograms from Iceland.

The scatter of polarizations

Shear-waves are very sensitive to surface interactions, and stations that show the most irregular scatter in polarizations in Figure 4 are typically locations on steep slopes and/or close to irregular surface or subsurface topography (Crampin 1990a; Liu *et al.* 1997). Some sources of scatter in polarizations can be identified: GRI is on a steep-sided island; KRO is in the onshore rift of the Mid-Atlantic Ridge; and KRI has an irregular cliff-face less than a wavelength from the station. Small earthquakes are typically the result of some stress or tectonic anomaly which is likely to disturb measurements of shear-wave splitting, so the remarkable linearity of polarizations implying comparatively uniform microcrack- and stress-alignments at many of the stations is surprising.

The polarizations in Figure 4 are from arrivals within the *straight-line* shear-wave window defined by tan^{-1}(epicentral distance/focal depth) = 45°. The 45° limit is to allow for ray curvature due to low-velocity near-surface layers bending arrivals within the theoretical 35.26° shear-wave window for a Poisson's ratio of 0.25. The occasional orthogonal polarizations are expected when the source mechanism radiates shear-waves in the direction of the station which happen to be polarized parallel to the slower split shear-wave so the faster wave is not excited (Crampin *et al.* 1986). Rose diagrams of shear-wave polarizations else-

(b)

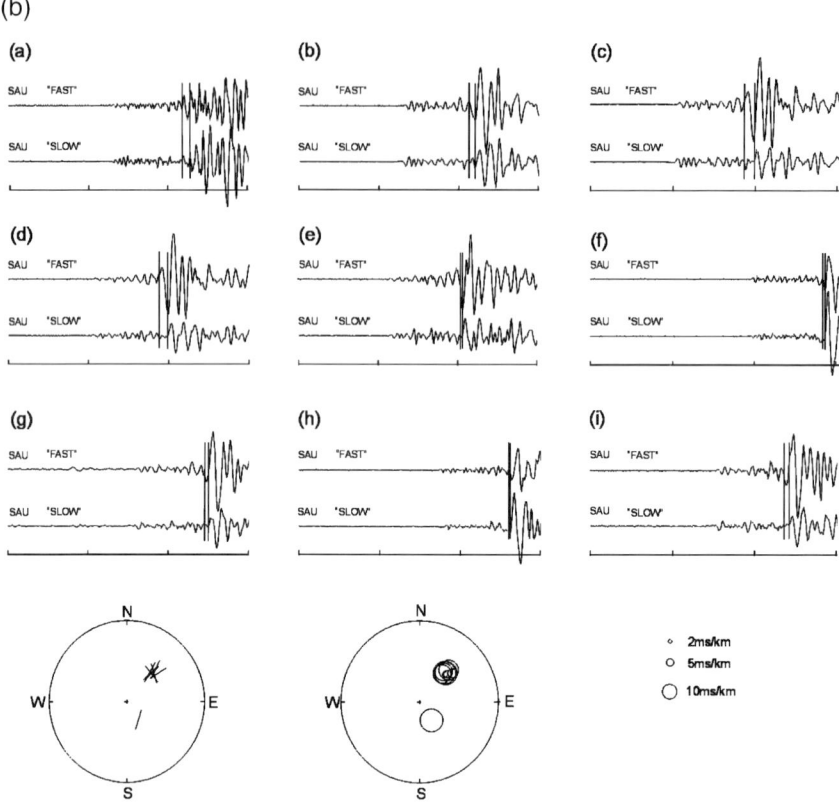

Fig. 5. Continued.

where typically display a scatter of at least ±20° regardless of whether the shear-waves are read automatically or visually.

Note that no corrections are made for the free-surface interactions. All observations are within the effective shear-wave window, where the effects are minimal, and in any case free-surface interactions would only marginally affect the rose diagrams in the roundels. Free-surface corrections do not affect time-delays.

Note that Menke *et al.* (1994), who previously examined shear-wave splitting in SW Iceland, generally found a larger scatter of polarizations at the stations common to Figure 4 – HEI and SOL – and especially BJA, HAU and SAU which are particularly linear in Figure 4. This is probably because Menke *et al.* used arrivals from earthquakes with epicentral distances up to 60 km from stations, which the assumed crustal structure suggested would still arrive within the shear-wave window. The scatter in Menke *et al.* (1994) suggests that the arrivals are either outside the effective shear-wave window, or that complications along the long ray-paths disturbed the wave trains. As such long ray-paths would greatly complicate time-delays between the split shear-waves, which we interpret as indicating stress changes in Volti & Crampin (2003), we restrict measurements to arrivals within the straight-line shear-wave windows of Figure 4.

The scatter of time-delays

Since one of the most puzzling aspects of shear-wave splitting is the scatter of time-delays, we discuss this scatter in some detail. Measured time-delays in figure 1 of Volti & Crampin (2003), show a large scatter about the nine-point moving averages and about the least-squares lines. The degree of scatter, sometimes as large as ±80%, is similar to that observed whenever time-delays above earthquakes are measured (Crampin 1999). Thus a large scatter appears to be an inherent feature of measurements of time-delays above small earthquakes. The only exceptions, where measurements of shear-wave splitting show considerably less scatter, are observations above isolated swarms of small earthquakes as in Arkansas (Booth *et al.* 1990) and Hainan Island (Gao *et al.* 1998), where the shear-waves travel along a restricted range of ray-paths from very tight clusters of foci.

Possible sources of scatter in a conventional (non-critical) crust

There are at least five possible sources of scatter of time-delays in a conventional (non-critical) crust where the usual deterministic geophysics applies.

(1) *Possible source*: Variations in time-delays inherent to geometry of time-delays along ray-paths in Band-1 assuming distributions of nearly vertical, nearly parallel cracks. Line-singularities (Crampin & Yedlin 1981) cross Band-1 directions in shear-wave windows in parallel vertical cracks. Consequently, time-delays in the solid angle of directions in Band-1 may be expected to vary from zero to nearly the maximum value (e.g. Crampin *et al.* 1990; Crampin 1993*b*). This means that time-delays along any arbitrary choice of directions in Band-1 will naturally have a wide variation. (We have found that attempting to correct for variations in time-delays for different directions within Band-1, assuming parallel vertical cracks, does not significantly reduce the scatter. Since this makes estimated values dependent on structure and earthquake locations without improving scatter, we have not made such corrections.)

Arguments against: In contrast, under the same assumptions, the theoretical time-delays in Band-2 should vary very little. Since in general, the scatter of time-delays in Band-2 is observed to be very similar to those in Band-1 (figure 1 of Volti & Crampin 2003), a similar source of scatter may be expected. Consequently, although variations in Band-1 are expected, the wide variation of theoretical values is unlikely to be the major source of the scatter. The ±80% range is in Band-2 directions where there are no shear-wave singularities.

(2) Possible errors in location of earthquake foci provide two sources of scatter.

(2a) *Possible source*: Ray-path directions in Band-1 of the shear-wave window through distributions of nearly vertical, nearly parallel cracks are expected to include directions of shear-wave line-singularities where shear-wave splitting is irregular and time-delays may be zero. This means that time-delays in Band-1 may be expected to vary widely for small changes in ray-path directions (Crampin 1991*b*; Crampin *et al.* 1990).

Arguments against: Although variations within the shear-wave window can certainly cause scatter, there are several reasons why this is thought not to be the major source of scattering.

(i) Time-delays in Band-2 are not expected to include singularities from geometrical considerations, and yet the scatter in Band-2 as is similar to that in Band-1, where there is a line-singularity.

(ii) All four stations, but particularly KRI and GRI, include local activity where clustering is common: many events within the shear-wave window within a short period of time. Such repeated seismicity is likely to have similar foci, similar ray-paths, and consequently similar time-delays. However, these 'stripes' of time-delays (Fig. 5b, and Fig. 1 of Volti & Crampin 2003) typically show a scatter as large as, and in some cases larger than, at other times for the particular station (Fig. 5b).

(iii) Böðvarsson relocated a selection of events within the shear-wave window with a multi-event relative-location technique (Slunga *et al.* 1995). This improved locations, as was shown by increased spatial clustering of foci, more planar locations, and reduction in location errors, but the scatter in time-delays was not significantly reduced.

(2b) *Possible source*: Time-delays are normalized by hypocentre path length to milliseconds per kilometre so that time-delays along different paths can be compared, and if the measured path lengths are in error, the normalized time-delays will also be in error.

Arguments against: Possible errors in location are usually much too small (see error bars in time-delay plots in Fig. 3 and in Fig. 1 of Volti & Crampin (2003), and the tight focal zone in Fig. 5b) to cause the ±80% scatter in time-delays.

(3) *Possible source*: Errors in picking and measuring time-delays.

Arguments against: Most measurements are checked by more than one analyst and any shear-wave measurements that appear in any way doubtful are rejected. Consequently, it is unlikely that errors large enough to cause the ±80% scatter are common.

(4) *Possible source*: Complicated rock structure beneath the station causing variations in velocity structure, so that small differences in source location result in different ray-paths with different degrees of anisotropy.

Arguments against: Although clearly a possible source of scatter, it would not be expected to cause such significant scattering on almost all occasions at all stations, particularly along ray-paths from the comparatively tight clusters of foci in the stripes of KRI and GRI in figure 1 of Volti & Crampin (2003) and at SAU in Figure 5b.

(5) *Possible source*: Complicated stress-aligned EDA-crack structure beneath the station.

Arguments against: There are now a large number of measurements of shear-wave splitting in controlled source exploration experiments in mostly sedimentary rocks (Crampin 1994, 1996; Winterstein 1996). Controlled-source experiments typically allow detailed high-precision measurements, particularly in VSP experiments (Li & Crampin 1991*a*, *b*; Yardley & Crampin 1993). These measurements typically show regular well-ordered splitting with no sign of the ±80% scatter in time-delays above small earthquakes. It should be noted that exploration experiments are typically in sedimentary strata, whereas ray-paths above small earthquakes are usually, but not always, in igneous and metamorphic rocks. However, in view of the observed similarities in shear-wave splitting in all types of rock (Crampin 1994, 1996), it seems unlikely that the scattering can be wholly attributed to igneous and metamorphic rocks.

The five possible sources of the scatter, listed above, have each been rejected as the primary source. It is possible that each source contributes so that the total scatter is a combination of the effects of several sources of scatter, but in view of the consistency of exploration experiments in Item 5, above, this does not seem likely. Remarkably, the variations before earthquakes and volcanic disturbances appear reasonably well established despite this large scatter. This suggests that the scatter is in some sense a random fluctuation about the nine-point moving average and is independent of the stress-induced mean-value modifications that we report in Volti & Crampin (2003). This is confirmed by the seismograms in Figure 5b where a scatter of ±80% of time-delays is observed along similar ray-paths in a short period of time when there is no known major change of conditions.

Possible source of scatter in a critical crust

The anisotropic poro-elasticity (APE) model, which successfully matches so much of the behaviour of shear-waves and cracks in the crust (Crampin 1999, 2000), is a mean field theory (in the sense of Jensen 1998). APE assumes that the cracks in the crust are a critical system held close to (fracture) criticality (Crampin 1998; Crampin & Chastin 2001). Mean field theories average over the clustering of heterogeneities at all scale lengths and all dimensions inherent in the self-similar, fractal, or power-law distributions of many phenomena associated with cracks and earthquakes (Crampin 1999), including the well-known Gutenberg-Richter relationship between numbers and magnitudes of earthquakes.

APE-modelling suggests, and both field (Crampin 1999) and laboratory observations (Crampin *et al.* 1999*b*) confirm, that stress-aligned fluid-saturated cracks are sensitive to even minor modifications of stress, pressure and other conditions. The driving mechanism for deformation is fluid migration by flow or diffusion along pressure gradients between neighbouring grain-boundary cracks and low aspect-ratio pores at different orientations to the stress-field (Zatsepin & Crampin 1997; Crampin & Zatsepin 1997; Crampin 1999). These distributions necessarily contain clusters of cracks with different numbers of cracks, different aspect ratios, and different crack densities. This means that although the averages may be stable (witness the wide range of phenomena, where APE-modelling appears to be a good, or at least satisfactory, model of observations; Crampin 1999), any individual observation may deviate substantially from the mean.

Note that the very wide scatter often occurs in comparatively short periods of time, in foreshock and aftershock sequences and short-lived series of small earthquakes, at all stations (Fig. 5b, and Fig. 1 of Volti & Crampin 2003). This suggests that the mechanism that drives the phenomena responds very quickly to changes, with time constants of perhaps at most a few hours. The time constants cannot be too small, much less than an hour, say, or the effects would disturb the smooth variations in measurements of shear-wave splitting in reflection profiles and VSPs as in Li & Crampin (1991*a*, *b*) and Yardley & Crampin (1993), and would have been recognized previously.

Thus, although in principle some scatter is expected in a critical crust, a better understanding of the mechanism that modifies the distributions is required. One of the first aims of the proposed controlled-source stress-monitoring site, discussed in Volti & Crampin (2003), will be to examine the stability of shear-wave splitting and investigate the cause of the scatter.

We can speculate that the driving mechanism for this scattering is the stress fluctuations introduced by Earth and ocean tides. Tatham *et al.* (1993) found approximately 0.5% variations in both horizontally and vertically propagating *P*-wave velocities correlating with in particular the diurnal Earth tides in transmission and reflection experiments in Illinois, USA. Tatham *et al.* attributed these variations to 'the opening and closing of cracks, pores and/or pore throats'. Since the sensitivity of shear-wave splitting to comparatively minor modification of crack distributions is much greater than that of *P*-waves, we can expect these effects to cause significant variations in time-delays in shear-wave splitting. The sensitivity of fluids in the crust to Earth tides is well-known; for example, Gupta *et al.* (2000) observe tidal signals in water-level fluctuations in wells around the Koyna Dam in India; Kümpel (1997) reviews water-level changes in wells, and reports distinctive fluctuations of water level correlating both with diurnal tides and with barometric pressures. Stresses from such phenomena produce changes in water levels by modifying the aspect-ratios of water-filled cracks in the surrounding rock, so that water is squeezed in and out of the rock mass. This is exactly the process that is monitored by shear-wave splitting and is expected to be one of the causes of the scatter in shear-wave time-delays.

Conclusions

Shear-wave splitting in Iceland, particularly above the persistent seismicity of the onshore transform zone of the Mid-Atlantic Ridge in SW Iceland, has been monitored for four years. The seismicity and seismic network of Iceland seem exceptionally favourable for measuring shear-wave splitting having high seismicity within the shear-wave windows of a number of SIL stations, and good recording and location techniques (particularly of small earthquakes), as well as seismic records being available on the Internet. All shear-wave arrivals within the shear-wave window, with very few exceptions, display shear-wave splitting. The shear-wave splitting is consistent with propagation through distributions of nearly vertical, nearly parallel fluid-saturated cracks striking approximately parallel to the direction of maximum horizontal stress. The shear-wave polarizations, and inferred directions of maximum compressional stress, in the southern half of Iceland average about NE–SW, whereas the polarizations and stress directions in northern Iceland average about NNE–SSW.

The suggested importance of these studies of microcrack deformation monitored by shear-wave splitting is that in principle any modification in crack geometry can be directly interpreted in terms of changes in direction and magnitude of stress. We suggest this opens a new window for examining and interpreting stress in *in situ* rock.

As suggested in the Forewarnings section at the beginning of this chapter, one of the major problems raised by this chapter and Part 2 (Volti & Crampin 2003), and by almost all measurements of shear-wave splitting elsewhere (reviewed by Crampin 1999), is the cause of the large scatter in measured time-delays between split shear-waves. They appear to be too large to be caused by conventional (non-critical) geophysics. The dense distributions of cracks suggested by the levels of recorded shear-wave splitting strongly suggest that cracks are a critical system verging on (fracture) criticality. This suggests that the large scatter is due to wave propagation through the heterogeneities and clustering in time and space inherent in critical systems (Bruce & Wallace 1989) of cracks on the verge of criticality. One of the major aims of the controlled-source stress-monitoring site currently being developed in northern Iceland will be to investigate the source of the scatter.

This work was partially supported by European Commission PRENLAB Projects, Contracts ENV4-CT96–0252 and ENV4-CT97–0536, and SMSITES Project, Contract EVR1-CT1999–40002. We thank Ragnar Stefánsson, of the Iceland Meteorological Office, without whose collaboration this investigation would not have been possible. We also thank: Reynir Böðvarsson, of Uppsala University, who recalculated earthquake locations; Yuan Gao, of the Chinese Seismological Bureau, for the use of his CCF technique for automatic measurement of shear-wave splitting; and many colleagues at home and abroad, too numerous to mention individually, who have each contributed to the arguments in one way or another.

References

ANDO, M., ISHIKAWA, Y. & WADA, H. 1980. *S*-wave anisotropy in the upper mantle under a volcanic area in Japan. *Nature,* **286,** 43–46.

ASTER, R. C., SHEARER, P. M. & BERGER, J. 1990. Quantitative measurements of shear wave polarizations at the Anza seismic network, southern California: Implications for shear wave splitting and earthquake prediction. *Journal of Geophysical Research,* **95,** 12449–12473.

ASTER, R. C., SHEARER, P. M. & BERGER, J. 1991. Reply to CRAMPIN, S., BOOTH, D. C., EVANS, R., PEACOCK, S. & J. B. FLETCHER. *Journal of Geophysical Research,* **96,** 6415–6419.

BÖÐVARSSON, R., RÖGNVALDSSON, S. TH., SLUNGA, R., KJARTANSSON, E. 1999. The SIL data acquisition system – at present and beyond year 2000. *Physics of the Earth and Planetary Interiors,* **113,** 89–101.

BOOTH, D. C. & CRAMPIN, S. 1985. Shear-wave polarizations on a curved wavefront at an isotropic free-surface. *Geophysical Journal of the Royal Astronomical Society,* **83,** 31–45.

BOOTH, D. C., CRAMPIN, S., LOVELL, J. H. & CHIU, J.-M. 1990. Temporal changes in shear wave splitting during an earthquake swarm in Arkansas. *Journal of Geophysical Research,* **95,** 11151–11164.

BOWMAN, J. R. & ANDO, M. 1987. Shear-wave splitting in the upper-mantle wedge above the Tonga subduction zone. *Geophysical Journal of the Royal Astronomical Society,* **88**, 25–41.

BRODIE, K. H. & RUTTER, E. H. 1985. On the relationship between deformation and metamorphism, with special reference to the behaviour of basic rock. *In:* THOMPSON, A. B. & RUBIE, D. C. (eds) *Metamorphic Reactions: Kinetics, Textures, and Deformation.* Springer, New York, Advances in Physical Chemistry, **4**, 138–179.

BRUCE, A. & WALLACE, D. 1989. Critical point phenomena: universal physics at large length scale. *In:* DAVIS, P. (ed.) *The New Physics,* Cambridge University Press, 236–267.

CHEN, T.-C., BOOTH, D. C. & CRAMPIN, S. 1987. Shear-wave polarizations near the North Anatolian Fault – III. Observations of temporal changes. *Geophysical Journal of the Royal Astronomical Society,* **91**, 287–311.

CRAMPIN, S. 1981. A review of wave motion in anisotropic and cracked elastic-media. *Wave Motion,* **3**, 343–391.

CRAMPIN, S. 1985. Evaluation of anisotropy by shear-wave splitting. *Geophysics,* **50**, 142–152.

CRAMPIN, S. 1990*a.* The scattering of shear waves in the crust. *Pure & Applied Geophysics,* **132**, 67–91.

CRAMPIN, S. 1990*b.* Alignment of near-surface inclusions and appropriate crack geometries for geothermal hot-dry-rock experiments. *Geophysical Prospecting,* **38**, 621–631.

CRAMPIN, S. 1991*a.* An alternative scenario for earthquake prediction experiments. *Geophysical Journal International,* **107**, 185–189.

CRAMPIN, S. 1991*b.* Effects of point singularities on shear-wave propagation in sedimentary basins. *Geophysical Journal International,* **107**, 531–543.

CRAMPIN, S. 1993*a.* Do you know of an isolated swarm of small earthquakes? *EOS, Transactions, American Geophysical Union,* **74**, 451 and 460.

CRAMPIN, S. 1993*b.* A review of the effects of crack geometry on wave propagation through aligned cracks. *Canadian Journal of Exploration Geophysicists,* **29**, 3–17.

CRAMPIN, S. 1994. The fracture criticality of crustal rocks. *Geophysical Journal International,* **118**, 428–438.

CRAMPIN, S. 1996. *Anisotropists Digest 149, 150.* anisotropists@sep.stanford.edu

CRAMPIN, S. 1998. Shear-wave splitting in a critical crust: the next step. *In:* RASOLOFOSAON, P. N. J. (ed.) *Proceedings of Eighth International Workshop on Seismic Anisotropy, Boussens, 1998, Revue de l'Institut Français du Pétrole,* **53**, 749–763.

CRAMPIN, S. 1999. Calculable fluid–rock interactions. *Journal Geological Society,* **156**, 501–514.

CRAMPIN, S. 2000. The potential of shear-wave splitting in a stress-sensitive compliant crust: a new understanding of pre-fracturing deformation from time-lapse studies. *70th Annual International Meeting of the Society of Exploration Geophysicists, Calgary, Expanded Abstracts,* **2**, 1520–1523.

CRAMPIN, S. & CHASTIN, S. 2001. Shear-wave splitting in a critical crust: II – compliant, calculable, controllable fluid–rock interactions. *In:* IKELLE, L. T. GANGI, T. (eds) *Anisotropy 2000: Fractures converted waves and case studies, Proceedings of the Ninth International Workshop on Seismic Anisotropy, Camp Allen, 2000.*

Society of Exploration Geophysicists, Open File Publication, **6**, 21–48.

CRAMPIN, S. & YEDLIN, M. 1981. Shear-wave singularities of wave propagation in anisotropic media. *Journal of Geophysics,* **49**, 43–46.

CRAMPIN, S. & ZATSEPIN, S. V. 1995. Production seismology: the use of shear waves to monitor and model production in a poro-reactive and interactive reservoir. *65th Annual International Meeting of the Society of Exploration Geophysicists, Houston, Expanded Abstracts,* 199–202.

CRAMPIN, S. & ZATSEPIN, S. V. 1997. Modelling the compliance of crustal rock, II – response to temporal changes before earthquakes. *Geophysical Journal International,* **129**, 495–506.

CRAMPIN, S., EVANS, R., ÜÇER, B., DOYLE, M., DAVIS, J. P., YEGORKINA, G. V. & MILLER, A. 1980. Observations of dilatancy-induced polarization anomalies and earthquake prediction. *Nature,* **286**, 874–877.

CRAMPIN, S., EVANS, R. & ATKINSON, B. K. 1984. Earthquake prediction: a new physical basis. *Geophysical Journal of the Royal Astronomical Society,* **76**, 147–156.

CRAMPIN, S., EVANS, R. & ÜÇER, S. B. 1985. Analysis of records of local earthquakes: the Turkish Dilatancy Projects (TDP1 and TDP2). *Geophysical Journal of the Royal Astronomical Society,* **83**, 1–16.

CRAMPIN, S., BOOTH, D. C., KRASNOVA, M. A., CHESNOKOV, E. M., MAXIMOV, A. B. & TARASOV, N. T. 1986. Shear-wave polarizations in the Peter the First Range indicating crack-induced anisotropy in a thrust-fault regime. *Geophysical Journal of the Royal Astronomical Society,* **84**, 401–412.

CRAMPIN, S., BOOTH, D. C., EVANS, R., PEACOCK, S. & FLETCHER, J. B. 1990. Changes in shear wave splitting at Anza near the time of the North Palm Springs Earthquake. *Journal of Geophysical Research,* **95**, 11 197–11 212.

CRAMPIN, S., BOOTH, D. C., EVANS, R., PEACOCK, S. & FLETCHER, J. B. 1991. Comment on "Quantitative measurements of shear wave polarizations at the Anza Seismic Network, Southern California: Implications for shear wave splitting and earthquake prediction" by ASTER, R. C., SHEARER, P. M., BERGER, J., *Journal of Geophysical Research,* **96**, 6403–6414.

CRAMPIN, S., VOLTI, T. & STEFÁNSSON, R. 1999*a.* A successfully stress-forecast earthquake. *Geophysical Journal International,* **138**, F1–F5.

CRAMPIN, S., ZATSEPIN, S. V., ROWLANDS, H. J., SMART, B. J. & SOMERVILLE, J. McL. 1999*b.* APE-modelling of fluid/rock deformation of sandstone cores in laboratory stress-cells. *61th Conference, European Association of Geoscientists & Engineers, Helsinki, Extended Abstracts, 1,* 2–08.

DAI, H. & MACBETH, C. 1994. Split shear-wave analysis using an artificial neural network? *First Break,* **12**, 605–613.

GAO, Y., WANG, P. ZHENG, S., WANG, M. & CHEN, Y.-T. 1998. Temporal changes in shear-wave splitting at an isolated swarm of small earthquakes in 1992 near Dongfang, Hainan Island, Southern China. *Geophysical Journal International,* **135**, 102–112.

GUPTA, H. K., RADHAKRISHNA, I., CHADHA, R. K., KÜMPEL, H.-J. & GRECKSCH, G. 2000. Pore pressure studies initiated in area of reservoir-induced earthquakes in

India. *EOS, Transactions, American Geophysical Union*, **81**, 14, 145, 151.

JENSEN, H. J. 1998. *Self-organized Criticality*. Cambridge University Press.

KÜMPEL, H.-J. 1997. Tides in water saturated rock. *Lecture Notes in Earth Sciences*, **66**, 277–291.

LI, X.-Y. & CRAMPIN, S. 1991a. Complex component analysis of shear-wave splitting: theory. *Geophysical Journal International*, **107**, 597–604.

LI, X.-Y. & CRAMPIN, S. 1991b. Complex component analysis of shear-wave splitting: case studies. *Geophysical Journal International*, **107**, 605–613.

LIU, Y., CRAMPIN, S. & MAIN, I. 1997. Shear-wave anisotropy: spatial and temporal variations in time delays at Parkfield, Central California. *Geophysical Journal International*, **130**, 771–785.

MENKE, W., BRANDSDÓTTIR, B., JAKOBSDÓTTIR, S. & STEFÁNSSON, R. 1994. Seismic anisotropy in the crust at the mid-Atlantic plate boundary in south-west Iceland. *Geophysical Journal International*, **119**, 783–790.

NADEAU, R. M. & MCEVILLY, T. V. 1997. Seismological studies at Parkfield V: Characteristic microearthquake sequences as fault-zone drilling targets. *Bulletin of the Seismological Society of America*, **87**, 1473–1483.

PEACOCK, S., CRAMPIN, S., BOOTH, D. C. & FLETCHER, J. B. 1988. Shear-wave splitting in the Anza seismic gap, Southern California: temporal variations as possible precursors. *Journal of Geophysical Research*, **93**, 3339–3356.

RUTTER, E. H. & BRODIE, K. H. 1991. Lithosphere rheology – a note of caution. *Journal of Structural Geology*, **13**, 363–367.

SHIH, X. R., MEYER, R. P. & SCHNEIDER, J. F. 1989. An automated, analytical method to determine shear-wave anisotropy. *Tectonophysics*, **165**, 271–278.

SILVER, P. G. & CHAN, W. W. 1991. Shear wave splitting and subcontinental deformation. *Journal of Geophysical Research*, **96**, 16429–16454.

SLUNGA, R., RÖGNVALDSSON, S. TH. & BÖÐVARSSON, R. 1995. Absolute and relative location of similar events with application to microearthquakes is southern Iceland. *Geophysical Journal International*, **123**, 409–419.

STEFÁNSSON, R., BÖÐVARSSON, R., SLUNGA, R. *et al.* 1993. Earthquake prediction research in the South Iceland Seismic Zone and the SIL Project. *Bulletin of the Seismological Society of America*, **83**, 696–716.

TATHAM, R. H., VARLAMOV, D. A., PURNALL, G. W., MILLER, R. L. & GOODWIN, J. E. 1993. Observations of Earth-tide effects on seismic velocity and electrical conductivity. *63rd Annual International Meeting of the Society of Exploration Geophysicists, Washington, Expanded Abstracts*, 445–448.

VOLTI, T. & CRAMPIN, S. 2003. A four-year study of shear-wave splitting in Iceland: 2. Temporal changes before earthquakes and volcanic eruptions. *In*: NIEUWLAND, D. A. (ed.) *New Insights into Structural Interpretation and Modelling*. Geological Society, London, Special Publications **212**, 135–149.

WINTERSTEIN, D. L. 1996. *Anisotropists Digest, 147*. anisotropists@sep.stanford.edu

YARDLEY, G. S. & CRAMPIN, S. 1993. Shear-wave anisotropy in the Austin Chalk, Texas, from multi-offset VSP data: case studies. *Canadian Journal of Exploration Geophysicists*, **29**, 163–176.

ZATSEPIN, S. V. & CRAMPIN, S. 1997. Modelling the compliance of crustal rock: I – response of shear-wave splitting to differential stress. *Geophysical Journal International*, **129**, 477–494.

Note added in proof

Explanation for the ±80% scatter observed in time-delays

As anticipated the European Commission funded SMSITES Project appears to have found the explanation for the problematic ±80% scatter in observed time-delays. Three new seismic stations were installed by the project immediately above the major WNW-to-ESE-trending Húsavík-Flatey Fault (HFF) in northern Iceland. Shear-wave polarisations at these stations from small earthquakes on the fault are approximately fault-parallel and perpendicular to the dominant polarisations at all other stations in Iceland (see Figure 4). (Similar phenomena are also found at two places immediately above the major San Andreas Fault in California.) The HFF is seismically active and such active faults are expected to be pervaded by high pore-fluid pressures, in order to relieve frictional forces and allow pressure-clamped faults at depth to slip as earthquakes occur.

Crampin *et al.* (2002) use APE-modelling to show that shear-wave polarisations in ray paths through cracks saturated by critically high pore-fluid pressures undergo 90°-flips such as those observed at the new stations. The 90°-flips are caused by small changes in the three-dimensional crack distributions.

Most earthquakes showing the scatter occur on small faults. We suggest that small active faults will also be pervaded by high pore-fluid pressures. Consequently, 90°-flips (with negative time-delays, say) are expected near such faults, but these will revert to the normal polarisations (positive time-delays) for the remainder of the paths to the surface. Stress and fluid-pressures will be redistributed after the stress released at each earthquake, and work in progress demonstrates that minor modifications to the proportions of the flipped and normally polarised segments of the ray paths can easily cause the observed ±80% scatter in shear-wave time-delays.

Note that we cannot obtain sufficient information about the structure to eliminate or quantify the scatter. However, the scatter serves to remind us that all seismically-active faults, however small, are pervaded by high pore-fluid pressures.

CRAMPIN, S., VOLTI, T., CHASTIN, S., GUDUNDSSON, A. & STEFÁNSSON, R. 2002. Indication of high pore-fluid pressures in a seismically-active fault zone. *Geophysical Journal International*, **151**, F1–F5.

A four-year study of shear-wave splitting in Iceland: 2. Temporal changes before earthquakes and volcanic eruptions

T. VOLTI[1,2] & S. CRAMPIN[1,3]

[1]*School of Geosciences, University of Edinburgh, Grant Institute, West Mains Road, Edinburgh, EH9 3JW, UK (e-mail: tvolti@glg.ed.ac.uk; scrampin@ed.ac.uk)*
[2]*Japan Marine Science and Technology Centre, Yokohama, 237–0061, Japan*
[3]*Edinburgh Anisotropy Project, British Geological Survey, Edinburgh EH9 3LA, UK*

Abstract: This chapter reports temporal variations in the time-delays between split shear-waves before both earthquakes and volcanic eruptions in Iceland. The hypothesis is that during a build-up of stress, crack distributions in a large volume surrounding the immediate source zone are modified until the level of cracking reaches fracture criticality, when shear strength is lost, rocks fracture, and earthquakes, or some types of eruptions occur. In one two-year period, when volcanic and magmatic activity appeared to be low, changes in shear-wave splitting in SW Iceland were observed routinely before earthquakes with magnitudes between $M3.5$ and $M5.1$. Assuming a linear relationship between earthquake magnitude and the rate of increasing crack aspect-ratio in this comparatively narrow amplitude range, the time and magnitude of a $M5$ earthquake was successfully stress-forecast. These results confirm a new understanding of pre-fracturing deformation of *in situ* rock that has implications over a wide range of situations where the crust undergoes changes at low levels of deformation below those at which rocks fracture. Potential applications of this new understanding include monitoring hydrocarbon production, as well as stress-forecasting earthquakes and some volcanic eruptions.

This chapter is the second part of a report of a four-year study of shear-wave splitting in Iceland as part of the European Commission funded PRELAB Projects, 1996 to 2000. The first part (Volti & Crampin 2003) outlines the current understanding of shear-wave splitting in crustal rocks and reports preliminary observations of shear-wave splitting in Iceland. Shear-wave splitting is caused primarily by propagation through the fluid-saturated stress-aligned grain-boundary cracks and pore-throats pervading most rocks in the crust (Crampin 1994; Volti & Crampin 2003). The evidence suggests that these 'crack' distributions are critical systems (Crampin 1998*a*, 2000; Crampin & Chastin 2001; Volti & Crampin 2003) and are consequently very sensitive to minor variations of stress in *in situ* rocks.

Since rock is weak to shear-stress, partly as a consequence of criticality, the build-up of stress before a large earthquake is extensive (Crampin 1999*a*, 2000) and shear-wave splitting monitors this build-up of stress in the rock mass. Such measurements are free of the complications and heterogeneities of the immediate source zone. It appears that the build-up of stress can be identified by monitoring with shear-wave splitting almost anywhere in the large volume of stressed rock, so that the build-up and progress towards fracture criticality, and fracturing, faulting and earthquakes, may be recognized at substantial distances from eventual epicentres.

Volti & Crampin (2003) describe shear-wave splitting recorded by the South Iceland Lowland (SIL) seismic network, and suggests that the scatter in time-delay measurements and several other phenomena indicate that the cracks in the crust are critical systems, which necessarily lead to scattered shear-wave splitting. Figure 4 of Volti & Crampin (2003) maps the polarizations of shear-wave splitting at stations principally in the transform zones of the Mid-Atlantic Ridge in SW Iceland and northern Iceland.

Note that earthquake magnitudes referred to by '*M*' in this chapter are magnitudes reported by the SIL seismic network in Iceland, which are approximately equivalent to the body-wave magnitude *mb*.

From: Nieuwland, D. A. (ed.) *New Insights into Structural Interpretation and Modelling.* Geological Society, London, Special Publications, **212**, 135–149. 0305-8719/03/$15

Temporal variations in shear-wave splitting

Monitoring temporal changes in shear-wave splitting before larger earthquakes, using small earthquakes as the signal source, requires six stringent conditions:

(1) persistent swarms of small-scale seismic activity with nearly continuous activity to provide source signals;
(2) seismic stations within the shear-wave window of the swarm activity;
(3) rapid earthquake parameter determination;
(4) three-component seismograms with adequate digital sampling rates;
(5) minimal volcanic/magmatic activity so that build-up of stress before the earthquake is not disturbed by magmatic activity;
(6) a large or larger earthquake (or eruption) nearby.

These restrictions are severe, and any absence of any one hinders or prevents monitoring temporal variations.

Consequently, although figure 4 of Volti & Crampin (2003) shows extensive observations of shear-wave splitting throughout Iceland, earthquakes are typically clustered in time and space and only stations BJA, GRI, KRI and SAU have sufficiently persistent swarm-type activity within the shear-wave window to display temporal variations over most of the four-year period. In Figure 1 we present the variations of time-delays for these four stations for the four years of data, 1996 to 1999. It is recognized theoretically (Crampin & Zatsepin 1997), and confirmed observationally (Crampin et al. 1990, 1991; Crampin 1999a), that increases of stress increase the average aspect-ratio of distributions of parallel stress-aligned cracks. Such increasing aspect-ratios increase the *average* time-delay in Band-1 of the shear-wave window. Band-1 is the double-leafed solid angle of ray path directions from 15° to 45° either side of the average plane of the vertical cracks. Time-delays in Band-2, with directions ±15° to the average crack plane, are sensitive principally to changing crack density. The middle and upper diagrams in Figure 1 represent time-delays, normalized by path length to milliseconds per kilometre, in Band-1 and Band-2, respectively.

We examine only arrivals from earthquakes at more than 5 km depth in order to avoid the more severe shallow irregularities. The most noticeable feature of the measured time-delays is the large scatter of up to ±80% about any mean value. The scatter is discussed in more detail in Volti & Crampin (2003), where propagation in a crack-critical crust is suggested as the most likely cause. We analyse this scatter by plotting nine-point moving averages through the data points, where nine-point averages were chosen as giving appropriate smoothing to the data sets which show broad minima correlating with larger earthquakes. This was an empirical choice but leads to useful results for visual appreciation where other averages were not so informative. The bottom diagrams for each station in Figure 1 are earthquakes with magnitudes $M2$ located within 20 km of the station. Note that satisfactory statistical analyses of data sets, which display sporadic but meaningful abrupt changes of values, as in Band-1 of Figure 1, do not yet exist. Consequently, the analyses are to some extent empirical, and are justified by displaying temporal variations, which can be correlated with earthquakes.

Note that, although the top diagrams (Band-2 directions) vary with time, we have not identified any meaningful association or correlation of activity in Band-2 with any earthquakes or volcanic activity. The variations in Band-2 are presented for information only.

Variations in shear-wave time-delays in Band-1 are believed to monitor stress-induced changes in crack geometry (Crampin 1994, 1998a, 1999a; Volti & Crampin 2003). Increases in stress, leading to increases of average time-delays in Band-1, are expected to occur before both earthquakes and at least some types of volcanic or magmatic activity. Volcanic activity appears to be dominant in 1996 as a result of the build-up of stress before the large Vatnajökull (Gjàlp) eruption, whereas earthquake activity dominates 1997 and 1998 and both volcanic and earthquake activity are present in 1999. We first discuss observations sequentially: 1996; 1997 and 1998; and 1999.

Changes in time-delays during 1996

Changes before the Vatnajökull eruption

The first distinctive variation of time-delays in Band-1 at all four stations in Figure 1 is the increase of delays (marked by dashed least-squares lines) from about May 1996 (June at BJA) to the end of September 1996, when the Vatnajökull eruption occurred. The eruption was a massive event that injected approximately 0.4 km³ magma into the Earth's crust (Gudmundsson et al. 1997) along a 10 km long fissure (with a near-vertical planar surface of between 50 and 100 km²), between the Bárðarbunga and Grímsvötn volcanoes (see map in fig. 4 of Volti & Crampin 2003). About 0.7 km³ of volcanic products were ejected, making Vatnajökull the largest eruption in Iceland since the island of Surtsey was created in 1963. Increases of time-delays in Band-1 are

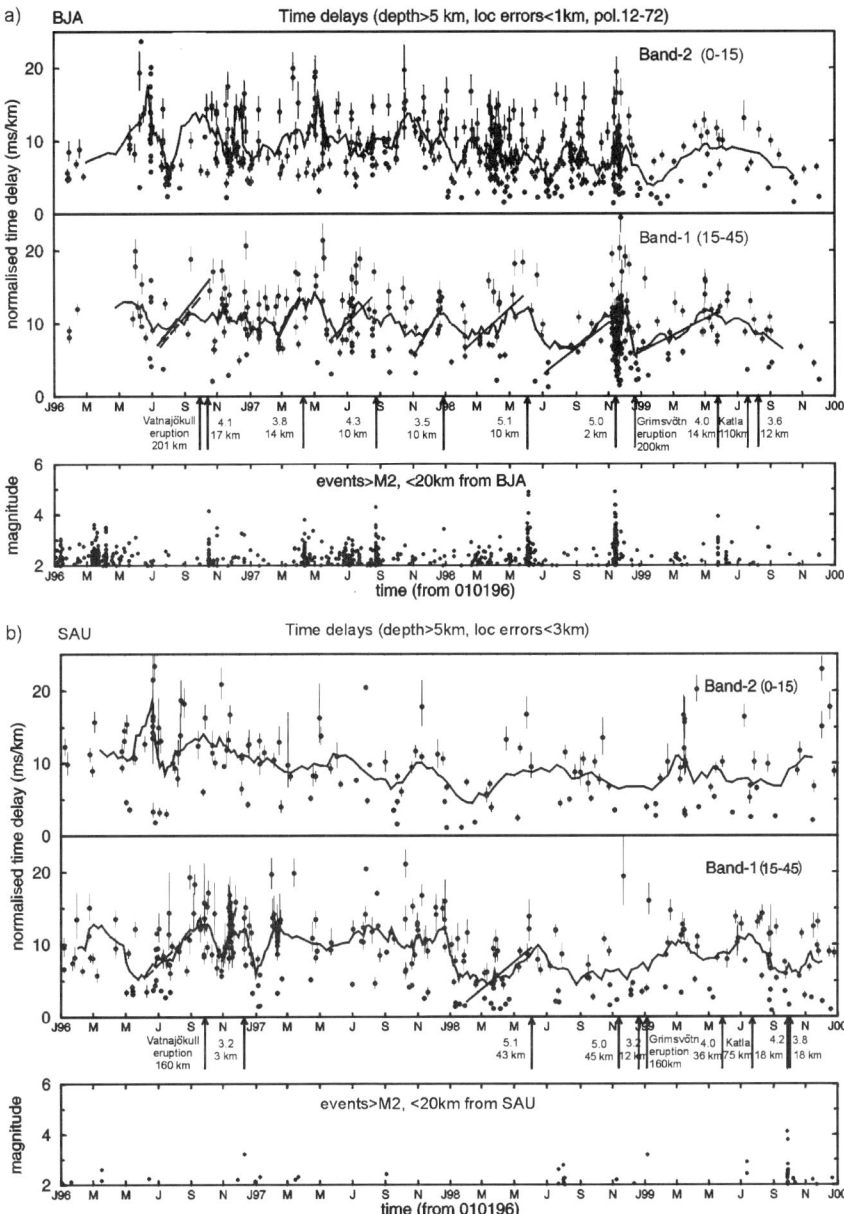

Fig. 1. Variations of time-delays between split shear-waves for 1 January 1996 to 31 December 1999 at stations (**a**) BJA, (**b**) SAU, (**c**) KRI and (**d**) GRI. The middle and upper diagrams show the variation of time-delays with time for ray-paths in Band-1 and Band-2, which are solid-angles ±15 to 45° and ±15° to the average crack plane, respectively. The time-delays (in ms) are normalized to a 1 km path-length. Error bars are time-delays divided by the errors in ray-path distance from locations. The irregular lines are nine-point moving averages. The straight lines in Band-1 are least-square estimates beginning just before a minimum of nine-point average and ending at a larger earthquake or eruption. Dashed lines refer to changes before the Vatnajökull eruption. The arrows below each plot indicate the times of these larger events with magnitudes and epicentral distances indicated. The bottom diagrams show the magnitudes and times of earthquakes greater than *M*2 within 20 km of the recording station.

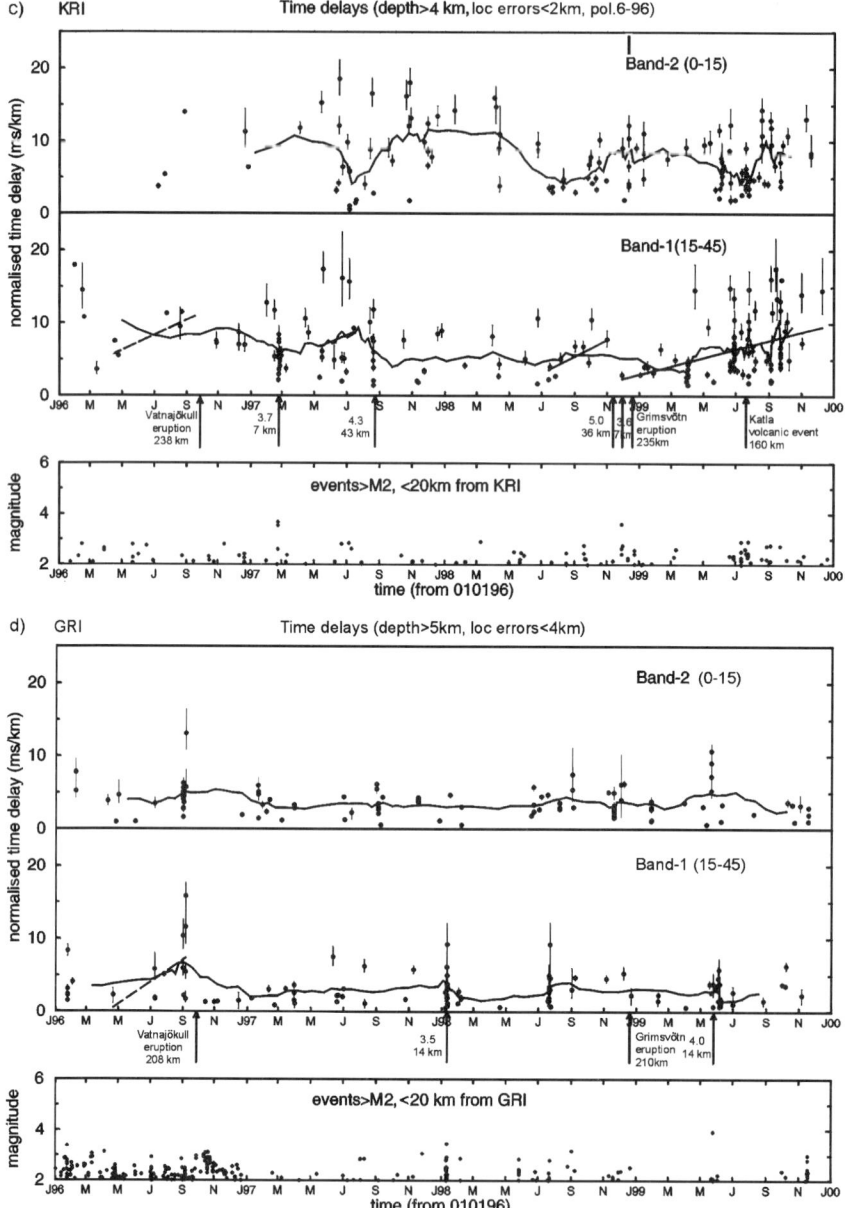

Fig. 1. Continued.

thought to indicate increasing stress (Crampin & Zatsepin 1997), and the changes identified here are interpreted as being caused by the build-up of stress as the crust is pressurized by magma from below. The interpretation is that the pressure increases until fracture criticality is reached, when the fissure opens against the minimum compressional stress. This is presumed to be similar to the process of hydraulic fracturing in the oil indus-

try. The observations suggest that the behaviour and fluid pressures for hydraulic and magmatic fracturing may be similar despite the different fluid properties of water and magma. The major difference is that hydraulically induced cracks immediately begin to close once the injected pressure opens the crack and the pressure is released (Meadows & Winterstein 1994), whereas with a magma-induced fracture, the dyke remains open and the strain remains, although as in this case the stress persists until movement of the surrounding rock mass (in this case, the offset of the Mid-Atlantic Ridge) accommodates the strain.

The behaviour of the time-delays in Band-1 associated with the Vatnajökull eruption show several notable features.

(1) The increase in time-delays at all four stations in Figure 1 begins in May (June at BJA) and finishes at the end of September 1996, which is consistent with a four or five month increase of stress. This is assumed to be the time taken by the magma to build up sufficient pressure to fracture the brittle crust.

(2) The increases in time-delays before the eruption are observed at distances of about 240 km at KRI, 200 km at BJA, 160 km at SAU, and 210 km at GRI. This is almost the full westward and northward extent of Iceland, and these are by far the greatest distances that stress-induced changes in shear-wave splitting have yet been observed.

(3) Following earthquakes (see next subsection) the increase of time-delays stops, and rapidly returns to approximately the level at the start of the increase, implying an abrupt release of stress at the time of the earthquake or shortly before (Crampin 1999a; see also below). Following the Vatnajökull eruption, however, the increase in time-delays stops at all stations. However, instead of immediately decreasing, the level of time-delays shows a slow reduction of about 2 ms km^{-1} per year over a period of about two years in both bands at Stations BJA and SAU, and in Band-1 at KRI (the data in Band-2 at KRI and both bands at GRI, are too sparse to show consistent variations). The changes associated with earthquakes in the next sections are superimposed on this gradual decrease following the eruption.

Changes before earthquakes during 1996

The lower diagrams in Figure 1 show that there are only two earthquakes larger than $M3$ during 1996 (except for a swarm of events in March to May at BJA when there was a gap in the source earthquake sequence). These are an $M4.1$ earthquake at BJA and an $M3.2$ earthquake at SAU.

The $M4.1$ earthquake 17 km from BJA in Figure 1a, in December 1996, is where the trend of the solid least-squares line for the earthquake lies close to the dashed line for the increase before the Vatnajökull eruption. It is impossible to separate the effects of the eruption and earthquake (the separation in the figure is for clarity), but there are two

phenomena that suggest that increase at BJA is primarily associated with the eruption rather than the $M4.1$ earthquake. The effects of the eruption are widely observed at comparatively large distances, including Stations KRI and SAU, which bracket BJA, so the effects of the eruption are expected to be seen at BJA. The second phenomenon is that following the Vatnajökull eruption the time-delays at BJA show a slow decrease over about two years (see Item 3, in previous subsection), whereas following earthquake time-delays typically show an abrupt decrease. We conclude that the effects of the eruption dominate the effects before the $M4.1$ earthquake at BJA.

The time-delays in Band-1 for the $M3.2$ earthquake 3 km from SAU in December 1996 show the steep increase expected for $M3$ earthquakes, similar to the behaviour at BJA in 1997 and 1998, below. However, the peak value of about 11 to 14 ms km^{-1} is high and may reflect the raised values of time-delays following the Vatnajökull eruption (see previous section) some two months earlier.

Changes in time-delays during 1997 and 1998

There was only one volcanic eruption during 1997 and 1998. This was a comparatively small eruption at the Grímsvötn Volcano in December 1998, which does not appear to have affected time-delays at any of the seismic stations with time-delay data in Figure 1.

The lack of significant volcanic (magmatic) activity in 1997 and 1998 makes these two years particularly important, as earthquake-induced effects are not disturbed by movement of magma in the crust. The time-delays are merely superimposed on the slow 2 ms km^{-1} per year reduction in both Band-1 and Band-2 following the Vatnajökull eruption. We shall examine the effects at the four stations separately.

Changes in shear-wave splitting before earthquakes at Station BJA during 1997 and 1998 (Fig. 1a)

Station BJA in Figure 1a displays what we assume to be the standard or type-behaviour, where nine-point moving averages of the time-delays show long-term and short-term variations with time in Band-1 of the shear-wave window. The nine-point moving averages in Band-1 show broad peaks and troughs, where the peaks appear to be associated with the larger earthquakes at the specified epicentral distances marked next to the arrows indicating the earthquakes below the diagram. We have not identified any meaningful association or correlation of variations in Band-2 with any other phenomenon apart from the *c.* 2 ms km⁻¹ per year decrease following the Vatnajökull eruption.

Between January 1997 and December 1998 there are five broad minima in the moving averages in Band-1, each associated with a later larger event (with a sixth minimum at the end of December 1998 associated with activity in 1999; see next section). Least-squares lines are fitted to the data starting immediately before each broad minimum. These least-squares lines show time-delays increasing until a larger earthquake occurs, when the time-delays drop abruptly to a level close to the level before the increase began. This level is usually approximately half the peak value. At BJA the earthquakes occur when the least-squares lines reach levels of shear-wave splitting between about 11 and 14 ms km⁻¹, which is assumed to be the level of fracture criticality in the rock mass along the ray-paths. This behaviour is broadly similar to the variations seen before the four other earthquakes worldwide where changes before earthquakes have been observed (Crampin 1999*a*). The variations in time-delays before earthquakes in Figure 1a are superimposed on the general reduction of about 2 ms km⁻¹ per year following the Vatnajökull eruption. This general reduction is visible in the moving averages in both Band-1 and Band-2 at stations BJA and SAU, and in Band-1 at KRI. The presumed level of fracture criticality at which the earthquakes occur also appears to decline from about 14 ms km⁻¹ at the beginning of 1997 to about 10 ms km⁻¹ at the end of 1998, consistent with the 2 ms km⁻¹ per year decrease following the Vatnajökull eruption.

The least-squares lines for all the earthquakes, with the exception of the *M*4.3 earthquake in August 1997, are comparatively robust in that the slopes and durations are not very sensitive to minor variations in the time of starting or ending the least-squares line. The slope and the duration of the line for the *M*4.3 event, however, are both extremely sensitive to the starting and particularly the ending times of the least-squares line. This sensitivity could be due to complications caused by a swarm of local events including seven events with magnitudes greater than *M*3, but it is also associated with a possible reduction in time-delays immediately before the earthquake. Such reductions are discussed below.

Note that levels of time-delays in Band-1 at which earthquakes occur, the assumed levels of fracture criticality, are observed to vary with location from about 5 ms km⁻¹ on the San Andreas Fault near Anza (Crampin *et al.* 1990) to the 11 to 14 ms km⁻¹ at BJA in Iceland. The presumed levels of fracture criticality at other stations in Iceland

also vary: 10 to 15 ms km⁻¹ at SAU; 9 to 11 ms km⁻¹ at KRI; and 8 ms km⁻¹ at GRI.

Note also that the relationship of shear-wave velocity to levels of fracture criticality of crack distributions varies with conditions. For example, the effect of a given crack distribution on shear-wave velocity anisotropy is strongly dependent on Poisson's ratio and hence on the *Vp/Vs* ratio (Crampin 1993*b*). Thus different levels of fracture criticality in different locations may reflect different relationships between shear-wave velocity anisotropy and crack density, due to variations in Poisson's ratio and other parameters, rather than fundamentally different crack distributions and crack densities. It is also possible that different combinations (ratios) of triaxial stresses may also lead to different values of fracture criticality.

The changes in splitting in Figure 1a are seen before earthquakes with magnitudes ranging from *M*3.5 to *M*5.1. Figure 2a shows the magnitude of these earthquakes plotted against the duration of the increase in days, and Figure 2b the magnitude plotted against the slope of the increase in ms km⁻¹ per day. The parameters for *M*4.3 earthquake of August 1997 are unreliable and although they are

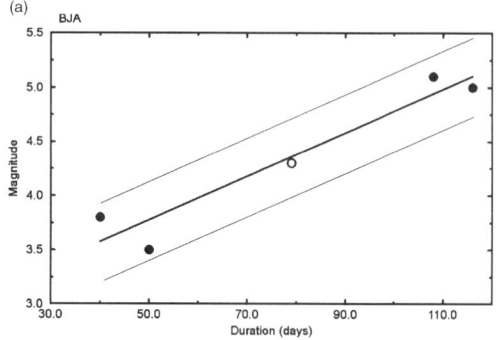

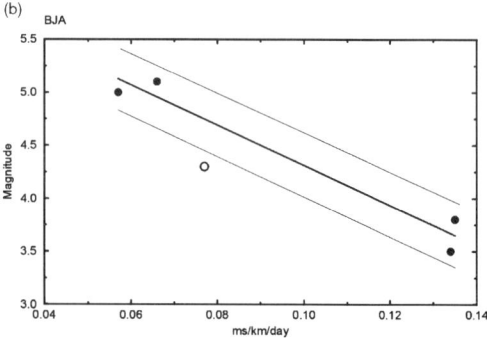

Fig. 2. Earthquake magnitude plotted against (a) durations and (b) slopes of increasing time-delays before five larger earthquakes in Band-1 in Figure 1a at BJA. The least-squares lines and standard errors omit the *M*4.3 event of August 1997 (open circle) as being unstable.

plotted, the values are omitted from the least-squares lines as being too subjective. Figure 2 shows that in this limited 1.6 magnitude range, if the relationships are assumed to be linear, magnitudes are proportional to the duration of the increase and inversely proportional to the rate of increase in time-delays. Although linearity is an unproven assumption, it is supported by the values of $M4.3$ earthquake even though they are unreliable. The assumption of linearity means that the time of an impending event can be estimated from the time when the increase reaches fracture criticality, and the magnitude can be estimated from the slope or duration of the increase. We call such estimates 'stress-forecasts'.

Note that because there are uncertainties in determining both rates of increase in time-delays and levels of fracture criticality, stress-forecasts are defined in smaller–earlier to larger–later (SELL) windows for impending events, where smaller and larger refer to magnitude, and earlier and later to time of occurrence. There is an example of a SELL window in the successful stress-forecast in Table 1 (Crampin *et al.* 1999).

In October 1998 we recognized that there was an increase time-delays in Band-1 (that had started in July), and the time and magnitude of a $M5$ earthquake on 13 November 1998 was successfully stress-forecast (Crampin *et al.* 1999). Table 1 lists the exchange of e-mail messages between Edinburgh University and the Icelandic Meteorological Office between 27 October and 13 November 1998. In summary, on 27 and 29 October, Edinburgh suggested that a significant event would happen 'soon'. On 28 October, Ragnar Stefánsson of the Icelandic Meteorological Office suggested that this event might be linked to the previous $M5.1$ earthquake, where low-level seismicity was still continuing. In the final stress-forecast on 10 November 1998, Edinburgh suggested (a SELL window) that 'an event could occur any time between now', when it would be '($M = 5$), and end of February, when it would be ($M = 6$)'. Three days later, on 13 November, the Meteorological Office reported that 'there was a magnitude 5 earthquake … near to BJA … this morning at 10 38 GMT', which was on the fault line suggested on 28 October. Thus the time and magnitude were successfully stress-forecast, and local information correctly predicted the approximate location (Crampin *et al.* 1999).

Note that there is typically a four to seven day delay before seismograms and earthquake location data are put on the Internet web site for accessing by Edinburgh. This means that the foreshock activity before the 13 November earthquake was

Table 1. *Timetable of e-mail communications*

1998	E-mails, facsimiles, and actions
(1) 27 Oct.	Edinburgh University (EU) e-mails Iceland Meteorological Office (IMO) reporting shear-wave time-delays in Band-1 increasing from July at stations BJA and KRI and suggests: '… *there was an 80% chance of something significant happening somewhere between BJA and KRI within three months.*'*
(2) 28 Oct.	EU faxes data for BJA and KRI to IMO. IMO suggests $M5.1$ earthquake near BJA in June 1998 may be linked to current increase in time-delays.
(3) 29 Oct.	EU updates current interpretation and suggests: '*Shear-wave splitting at both BJA and KRI indicate something is going to happen soon, probably within a month…*'*
(4) 30 Oct.	IMO sends notice to National Civil Defence Committee (NCDC) in Reykjavik suggesting a meeting.
(5) 31 Oct.– 4 Nov.	Faxes and e-mails updating information. EU refines data and interpretation. IMO increases local geophysical and geological investigations.
(6) 5 Nov.	IMO presents stress-forecast and other data from surrounding area to scientific advisors of NCDC, who conclude no further action was required of them (see comment in the Discussion).
(7) 6–9 Nov.	Exchange of various faxes and emails updating information and interpretation.
(8) 10 Nov.	EU concludes: '… *the last plot … is already very close to 10 ms km^{-1}. This means that an event could occur any time between now (M = 5) and end of February (M = 6).*'*†
(9) 11 Nov.	EU faxes updated data for KRI and BJA, with SAU now also suggesting increasing time-delays from September.
(10) 13 Nov.	IMO reports: '… *here was a magnitude 5 earthquake just near to BJA (preliminary epicenter 2 km west of BJA) this morning 10 38 GMT.*'*

After Crampin *et al.* (1999)
*Quotations in *italics* are exact texts from e-mails.
†This is an example of a Smaller-Earlier to Larger-Later (SELL) window.

not available to Edinburgh when the final stress-forecast was made (Crampin *et al.* 1999). Even if the data had been available, without recognizing the increase of stress, it would have been difficult to identify the activity as foreshocks, which are relatively uncommon in Iceland. It is the only clear foreshock sequence in the four years of data at the four stations in Figure 1.

Changes in shear-wave splitting before earthquakes at Station SAU during 1997 and 1998 (Fig. 1b)

The increase in time-delays of the *M*3.2 earthquake in December 1996 (see previous sections), is followed by another very similar increase. This peaks in February, and does not appear to be associated with any nearby earthquake (a *M*3.7 earthquake in February, at 81 km distant near KRI is presumably too far to be associated). This steep rise begins a 'plateau' in Band-1 time-delay values, averaging about 12 ms km^{-1}, which continues for approximately the whole of 1997, until there is an equally steep decline around 1 January 1998. We have no explanation for this plateau, but could speculate that it is caused by otherwise unrecognized subsurface volcanic (magmatic) activity.

The only temporal variation in Band-1 in 1998 that can be associated with earthquakes is a January to May increase, very similar to a simultaneous increase at BJA before a *M*5.1 earthquake 10 km from BJA. The earthquake is at a distance of 45 km from SAU and it is suggested that the two increases are associated. If this is the case, it is the largest distance that changes in shear-wave splitting have been observed *before earthquakes*. In contrast, note that the stress-forecast *M*5 earthquake of November 1998, near to BJA, which is close in epicentre and magnitude to the *M*5.1 event which does modify time-delays, does not appear to affect time-delays at SAU. We have no explanation for this.

Changes in shear-wave splitting before earthquakes at Station KRI during 1997 and 1998 (Fig. 1c)

Figure 1c shows there are insufficient small earthquakes within the shear-wave window at KRI to provide enough shear-wave sources to show temporal changes before the two *M*> 3 earthquakes within 20 km of KRI in 1997. The only clear increase in 1998 is before the stress-forecast earthquake of November 1998, at a distance of 36 km. This increase at KRI was one of the phenomena that was used to support the stress-forecast (see entries 1, 3 and 9 in Table 1) (Crampin *et al.* 1999).

In contrast to SAU which sees changes before the *M*5.1 June 1998 event but not before the *M*5 November 1998 event, KRI does not show changes before June 1998 but does before the November 1998 event.

Changes in shear-wave splitting before earthquakes at Station GRI during 1997 and 1998 (Fig. 1d)

Figure 1d shows there are insufficient (shear-wave source) earthquakes within the shear-wave window at GRI to establish temporal changes before the few *M*>3 earthquakes close to GRI in 1997 and 1998.

Changes in time-delays during 1999

During 1996 there was dominating volcanic (magmatic) activity. During 1997 and 1998, there was largely earthquake activity, superimposed on the 2 ms km^{-1} per year decrease following the Vatnajökull eruption. The behaviour before earthquakes followed the characteristic behaviour discussed in the previous section. Thus in those years, the volcanic and earthquake were largely separated in time. In contrast, during 1999 there was both earthquake and volcanic activity and it is difficult to separate the effects.

SAU is the closest (Fig. 4 in Volti & Crampin 2003) of the four stations in Figure 1 to the volcanoes of Katla, which erupted in July 1999, and Hekla, which erupted in March 2000 (just outside our four-year study period). The Katla eruption was small but the Hekla eruption was the largest eruption in Iceland since Vatnajökull at the end of September, 1996. Again we examine the four critical stations separately.

Changes in shear-wave splitting at Station BJA during 1999 (Fig. 1a)

Shear-wave splitting time-delays in Band-1 show similar behaviour to the previous earthquakes in 1997 and 1998 before a *M*4 earthquake in May 1998 at 14 km distance. However, the slope and duration of the increase are not compatible with those expected for *M*4 earthquakes in the previous three years, and the time-delays do not show the abrupt decrease following the event expected for earthquakes. These anomalies may be due to complications caused by the stress changes before the volcanic eruptions of Katla in July 1999 (and possibly the much larger eruption of Hekla in February and March 2000). Note that Band-2 for this period shows similar variations to those in Band-1 at BJA.

Changes in shear-wave splitting at Station SAU during 1999 (Fig. 1b)

The nine-point moving averages in Band-1 show two broad peaks. The second peak in July may be correlated with the minor eruption at Katla 75 km to the SW, but the first peak in March does not appear to be correlated with any known seismic or volcanic activity. These $M4.2$ and $M3.8$ earthquakes were within 18 km of SAU in September 1999, but the time-delays do not show any recognizable correlation with either earthquakes or the eruptions which were comparatively close to SAU: Katla at 75 km, and Hekla at 35 km. Again, the absence of a simple relationships may well be due to stress-fields showing a combination and interaction of the effects of build-up of stress before earthquakes and volcanic eruptions.

Changes in shear-wave splitting at Station KRI during 1999 (Fig. 1c)

The shear-wave splitting in Band-1 at KRI display a large scatter but show a consistent increase starting in December 1998. This increase lasted the whole of 1999 and ended in three very irregular peaks in February, April and June 2000; since then it slowly decreased over a period of some six months. This increase is the most distinctive phenomenon visible in the four years of recordings at the four stations KRI, BJA, SAU and GRI. The duration and rate of increase suggest a large event. The difficulty in interpreting this increase is that it is only apparent at KRI; we have no immediate explanation. This is the subject of further investigations.

Changes in shear-wave splitting at Station GRI during 1999 (Fig. 1d)

There was a $M4$ earthquake in May 1999 but insufficient shear-wave source earthquakes to show any variations.

Shear-wave splitting showing 'precursory' changes immediately before earthquakes

In general, the changes in shear-wave splitting observed in Iceland, and elsewhere, show increasing time-delays in Band-1 before earthquakes and an abrupt reduction to approximately the original level at the time of the earthquake. However, in those few occasions where there is adequate information (that is, sufficient small source earthquakes immediately before the larger earthquake in the shear-wave windows below the stations) to reliably

delineate the behaviour, the reduction appears to begin shortly *before* the earthquake.

The first time this 'precursory' decrease was observed as the M_L 3.8 earthquake in the swarm at Enola in Arkansas, where the reduction in time-delays began four hours before the event (Booth *et al.* 1990; Crampin 1999*a*). The Arkansas data set was extremely sparse so the 'precursory' decrease was unreliable.

Figure 3, a time-expanded section of Figure 1a, shows a well-established example of such precursory changes at BJA for the stress-forecast $M5$ event of November 1998. There is an abrupt decrease on 9 November, where the moving average through the normalized time-delays drops from close to fracture criticality, *c.* 11 ms km^{-1}, to approximately 8 ms km^{-1}. This is about four days before the $M5$ forecast earthquake. Note that the nine-point moving average smoothes over the abrupt drop in time-delays. Note also that there is some sign of 'precursory changes' before the $M4.3$ earthquake in July 1997 at BJA (see Fig. 1a), although the data are too irregular to be reliable. It is possible that if appropriate observations could be made, as in the controlled source seismics discussed below, such short-term precursors could be used to signal the proximity of most or many large or larger earthquakes.

Interpretation

At Station BJA, all the broad peaks and troughs of the nine-point averages in Band-1 are associated

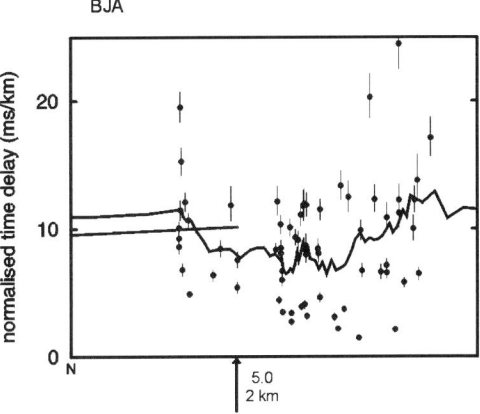

Fig. 3. Time-expanded section of Figure 1a. Changes in time-delays in Band-1 of the shear-wave window at Station BJA for the month of November 1998, showing precursory changes in time-delays at BJA beginning four days before the earthquake of 13 November 1998.

with earthquakes or volcanic eruptions. In particular, during 1997 and 1998 when volcanic activity appeared to be at a minimum, five increases in time-delays in Band-1 fitted by least-squares lines could be interpreted as increases of aspect-ratios indicating the build-up of stress before earthquakes. Based on the last of these increases, a $M5$ earthquake was successful stress-forecast (Crampin et al. 1999). Although the time, magnitude and location of future earthquakes may well be impossible to predict deterministically (Geller 1997; Kagan 1997; Leary 1997), a large earthquake can only occur when sufficient stress has accumulated in the surrounding rock mass. Rocks are comparatively weak under shear stress, so that the volume where stress has accumulated before a large earthquake necessarily has to be large, probably tens to hundreds of millions of cubic kilometres before a $M8$ earthquake.

It is suggested that increasing stress modifies the highly compliant fluid-saturated microcrack geometry throughout a large part of the stressed volume until fracture criticality is reached, when an earthquake or eruption occurs (Zatsepin & Crampin 1997; Crampin & Zatsepin 1997). The evidence suggests that the approach to criticality can be recognized probably over a significant part of the stressed volume. These results suggest that shear-wave splitting can monitor the stress-induced increase in crack aspect ratios and the approach of fracture criticality, and that the time and magnitude of large, or larger, impending earthquakes (or the times of eruptions) can be stress-forecast.

Interpretation of changes before earthquakes

Figure 2 shows that for the very limited range of earthquakes available ($M3.5$ to $M5.1$) the magnitudes are approximately proportional to the duration of the increasing stress and inversely proportional to the rate of increase. This is what would be expected if an approximately constant rate of increasing stress accumulated in a heterogeneous crust. If stress accumulates over a small volume, the rate of increase will be relatively fast over a short period of time and the eventual earthquake will be relatively small. However, if stress accumulates over a larger volume, the rate of increase will be relatively slow and last for a longer period of time, but the eventual earthquake will be relatively large. For the very limited range of earthquakes in Iceland ($M3.5$ to $M5.1$) before which changes in shear-wave splitting have been identified, the effects are assumed to be approximately linear. They are not expected to be linear for larger ranges of magnitudes.

This means that the time of occurrence of the earthquake (or eruption) can be estimated by the time that the increase reaches the level of fracture criticality. Similarly, the magnitude of the event can be estimated by rate of increase of time-delays or the duration of the increase of stress in a SELL window. We call this stress-forecasting. Note that these changes in shear-wave splitting do not indicate the location of the event. In principle, the release of stress could occur anywhere in the large stressed volume. However, if a larger earthquake is stress-forecast, anticipation of a large forthcoming event enables otherwise minor or easily overlooked precursory anomalies to be interpreted realistically, as was done successfully in the stress-forecast event of November 1998 (Crampin et al. 1999). Note, however, that this was possible only for the years 1997 and 1998 when volcanic (magmatic) activity appeared to be negligible.

Consistent changes in shear-wave splitting observed above small earthquakes, indicating that stress-induced changes to crack aspect ratios can be monitored by shear-wave splitting along ray-paths in Band-1, have been observed in four earthquakes worldwide (Crampin 1999a), and now in five earthquakes (Fig. 1) at several stations in SW Iceland. Below we shall suggest that similar observations can be made in stress-monitoring sites using controlled-source cross-hole seismics.

Interpretation of changes before volcanic eruptions

Iceland is situated over a mantle plume in an actively spreading ridge system, in which crust is being formed at the rate of about 2 cm per year (Einarsson 1991). Volcanism is generally concentrated into the central volcanoes such as Bárðarbunga and Grímsvötn, and their associated fissure systems, although there is also significant volcanism in outlying areas such as Katla and Hekla (Fig. 1). Magma is thought to accumulate into small crustal chambers, followed by lateral migration of this magma away from the chamber into dykes along fissure swarms. Five months' duration of the increase in stress before the 1996 Vatnajökull eruption is indicated by changes in time-delays at all of the stations in figure 3 (Volti & Crampin 2003). This suggests that the increase of pressure necessary for opening the fissure by fluid fracture in a magmatic fracture (analogous to the hydraulic fractures of the oil industry) took about five months to accumulate.

The eruption beneath the Vatnajökull glacier in Central Iceland started in the late evening of 30 September 1996 and was preceded by a sequence of earthquakes a few hours before the largest event of $M4.3$ immediately preceding the eruption. Based on the amount of magma ejected, this was the fourth largest eruption in Iceland in the twentieth century, and Vatnajökull is the largest ice cap in Europe (Gudmundsson *et al.* 1997).

The observations in Figure 1 suggest that the effects of increasing stress before the Vatnajökull fissure eruption are visible at extensive distances (essentially over the whole of Iceland) before the impending eruption. Following the eruption, the stresses decayed to close to their original level over a period of about two years. This can be interpreted as the Vatnajökull eruption, above the Iceland plume, initiating a spreading cycle of the Mid-Atlantic Ridge, where the two-year decay in stress is the time taken by the two transform zones, south and north of Iceland, to adjust to the plume-driven movement of Iceland. This comparatively slow response suggests that the spreading of the ridge is driven by the magma injection above the plume, rather than the magma merely accommodating the potential space left by the tectonic plates separating at approximately 2 cm per year. This suggests that, at least around Iceland, the spreading takes place in sporadic episodes driven by the magmatic injections. Einarsson (1991) reviews present-day tectonics in Iceland.

The details of stress changes before fissure eruptions may well be different from the more usual eruptions along previously established, more symmetrical vents. The build-up of stress before the Vatnajökull eruption implied by the changes in shear-wave splitting, as interpreted here, suggests a stress-induced break in a seal, similar to the familiar pressure-induced hydraulic fractures used in oil field operations. Since rising magma is fluidized by volatiles with an increase of volume and pressure, it is likely that this will increase horizontal stress before vent eruptions in a similar way to that suggested before the Vatnajökull fissure eruption.

Recent earthquake in SW Iceland which was not stress-forecast

During the preparation of this chapter, which reports the period from 1 January 1996 to 31 December 1999, there was a cluster of three larger earthquakes in SW Iceland, two on 17 June and one on 21 June 2000. This cluster of earthquakes is outside the period covered by the study but is relevant to the report as it includes the largest earthquakes recorded in Iceland since January, 1996. The earthquakes had SIL magnitudes $M5.6$, 5.0 and 5.3 (approximately equivalent in this magnitude range to $Ms = c.$ 6.6, 6.0 and 6.3, respectively) with epicentral distances from Station SAU of approximately 3 km, 7 km and 14 km, respectively. There were changes in the time-delays between the split shear-waves before the earthquakes, but these earthquakes were not stress-forecast.

Figure 4 shows time-delays from station SAU for the eight months from November 1999 to June 2000. There were no associated anomalies recognized at other seismic stations. The normalized time-delays in Band-1 in Figure 4 show an increasing delay following a swarm of seismicity at the of February 2000. We currently have no experience of earthquakes larger than $M5.1$, or from clusters of earthquakes. However, judging from the rates of increase and durations during 1996 to 1999, the rate of increase of the least-squares fit of time-delays in Band-1 in Figure 4 appears to be consistent with an $M5.6$ ($Ms6.6$) earthquake, but the duration of about four months is certainly too short.

The reason for this inconsistency, and the reason why the increase was not recognized and the earthquake cluster stress-forecast, was that before the swarm of activity in February 2000 there was a period of about seven weeks without any suitable shear-wave source earthquakes near SAU. Consequently, there could be no time-delay measurements. This means that the least-squares fit was dominated by the swarm, and following the swarm, the average time-delays in Band-1 actually decreased until the foreshock activity beginning seven days before the $M5.6$ earthquake. Routine delays in putting data on the web site and accessing the data meant that the increase was only recognized with hindsight. However, had there been some small-scale shear-wave source activity during January and February so that the duration of the increase was longer, it is likely that the duration and rate of increase of time-delays before the first of the earthquakes would have correctly stress-forecast the $M5.6$ earthquake (within the usual SELL window).

Note the reliance on and need for nearly continuous small-scale activity to provide a source of shear-waves to monitor the rock mass and stress-forecast earthquakes. A stress-monitoring site using controlled-source cross-hole seismology to routinely monitor the rock mass with shear-wave splitting (see next section, and Crampin 2001) is currently being developed near Húsavík, northern Iceland (Crampin *et al.* 2000). Stress-monitoring sites would use shear-waves to monitor the rock mass, and would remove the need for continuous small-scale seismic activity to stress-forecast earthquakes.

Development of stress-monitoring sites

The possibility of stress-forecasting the times and magnitudes of future large earthquakes by monitoring shear-wave splitting suggested by these results may well be the best way, perhaps the only way, to provide reliable warnings of imminent earthquakes and eruptions by monitoring the build-up

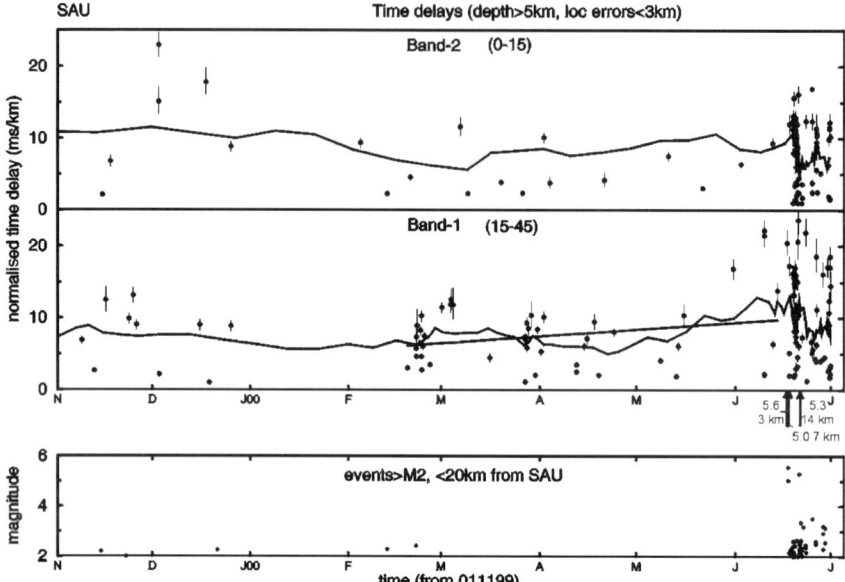

Fig. 4. Shear-wave splitting time-delays at SAU for the period November 1999 to June 2000, showing changes of time-delays in Band-1 before the $M5.6$, 5.0 and 5.3 earthquakes of 17 June (15:40 and 15:42 hours) and 21 June, respectively. Note that in this magnitude range, the $M5.6$, 5.0 and 5.3 magnitudes are approximately equivalent to $Ms6.6$, 6.0 and 6.3. Format and notation as in Figure 1.

of stress. Such stress-forecasting using small earthquakes as the source of shear-waves, as discussed in this chapter, is only possible when there is a more or less continuous source of small shear-wave source earthquakes. Such persistent seismicity is extremely rare. In ten years of searching for appropriate data (Crampin 1993a), we found only four examples in a worldwide search where seismicity and source and receiver geometry were appropriate to estimate variations of time-delays before large or larger earthquakes (Crampin 1999a). Only in the persistent seismicity of Iceland have we found more examples of changes before earthquakes (Crampin et al. 1999; Volti & Crampin 2003).

To stress-forecast earthquakes near vulnerable cities or vulnerable installations, without such persistent seismicity, requires cross-well seismology using a borehole source transmitting orthogonally polarized shear-waves along appropriate ray-paths in Band-1 to three-component borehole receivers in other boreholes. This proposition was first discussed by Crampin & Zatsepin (1997b) and Crampin (1998b) who proposed using cross-hole seismics between (expensive) deviated wells with an airgun as a source of shear-waves. Recently, a borehole shear-wave source, the Downhole Orbital Vibrator (DOV, previously known as the Conoco Orbital Vibrator or COV; Cole 1997; Daley & Cox 2001), has been commercialized by Geospace Engineering Resources, Inc. The DOV has an

eccentric cam, which is swept in both clockwise and counter-clockwise directions, that exerts a radial force on the borehole wall. The radiated signal from a vertical well can be processed to simulate two orthogonally polarized shear-waves at specified orientations at a frequency peaking at about 200 Hz. There is also minimal tube wave generation. This source allows shear-wave propagation along appropriate ray-paths in Band-1 to be generated using conventional vertical wells. The receiver boreholes need to be at specific stress-related offsets and azimuths with respect to the source borehole to get observations within the solid angle of directions in Band-1 (Crampin 2001).

Note that subsurface cross-well seismics is required in order to avoid the severe attenuation and scattering that occurs at shallow depths above some critical depth between 500 m to 1000 m, for example (Crampin 1990, 1994). This is not such a serious problem for shear-waves above small earthquakes as the much lower frequencies and much longer wavelengths are less sensitive to the near-surface anomalies.

Many townships in Iceland sink boreholes to use geothermal heat. The small port of Húsavík, 2 km south of where the Húsavík–Flatey Fault of the Tjörnes Fracture Zone runs ashore at the base of the Tjörnes Peninsula, has wells in the appropriate stress-oriented geometry that may be used as a stress-monitoring site, courtesy of Hreinn Hjartarson of Orkuveita Húsavíkur, the municipal service company (Crampin *et al.* 2000). The Tjörnes Fracture zone is the offshore northern leg of the transform zone of the Mid-Atlantic Ridge, which is largely onshore in SW Iceland. The European Commission is funding the SMSITES Project to develop a stress-monitoring site (SMS) at Húsavík in order to develop techniques for stress-forecasting the times and magnitudes of impending large earthquakes wherever required in Europe and worldwide (Crampin 2001; Crampin *et al.* 2000).

Note that the need to monitor the rock mass with controllable sources was particularly evident in June 2000 (see previous section). The $M5.6$ ($M_S6.6$) earthquake on 17 June 2000, in SW Iceland, was the largest event in Iceland for several years. It was not stress-forecast, because there was a seven week gap in small-scale seismicity beneath the nearest station SAU. Since the gap was near the beginning of the increase in time-delays, the increase was not interpreted correctly. A SMS would have resolved the difficulty.

Other applications

It might have been thought that crystalline igneous and metamorphic rocks, generally considered to be comparatively hard and brittle, would be relatively insensitive to small changes of stress. In fact, shear-wave splitting appears to be particularly sensitive in the basalts of Iceland with levels of fracture criticality for normalized time-delays ranging between about 8 and 15 ms km^{-1} at different stations in SW Iceland, as opposed to less than 8 ms km^{-1} in other locations elsewhere (Crampin 1999*a*). If the compliance of fluid-saturated cracks can be recognized in crystalline rocks, the effects in softer sedimentary rocks is likely to be even more sensitive. The widespread observations of shear-wave splitting (Crampin 1994, 1996; Winterstein 1996) suggest that almost all rocks – igneous, metamorphic and sedimentary rocks, including hydrocarbon reservoirs – respond to changes of stress in much the same way.

Since almost any change in rock mass conditions produces changes of stress, by implication, the rock mass is sensitive to changes in a great variety of phenomena. This suggests that shear-waves are sensitive indicators of a wide range of changes in hydrocarbon reservoirs. It has been observed by Davis *et al.* (1997) and confirmed by Angerer *et al.* (2000, 2002) that variations in shear-wave splitting are the most distinctive seismic signatures of a high-pressure CO_2 injection on a time-lapse three-component reflection survey in a carbonate reservoir. Angerer *et al.* (2000, 2002) used a modified version of APE-modelling, incorporating fixed large fractures as well as fluid-saturated

microcracks, to match (in effect predict with hindsight) the effects of the injection. Matching the pre-injection shear-wave splitting with a model of a crack geometry having a mixture of large cracks and microcracks, the effects of the post-injection crack geometry were predicted by inserting the appropriate high pore-fluid pressure into the APE model. The synthetic seismograms of shear-wave arrivals from the predicted response exactly matched the observed arrivals for two injection pressures, one an over-pressure (Angerer *et al.* 2002). This suggests that monitoring hydrocarbon reservoirs with shear-waves may be a powerful seismic technique for understanding fluid–rock interactions and the progress of fluid–fluid fronts in production fields (Crampin 1999*b*).

Conclusions

Shear-wave splitting in Iceland, particularly above the persistent seismicity of the onshore transform zone of the Mid-Atlantic Ridge in SW Iceland, has been monitored for the four years 1996 to 1999. A number of results have been confirmed or established for the first time.

Confirmation of existing ideas

1. Virtually all shear-wave arrivals within the shear-wave window display shear-wave splitting.
2. The shear-wave splitting is consistent with propagation through distributions of nearly vertical, nearly parallel fluid-saturated cracks striking approximately perpendicular to the direction of minimum horizontal stress.
3. The normalization of time-delays to milliseconds per kilometre assumes that anisotropy is approximately uniformly distributed along the hypocentral distance. The remarkable match of APE-modelling to a range of observations (Crampin 1999*a*) suggests that this is a good first approximation.
4. Temporal changes in the time-delays between split shear-waves, indicating increases of crack aspect ratio as a result of the stress build-up, have been routinely observed before five earthquakes in SW Iceland. These changes are in the double-leafed solid angle of ray-path directions known as Band-1 (15° to 45° to the average crack plane). Shear-wave splitting in Band-1 is sensitive to changes in crack aspect ratios, and changes in aspect ratio are the immediate response of crack distributions to changes of stress (Crampin & Zatsepin 1997).
5. The only station where there was sufficient local seismicity to monitor the detailed changes in shear-wave splitting immediately

before an earthquake, showed a precursory decrease in time-delays beginning some four days before the $M5$ earthquake. A similar precursory decrease began some four hours before a $M_D3.8$ earthquake in Arkansas (Booth *et al.* 1990).

6. Using variations in shear-wave splitting above small earthquakes to stress-forecast the time and magnitude of larger earthquakes is only possible when there is persistent swarm-type seismicity to provide a continuous source of shear-wave signals.

New results

1. The shear-wave polarizations, and hence directions of maximum compressional stress, in the southern half of Iceland are NE–SW, whereas the polarizations and stress directions in northern Iceland are NNE–SSW.

2. The maximum distance at which changes in time-delays have been observed from the source of the stress changes is about 45 km before a $M5.1$ earthquake, and about 240 km before the Vatnajökull eruption.

3. The time and magnitude of a $M5$ earthquake on 13 November 1998 were successfully stress-forecast.

4. In principle, such stress-forecasting can be made at stress-monitoring sites (SMSs) where cross-hole seismics using the DOV source monitor changes in aspect ratio along ray-path directions in Band-1. A SMS is currently being set up in Iceland.

5. Similar changes in shear-wave splitting to those observed before earthquakes have also been observed to begin five months before the 1996 Vatnajökull eruption at distances of some 240 km north and 240 km WSW.

6. Relaxation of stress is inferred from the 2 ms km^{-1} per year reduction in time-delays over a two-year period following the Vatnajökull eruption. The suggested interpretation is that these changes are the response of the tectonic plates adjusting to the plume-driven spreading cycle of the Mid-Atlantic Ridge.

7. The source of the large scatter in measurements of time-delays above small earthquakes is not yet understood, and cannot be explained by conventional non-critical geophysics. It may well reflect the temporal instability of shear-wave splitting as critical crack distributions respond to fluctuations in stress of Earth and ocean tides, barometric pressures, and possibly other phenomena.

These various findings confirm that low-level deformation, before fracture criticality is reached and fracturing occurs, can be monitored with shear-wave splitting, and that the evolution of stressed fluid-saturated cracks can be modelled and evaluated by anisotropic poro-elasticity (APE). The overall conclusion is that the rock mass is highly compliant, is a critical system, and responds to small changes in stress whose effects can be monitored in detail by shear-wave splitting, and modelled by APE.

This work was partially supported by European Commission PRENLAB Projects, Contracts ENV4-CT96–0252 and ENV4-CT97–0536, and SMSITES Project EVR1-CT1999–40002. We thank R. Stefánsson, of the Iceland Meteorological Office, without whose collaboration this investigation would not have been possible. We also thank many colleagues at home and abroad, too numerous to mention individually, who have contributed to the arguments in one way or another.

References

ANGERER, E., CRAMPIN, S., LI, X.-Y. & DAVIS, T. L. 2000. Time-lapse seismic changes in a CO_2 injection process in a fractured reservoir. *70th Annual International Meeting of the Society of Exploration Geophysicists, Calgary, Expanded Abstracts,* **2**, 1532–1535.

ANGERER, E., CRAMPIN, S., LI, X.-Y. & DAVIS, T. L. 2002. Processing, modelling, and predicting time-lapse effects of over-pressured fluid-injection in a fractured reservoir. *Geophysical Journal International,* **149**, 267–280.

BOOTH, D. C., CRAMPIN, S., LOVELL, J. H. & CHIU, J.-M. 1990. Temporal changes in shear wave splitting during an earthquake swarm in Arkansas. *Journal of Geophysical Research,* **95**, 11 151–11 164.

COLE, J. H. 1997. The orbital vibrator, a new tool for characterizing interwell reservoir space. *The Leading Edge,* **16**, 281–283.

CRAMPIN, S. 1990. Alignment of near-surface inclusions and appropriate crack geometries for geothermal hot-dry-rock experiments. *Geophysical Prospecting,* **38**, 621–631.

CRAMPIN, S. 1993*a*. Do you know of an isolated swarm of small earthquakes? *EOS, Transactions, American Geophysical Union,* 74, 451 and 460.

CRAMPIN, S. 1993*b*. A review of the effects of crack geometry on wave propagation through aligned cracks. *Canadian Journal of Exploration Geophysicists,* **29**, 3–17.

CRAMPIN, S. 1994. The fracture criticality of crustal rocks. *Geophysical Journal International,* **118**, 428–438.

CRAMPIN, S. 1996. Anisotropists Digest 149 and 150. *anisotropists@sep.stanford.edu.*

CRAMPIN, S. 1998*a*. Shear-wave splitting in a critical crust: the next step. *In:* RASOLOFOSAON, P. N. J. (ed.) *Proceedings of Eighth International Workshop on Seismic Anisotropy, Boussens, 1998, Revue de l'Institut Français du Pétrole,* **53**, 749–763.

CRAMPIN, S. 1998*b*. Stress-forecasting: a viable alternative to earthquake prediction in a dynamic Earth. *Transactions of the Royal Society of Edinburgh, Earth Sciences,* **89**, 121–133.

CRAMPIN, S. 1999*a*. Calculable fluid-rock interactions. *Journal of the Geological Society,* **156**, 501–514.

CRAMPIN, S. 1999*b*. Implications of rock criticality for reservoir characterization. *Journal of Petroleum Science & Engineering,* **24**, 29–48.

CRAMPIN, S. 2000. The potential of shear-wave splitting in a stress-sensitive compliant crust: a new understanding of pre-fracturing deformation from time-lapse studies. *70th Annual International Meeting of the Society of Exploration Geophysicists, Calgary, Expanded Abstracts,* **2**, 1520–1523.

CRAMPIN, S. 2001. Developing stress-monitoring sites using cross-hole seismology to stress-forecast the times and magnitudes of future earthquakes. *Tectonophysics,* **338**, 233–245.

CRAMPIN, S. & CHASTIN, S. 2001. Shear-wave splitting in a critical crust: II - compliant, calculable, controllable fluid-rock interactions. *In:* IKELLE, L. T. & GANGI, T. (eds) *Anisotropy 2000: Fractures converted waves and case studies, Proceedings of the Ninth International Workshop on Seismic Anisotropy, Camp Allen, 2000,* Society of Exploration Geophysicists, Open File Publication, **6**, 21–48.

CRAMPIN, S. & ZATSEPIN, S. V. 1997. Modelling the compliance of crustal rock, II – response to temporal changes before earthquakes. *Geophysical Journal International,* **129**, 495–506.

CRAMPIN, S., BOOTH, D. C., EVANS, R., PEACOCK, S. & FLETCHER, J. B. 1990. Changes in shear wave splitting at Anza near the time of the North Palm Springs Earthquake. *Journal of Geophysical Research,* **95**, 11197–11212.

CRAMPIN, S., BOOTH, D. C., EVANS, R., PEACOCK, S. & FLETCHER, J. B. 1991. Comment on "Quantitative measurements of shear wave polarizations at the Anza Seismic Network, Southern California: Implications for shear wave splitting and earthquake prediction" by ASTER. R. C., SHEARER, P. M. & BERGER, J., *Journal of Geophysical Research,* **96**, 6403–6414.

CRAMPIN, S., VOLTI, T. & STEFÁNSSON, R. 1999. A successfully stress-forecast earthquake. *Geophysical Journal International,* **138**, F1–F5.

CRAMPIN, S., VOLTI, T. & JACKSON, P. 2000. Developing a stress-monitoring site (SMS) Húsavík for stress-forecasting the times and magnitudes of future large earthquakes. *In:* THORKELSSON, B. & YEROYANNI, M. (eds) *Destructive Earthquakes: Understanding Crustal Processes Leading to Destructive Earthquakes, Proceedings of the Second EU–Japan Workshop on Seismic Risk, Reykjavik, 1999,* European Commission, Research Directorate General, 136–149.

DALEY, T. M. & COX, D. 2001. Orbital vibrator seismic source for simultaneous *P-* and *S-*wave crosswell acquisition. *Geophysics,* **66**, 1471–1480.

DAVIS, T. L., BENSON, R. D., ROCHE, S. L. & TALLEY, D. 1997. 4-D 3-C seismology and dynamic reservoir characterization - a geophysical renaissance. *67th Annual International Meeting of the Society of Exploration Geophysicists, Dallas, Expanded Abstracts,* 1, 880–882, 883–885, 886–889.

EINARSSON, E. 1991. Earthquakes and present-day tectonism in Iceland. *Tectonophysics,* **189**, 261–279.

GELLER, R. J. 1997. Earthquake prediction: a critical review. *Geophysical Journal International,* **131**, 425–450.

GUDMUNDSSON, M. T., SIGMUNDSSON, F. & BJÖRNSSON, H. 1997. Ice-volcano interaction of the 1996 Gjálp subglacial eruption, Vatnajökull, Iceland. *Nature,* **389**, 954–957.

KAGAN, Y. Y. 1997. Are earthquakes predictable? *Geophysical Journal International,* **131**, 505–525.

LEARY, P. C. 1997. Rock as a critical-point system and the inherent implausibility of reliable earthquake prediction. *Geophysical Journal International,* **131**, 451–466.

MEADOWS, M. A. & WINTERSTEIN, D. F. 1994. Seismic detection of a hydraulic fracture from shear-wave VSP data at Lost Hills Field, California. *Geophysics,* **59**, 11–26.

VOLTI, T. & CRAMPIN, S. 2003. A four-year study of shear-wave splitting in Iceland: 1. Background and preliminary analysis. *In:* NIEUWLAND, D. A. (ed.) *New Insights into Structural Interpretation and Modelling.* Geological Society, London, Special Publications.

WINTERSTEIN, D. L. 1996. *Anisotropists Digest* **147**. anisotropists@sep.stanford.edu

ZATSEPIN, S. V. & CRAMPIN, S. 1997. Modelling the compliance of crustal rock: I – response of shear-wave splitting to differential stress. *Geophysical Journal International,* **129**, 477–494.

Note added in proof

Confirmation of the compliant rock-mass

Measurements at the European Commission funded SMSITES Project developing a Stress-Monitoring Site in Northern Iceland have demonstrated that the rock mass (in this case basalts) is extremely compliant to small changes. Crampin *et al.* (2003) report significant well-recorded changes in *P- SV-,* and *SH-*wave travel times, and *SV-SH* at 500 m-depth between wells at 315 m-offset, water well level changes, and changes in Global Position System strain measurements all correlating with minor seismicity (equivalent energy release to one *M*4 earthquake) at 70 km-distance. This extraordinary sensitivity reinforces that demonstrated in this chapter (compare with Item 2 in *New results* in Conclusions) and confirms the critical nature of the fluid-saturated microcracked rock mass. These results confirm both science and technology for using Stress-Monitoring Sites for stress-forecasting large earthquakes.

CRAMPIN, S., CHASTIN, S. & GAO, Y. 2003. Shear-wave splitting in a critical crust: III – preliminary report of multi-variable measurements in active tectonics. *In:* GAJEWSKI, D., VANELLE, C. & PSENCIK, I. (eds) *Proceedings of Tenth International Workshop on Seismic Anisotropy, Tutzing, 2002, Journal of Applied Geophysics* (in press).

3D analogue models of variable displacement extensional faults: applications to the Revfallet Fault system, offshore mid-Norway

T. DOOLEY[1], K. R. MCCLAY[1] & R. PASCOE[2]

[1]*Fault Dynamics Research Group, Geology Department, Royal Holloway University of London, Egham, Surrey, TW20 0EX, UK*

[2]*BHP, 1360 Post Oak Boulevard, Suite 500, Houston, TX 77056, USA*

Abstract: The Revfallet Fault system, Heidrun area, offshore mid-Norway, illustrates considerable variation in structural style along strike which is attributed to variation in coupling between basement and cover (separated by Triassic salt), due to the variable displacement along the fault system and to distinct phases of shallow level as well as basement-involved extension. Deformation in the southern part of the fault is partitioned over a number of cover-graben and associated with the development of a monoclinal flexure above the basement fault system. With an increase in displacement northwards, cover-graben become grounded on the pre-salt basement resulting in the rotation and breaching of the extensional drape fold and, eventually, partial and full coupling between basement and cover fault systems.

Analogue models were designed to simulate the 3D evolution of cover deformation above a variable displacement extensional fault separated from the cover by a ductile basal unit and to validate the structural interpretations and evolutionary models for the Revfallet Fault system. Analogue model results reveal the changing geometry along strike from a gentle extensional drape fold with minimal brittle deformation at low extension, to a breached drape fold system with significant rotation and collapse of the cover across the basement fault as displacement increases. Models conducted with a two-phase history – shallow extension (orthogonal and oblique) followed by basement-involved extension illustrate variable reactivation of cover-graben along strike during the basement-involved event, and collapse and rotation of the extensional footwall to the basement fault as displacement increases to a maximum. Coupling between cover faults and the basement fault only occurred where an oblique, precursor cover-graben cuts across the basement fault from hanging wall to footwall. Closer examination of the structural and stratigraphic relationships reveals major discrepancies between throw on cover and basement fault segments. The analogue model results compare favourably to seismic data from the Revfallet Fault system and validate the structural model for this fault system.

The hydrocarbon province of mid-Norway is situated north of the Triassic–Jurassic extensional systems of the North Sea and SE of the Late Cretaceous–early Tertiary volcanic, oceanic margin of the Vøring Marginal High (Fig. 1). This study focuses on the NE–SW and NNE–SSW trending Revfallet Fault system and its extensional footwall, the Nordland Ridge (Figs 1 & 2), and follows on from a 3D seismic analysis of the structure by Pascoe *et al.* (1999). The Revfallet Fault/Nordland Ridge system is a major Upper Jurassic to Cretaceous extensional fault system that displays variations in coupling between basement and the Jurassic–Cretaceous cover sequences due to intervening Triassic evaporite sequences, and variations in displacement along the Revfallet Fault system (Pascoe *et al.* 1999; Fig. 2). Withjack *et al.* (1990) and Withjack & Callaway (2000) documented decoupled deformation styles in the form of extensional forced folding in the cover above the basement faults from 2D analogue models and from the Haltenbank area. Pascoe *et al.* (1999) put forward a new model for the development of this fault system based on new 3D seismic data as well as comparison to published analogue models (Withjack *et al.* 1990; Vendeville & Jackson 1992) and in-house structural studies (Dooley & McClay 1997). This model differed from previous studies as it concluded that the main extensional event was in Aptian to Cenomanian times rather than the Late

From: NIEUWLAND, D. A. (ed.) *New Insights into Structural Interpretation and Modelling.* Geological Society, London, Special Publications, **212**, 151–167. 0305-8719/03/$15

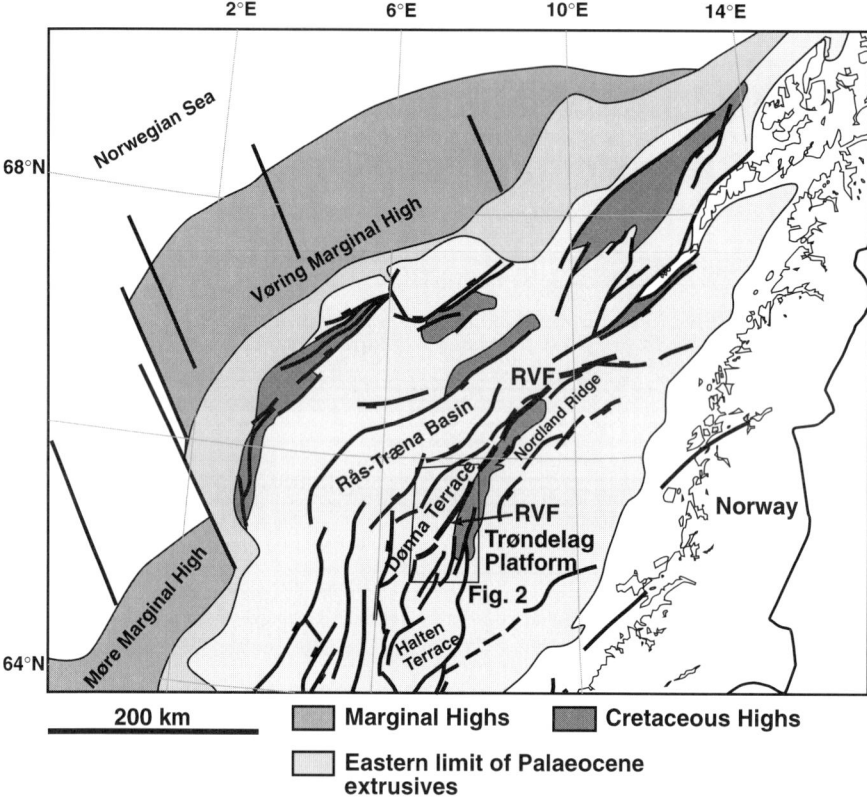

Fig. 1. Major tectonic features of the mid-Norway margin and the study area (modified after Pascoe *et al.* 1999). RVF, Revfallet Fault system.

Jurassic to earliest Cretaceous event documented by previous workers (Blystad *et al.* 1995; Koch & Heum 1995).

A series of 3D analogue models was carried out to investigate the relationship between basement–cover coupling and cumulative displacement on a basement master fault with variable displacement along strike, and to validate the structural model of Pascoe *et al.* (1999). 2D models such as those by Vendeville & Jackson (1992), Vendeville *et al.* (1995), Withjack *et al.* (1990) and in-house structural modelling (Dooley & McClay 1997) illustrate the buffering effect of salt on cover deformation, with initial displacements accommodated by a cover monocline that becomes progressively steeper and dissected by cover faults with increased displacement on the basement fault (Fig. 3; Dooley & McClay 1997). These models illustrate that despite large displacements on the basement system, cover deformation is largely decoupled from the basement fault and vertical displacements are accommodated over a diffuse zone consisting of a footwall graben and complex, steeply dipping

fault segments that cut the hanging wall syncline (Fig. 3). Recent modelling by Withjack & Callaway (2000) also indicated the importance of strain rate, layer thickness and cohesive strength of the overburden on the amount of coupling between basement and cover. The models presented in this chapter differ from previously published results for two reasons: (1) the deformation apparatus models a single extensional fault with variable displacement along strike, i.e. displacement increases from zero to a maximum in the middle of the fault trace; and (2) models were initially deformed under gravity to generate decoupled cover-graben and thus to investigate the effects of pre-existing structures on the amount of coupling between basement and cover during a major basement-involved extensional event.

Revfallet Fault/Nordland Ridge system

Regional setting

Figures 1 and 2 illustrate the setting and local structure of the study area. The Revfallet Fault sys-

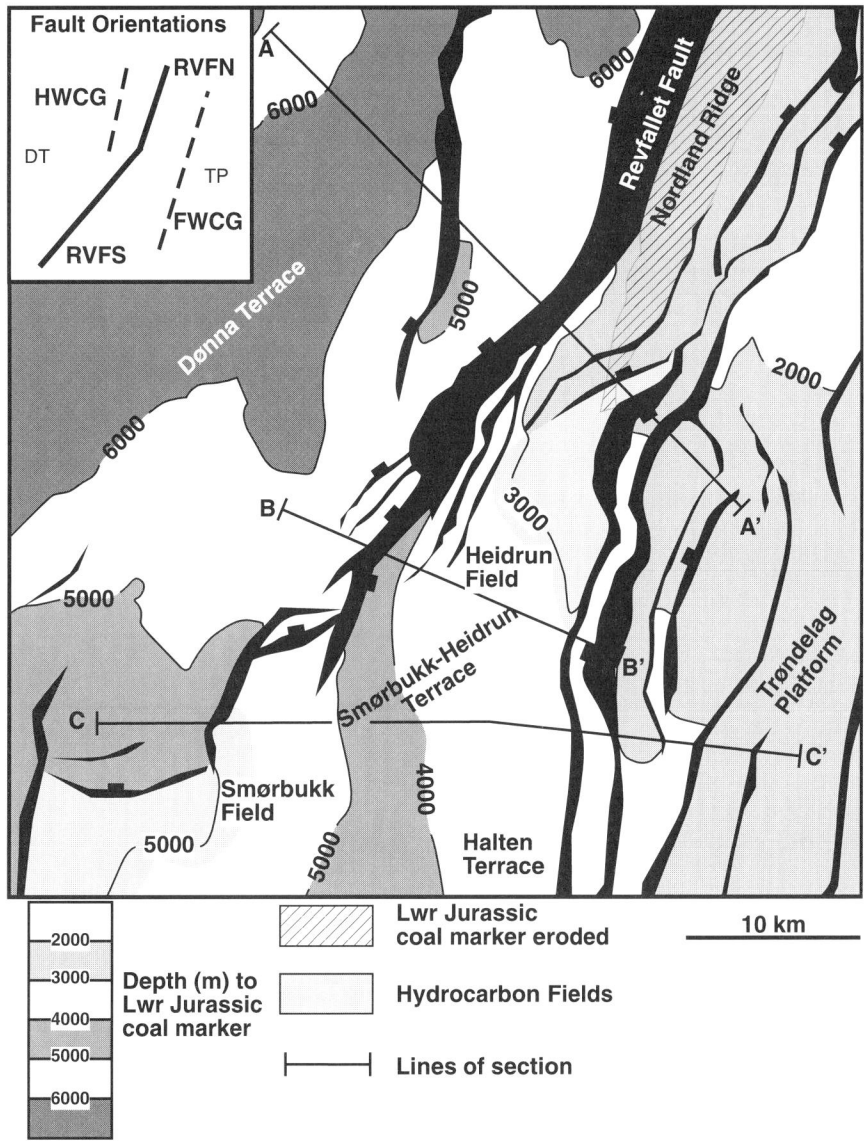

Fig. 2. Structure map of the area on the intra-Lower Jurassic coal marker. Lines of section are shown in Figure 5. Inset figure illustrates orientations of the Revfallet Fault system (RVFS, Revfallet Fault south; RVFN, Revfallet Fault north) and orientations of cover-graben in hanging wall and footwall of this extensional fault (HWCG, hanging wall cover-graben; FWCG, footwall cover-graben). DT, Dønna terrace; TP, Trøndelag Platform.

tem and associated extensional footwall block, the Nordland Ridge, are situated between the mainland and the deep water Rås–Træna Basin, inboard of the Vøring Margin (Fig. 1). The Revfallet Fault/Nordland Ridge system separate the shallow Trøndelag Platform from the Dønna Terrace, and it consists of a complex system of interlinked fault segments that can be defined for over 200 km strike length (Figs 1 & 2; Pascoe *et al.* 1999). In the south

the Revfallet Fault tips out into a broad zone of distributed extension, as displacement is transferred to the south across the Smørbukk–Heidrun and Halten Terraces through the Smørbukk–Revfallet Transfer system (Fig. 2; Pascoe *et al.* 1999). In the cover this generates a relatively unfaulted relay in which the Heidrun and Smørbukk hydrocarbon fields are situated (Fig. 2; Pascoe *et al.* 1999). Throw on the Revfallet Fault

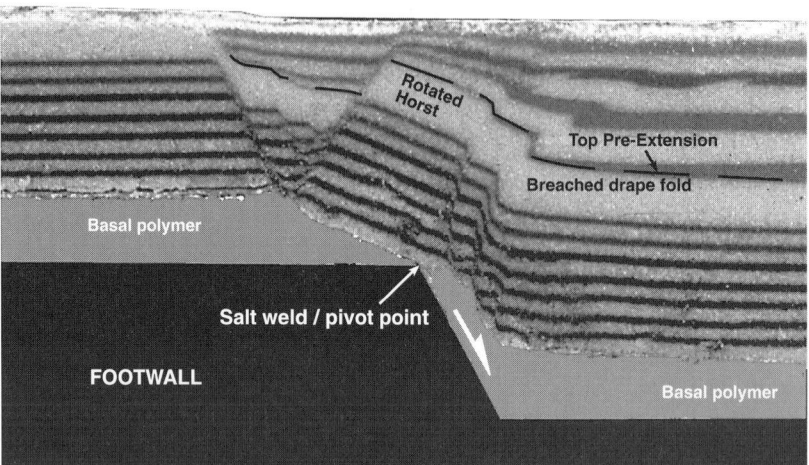

Fig. 3. 2D model of deformation in the cover above a ductile horizon above a 60° extensional fault (from Dooley & McClay 1997). Strata below the marker 'Top Pre-Extension' were originally horizontal, pre-kinematic, strata overlying an horizontal ductile basal unit. The hinge point of the extensional monocline is marked by the major synthetic fault in the footwall graben.

gradually increases northwards from the transfer system, up to a maximum of approximately 7 km, along the north–south trending segment north of line A–A′ (Fig. 2; Pascoe *et al.* 1999).

The stratigraphy of the area consists of up to 5 km of marine and marginal non-marine Triassic to Middle Jurassic rocks (including 800 m of Triassic evaporites), up to 3 km of shelfal to bathyal Upper Jurassic and Cretaceous rocks and approximately 2 km of abyssal to deltaic Tertiary rocks (Fig. 4). Figure 4 also highlights three major Mesozoic extensional events: Camian–Scythian, Bathonian–Oxfordian and Aptian–Cenomanian (Pascoe *et al.* 1999). Triassic evaporites play a pivotal role in the structural model presented by Pascoe *et al.* (1999), and in the models presented in this chapter. A well on the Trøndelag platform recorded two Middle to Upper Triassic evaporitic intervals of approximately 400 m thickness each and these two salt layers can be traced over most of the Trøndelag Platform and Halten Terrace (Fig. 5; Pascoe *et al.* 1999). These evaporites are extremely thin to absent along portions of the Nordland Ridge (Pascoe *et al.* 1999).

Structural geometry

Pascoe *et al.* (1999) presented a detailed discussion of the structural geometry and along-strike changes along the Revfallet Fault/Nordland Ridge system, and thus only a short summary is included here. Figure 5 illustrates three transects across the study area demonstrating along-strike variations in struc-

tural geometry from south to north. Note that the Revfallet Fault system deviates from its NNE–SSW trend south of the Nordland Ridge to a NE–SW trend crossed by the three regional section lines (Fig. 2).

Line C–C′ (Fig. 5c) illustrates the general structure to the south of the termination of the Revfallet Fault. Here, deformation is partitioned across a broad zone and cover faulting detaches on or within the Triassic salt (Fig. 5c; Pascoe *et al.* 1999). The elevation drop between the Trøndelag Platform and the Smørbukk–Heidrun terrace is accommodated by an extensional drape fold and associated small graben in the Jurassic rocks and a series of faults in the basement (Fig. 5). The Upper Jurassic Viking Group shows a gradual westward thickening into the hanging wall syncline of the extensional drape fold and localized accumulation.

Approximately 10 km to the north a distinct Revfallet Fault is evident in the basement with approximately 3 km of throw (line B–B′, Fig. 5b; Pascoe *et al.* 1999). Cover deformation along this transect consists of an extensional drape fold with a steep front limb across the basement Revfallet Fault generating a hanging wall syncline (line B–B′, Fig. 5b). Major thickening of the Triassic evaporite sequence occurs across the basement fault and grounding of the Jurassic cover onto the Triassic pre-salt basement occurs. Pronounced erosion of the Middle to Upper Jurassic Viking Group has occurred on the outer arc of the monocline, whereas a complete Jurassic sequence is present on the Smørbukk–Heidrun terrace and westward

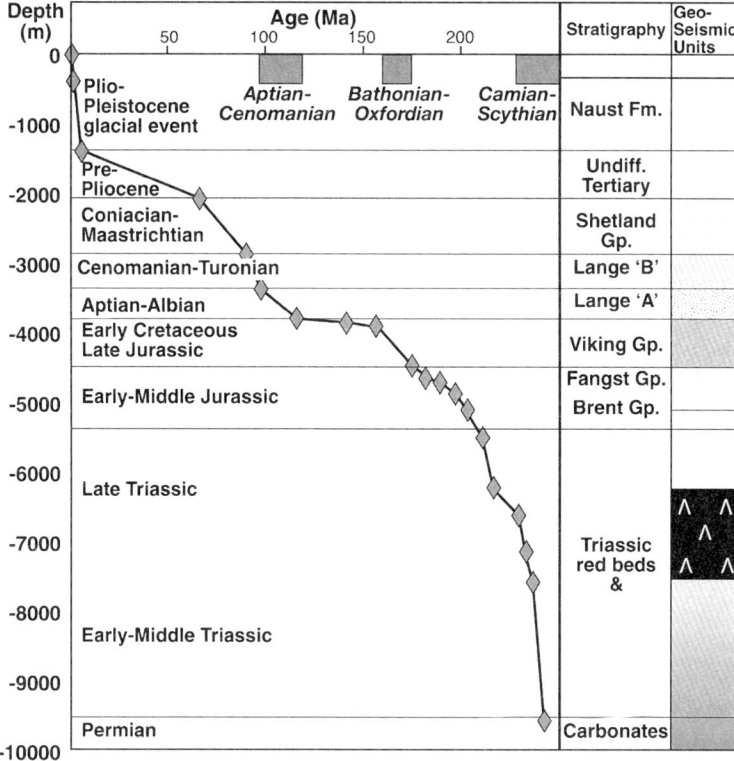

Fig. 4. Regional stratigraphic column and net sediment accumulation data from the study area. The three major extensional areas discussed in the text are marked by the grey bars at the top of the figure. Diagram modified from Pascoe *et al.* 1999.

thickening of the Upper Jurassic Viking Group is seen in the crestal graben and immediately west of the basement Revfallet Fault (Fig. 5b).

Line A–A′, to the north of the Heidrun Field, illustrates a strikingly different structural style with the Nordland Ridge and Dønna Terrace clearly distinct features at both Jurassic–Cenomanian and basement levels (Fig. 5a; Pascoe *et al.* 1999). In this location there is approximately 5 km throw on the Revfallet Fault at the pre-salt surface and coupling between basement and a fault in the cover has occurred (Fig. 5a). Preservation of Upper Jurassic strata in a graben in the footwall and the eroded tilt blocks on the Dønna terrace indicate that major offset across the Revfallet Fault occurred after the Jurassic (Fig. 5a; Pascoe *et al.* 1999). In addition, major thickness increase of the Upper Jurassic Viking Group occurs immediately across the Revfallet Fault, demonstrating coupling between basement and cover during the Aptian–Cenomanian extensional event (Figs 4 & 5). The Aptian–Albian Lange 'A' has a wedge-shaped downlapping geometry along this line, interpreted as synextensional sedimentation, onlapped by the

younger Cretaceous strata that appear to be passive infill deposits (Lange 'B', Figs 4 & 5; Pascoe *et al.* 1999). The shallow cover-grabens observed in the seismic lines above the salt detachment, have a NNE–SSW trend that roughly parallels the fault systems on the Trøndelag platform (NNE–SSW to north–south; Fig. 2). These structures are thus oblique to the main trace of the Revfallet Fault system to the south of the Nordland Ridge (Fig. 2; Pascoe *et al.* 1999).

Analogue modelling

Model configurations

The physical models presented in this chapter underwent a dual-phase deformation history consisting of an initial shallow extensional event and a subsequent basement-involved extensional event, similar to that described along the Revfallet Fault system (Pascoe *et al.* 1999). During the first phase no extension occurred on the basement fault. Models were 'seeded' (see below) with cover-graben both parallel and oblique to the basement fault sys-

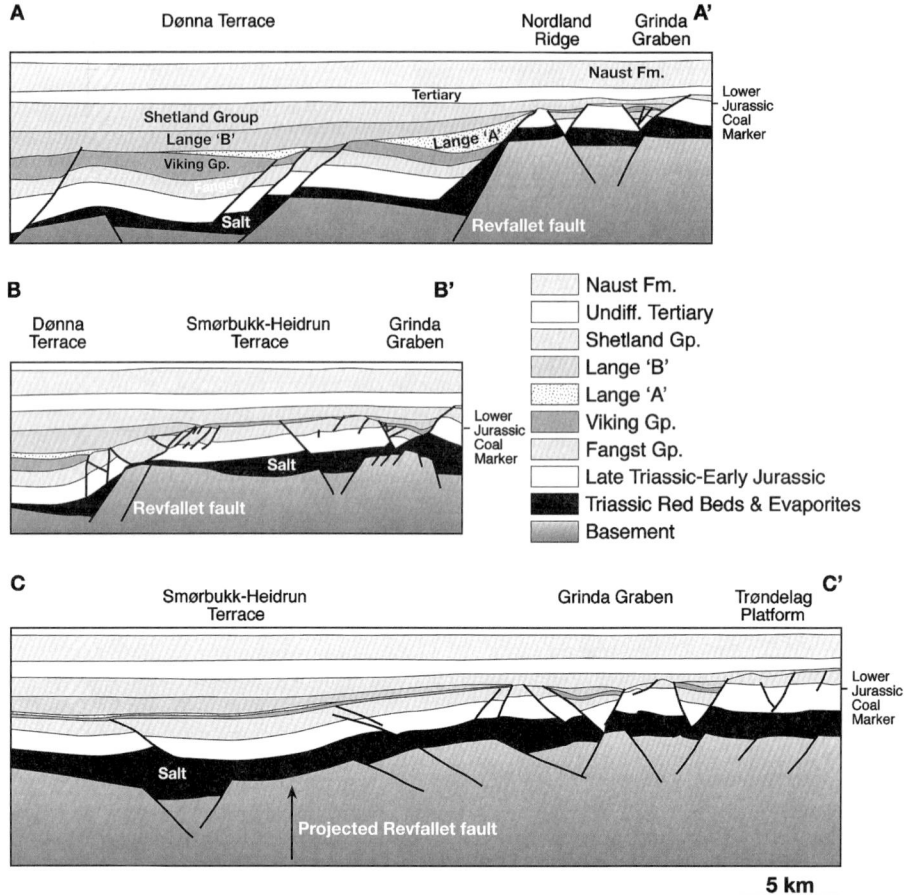

Fig. 5. (a, b, c) Semi-regional transects across the southern Revfallet Fault system. Location marked on Figure 2. Note the dramatic change in cover deformation style from south to north. Lower Jurassic coal marker is horizon mapped in Figure 2.

tem (Fig. 6). The bend in the Revfallet Fault to a NNE–SSW trend to the south of the Nordland Ridge (Fig. 2), results in cover-graben crossing the trace of the basement fault and thus moving from upthrown to downthrown block (Pascoe *et al.* 1999). These cover-graben have north–south trends that parallel the fault systems to the east of the Nordland Ridge–Halten Terrace on the Trøndelag Platform (see Fig. 2). The models run with gravity-induced precursor graben oblique to the basement fault system thus investigate the effects of pre-Cretaceous cover-graben moving from hanging wall to footwall across the trace of the Revfallet Fault zone.

Models were constructed in a deformation rig specifically designed to model a single, 70 cm long, extensional fault with an elliptical displacement curve (Fig. 6). The basal salt section was modelled using a viscous silicone polymer (SGM-36, Dow

Corning; Weijermars *et al.* 1993) and covered with layered, dry, quartz sand with an average grain size of 190 μm to simulate the brittle sedimentary sequence. The models are dynamically scaled such that 1 cm in the model approximates to 1 km in nature (see Brun *et al.* (1994) and McClay (1990) for discussions on scaling). In all the models the polymer layer was 0.5 cm thick to model a proto-type stratigraphy of 500 m thick Triassic evaporitic units. Basement extension was generated in the overburden by a stepper motor with a displacement rate of 1.1×10^{-4} cm s^{-1} linked to the triangular-shaped hanging wall block in the deformation rig (Fig. 6). This displacement rate simulates an extensional fault system with a relatively high displacement rate ($\geq 10^{-9}$ cm s^{-1}; Withjack & Callaway 2000) such as that described for the Revfallet Fault system in Mid-Cretaceous times (Pascoe *et al.* 1999) or where the salt had a high viscosity

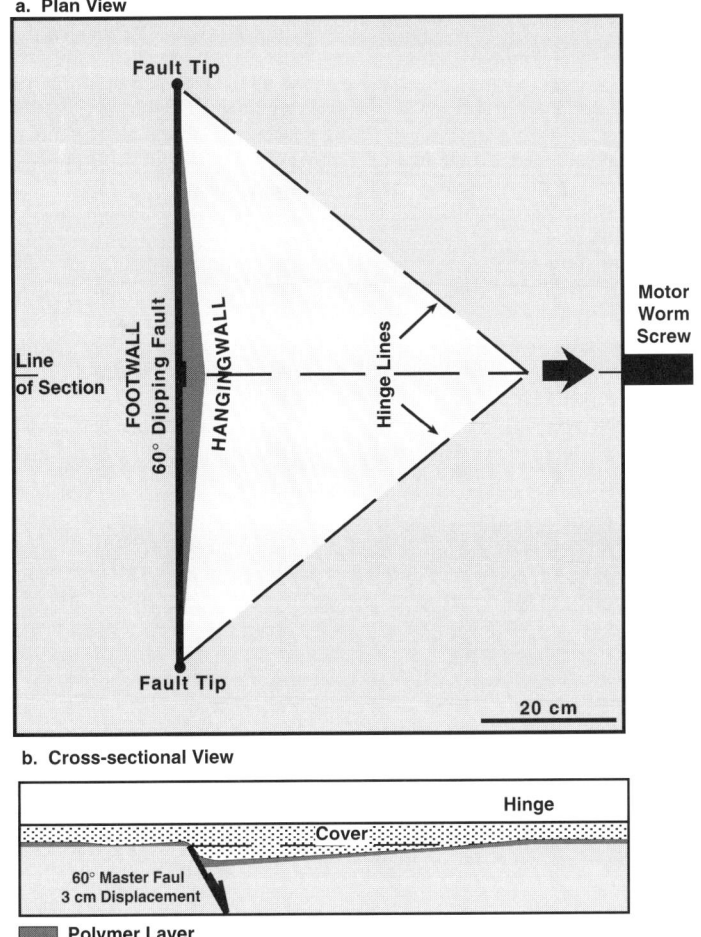

a. Plan View

Fault Tip

FOOTWALL
60° Dipping Fault

HANGINGWALL

Hinge Lines

Line of Section

Motor Worm Screw

Fault Tip

20 cm

b. Cross-sectional View

Hinge

Cover

60° Master Faul
3 cm Displacement

Polymer Layer

Fig. 6. Summary diagram illustrating the geometry of the variable displacement deformation apparatus. Lateral hinges separate deformed from non-deformed, zero-displacement, strata.

(possibly due to previous thinning; see Withjack & Callaway 2000). The model series was conducted to test and validate deformation processes and structural models of deformation above variable displacement extensional faults and to test the influence of precursor cover-grabens on the degree of coupling along the fault system and not specifically to model any one portion of the Revfallet Fault system. Thus, synkinematic sedimentation rates during the basement extension phase were not dynamically scaled.

The two models presented in detail in this chapter were identical in terms of polymer thickness, overburden thickness and displacement rates. The only variable was the orientation of the precursor cover-graben with respect to the basement fault system. These pre-basement extension cover-

graben were 'seeded' by removing 2 mm thick lines of sand from an initial, uniform thickness, 4 mm sand layer overlying the polymer. The model was left for 16–18 hours, allowing polymer to well up along these seed lines due to the differential loading. The remaining 2.1 cm sandpack was then added and the rig was placed on a 2° tilt. This tilting generated graben in the cover exclusively above the 'seeds' (see Figs 7a & 9a). Cover-graben were progressively infilled every 30 minutes and allowed to extend for 4.5 hours, prior to the basement-involved extensional event. The model was then extended in 5 mm increments and a different synkinematic layering added to infill extensional features after each 5 mm displacement on the basement fault. Models were gelled after each experiment and serially sectioned. The two models

presented in detail in this chapter are from a series
of 16 models run during the experimental study.
Models were repeated to test reproducibility.

Model results

Model 1: fault-parallel cover-graben At the end of
the 2° tilt phase a cover-graben system had
developed above the 'seed' in the immediate footwall of the basement extensional fault system in
this thin-skinned extensional event (Fig. 7a). This
extensional system consisted of two symmetrical
graben separated by a central horst block (Fig. 7a).
The down-dip graben partially overlaps the projected surface trace of the basement fault (Fig. 7a).
During the basement-involved extensional phase
the down-dip cover-graben was reactivated along
the majority of the strike length of the basement

fault, gradually losing displacement along strike
and tipping out approximately 10 cm from the lateral tips of the basement fault (Fig. 7b). Initially a
monocline developed in front of the reactivated
cover-graben and this progressively steepened as
displacement increased. The monocline is bounded
down-dip by a fault with apparent reverse separation (Fig. 7b).

Serial sections through the model reveal the
complex, multiphase history of this experiment
(Fig. 8). These sections show a sequence from zero
extension on the basement fault (section 2, Fig. 8)
to section 19 in the centre of the model with 3.2
cm of displacement on the basement fault (Fig. 8).
Section 2 reveals cover-graben formed during the
tilt phase that have not been reactivated during subsequent basement extension, and thus lack synext-
ension sediments. With a gradual increase in dis-

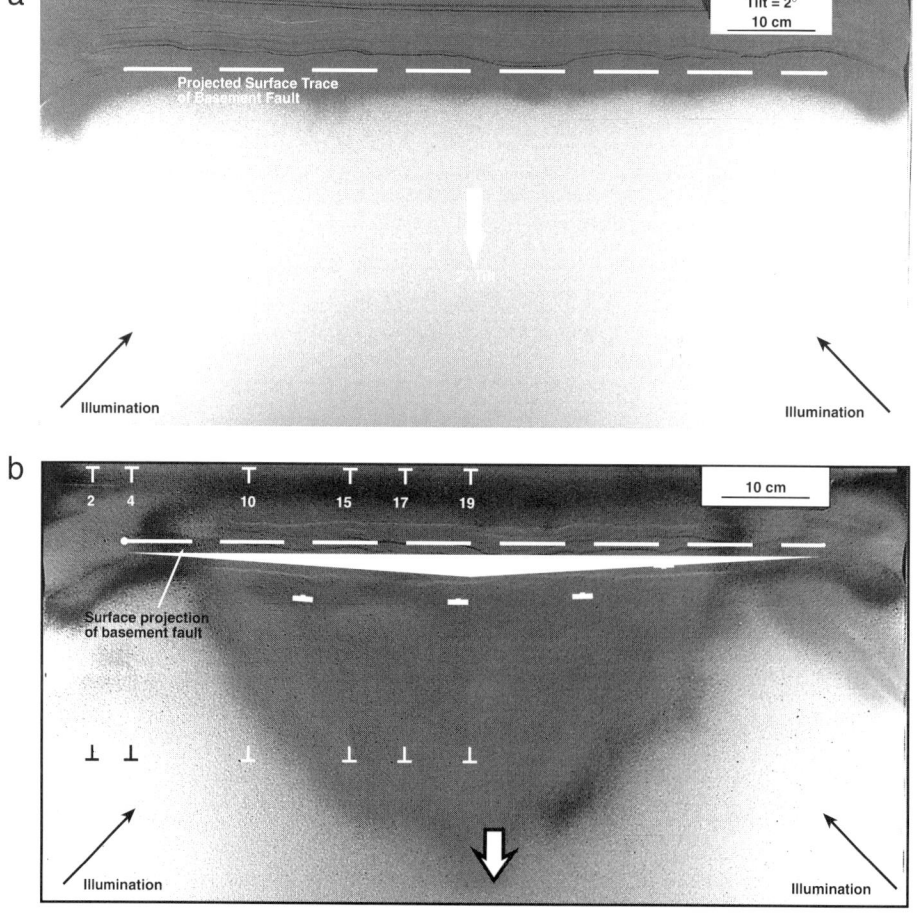

Fig. 7. (**a**) Overhead view of Model 1 after 6 hours on a tilt of 2°. Note the cover-graben system in the immediate
footwall to the basement structure. (**b**) Model 1 after 3.2 cm maximum displacement on the basement structure. Note
partial reactivation of the tilt-graben and the breached hanging wall syncline.

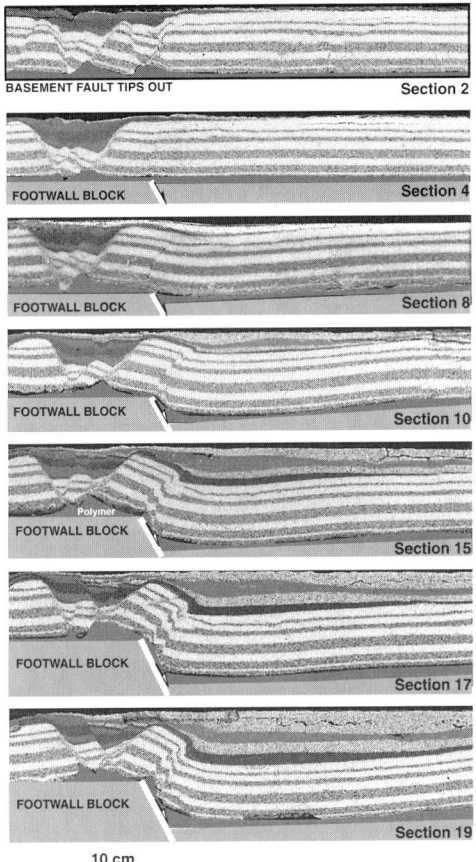

Fig. 8. Serial sections through Model 1. Locations are indicated on Figure 7. Tilt-graben are reactivated as displacement exceeds 1 cm in this model and with increasing displacement the antithetic faults in the cover-graben are rotated in an clockwise about a pivot point on the crest of the footwall block.

placement on the basement fault from section 4 to section 19, the antithetic faults of the cover-graben are seen to have rotated clockwise as these graben collapse across the basement structure (Fig. 8). Rotation of the footwall increases as displacement along the fault increases resulting in a steep limb in the centre of the model that is bounded by a convex-upward fault system that accommodates the rotation and translation of the footwall to the tilt-graben and vertical motion of the hanging wall to the basement fault (Fig. 8). Despite significant displacement on the basement fault in the centre of the model there is no throughgoing fault linking basement to cover and the deformation is partitioned between reactivation of the tilt-graben, rotation of the cover and activity along convex-upward fault zones (Fig. 8).

Model 2: oblique cover-graben The experimental set-up was identical to that of Model 1 except that the seeded cover-graben were 25° oblique to the basement fault trace (Fig. 9a). At the end of the tilt phase a series of extensional graben traverse the model but only one of these intersects the projected trace of the basement fault (Fig. 9a). During the basement-involved extensional phase of this model this cover-graben was partially reactivated in the immediate vicinity of the projected trace of the basement fault (Fig. 9b). Along the majority of the fault the predominant structural features consist of a tapering graben with more prominent antithetic fault segments, and a monocline bounded in the hanging wall of the basement fault by a fault with apparent reverse separation (Fig. 9b). The reactivated portion of the tilt-graben forms an S-shape as it links the base of the drape fold with a basement-fault-parallel segment immediately in the footwall of the basement fault segment (Fig. 9b).

Figure 10 illustrates a series of sections through the model demonstrating the change in deformation geometry along the strike of the variable displacement system. Synextensional structures consist primarily of drape folds that become progressively steeper and are breached by convex-upward fault systems as displacement increases (Fig. 10). All sections show ponding of the polymer in the immediate hanging wall of the basement structure (Fig. 10). Section 8 reveals a well-developed extensional drape fold with no breaching, and a tilt-graben preserved some distance into the hanging wall of the basement fault (Fig. 10). The drape fold is very well developed in this section, as there is no tilt-graben at this location to influence the deformation geometry (compare to Section 8 in Figure 8). Sections 17–22 shows the structural geometry where a tilt-graben is observed to migrate through the hanging wall, to the immediate hanging wall and juxtaposed above and in the immediate footwall of the basement fault. Section 17 shows the typical geometry of a high displacement section with no reactivation of the tilt-graben.

Deformation in the footwall is characterized by an asymmetric graben with partial collapse of this structure across the crest of the basement fault and subsequent rotation of the antithetic cover faults (Fig. 10). Immediately in the hanging wall of the basement fault a vertical to overturned fault system marks the boundary between footwall and hanging wall with major thickening of the synextensional sediments across this structure (Fig. 10). Sections 18–20 illustrate a degree of coupling between basement and cover due to the reactivation of the tilt-graben (Fig. 10). On section 18 reactivation of a synthetic fault that bounds the pre-extensional tilt-graben has occurred along with the formation of a weak graben in the footwall of the basement struc-

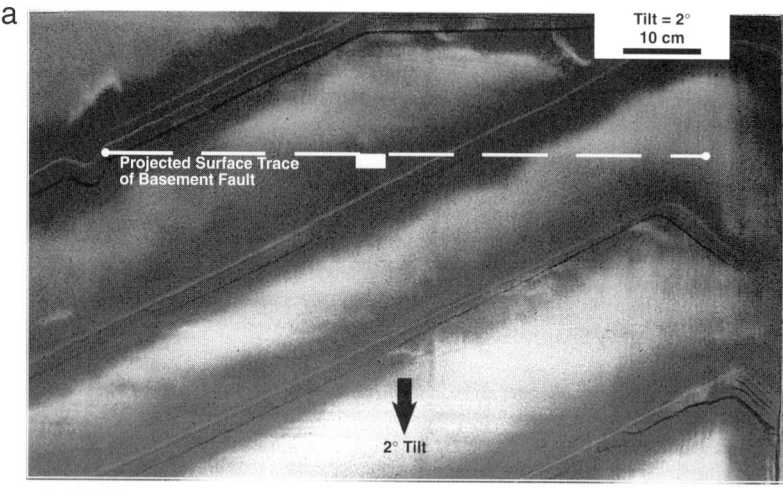

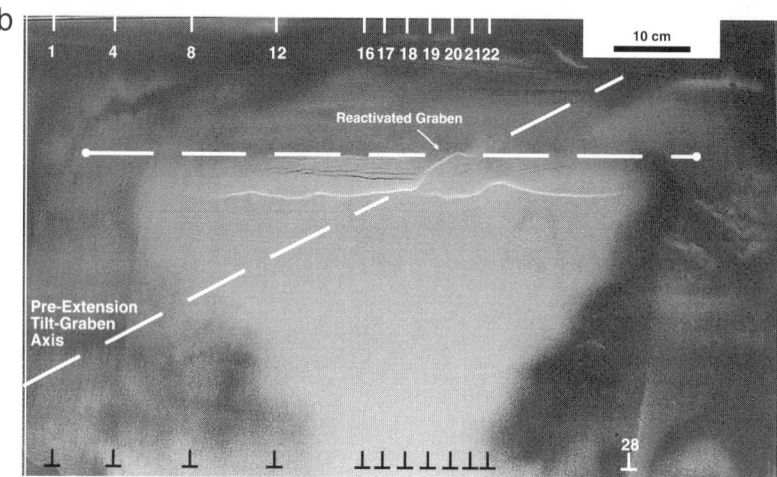

Fig. 9. (**a**) Overhead view of Model 2 after 6 hours at a tilt of 2°. Cover-graben are developed above the 'seed' lines placed in the model. One of the graben intersects the basement fault at the site of maximum displacement. (**b**) Overhead view of Model 2 after 4.2 cm maximum displacement on the basement fault. Note the reactivated cover-graben in the central part of the basement fault, keystone graben and breached drape fold.

ture and a convex-upward fault surface in the hanging wall (Fig. 10). On section 19 the tilt-graben synthetic fault coincides with the basement fault and clear coupling has occurred (Fig 10). The antithetic fault of the tilt-graben has also been reactivated as a convex-upward structure bounding the triangular zone of deformation above the basement fault (Fig. 10). Closer examination reveals that there is less than 1 cm displacement on the cover fault during the basement event as opposed to 3.33 cm displacement on the basement fault itself, and thus the appearance of a hard link between basement and cover structures is partially misleading (Section 19, Fig. 10).

On Section 20, where the tilt-graben is situated in the footwall of the basement structure, reactivation of both synthetic and antithetic tilt-graben extensional faults in the immediate footwall to the basement structure has occurred, resulting in enhanced stretching across the graben and a resultant growth of the reactive diapir beneath (Fig. 10). The footwall to the tilt-graben has grounded on the crest of the basement structure and illustrates significant clockwise rotation across this cusp (Fig. 10). Again, there appears to be a clear link between the basement fault and a reactivated tilt-graben fault in the cover but closer inspection shows that this is not the case (Fig. 10). The reactivated syn-

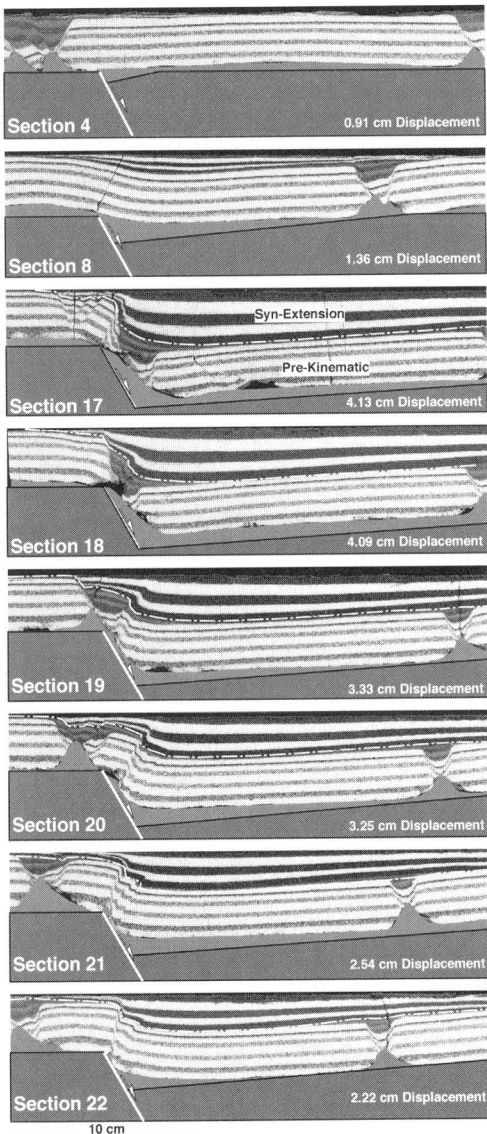

Fig. 10. Serial sections through Model 2. Locations are indicated on Figure 9. Note the extensional drape fold geometry on Section 8 and the gradual reactivation of the tilt-graben as it approaches and cross-cuts the basement structure.

thetic fault has approximately 6–8 mm of synbasement extension displacement (blue and white layers) but this fault does not cut down-section further than the top of the polymer (Fig. 10). The sharp, steep, sidewall to the reactive diapir gives the impression of up-section linkage between the basement and cover that is misleading (Fig. 10). Sections 21 and 22 illustrate grounding of the cover

on the crest of the basement block and rotation of the tilt-graben footwall across this hinge point, due to reactive rise of polymer beneath the tilt-graben during the tilt phase and consequential thinning of the surrounding polymer. Little or no reactivation of the tilt-graben is observed in these sections.

Discussion

Analogue models

2D models of cover extension above a ductile basal unit generate a footwall graben that gradually widens by outer arc extension as displacement increases on the basement structure (Fig. 3). The footwall to the cover-graben rotates about a pivot or weld point on the basement footwall cut-off line, allowing steepening and collapse of the forelimb of the extensional drape fold (Fig. 3). This decoupled geometry is similar to that documented by Withjack et al. (1990) and to models described in Withjack & Callaway (2000). The 3D models presented in this chapter have the added difference of variable displacement and thus represent an evolutionary spectrum from low to high strains in one model. The addition of precursor cover-graben allows evaluation of the influence of pre-existing structures on the final geometry, and their role in the variability in coupling between basement and cover.

In general, models evolve from non-deformed (i.e. outside the influence of the basement fault) through extensional drape folds to breached drape folds with significant rotation of a cover horst block about a pivot point on the footwall cut-off line as displacement increases along the basement fault (Figs 8, 10 & 11a). Synextensional sediments thin onto the fold hinge along sites of limited breaching (Figs 8, 10 & 11a). Vertical sections through the models reveal that up to 1.5 cm displacement on the basement fault can be taken up by extensional folding of the cover above the 'salt' analogue prior to breaching of the cover by brittle structures (Figs 8, 10 & 11). Model 1, run with tilt-graben that parallel the basement structure, developed drape folds at low displacements with no reactivation of the tilt-graben. Along strike, as displacement increased on the basement fault, the tilt-graben is gradually reactivated and the cover collapses across the basement fault as the extensional footwall is oversteepened and rotates around the weld point. In the central portion of the basement structure, cover faults in the precursor tilt-graben are reactivated and antithetic faults are rotated to shallower angles due to the collapse of the cover. New convex-upward fault systems mark the boundary from this extensional rotation and

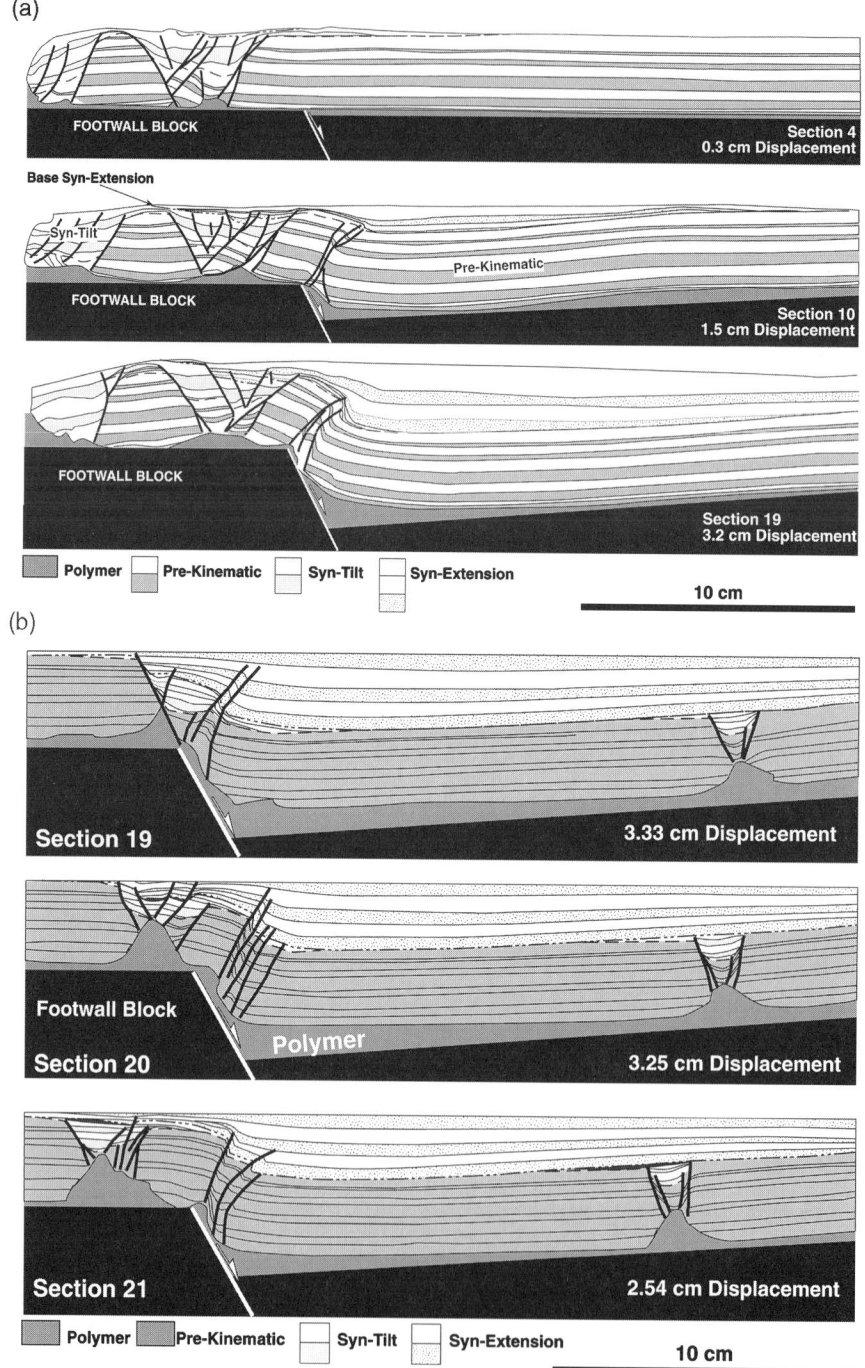

Fig. 11. (**a**) Line drawing interpretations of selected sections from Model 1. Section locations located on Figure 7b. Note the gradual collapse of the cover across the crest of the footwall block and the rotation of antithetic cover faults. (**b**) Line drawing interpretations of selected sections from Model 2 illustrating partial coupling between cover and basement as the tilt-graben cross-cuts the basement structure. Location of sections illustrated on Figure 9b.

collapse zone and the passive, translating, hanging wall (Fig. 11a).

Model 2, with oblique precursor cover-graben, generated broadly similar geometries to those with parallel precursor graben. The lack of precursor graben along much of the basement fault trace in this model allowed the development of classic extensional drape folds at sites of displacements up to 1.5 cm (e.g. Section 8, Fig. 10). With increased displacement on the basement fault along strike, small keystone cover-graben form in the immediate footwall of the basement structure and the hanging wall dipping panel is truncated by a convex-upward fault system above the basement fault (Sections 17 and 20, Figs 10 & 11b). Four sections from the central portion of the model illustrate reactivation of the tilt-graben (Figs 10 & 11b). Despite the appearance of full coupling between basement and cover in these sites, closer examination of the sections reveals major discrepancies between displacement on the basement and cover faults, with deformation in the cover distributed across several structures rather than on one cover fault (Figs 10 & 11b). The symmetric nature of the diapirs in sections 19 and 20 (Figs 10 & 11b) indicates that extension in the cover above these diapirs was due to outer arc extension across the reactivated tilt-graben and subsequent reactive rise of the polymer beneath these graben.

Comparison with the Revfallet Fault system

Figure 12a illustrates detail of an interpreted seismic section from the southern end of the Revfallet Fault system near line B–B' (Figs 2 & 5). This section illustrates spectacular decoupled geometries from this position of the Revfallet Fault zone. Clear structural pull-up of the Jurassic and Cretaceous beds is observed in the hanging wall of the Revfallet Fault above a major accumulation of salt adjacent to the fault plane (Fig. 12a). This 'pull-up' of the hanging wall strata reflects initial extensional drape folding across the basement structure (cf. Pascoe et al. 1999), and is almost identical to geometries seen from high displacement sites in Model 1 (Figs 8 & 11a). On the crest of the basement block a major cover-graben is present with steep northwesterly dipping cover faults countered by a moderate to shallow southeasterly dipping fault associated with localized thickening of Jurassic strata during the early shallow extensional event (Fig. 12a). This shallow-dipping antithetic fault is associated with a weld beneath the cover-graben and is analogous to geometries observed in Model 1 whereby an original cover-graben collapses across the basement fault as displacement on this structure increases resulting in outer arc extension across the graben and

rotation of the antithetic fault segments (Figs 8, 11a & 12a). Erosion of the part of the Upper Jurassic section occurred during amplification and eventual breaching of the monocline during the Cretaceous event. The steep to overturned faults in the immediate hanging wall of the main Revfallet Fault are analogous to the convex-upward fault segments observed in the models, marking the transition from the steep forelimb of the extensional fold to the flat-lying segments of the system (Fig. 12a).

Coupled deformation geometries along the Revfallet Fault system are seen to the north of the Heidrun Field (Figs 2 & 12b). This takes the form of tilted fault blocks formed predominantly during the main phase of structural development during the Cretaceous extensional event (cf. Pascoe et al. 1999). Jurassic strata in the hanging wall are terminated abruptly by the steeply dipping Revfallet Fault in this location (Fig. 12b). Bending of some of the Jurassic strata is observed and may relate to an early phase of folding prior to breaching as displacement increased, similar to that seen along low-displacement sites of the two models (Fig. 11). There has been significant erosion and/or non-deposition on the crest of this fault block. Despite breaching of the cover by the basement system, structures observed above Salt 2 on the crest of the tilt-block indicate that extension during the major Cretaceous event was still partitioned across a series of cover-graben in the footwall of the Revfallet Fault (line A–A', Fig. 5; Fig. 12b). Serial sections through Model 2 in the vicinity of the reactivated tilt-graben display similar geometries whereby large displacements are observed on the basement fault, but in the thin cover on the crest and footwall displacement is partitioned across the reactivated graben (Figs 10, 11b & 12).

Structural development

Analogue model results indicate that with increasing displacement hanging wall deformation evolves from extensional folding with minor footwall keystone graben formation through to breached hanging wall folds with varying degrees of cover-graben reactivation depending on the location of the cover-graben with respect to the basement fault system. Cross-sections across the Revfallet Fault system from south to north document a similar evolution from decoupled to coupled deformation geometries as displacement increases on the Revfallet Fault system (Fig. 5). Restoration of cross-section B–B' immediately south of the Heidrun Field by Pascoe et al. (1999; Fig. 13) reveals the time–structure evolution of this portion of the fault system and the variation in geometry is very similar to that observed in Model 1 (Fig. 8), as well

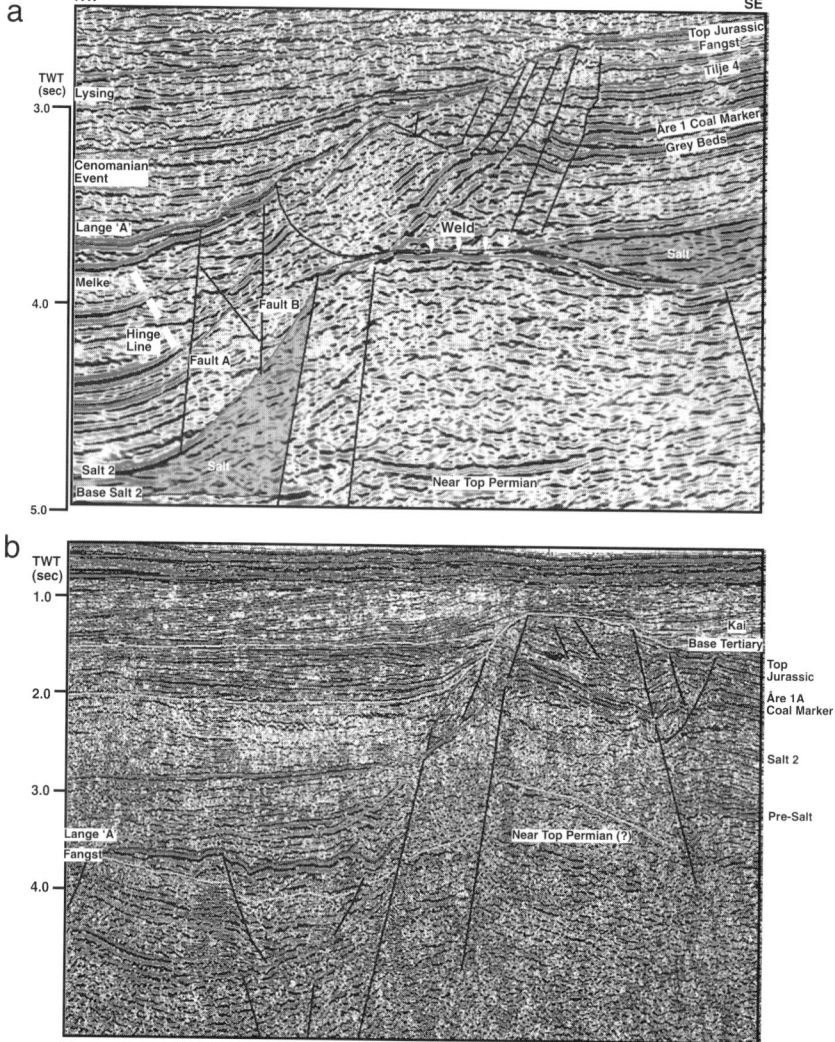

Fig. 12. (**a**) Interpreted seismic line near the location of line B–B′ in Figure 5. Note the collapse of the cover around a weld point on the crest of the footwall block and outer-arc extension across the footwall graben. Faults labelled A and B in the hanging wall may be interpreted as convex-upward faults like those observed in the models. (**b**) Interpreted seismic line near the location of line A–A′. Note the steeply dipping Top Jurassic reflector, and the partitioning of deformation across a series of structures in the footwall to the main fault. See text for details.

as the change in geometry from south to north along the Revfallet Fault (Fig. 5). In Early to Mid-Jurassic time extension across the Revfallet system in this region was accommodated by a decoupled graben system. With an increase in extension rate in the late Jurassic some coupling is believed to have occurred to the north along the Revfallet Fault but in the region of B–B′ (Fig. 5b) the structure is largely decoupled from the basement fault system, with a minimal footwall uplift component (Pascoe *et al.* 1999).

The major extension phase occurred in the Aptian–Cenomanian, generating a major extensional fold in the Mesozoic cover (Fig. 13). Deposition of Cretaceous Aptian–Albian Lange 'A' deposits was limited to the hanging wall of the main Revfallet system as the precursor graben sit in the footwall of the Revfallet system. Precursor graben were reactivated by outer-arc extension across the Revfallet system as the growth fold amplified (Fig. 13). During this event significant footwall uplift took place, resulting in the erosion

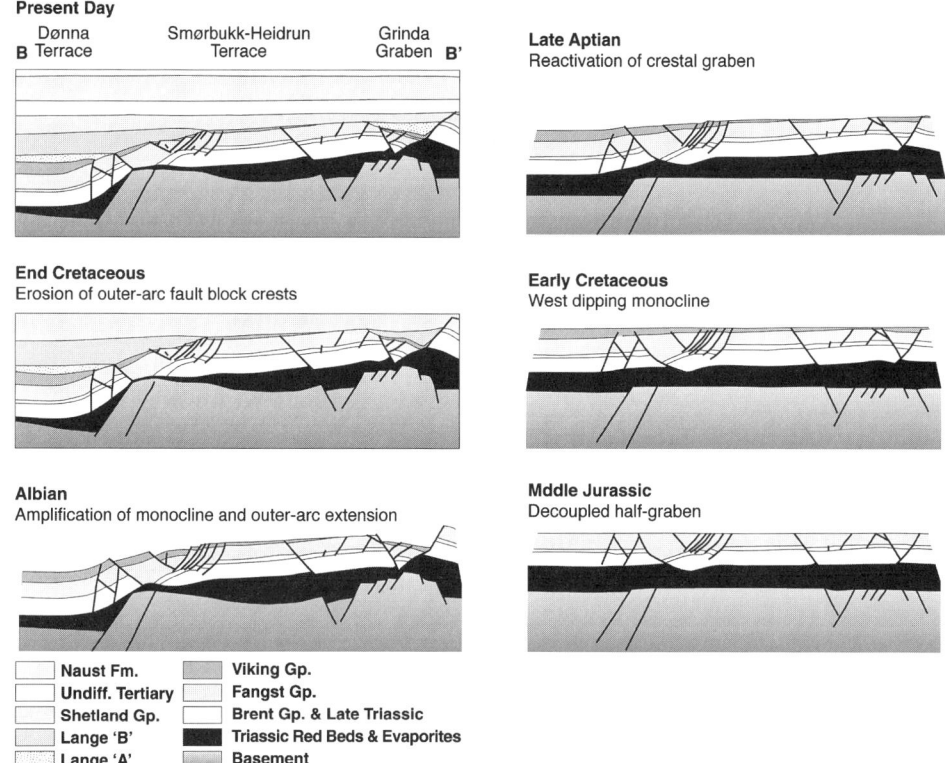

Present Day

Dønna Terrace   Smørbukk-Heidrun Terrace   Grinda Graben

End Cretaceous
Erosion of outer-arc fault block crests

Albian
Amplification of monocline and outer-arc extension

Late Aptian
Reactivation of crestal graben

Early Cretaceous
West dipping monocline

Mddle Jurassic
Decoupled half-graben

Legend:
- Naust Fm.
- Undiff. Tertiary
- Shetland Gp.
- Lange 'B'
- Lange 'A'
- Viking Gp.
- Fangst Gp.
- Brent Gp. & Late Triassic
- Triassic Red Beds & Evaporites
- Basement

Fig. 13. Combined forward and reverse structural model for transect B–B′. See text for details. Modified after Pascoe *et al.* (1999).

of Upper Jurassic strata (Viking Group) and tilting of pre-existing fault planes. Although major displacement of the basement marker occurs across the Revfallet Fault system in this location, the deformation in the cover is still largely decoupled and distributed over a broad zone of brittle and ductile deformation styles (Fig. 13).

Thus, for areas along the Revfallet Fault system from the immediate vicinity of line B–B′ (Fig. 2) and to the south the evolution of the Revfallet Fault system can be summarized as consisting of Early to Late Jurassic minor extensional events that generated detached cover-graben and a major Cretaceous event that resulted in the reactivation of precursor cover-graben, amplification of these graben, and a limited amount of coupling between basement and cover across the Revfallet Fault system. 3D physical models validate this structural interpretation, revealing deformation pathways that evolve from extensional drape folds through reactivated cover-graben to breached drape folds depending on the displacement on the basement fault and the presence or lack of precursor cover

structures (Fig. 14). These cover-graben rotate about a weld point on the crest of the footwall block generating outer-arc extension and shallowly dipping antithetic faults (Fig. 14). Steep frontal limbs to the breached drape fold are present in the models although these slopes are maintained by the addition of synkinematic sediments. These frontal limbs are potential slump sites and sites of low seismic resolution as observed in the Heidrun area (Pascoe *et al.* 1999).

As documented above and by Pascoe *et al.* (1999), coupling increases to the north of the Heidrun Field. There appears to be a hard link between faults in the cover and basement where there is approximately 5 km of displacement on the basement structure (section A–A′, Fig. 5). Hard linkage between basement and cover is estimated to have occurred prior to Upper Jurassic times, due to the five-fold increase in thickness of the Upper Jurassic Viking Group across the fault (Pascoe *et al.* 1999). Withjack & Callaway (2000) document increased coupling between basement and cover with increased displacements, similar to the models in

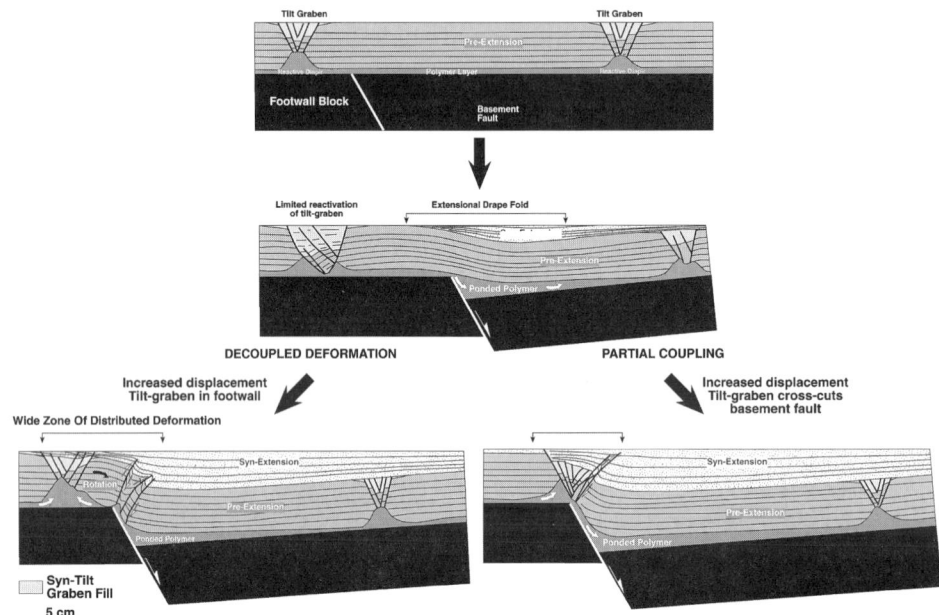

Fig. 14. Synoptic diagram illustrating deformation paths for decoupled and partially coupled cover sequences. Coupled geometries possess a much narrower zone of distributed deformation as displacement is taken up on pre-existing structures. Decoupled deformation geometries illustrate complex rotations and major steepening on the frontal limb of the extensional fold.

this study, but they also document an increase in coupling with an increase in displacement rate. According to Pascoe *et al.* (1999) more than 80% of post-Triassic extension occurred during the 10 Ma mid-Cretaceous event whereas the Jurassic events that account for only 20% total extension were spread over a 50 Ma period and largely decoupled from the basement. In this scenario one would expect full coupling to have occurred during the main, and very rapid, extensional event when the fault system accumulated a further 4 km displacement (cf. Withjack & Callaway 2000). Analogue models indicate that up to 1.5 cm, or 1.5 km, displacement is readily accommodated by extensional drape folding and that full coupling does not occur up to 3.5 cm or 3.5 km of displacement. Seismic line B–B' (Figs 5 & 13) illustrates that at 3 km displacement in this location the deformation geometry is primarily decoupled. Coupling along line A–A' to the north of the Heidrun Field initially occurred during the Jurassic extensional events at less than 1 km displacement on the basement structure. Basement–cover coupling in this location may have been aided by a Jurassic cover-graben cross-cutting the basement structure from hanging wall to footwall, as the trend of these pre-Cretaceous structures is NE–SW, approximately 25° counter-clockwise to the trend of the Revfallet Fault in this region (Fig. 2; Pascoe *et al.* 1999; R. Hooper, pers. comm. 2000).

Analogue models with oblique, cross-cutting cover-graben illustrate a significant increase in coupling or apparent coupling where cover-graben faults are reactivated above the basement structure (Figs 9, 10 & 14). Cover-graben are reactivated at relatively low displacements (<1.5 cm), at sites where they cut across the basement structure (Dooley & McClay 1997). Coupling along line A–A' along the Revfallet Fault (Fig. 5) is proposed to have been promoted by the intersection between a cover-graben and the basement structure during Jurassic rifting. This was further enhanced during the rapid extensional event in the Cretaceous where major displacement on the basement system promoted hard linkage between basement and cover (Fig. 14; cf. Withjack & Callaway 2000). Analogue models indicate that although reactivation and linkage between basinward-dipping precursor graben-bounding faults and the basement structure occur, displacement on the basement structure is still partitioned across a broad deformation belt consisting of a breached extensional fold and footwall grabens.

Conclusions

This study has shown that 3D physical models of variable displacement extensional faults generate extensional folds in the brittle cover at low to moderate displacements. High displacement sites display breached fold geometries and rotated keystone graben in the footwall of the basement structure. Precursor cover-graben that parallel the basement structure display reactivation along the fault as displacement increases. High displacement sites result in the basinward rotation of cover-graben antithetic faults as these structures ground on the crest of the basement structure and collapse into the developing half-graben. Models run with cross-cutting cover-graben demonstrate reactivation of these structures, in the immediate vicinity of the basement fault, at relatively low displacements resulting in partial coupling between basement and cover.

The Revfallet Fault system demonstrates considerable variation in structural geometry along strike from south to north due to increasing displacement on the basement fault system as well as a rapid increase in strain rate in the Cretaceous extensional event. The structural model proposed by Pascoe *et al.* (1999) and this study of Jurassic extensional events generating primarily decoupled cover-graben and extensional drape folds followed by reactivation, amplification and rotation of these structures during the major Cretaceous extensional event is validated by the physical models. Displacements up to 3 km are accommodated by major extensional folds and outer-arc extension across the fault-parallel precursor graben. To the north of the Heidrun field coupling between basement and cover was enhanced by NNE–SSW Jurassic cover-graben cross-cutting the NW–SE trending basement structure, and its subsequent reactivation and hard linkage with the basement structure during the major Cretaceous extensional event.

K. McClay gratefully acknowledges support from BP. H. Moore constructed the deformation apparatus. C. Elders and R. Hooper are thanked for comments on the manuscript. The authors thank J. K. Blom and an anonymous reviewer for constructive comments on the original manuscript.

References

BLYSTAD, P., BREKKE, H., FŒRSETH, R. LARSEN, B., SKOGSEID, J. & TØRUDBAKKEN, B. 1995. Structural elements of the Norwegian continental shelf, Part II: The Norwegian Sea Region. *NPD Bulletin,* **8**.

BRUN, J., SOKOUTIS, D. & VAN DEN DRIESSCHE, J. 1994. Analogue modelling of detachment fault systems and core complexes. *Geology,* **22**, 319–322.

DOOLEY, T. & MCCLAY, K. R. 1997. *Heidrun Modelling Project.* Internal report to Norske Conoco AS.

KOCH, J. O. & HEUM, O. R. 1995. Exploration trends of the Halten Terrace. *In*: HANSLIEN, S. (ed.) *Petroleum Exploration and Exploitation in Norway*, NPF Special Publications, **4**, 235–251.

MCCLAY, K. R. 1990. Extensional fault systems in sedimentary basins. A review of analogue model studies. *Marine and Petroleum Geology,* **7**, 206–233.

PASCOE, R., HOOPER, R., STORHAUG, K. & HARPER, H. 1999. Evolution of extensional styles at the southern termination of the Nordland Rdge, Mid-Norway: a response to variations in coupling above Triassic salt. *In*: FLEET, A. J. & BOLDY, S. A. R. (eds) *Petroleum of Northwest Europe: Proceedings of the 5th Conference*, 83–90.

VENDEVILLE, B. C. & JACKSON, M. P. A. 1992. The rise and fall of diapirs during thin-skinned extension. *Marine and Petroleum Geology,* **9**, 331–353.

VENDEVILLE, B. C., GE, H. & JACKSON, M. P. A. 1995. Models of salt tectonics during basement extension. *Petroleum Geoscience,* **1**, 179–183.

WEIJERMARS, R., JACKSON, M. P. A. & VENDEVILLE, B. 1993. Rheological and tectonic modelling of salt provinces. *Tectonophysics,* **217**, 143–174.

WITHJACK, M. O. & CALLAWAY, S. 2000. Active normal faulting beneath a salt layer: An experimental study of deformation patterns in the cover sequence. *AAPG Bulletin,* **84**, 627–651.

WITHJACK, M. O., OLSON, J. & PETERSON, E. 1990. Experimental models of extensional forced folds. *AAPG Bulletin,* **74**, 1038–1054.

3D evolution of a pop-up structure above a double basement strike-slip fault: some insights from analogue modelling

W. P. SCHELLART* & D. A. NIEUWLAND

Faculty of Earth and Life Sciences, Tectonics, Vrije Universiteit, De Boelelaan 1085, 1081 HV Amsterdam, The Netherlands
*Present address: Australian Crustal Research Centre, School of Geosciences, PO Box 28E, Monash University, Melbourne, VIC 3800, Australia
(e-mail: wouter@mail.earth.monash.edu.au)*

Abstract: In analogue modelling, pop-up structures have previously only been described in models of basement-involved strike-slip deformation with a stepover or restraining bend configuration. We describe the results of an analogue experiment with an alternative model configuration to investigate the structural development of pop-up structures. This model consists of a cover sequence of sand overlying two parallel running basement faults, which experience the same sense and the same amount of slip during deformation. The ratio of basement fault separation (1 cm) to overburden thickness (4 cm) is small. The experiment has been recorded in an X-ray tomograph, by which it was possible to investigate the 3D structural evolution of the model. During deformation, an elongated rhombic pop-up structure developed centrally above the basement faults. The faults bounding the pop-up structure displayed convex-upward as well as straight to concave-upward shapes. During deformation the topography above the basement faults increased to a maximum of *c.* 4.8 mm, compared to a maximum subsidence of *c.* 1.5 mm on both sides of the basement faults. During progressive deformation, local maximum horizontal principal stresses displayed a considerable rotation, from 45° in the initial stages, to (sub)parallel in a more advanced stage and finally back to *c.* 45° to the strike of the basement faults.

Pop-up structures are an integral part of strike-slip tectonics (Sylvester 1988) and form in restraining bend or stepover strike-slip settings (Harding 1990), as well as in pure strike-slip settings (Dooley & McClay 1996). Pop-up structures form anticlinal uplifts or push-ups, commonly with doubly plunging arrangements of folds, are of limited strike extent and are rhomboidal or lozenge shaped in plan view (McClay & Bonora 2001).

Pop-up structures have previously been generated in sandbox experiments, where the basement fault displayed a strike-slip geometry, with a stepover or offset in the middle (a discontinuity connecting two parallel running, overlapping or underlapping basement faults: Mandl 1988; Richard *et al.* 1995; Dooley *et al.* 1999; McClay & Bonora 2001). Variations on this model were made by changing the length of the stepover spacing (compared to the overburden thickness) and the angle of the stepover (compared to the strike of the basement strike-slip fault segments). Both of these variations had an influence on the size and

geometry (e.g. rhombic or sigmoidal shaped) of the pop-up structure (McClay & Bonora 2001).

The boundary conditions of these models have been discussed previously, and although elegant, some of these conditions are not realistic compared to the natural prototype. Furthermore, the pop-up structures that develop in these models have a relatively large surface extent compared to basement fault separation and are bounded by shallow dipping faults (*c.* 30°), while these features are not necessarily general characteristics of pop-up structures, e.g. Pijnacker oil field structure (Racero-Baema & Drake 1996; Dooley *et al.* 1999), Confidence Hills pop-up structure (Dooley & McClay 1996).

An alternative model is proposed to investigate the possible occurrence of pop-up structures in a pure strike-slip setting. This will most likely result in less pronounced pop-up structures, which are oriented more in line with the trend of the fault zone and are bounded by steeper dipping faults. The model is made up of a cover sequence overly-

From: NIEUWLAND, D. A. (ed.) *New Insights into Structural Interpretation and Modelling.* Geological Society, London, Special Publications, **212**, 169–179. 0305-8719/03/$15
© The Geological Society of London 2003.

ing two parallel running basement faults, which experience the same sense and the same amount of slip during deformation and where the ratio of basement fault separation (S) to overburden thickness (T) is small (here $S/T = 0.25$). A small S/T ratio leads to interaction of the individual fault zones in the upper part of the overburden (e.g. Richard *et al.* 1995). Such an experimental configuration results in relatively long, overlapping Riedel shears (Richard *et al.* 1995), between which the horizontal principal stresses will be rotated considerably (e.g. Naylor *et al.* 1986; Mandl 1999), and might result in the generation of pop-up structures in these overlapping zones.

The experiment described below has been recorded in an X-ray computer tomograph (CT-scanner). The advantage of X-ray scanning is that it is non-destructive for the model. Therefore, the progressive 3D geometrical development of structures in the model can be studied in a single experiment (e.g. Mandl 1988, 1999; Richard *et al.* 1990; Colletta *et al.* 1991; Schreurs 1994). When the analysis of an experiment is done by wetting the sediment package with water or a dilute gelatine solution, sequential slicing and photographing, then only one stage in the 3D structural evolution of the experiment can be investigated.

Scaling theory and experimental material

Analogue or physical experiments are subjected to specific scaling rules. The theory of these rules applied to geological processes was first introduced by Hubbert (1937), and was later also discussed by Ramberg (1967), Horsfield (1977), Richard (1991) and Davy & Cobbold (1991). According to this theory, models should be properly scaled for stresses, where stresses scale down as the product of gravity, density and length vectors scale down:

$$\frac{\partial \sigma_{ij}^a}{\partial \sigma_{ij}^n} = \frac{\rho^a g_i^a \partial x_j^a}{\rho^n g_i^n \partial x_j^n} \quad (i, j = 1, 2, 3) \tag{1}$$

where superscript a denotes the analogue model and n denotes the natural prototype; σ is the stress tensor; ρ is the density; g is the acceleration due to gravity; and x is a length vector. The suffixes refer to Cartesian vector and tensor components in a fixed spatial frame.

When the experiment is executed in a normal field of gravity, then gravity is the same in both the model and in nature. Furthermore, when a modelling material is used with a density similar to the density of rocks in nature, then Equation 1 reduces to:

$$\frac{\partial \sigma_{ij}^a}{\partial \sigma_{ij}^n} = \frac{\partial x_j^a}{\partial x_j^n} \tag{2}$$

From Equation 2 it is concluded that stresses scale down as length vectors scale down. In the experiment, a length factor of 10^{-5} has been applied, thus 1 cm in the experiment simulates 1 km in nature. The material used in the experiment was dry fine sand (grain size = 0.075–0.125 mm) to simulate brittle faulting in upper crustal rocks. Sand is a frictional plastic material (Mandl 1988) and therefore deformation is strain rate independent. Sand has an angle of internal friction (ϕ) which varies from *c.* 30° (Hubbert 1951; Mandl 1988), *c.* 41° (Schellart 2000) and *c.* 30–45° (Krantz 1991), which is comparable with values of ϕ for rocks. The cohesion of sand ranges between zero and a few hundred Pa, depending on the normal stress (Schellart 2000). According to Horsfield (1977) cohesion has to be scaled down in a similar way that stresses scale-down. With a length factor of 10^{-5}, this would lead to values for cohesion of natural rocks up to a maximum of a few tens of megapascals. Values for cohesion of rocks range between 5 MPa for loose compacted sediments (Horsfield 1977) and some tens of megapascals to a maximum of *c.* 100 MPa for igneous, metamorphic and consolidated sedimentary rocks (Handin 1969; Jaeger & Cook 1977). Therefore, it can be concluded that sand is properly scaled for both cohesion and angle of internal friction. Thus, sand is a good analogue to model brittle behaviour of upper crustal rocks.

Model set-up

The experiments were executed in a rectangular sandbox with two basement faults (BF), running parallel to each other. These BF were created by inserting a fixed basement strip between the two basement plates of the sandbox (Fig. 1). During deformation, the two basement plates moved horizontally, parallel to the BF, resulting in the same

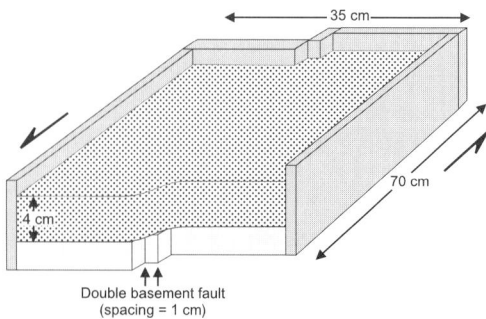

Fig. 1. Model configuration of strike-slip faulting above a double basement fault. Horizontal dimensions are 70 by 35 cm, sediment thickness is 4 cm, sediment used is fine sand (0.075–0.125 mm), double basement fault spacing is 1 cm and sense of shear is sinistral.

sense of slip (sinistral) and the same amount of dis-placement along the BF. The fault separation between the BF was 1 cm. The apparatus contained four side walls. The horizontal dimensions of the model were 70 by 35 cm. Several experiments were run with an overburden thickness varying between 3 and 4 cm. The individual experiments showed a similar structural evolution in plan view and cross-section (Schellart 1998), indicating that the experiments are reproducible. One experiment with an overburden thickness of 4 cm was recorded in a CT-scanner, from which the results are presented in this paper. In this experiment, three thin layers of zirprop (grain size = 0.28–0.50 mm) were inserted, with a spacing of 1 cm, to visualize internal layering in the cross-sections. These layers were only one grain size thick to minimize their influence on the homogeneity of the cover sequence.

Recording and interpretation

The experiment was recorded in an X-ray computer tomograph (CT-scanner), which measures density contrasts in cross-sectional slices with a thickness of 2 mm. 3D volumes were obtained by scanning crosslines (cross-sections, perpendicular to the BF) every 2 mm after every 0.5 cm of displacement between the two halves of the model (up to 4 cm of displacement). From the total area of the experi-ment (70 by 35 cm), only a small part (24.0 by 12.8 cm) was scanned. The volumes obtained were imported into 3D interpretation and visualization programs Charisma 3.8.7™ and VoxelGeo 2™. Here, the four horizons (the top surface and the three zirprop layers) and the faults were inter-preted. For the horizons, topographic elevation was plotted. For each of these horizons, 0 cm is the reference level at 0 cm of displacement (before deformation). The density contrast between sand and air is very high and therefore the volumes dis-played a sharp boundary for the top surface. For this reason the top surface could be interpreted very accurately. The density contrast between sand and zirprop is less and because these zirprop layers are also somewhat irregular, their resulting interpret-ations are less accurate.

In the crosslines, the BF can be observed (see square at the bottom in Fig. 4). These BF were made by insertion of an aluminium strip in between the basement halves, which produced some distor-tion in the scanned images in a small area around the strip. The zirprop layers are visible as horizon-tal layers with black and white dots. Faults can be recognized as narrow, white zones. They are vis-ible, because the material in each thin fault zone is less dense than its surroundings, due to dilatation

of the sand in the fault zone (Mandl *et al.* 1977; Krantz 1991; Schellart 2000).

Results

Surface observations and tomograph results

The evolution of strike-slip fault zones in sandbox modelling has been described previously by Naylor *et al.* (1986) and Richard *et al.* (1995). The termin-ology used here to describe faults in the experiment is largely in accordance with these authors.

With the onset of deformation, sinistral displace-ment is accommodated by bulk shearing of a 4 cm wide zone, which is situated centrally above the BF. The area affected by topographic change, how-ever, is much wider (at least as wide as the area of investigation) (Figs 2a, 3a, 4a). In Figures 3a and 4a a broad open anticlinal bulge or push-up can be observed with its maximum elevation situated centrally above the BF and minimum elevation situated on both sides of the central zone. At this stage no faults are recognized at the top surface or in the cross-sections.

The first faults that form are en echelon, right-stepping, synthetic Riedel shears that form at an angle of *c.* 15° to the strike of the BF after 0.8 cm of displacement and are regularly spaced above the BF (Fig. 2b). As the basement displacement increases, the Riedels grow outwards. In cross-sec-tion, the Riedels display concave-upward to straight shapes in the inner parts, but their shapes become progressively more convex upward towards their outer tips (especially for the longer Riedels; Fig. 4b). Here they can dip down to *c.* 45° near the top surface and *c.* 60° near the BF. Most of the Riedels display scissor fault deformation (Fig. 5a), faults where the sense of vertical dis-placement reverses at their midpoints (Naylor *et al.* 1986). Splay faults may form at the tips of the Rie-dels and usually strike at a greater angle to the strike of the BF (up to 35°).

After 1.0–1.2 cm of displacement, synthetic low-angle shears start to develop at angles of 0–10° to the strike of the BF between or from the tips of the Riedels and start connecting them. Initially, they terminate at the Riedels, but with ongoing defor-mation they cut through them to form a through-going fault zone. The low angle shears normally display slightly concave-upward to straight shapes with dips between 75 and 90°. Figures 2c and 3b show the structural pattern of the top surface after 1.5 cm of displacement, where both Riedels and low angle shears can be seen. The central high is more pronounced and some of the faults define topographic steps (Fig. 3b). With increasing depth it was observed that the strike of the Riedels and

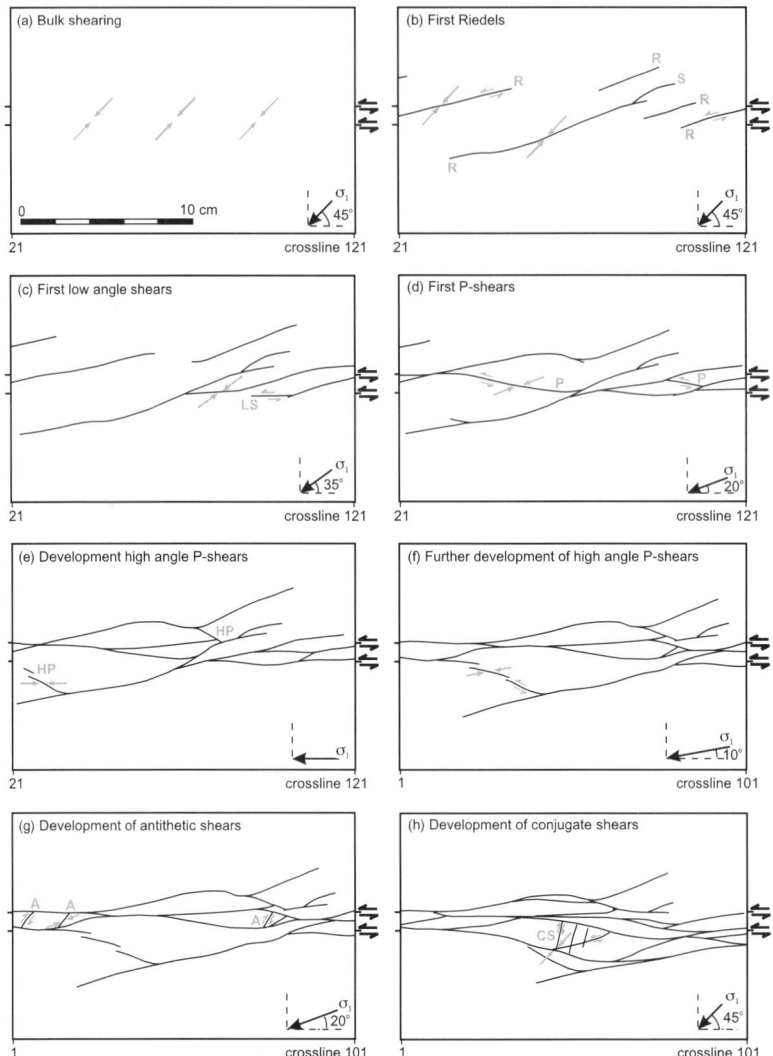

Fig. 2. Top view line drawings of development of fault pattern with related stress field during progressive stages of deformation, with (**a**) 0.5 cm, (**b**) 1.0 cm, (**c**) 1.5 cm, (**d**) 2.0 cm, (**e**) 2.5 cm, (**f**) 3.5 cm, (**g**) 4.0 cm and (**h**) 8.0 cm of displacement. During the first stages (a–e) the local maximum principal stress (σ_1) rotates clockwise, while in the following stages (f–h) σ_1 rotates anticlockwise. Parts a, c, d, e, f and g correspond to a, b, c, d, e and f in Figure 3, respectively. R = Riedel shear; S = splay fault; LS = low angle shear; P = P-shear; HP = high angle P-shear; A = antithetic shear; CS = conjugate shears.

low angle shears decreases to finally turn into one of the BF.

After *c.* 1.8 cm of displacement, synthetic P-shears shears start to develop between the Riedels and the low angle shears, at angles between −15° and −5°. The P-shears normally have straight to slightly concave-upward shapes with dips varying between 75° and 90° near the top surface (Fig. 4c). One of the P-shears, which connects two widely spaced Riedels (Figs 2d,e & 3c,d), displays a

scissor fault geometry, that is a mirror image of a Riedel shear geometry (Fig. 5b).

After *c.* 2.0 cm of displacement, a fault-bound rhombic elongated pop-up structure starts to develop centrally above the BF (Fig. 3c) and becomes more defined with ongoing deformation (Fig. 3d–f). The longer sides of this structure are bounded by Riedels, with concave-upward (near the BF) to convex-upward (near the tips) shapes. The shorter sides are bounded by synthetic shears,

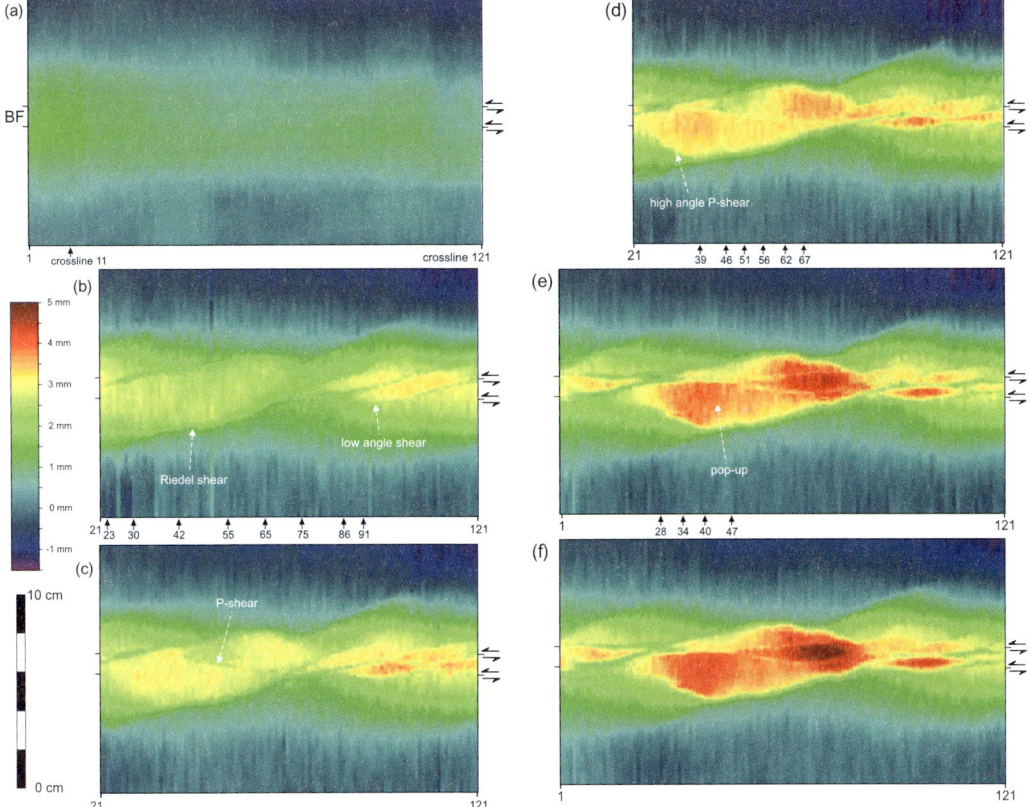

Fig. 3. Plan views of top surface of experiment. (**a**) After 0.5 cm of displacement with development of an anticlinal bulge above the basement faults. (**b**) After 1.5 cm of displacement with development of Riedels and low angle shears. (**c**) After 2.0 cm of displacement with development of P-shears between the earlier formed Riedels. (**d**) After 2.5 cm of displacement with development of high angle P-shears. (**e**) After 3.5 cm of displacement with development of a pop-up structure. (**f**) After 4.0 cm of displacement with the continuing rise in topography of the pop-up structure. Arrows with numbers refer to cross-sections in Figure 4. Interpretation of parts a, b, c, d, e and f are plotted in Figure 2a, c, d, e, f and g, respectively.

which strike $-35°$ to $-20°$ to the BF (we will call these shears 'high angle P-shears'). These shears start to develop at $c.$ 2.0 cm of displacement by growing from their bounding Riedels from the inner sides at the compressive ends of these Riedels towards the inner part of the central zone. They have straight to convex-upward shapes (Fig. 4d) with dips down to $60°$ near the top surface and $c.$ $90°$ near the BF. Furthermore, they display a relatively large amount of reverse dip-slip (up to $c.$ 2.5 mm). They continue to grow until $c.$ 3.5 cm of displacement (Figs 2d–f & 3c–e). Cross-sections through the pop-up structure display palm tree geometries with convex-upward shaped shears at one or at both sides of the structure (Fig. 4c).

From $c.$ 2.4 to 4.5 cm of displacement, straight to slightly curved antithetic shears develop at $c.$ $50°$ to the strike of the BF. They only occur between (sub)parallel synthetic shears, which are spaced

relatively close to each other (less than $c.$ 2 cm; Fig. 2g). With ongoing deformation, these antithetic shears rotate anticlockwise as in a bookshelf mechanism (Mandl 1987) up to a maximum of $c.$ $15°$. The new antithetic shears are not visible on the tomograph scans, possibly due to their short length, small amount of displacement and their relatively high angle to the scan line direction.

After 4.5 cm of displacement, a more or less through-going fault zone develops with several shear lenses running sub-parallel to the BF. At 6.2 cm of displacement, several new faults develop, which are conjugate shears that form at $c.$ $75°$ (antithetic shears) and $c.$ $15°$ (synthetic shears) to the strike of the BF and only form in the pop-up structure (Fig. 2h). New conjugate shears continue to develop until 7.7 cm of displacement. Deformation was stopped after 8.0 cm of displacement.

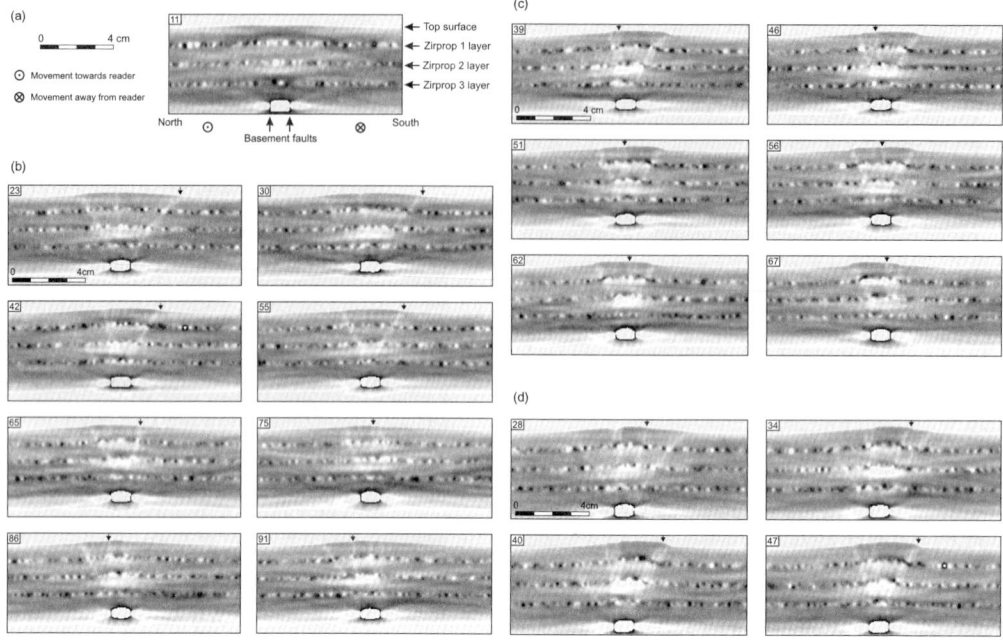

Fig. 4. Cross-sections perpendicular to the basement faults. (**a**) After 0.5 cm of displacement showing an anticlinal bulge centrally above the basement faults. (**b**) After 1.5 cm of displacement showing the geometry of a Riedel shear (arrow) with a convex-upward shape at its end and a concave-upward shape in the centre. (**c**) After 2.5 cm of displacement showing the geometry of a P-shear (arrow) with a concave-upward shape in the centre and a small reverse displacement along the fault. (**d**) After 3.5 cm of displacement showing the geometry of a high angle P-shear (arrow) with a convex-upward shape and a reverse displacement along the fault. For location of cross-sections see Figure 3a, b, d and e, respectively.

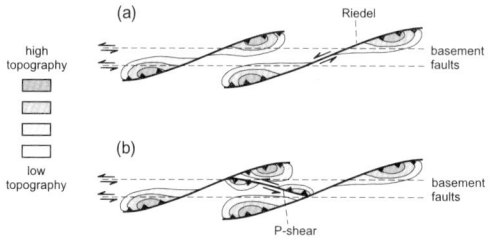

Fig. 5. Scissor faulting in (**a**) Riedels and (**b**) P-shears in sinistral strike-slip above a double basement fault. P-shears are the mirror image of Riedels.

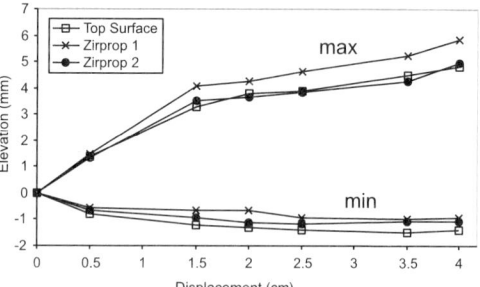

Fig. 6. Diagram illustrating the development of topographic relief during progressive stages of deformation. Maximum elevation (max) occurs in the central zone above the basement faults and minimum elevation (min) occurs at both sides of the central zone. Change in relief is most rapid in the first 1.5 cm of displacement.

Topographic change

In Figure 6, the maximum and minimum elevation of the top surface and the uppermost two zirprop layers have been plotted at different amounts of total displacement along the BF. From the maximum elevation it is clear that it increases with progressive displacement but that the amount of increase decreases with progressive deformation. Most of the positive relief develops between 0 and 1.5 cm displacement. The minimum elevation

graphs are more or less the mirror image of the maximum elevation graphs, suggesting that the increase in topography above the BF and decrease in topography at the sides share a similar origin. However, the amount of change (decrease in topography) is less. The maximum increase is *c.* 4.8 mm, while the maximum decrease is *c.* 1.5 mm.

Two explanations for the change in topography can be thought of.

(1) The increase in topography above the BF results from the dilatation of the fault zones during shearing, which produces extra horizontal compression in the overburden and pushes the fault wedges above the BF upwards, since the overburden is not allowed to expand horizontally outwards considerably. This extra horizontal compression could result in a small horizontal tectonic compaction of the overburden (e.g. Verschuren *et al.* 1996) at both sides of the fault zone. However, one would expect this to result in a slight increase rather then a slight decrease in topography at both sides of the fault zone.

(2) The increase in topography above the BF results from the geometry of and displacement along the fault zones during shearing, which pushes the fault wedges above the BF upwards, since the overburden is not allowed to expand horizontally outwards considerably. This upward extrusion of fault wedges could then give the surrounding undeformed overburden laterally more space, which would lead to a slight decrease in horizontal stress, perpendicular to the BF, and would then result in a slight decrease in topography at both sides of the fault zone.

The second explanation seems to be the most likely, since it explains both the increase and decrease in topography. However, to be verified, *in situ* stress measurements (as in Nieuwland *et al.* 2000) in similar strike-slip experiments are required, in which the horizontal stress in the overburden perpendicular to the BF is measured.

Low density zones

In cross-sections through the pop-up structure, fault-bound low-density zones inside the pop-up structure can be observed (Fig. 4). These low-density zones commonly develop just below the zirprop layers and are normally covered by sediments with a higher density in an arched geometry. These high-density zones are probably the result of some 'stress arching' in the overlying zirprop layers, which could be related to the bigger zirprop grain size. The most pronounced low-density zones occur below the second zirprop layer. This could be related to the geometry and amount of reverse slip along the bounding faults. The amount of reverse slip along the faults near the lowermost zirprop layer is about half the amount of reverse slip near the overlying zirprop layers and the top surface. This difference results in a small vertical extension in the lower region of the pop-up structure, which leads to an elongated low-density zone between the second and the third zirprop layer.

In nature, the development of such a zone could act as a pumping mechanism for fluids. The presence of such a low-density extensional zone inside the pop-up structure could act as a trap and seal pair for hydrocarbons. Here, the low-density zone could act as a reservoir, which synchronously develops with and is confined in 3D by the geometry of the pop-up structure. In hydrocarbon field development it may be necessary to take the orientation of σ_1 in and around the reservoir into account, for example, if a well needs to intersect or has to avoid fractures of a particular orientation. Alternatively, the sealing properties of faults may depend on their orientation relative to the *in situ* stress. The common assumption that in a strike-slip setting the orientation of σ_1 is at 45° to the strike of the BF can be misleading, especially on a local scale, as has been demonstrated by the experiment (Fig. 2, see also next paragraph), where the orientation of σ_1 in the pop-up structure changed considerably with progressive deformation.

Stress field development

The orientation of the principal stresses in a strike-slip environment can be approximated by measuring the orientation of a newly formed strike-slip fault and determining its sense of shear (sinistral or dextral). Furthermore, the angle of internal friction of the material in which the fault has formed should be known. The angle (α) between the fault and the maximum principal stress (σ_1) can then be determined by the following relationship (Naylor *et al.* 1986):

$$\alpha = 45° - \phi/2 \qquad (3)$$

If we assume that $\phi \approx 30°$ (Hubbert 1951; Mandl 1988), then $\alpha = 45° - \phi/2 \approx 30°$. Furthermore, we will assume that σ_2 is vertical. Therefore, σ_1 and σ_3 have horizontal orientations.

The first deformation is accommodated by a plastic bulk shearing of a 4 cm wide zone in the overburden. According to Mandl *et al.* (1977), σ_1 attains an orientation of 45° to the shear direction at the free surface during simple shearing of frictional plastic material. Therefore, we will assume σ_1 to be oriented at 45° to the BF in this first stage of deformation (Fig. 2a).

The first faults to form are Riedel shears, which develop at angles of *c.* 15° to the strike of the BF. With Equation 3 this leads to an orientation for σ_1 of *c.* 45° (Fig. 2b), which is the same as during the plastic bulk shearing.

In the following stage of deformation, low angle shears start to develop at angles of 0 to 10° to the strike of the BF. Thus, σ_1 has rotated some 5 to 15° clockwise (at least locally) and has obtained an orientation more parallel to the strike of the BF (Fig. 2c). With progressing deformation, P-shears develop at angles of −15 to −5° to the strike of the BF, indicating that σ_1 strikes c. 15 to 25° to the strike of the BF and has rotated even further (Fig. 2d). The low angle shears and P-shears form at a lower angle to the strike of the BF because they form in a local stress environment in between the earlier formed overlapping Riedels, where the local σ_1 has rotated (sub)parallel to the Riedels (e.g. Naylor et al. 1986).

Shortly after the formation of the P-shears, high angle P-shears start to develop at angles of −35 to −20° to the strike of the BF. Therefore, σ_1 has a local orientation between −5 and 10° to the strike of the BF (Fig. 2e,f). Thus, σ_1 has rotated c. 45° clockwise, from the beginning of the experiment until the formation of the high angle P-shears.

From 2.4 cm to 4.5 cm of displacement, several antithetic shears develop at angles of c. 50° to the strike of the BF, implying that σ_1 has a local orientation of c. 20° to the strike of the BF (Fig. 2g).

Finally, in a very late stage of deformation (6.2–7.7 cm of displacement), several conjugate shears develop in the pop-up structure at c. 75° and c. 15° to the BF respectively, indicating that σ_1 strikes 45° to the strike of the BF (Fig. 2h). Thus, σ_1 has rotated c. 35–45° anticlockwise in this region compared to its orientation during formation of the high angle P-shears. This rotation could be explained as follows. Further displacement of the pop-up structure along the P-shears and high angle P-shears leads to stacking of individual shear lenses, resulting in extra horizontal compressional stresses perpendicular to the BF. This results in an anticlockwise rotation of σ_1, where the angle between σ_1 and the BF progressively increases. The pop-up structure finally absorbs the externally applied stresses by internal faulting.

Discussion

With progressive deformation, the topography increases in the central part (above the BF) and decreases at both sides of the central part (Fig. 6). The three zirprop surfaces also display an increase in topographic elevation with progressing deformation in the central part and subsidence at both sides of the central zone, although the topographic change in the lowermost zirprop layer is about half the amount of the other two. The increase in topography above a basement fault has been reported earlier in other strike-slip experiments (Naylor et al. 1986; Mandl 1988; Richard et al. 1995;

Dooley & McClay 1996). However, the development of topographic lows at the sides has not been reported earlier in analogue strike-slip experiments. These topographic lows do occur in nature. Figure 7 displays a palm tree or pop-up structure with so called 'rim synclines' at both sides of this structure (from Nijman et al. 1992). These rim synclines have been filled syntectonically by coarse clastic sediments, derived from the adjacent pop-up structure. According to Nijman et al. (1992) these syntectonic sediments display a multistorey channel fill, where the mean maximum particle size decreases upwards. This decrease could be related to the decrease in subsidence rate of the rim synclines and the decrease in uplift rate of the central zone, as has been observed in the analogue experiment (Fig. 6).

During progressive deformation, σ_1 experienced considerable local rotation in the overburden above and close to the BF. The orientation of σ_1 started at an angle of c. 45° to the strike of the BF during formation of the Riedels and rotated clockwise during formation of the low angle shears and P-shears. Similar fault orientations and amounts of rotation of σ_1 have been reported earlier (Naylor et al. 1986). However, in the experiment described here, local stresses rotated even further with σ_1 running (sub)parallel to the BF in places where high angle P-shears developed. Also, in an advanced stage of deformation, σ_1 locally rotated back again to oblique angles to the BF (antithetic shears and conjugate shears).

In previously performed double basement fault strike-slip experiments with a comparable model set-up (Richard et al. 1995), pop-up structures have

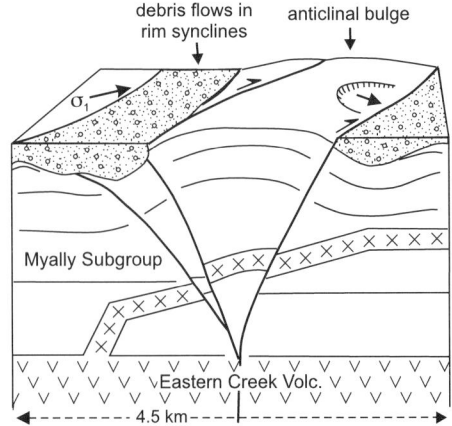

Fig. 7. Schematic sketch of palm tree structure with anticlinal bulge and rim synclinal deposition of coarse clastics in the Proterozoic of the Paroo range, Mount Isa Inlier, Queensland, Australia (modified from Nijman et al. 1992).

not been described. However, this could be related to the low ratio of BF separation to overburden thickness (*S/T*) in the experiment described here (*S/T* = 0.25) compared to higher ratios for the experiments described by Richard *et al.* (1995) (*S/T* = 0.5 and 2.0).

Comparison of the geometry of the pop-up structure with the pop-up structures described by McClay & Bonora (2001), who modelled pop-up structures with an offset strike-slip configuration with a different offset angle and *S/T* ratio, indicates that our results are most comparable with the 150° overlapping stepover model with a small *S/T* ratio (*S/T* = 0.5) (McClay & Bonora 2001, fig. 17i). This model shows a rhomboidal shaped pop-up structure with the sides striking at *c.* 15° and *c.* −30° to the strike of the BF, similar to the results described here (Fig. 3e, f). Furthermore, conjugate shears developed in the pop-up structure in a late stage of deformation, as has also been observed here (Fig. 2h). In cross-section, the pop-up structures described by McClay & Bonora (2001) are predominantly asymmetric, while symmetric geometries are only found in central sections. This is also observed in the pop-up structure described here.

In this pure strike-slip experiment, faults with concave and convex shapes occur equally. This is in contrast with findings of other researchers, who also describe analogue strike-slip experiments (Naylor *et al.* 1986; Mandl 1988; Richard *et al.* 1995). They claim that faults in comparable strike-slip experiments (with the only displacement being parallel to the basement fault(s)) predominantly have concave-upward shapes (tulip structures). According to these authors, the occurrence of convex-upward shaped faults (palm tree structures), when associated with an en echelon fault pattern in map view, would point to transpression or basement-involved reverse oblique-slip faulting (Naylor *et al.* 1986; Mandl 1988; Richard *et al.* 1995). However, the results of the experiment described here indicate that convex-upward faults are common in a pure strike-slip regime as well.

The convex-upward fault geometry can be considered a result of variation of the vertical stress (σ_v) with depth, which is small close to the surface and increases progressively downwards. The steep fault trace at depth and the lateral displacement along it implies that $\sigma_v = \sigma_2$, and σ_1 and σ_3 are oriented horizontally, while at the surface, the inclined fault trace and the reverse oblique-slip implies that $\sigma_v \neq \sigma_2$. This might explain why convex-upward fault segments form in a late stage of Riedel shear growth and in a relatively late stage of deformation (high angle P-shears). Early formation of concave-upward shear zones (Riedels, low angle shears, P-shears) has resulted in an increase

in horizontal compression in the overburden due to fault zone dilatation. This forced σ_3 out of the horizontal plane close to the surface, because vertical stresses are small here, and resulted in convex-upward oriented faults.

Distinction between different tectonic regimes for en echelon structures in map view in combination with surface uplift can be resolved in cross-sections through these structures. Here, cross-sections through transpressive structures normally display symmetrical palm tree geometries (e.g. Richard & Cobbold 1990) and cross-sections through basement-involved oblique reverse faulting display asymmetrical palm tree geometries with a step in the basement (e.g. Mandl 1988; Richard 1991). In contrast, cross-sections through elongated pop-up structures related to a pure strike-slip setting display mainly asymmetrical but also symmetrical palm tree geometries with a relatively flat basement. Furthermore, restraining bend pop-ups normally have much shallower dipping convex-upward faults (*c.* 30°) in cross-section (Richard *et al.* 1995; Dooley *et al.* 1999; McClay & Bonora 2001), compared to *c.* 60° in the experiment described here. Also, the sides of restraining bend pop-up structures in map view are of comparable length, while the pop-up structure described here is highly elongated, oriented (sub)parallel to the fault zone, with differing lengths for the sides of the pop-up structure.

A natural example of a palm tree structure, which could be related to strike-slip above a double basement fault, is illustrated in Figure 8. This migrated seismic section shows a palm tree struc-

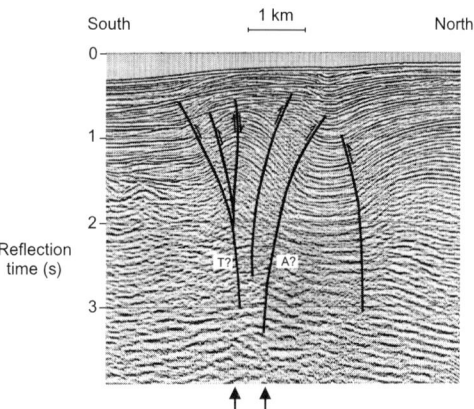

Fig. 8. Interpreted seismic section displaying a palm tree structure along the subsea extension of the Oak Ridge fault in the Santa Barbara Channel, Southern California, USA (from Sylvester 1988). The section has been migrated. A = movement away from the observer, T = movement towards the observer. The two arrows indicate the possible location of two basement faults.

ture along the subsea extension of the Oak Ridge fault in the Santa Barbara Channel, southern California, USA (Sylvester 1988). A plan view example of an elongated pop-up structure, which is more likely to be related to pure strike-slip deformation rather than to transpression, reverse oblique-slip faulting or restraining bend geometries, is the Pijnacker oil field in the West Netherlands Basin (Racero-Baema & Drake 1996; Dooley *et al.* 1999; McClay & Bonora, 2001). The Pijnacker oil field is an elongated pop-up structure that developed in a strike-slip setting, bounded by convex- and concave-upward strike-slip faults with a reverse displacement. Another example is the Confidence Hills pop-up structure in the southern Death Valley fault zone in California, USA (Dooley & McClay 1996). This structure shows elongated fault-bound pop-up segments with doubly plunging anticlines that run (sub)parallel to the bounding faults, similar to the structures observed in the experiment.

Conclusions

Four main conclusions can be drawn from this work.

1. Pop-up and palm-tree structures can form not only in restraining bends of strike-slip systems, transpressive regimes and reverse oblique-slip regimes, but can also form in pure strike-slip settings. In the experiment described here an elongated rhombic pop-up structure formed in the brittle overburden overlying two basement faults during sinistral strike-slip motion. This pop-up structure was bounded on the longer sides by Riedels striking *c.* 15° to the basement faults, displaying concave to straight to convex-upward shapes in cross-section, and on the shorter sides by convex-upward high-angle P-shears striking *c.* −20 to −35° to the strike of the basement faults. From comparison with experimental results of pop-up structures in transpression or restraining bend geometries it can be concluded that elongated pop-up structures bounded by relatively steeply dipping convex- and concave-upward faults are more likely to be the result of pure strike-slip faulting than transpression or restraining bend geometries.

2. During deformation, the topography of the pop-up structure, as well as other fault segments, increased from 0 mm up to 4.8 mm above the basement faults and decreased to −1.5 mm on both sides of the basement faults, corresponding to some 480 m and −150 m in nature. This change in topography took place during 4 cm of lateral displacement, corre-

sponding to some 4 km in nature. Most topographic change took place during the first 1.5 cm of displacement, with an increase of up to *c.* 3.5–4 mm and a decrease to *c.* −1 mm, indicating that in strike-slip settings topography can change rapidly.

3. In the lower part of the pop-up structure, an elongated low-density zone formed during formation of the pop-up structure. Such a structure is important for hydrocarbon exploration, since the low-density zone could act as a hydrocarbon reservoir, confined in 3D by the pop-up structure.

4. During deformation, the orientation of the local maximum horizontal principal stress above the basement faults varied significantly, from striking *c.* 45° to (sub)parallel and back to *c.* 45° to the strike of the basement faults from initiation to an advanced stage of deformation. Thus, under a static far-field stress field, local stresses can vary significantly during progressive development of a strike-slip fault zone.

We thank P. Richard, T. Dooley and an anonymous reviewer for helpful reviews that greatly improved this manuscript. This study was part of a research programme in strike-slip tectonics, undertaken at Shell International Exploration and Production B. V. in Rijswijk, the Netherlands. We thank Shell Research B. V. for permission to publish this paper. W. P. S. did the experimental work as part of the requirements for a MSc degree at the Free University, Amsterdam.

References

COLLETTA, B., LETOUZEY, J., PINEDO, R., BALLARD, J. F. & BALE, P. 1991. Computerized X-ray tomography analysis of sandbox models; examples of thin-skinned thrust systems. *Geology*, **19**, 1063–1067.

DAVY, PH. & COBBOLD, P. R. 1991. Experiments on shortening of a 4-layer model of the continental lithosphere. *Tectonophysics*, **188**, 1–25.

DOOLEY, T. P. & MCCLAY, K. R. 1996. Strike-slip deformation in the Confidence Hills, southern Death Valley fault zone, eastern California, USA. *Journal of the Geological Society, London*, **153**, 375–387.

DOOLEY, T., MCCLAY, K. & BONORA, M. 1999. 4D evolution of segmented strike-slip fault systems; applications to NW Europe. *In*: FLEET, A. J. & BOLDY, S. A. R. (eds) *Petroleum Geology of Northwest Europe: Proceedings of the 5th Conference*. Geological Society, London, 215–225.

HANDIN, J. 1969. On the Coulomb-Mohr failure criterion. *Journal of Geophysical Research*, **74**, 5343–5348.

HARDING, T. P. 1990. Identification of wrench faults using subsurface structural data; criteria and pitfalls. *AAPG Bulletin*, **74**, 1590–1609.

HORSFIELD, W. T. 1977. An experimental approach to basement-controlled faulting. *Geologie en Mijnbouw*, **56**, 363–370.

HUBBERT, M. K. 1937. Theory of scale models as applied to the study of geological structures. *Bulletin of the Geological Society of America,* **48**, 1459–1520.

HUBBERT, M. K. 1951. Mechanical basis for certain geological structures. *Bulletin of the Geological Society of America,* **62**, 355–372.

JAEGER, J. C. & COOK, N. G. W. 1976. *Fundamentals of Rock Mechanics.* John Wiley & Sons, New York.

KRANTZ, R. W. 1991. Measurements of friction coefficients and cohesion for faulting and fault reactivation in laboratory models using sand and sand mixtures. *Tectonophysics,* **188**, 203–207.

McCLAY, K. R. & BONORA, M. 2001. Analog models of restraining stepovers in strike-slip fault systems. *AAPG Bulletin,* **85**, 233–260.

MANDL, G. 1987. Tectonic deformation by rotating parallel faults – the 'bookshelf' mechanism. *Tectonophysics,* **141**, 277–316.

MANDL, G. 1988. *Mechanics of Tectonic Faulting, Models and Basic Concepts.* Elsevier, Amsterdam.

MANDL, G. 1999. *Faulting in Brittle Rocks; An Introduction to the Mechanics of Tectonic Faults.* Springer, Berlin.

MANDL, G., DE JONG L. N. J. & MALTHA, A. 1977. Shear zones in granular materials. *Rock Mechanics,* **9**, 95–144.

NAYLOR, M. A., MANDL, G. & SIJPESTEIJN, C. H. K. 1986. Fault geometries in basement induced wrench faulting under different initial stress states. *Journal of Structural Geology,* **8**, 737–752.

NIEUWLAND, D. A., URAI, J. L. & KNOOP, M. 2000. In-situ stress measurements in model experiments of tectonic faulting. *In:* LEHNER, F. K. & URAI, J. L. (eds) *Aspects of Tectonic Faulting.* Springer, Berlin, 155–166.

NIJMAN, W., VAN LOCHEM, J. H., SPLIETHOFF, H. & FEIJTH, J. 1992. Deformation model and sedimentation patterns of the Proterozoic of the Paroo Range, Mount Isa Inlier, Queensland, Australia. *In:* STEWART, A. J. & BLAKE, D. H. (eds) *Detailed Studies of the Mount Isa Inlier.* Australian Geological Survey Organisation, Bulletin, **243**, 29–73.

RACERO-BAEMA, A. & DRAKE, S. J. 1996. Structural styles and reservoir development in the West Netherlands oil province. *In:* RONDEEL, H. E., BATJES, A. J. & NIEUWENHUIS, W. H. (eds) *Geology of Gas and Oil under the Netherlands.* Kluwer, Amsterdam, 211–227.

RAMBERG, H. 1967. Model experimentation of the effect of gravity on tectonic processes. *The Geophysical Journal of the Royal Astronomical Society,* **14**, 307–329.

RICHARD, P. 1991. Experiments on faulting in a two-layered cover sequence overlying a reactivated basement fault with oblique-slip. *Journal of Structural Geology,* **13**, 459–469.

RICHARD, P. & COBBOLD, P. R. 1990. Experimental insights into partitioning of fault motions in continental convergent wrench zones. *Annales Tectonicae,* **4**, 35–44.

RICHARD, P. D., BALLARD, J. F., COLLETTA, B. & COBBOLD, P. R. 1990. Genese et propagation de failles au dessus d'un decrochement de socle; modelisation analogique et tomographie. *In: 13e reunion des sciences de la terre* **13**, Grenoble, France, 107.

RICHARD, P. D., NAYLOR, M. A. & KOOPMAN, A. 1995. Experimental models of strike-slip tectonics. *Petroleum Geoscience,* **1**, 71–80.

SCHELLART, W. P. 1998. *Analogue modelling of basement involved strike-slip faulting.* MSc thesis, Free University, Amsterdam.

SCHELLART, W. P. 2000. Shear test results for cohesion and friction coefficients for different granular materials: scaling implications for their usage in analogue modelling. *Tectonophysics,* **324**, 1–16.

SCHREURS, G. 1994. Experiments on strike-slip faulting and block rotation. *Geology,* **22**, 567–570.

SYLVESTER, A. G. 1988. Strike-slip faults. *Bulletin of the Geological Society of America,* **100**, 1666–1703.

VERSCHUREN, M., NIEUWLAND, D. & GAST, J. 1996. Multiple detachment levels in thrust tectonics: Sandbox experiments and palinspastic reconstruction. *In:* BUCHANAN, P. G. & NIEUWLAND, D. A. (eds) *Modern Developments in Structural Interpretation, Validation and Modelling.* Geological Society, London, Special Publications, **99**, 227–234.

Segment linkage during evolution of intracontinental rift systems: insights from analogue modelling

T. TENTLER[1] & S. TEMPERLEY[2]

[1]Hans Ramberg Tectonic Laboratory, University of Uppsala, SE-752 36 Uppsala, Sweden
(e-mail: tatiana.tentler@geo.uu.se)
[2]Department of Geology, University of Leicester, UK

Abstract: On the basis of the effective scale-independence of brittle structures, from microcracks to regional fault systems, we have used analogue centrifuge models to provide insights into the initiation and evolution of continental active rifts, with a single dilational fracture segment representing a prototype rift segment. In the models, which are physically and dynamically scaled, a semi-brittle compound material, with flexural rigidity, was devised to simulate the lithosphere as a single layer. This is justified by the fact that, at the largest scale, the lithosphere behaves as a single viscoelastic, flexurally rigid unit. Silicon polymers of two different viscosities and densities represent the asthenosphere. A parallelepiped of lower viscosity and lower density material is embedded within and near the base of the second polymer that fills up most of the model box. The former is activated as a plume-like diapir that rises, spreads laterally and ultimately generates extensional stress in the semi-brittle layer on top. In terms of model fracture distributions, both narrow and wide failure modes were achieved, analogous to narrow and wide modes of rifting. The processes of fracture initiation, propagation and coalescence are the same for each mode. During the early stages of model runs, fracture initiation is more important than the propagation of existing segments in relieving stress, although the latter dominates during later stages. The key parameters of overlap, offset, obliquity and propagation angle, defined and illustrated, are used to distinguish different ways in which pairs of fractures coalesce. We recognize three distinct types of coalescence (type 1, type 2 and type 3) involving both tip-to-tip and tip-to-sidewall linkages. These are described and both graphically and statistically discriminated. Whether narrow or wide mode, an intracontinental rift system consists of a number of discrete fault-bound rift segments. Similar types of interactions to those in the models have been identified between pairs of rift segments from the Cenozoic Baikal and East African rift systems, and the Mesozoic to early Tertiary Central African rift system. The factors that control the type of coalescence between natural rift segments would appear to be precisely those that govern linkages of fracture segments in the models.

Intracontinental rift systems are the result of horizontal extension and associated vertical thinning of continental lithosphere. In terms of causative processes, rifting may be 'active' or 'mantle-activated' when initiated by an upwelling asthenospheric diapir, or else may be 'passive' or 'lithosphere-activated' when generated directly by far-field stresses associated with plate tectonic motions (Sengor & Burke 1978; Turcotte & Emerman 1983; Keen 1985). Irrespective of the fundamental underlying cause, the near-surface expression of a rift is as a composite, elongate graben, bound and dissected by normal faults and variably infilled by volcanic and sedimentary material. Surface heat flow in and around the rift zone is generally high due to plutonic, volcanic and hydrothermal activity, whilst heat flow across the Moho (known as reduced heat flow) is also relatively high, due to the presence of relatively shallow, anomalously hot asthenosphere beneath thinned mantle lithosphere.

Here we define a *rift system* as a series of discrete, variably interlinked *rift segments*. This definition generally holds true whether extension is asthenosphere- or lithosphere-activated, or whether the mode of rifting is wide or narrow (England 1983; Buck 1991; Buck *et al.* 1999). It is, moreover, applicable irrespective of the degree of asymmetry of the fault and shear-zone hierarchies which constitute a rift zone, yielding bulk pure-shear to

From: NIEUWLAND, D. A. (ed.) *New Insights into Structural Interpretation and Modelling.* Geological Society, London, Special Publications, **212**, 181–196. 0305-8719/03/$15

bulk simple-shear model end members (McKenzie 1978; Wernicke 1985; Lister & Davis 1989; Lister *et al.* 1991). In this study, modelling intracontinental active rifting, we are interested in the ways and means by which individual rift segments evolve and interact and, in particular, how the propagation and coalescence of segments result in the development of a through-going intracontinental rift system which, in turn, can lead to continental break-up via a rift-drift transition.

An individual rift segment has finite length and width at the Earth's surface, and may be considered as a zone of extensional deformation accommodated by a population of normal faults and dilational fractures, with one or more zones of ductile flow, although some extension may be accommodated by magmatic accretion and dyking. The scaling characteristics of fault/fracture populations in such zones represents a field of considerable research and debate over the last two decades, with particular emphasis on frequency–length scaling and displacement–length scaling. Studies of cumulative frequency versus fracture length have revealed both power-law and exponential relationships (Poulimenos 2000). The modelling data of Spyropoulos *et al.* (1999) and the field data of Gupta & Scholz (2000) imply a frequency–size transition from power-law to exponential as bulk, brittle fracture-related strain accumulates. The nature of displacement–length scaling represents a more contentious issue. Workers have found both linear and non-linear scaling, with the latter either power-law or exponential (e.g. Cowie & Scholz 1992). However, the most important consequence of scaling in the context of this study is simply that faulting is a self-similar process that has been demonstrated to operate over at least eight orders of magnitude. Because of the fractal characteristics of faulting, small-scale observations of fracture processes, under controlled laboratory conditions, have the potential to yield significant insights into the evolution of much larger-scale, natural fault systems.

How fracture coalescence models for fault zone evolution can throw light on the much larger-scale evolution of an entire rift system remains an important but, as yet, unanswered question. Energy considerations are such that a large-scale fault cannot form all at once, but evolves from interactions at ever-increasing scales. Thus, a graben-defining fault, say between 10 and 100 km in length, will result from the coalescence of smaller-scale segments, which themselves grow by coalescence of even smaller-scale fractures, and so on down to the micro-scale. A system of graben-defining faults constitutes a rift segment, while several, variably linked rift segments define a rift system. The development of an oceanic basin is impossible without prior development of a through-going rift system, the latter demanding progressive coalescence of once-isolated rift segments. When viewed on a continental scale, a rift segment propagates as a single system. Studies of mid-ocean ridges (MOR) at this scale have demonstrated that interactions of ridge rift segments are subject to the same controls as for single faults and small-scale fractures.

Being extremely large-scale, long-lived features, rift systems are not easy to study directly: lengths and widths of individual segments can exceed 1000 and 100 km, respectively, and they evolve over time scales of the order of 10–100 Ma. For these reasons we have turned to physical (analogue) modeling in order to provide insights into rift system evolution. Here we utilize centrifuge-modelling techniques on scaled analogue materials in a three-layer configuration, representing lithosphere, 'normal' asthenosphere and anomalous (i.e. hot) asthenosphere. The scale difference is such that an entire rift segment in nature is represented by a single or composite extensional (mode I) fracture of the lithospheric analogue. The majority of fractures have no discernible shear component and so are purely extensional, opening in a direction perpendicular to their walls, which is the defining characteristic of opening mode, or Mode I cracks (Atkinson 1987). This rationale is directly comparable to that for modelling-based studies of the propagation and linkage of MOR segments (Oldenburg & Brune 1975; Macdonald & Fox 1983; Pollard & Aydin 1984; Shemenda & Grocholsky 1991; Shemenda & Grocholsky 1994; Dauteuil & Brun 1993; Mart & Dauteuil, 2000).

How isolated, individual fracture segments initiate, propagate and coalesce to form a through-going system is illustrated, described and quantified. We present evidence for three types of segment coalescence, and describe and quantify the geometrical factors controlling each type. The results are then compared to natural rift systems representing various stages in rift evolution, where an entire rift basin represents a single segment. Examples chosen for comparison include the Cenozoic Baikal and East African rift systems, together with the Mesozoic to early Tertiary Central African rift system.

Method

Analogue material and scaling

For each experiment, a ductile/brittle model simulating lithosphere and asthenosphere was prepared. Non-linear materials that mimic natural rock deformation and stress distribution were used to construct the model. A strain-softening, semi-brittle material was used to imitate lithosphere. It was prepared as a homogenized mixture of (by weight): 40% paraffin, 33% Vaseline and 27% plaster of Paris, which was heated, stirred when molten and then cooled. Non-linear silicone polymers, with power-law exponents (n) between 1.3 and 4.0, were used to represent the asthenosphere. It is commonly assumed that the asthenosphere behaves as a linear (Newtonian) viscous fluid, particularly at the low deviatoric stress values associated with mantle convection (Lliboutry 1999). However, recent evidence from laboratory experiments suggests that at slightly higher stresses the relationship becomes moderately non-linear, when the effective viscosity would be strongly dependent on temperature, pressure and water content (Davies 1999). For this reason we chose to use non-Newtonian material for simulation of asthenosphere. Polymers of two different rheologies were used. That with the higher viscosity and density is representative of 'typical' subcontinental asthenosphere, while the softer, less dense version represents hotter and more buoyant asthenosphere.

The properties of modelling materials and the dimensions of each material unit of the models are listed in Table 1, together with comparable data for equivalent prototype (natural) units. The rheological properties of the analogue materials were tested in a uniaxial pressure vessel at a constant strain-rate and temperature. The materials were designed or chosen in order to ensure a dynamic equivalence to the physical properties of the natural system. The properties of the analogue materials must therefore satisfy proportional scaling rules, such that the ratios of their stresses, rheologies and densities correspond as closely as possible to those in nature, derived from mechanical, petrological and geophysical data (Hubbert 1937; Ramberg 1981; Weijermars & Schmeling 1986). According

to the principles of dynamics and scaling, the following condition must be satisfied:

$$\sigma_r = \rho_r g_r l_r \tag{1}$$

where σ is stress, ρ is density, g is centrifugal (model) or gravitational (natural) acceleration and l is the length-scale. The subscript r indicates that these values are model/prototype ratios (Dixon & Summers 1985; Davy & Cobbold 1991; Brun 1999). The lower densities of the modelling materials compared to natural densities ensures that $\rho_r < 1$. The density of upper asthenospheric mantle varies between 3.38×10^3 and 3.99×10^3 kg m^{-3} (Dziewonski & Anderson 1981). Density values of 3.7×10^3 kg m^{-3} and 3.1×10^3 kg m^{-3} for normal and hot asthenosphere, respectively, yield g_r values of 0.34 and 0.37 (Table 1). While the density of the lithospheric mantle is about 3.3×10^3 kg m^{-3}, that of the crust is lower, between 2.6×10^3 and 3.0×10^3 kg m^{-3}. Rotation of the model increases acceleration, resulting and ensuring that $g_r \gg 1$. The model was rotated in a centrifuge at a speed of 600 rotations per minute, so the acceleration was scaled as $200g$ for model equivalent to $1g$ for prototype. In terms of length scale, 1 mm in the models corresponds to 30 km in nature, and so l_r is extremely small, at 3.3×10^{-8}.

Provided that the model and prototype viscosities are also proportionally scaled, the buoyant forces in the models should correspond to the forces generated in the asthenosphere beneath active rift zones. As a result, the spreading rates in the models should be appropriately scaled to those of natural systems. Assuming non-linear behaviour, viscosity in the mantle appropriate to the time-scale of convection is both temperature- and stress-dependent. The viscosity of the asthenosphere is

Table 1. *Characteristics of analogue materials and model ratios used in the experiments*

Earth unit	Thickness	Density (kg m^{-3})	Viscosity (Pa s) (or other value as stated)
Lithosphere			
Nature	60 km	2.7×10^3	Flexural rigidity: 10^{22}–2×10^{23} Nm
Model	2 mm	1.0×10^3	Strength: 2.5×10^4 Pa
Model ratio	0.03	0.37	
Abnormal asthenosphere			
Nature	300 km	3.1×10^3	10^{20}
Model	10 mm	1.1×10^3	3×10^4
Model ratio	0.03	0.35	3×10^{16}
Normal asthenosphere			
Nature	300 km	3.7×10^3	10^{21}
Model	10 mm	1.25×10^3	7.25×10^5
Model ratio	0.03	0.34	7.25×10^{16}

Model ratio is taken as ratio: (parameter in model)/(parameter in nature)

estimated to be between 10^{20} and 10^{21} Pa s (Mitrovica 1996; Lambeck & Johnston 1998), with the lower value more likely to be representative of hotter 'abnormal' asthenosphere. Time-dependent strength variations of the lithosphere imply that it is a viscoelastic unit (a Kelvin–Voight solid) over the time scale appropriate to rifting, with a flexural rigidity of between 10^{22} and 2×10^{23} Nm (Walcott 1970; Cathles 1975).

Modelling procedure and model evolution

Models were constructed so that the density difference in the asthenospheric analogue material would, under the action of centrifugal force, generate a gravitational instability. The technique of centrifuge modelling allows body and surface forces to be scaled with the same ratio, necessary to achieve dynamic similarity of non-Newtonian flows (Mulugeta 1988). First, a rectangular parallelepiped of the lower density polymer was placed centrally at the base of the model box. The dimensions of this unit are illustrated in Figure 1a. Next, the box was filled to within 0.2 cm of the top with the polymer of higher density and viscosity. Finally, an upper planar layer of the semi-brittle material was placed on top. The lateral dimensions of this layer were deliberately made less than those of the confining walls of the box, permitting limited spreading.

The centrifuge technique is particularly suited to deformation of models with small dimensions consisting of high viscosity fluids which can preserve finite deformation and that have fast instability amplification rates. Centrifugal force is employed to increase the apparent gravity field in the analogue materials, and represents the ultimate driving force generating stresses and allowing rapid deformation (Ramberg 1981; Weijermars & Schmeling 1986). Figure 1 illustrates a typical model run. In the centrifuge box, the model was laterally confined from four sides and the base, with the upper surface open. The small gap left around the margins of the upper layer, 1 cm wide, was infilled either with Plasticine or silicon polymer, materials of similar rheology but different strength. They act as soft buffers for the upper layer, limiting the amount of finite extension. Rotation of the model in the centrifuge was interrupted at regular intervals, in order to record the stage of deformation reached, and so provide a semi-continuous history of the development of deformation. After each experiment, the deformed model was removed from the centrifuge and cut vertically, in order to observe and photograph the structure in cross-section.

Stretching of the upper layer was activated by diapiric uprise of the lowest viscosity/lowest density material, analogous to the uprise of a plume of abnormally hot asthenosphere in the mantle. The parallelepiped of this material transforms into a linear plume, which in profile is mushroom shaped (Fig. 1b). Diapiric upwelling leads to divergent lateral flow of the higher viscosity silicon polymer above the plume's head. This in turn generates an extensional stress regime in the upper layer, with $\sigma 1$ vertical and $\sigma 3$ perpendicular to the linear axis of the plume. Amplification of perturbations on the upper surface of the diapir leads to the development of flame-like protrusions that begin to impinge on the lower surface of the upper layer and are responsible for stress concentrations in the latter (Fig. 1c). Thus not only is the regional stress configuration generated from below, but also the local stress anomalies that control positioning of sites of fracture initiation in the semi-brittle layer. Thus the process is analogous to intracontinental active rifting. The ductile, divergent flow above the plume permits further diapiric upwelling that ultimately enhances lateral stretching of the upper layer. As the plume head approaches the base of the upper layer it spreads laterally, while the neck and base of the plume feed progressively into the head, eventually leaving only a short, tapering tail (Fig. 1d). Further stress concentration in the analogue lithosphere results in continued fracturing. The fractures have a significant dilational component, with space between the walls progressively infilled by polymer flowing up towards the model surface (Fig. 1d). The fluid pressure exerted by the polymer on the walls promotes fracture propagation by enhancing dilation and increasing the stress concentration at fracture tips.

The limit of lateral spreading of the upper layer is constrained by two factors: the difference in lateral dimensions between the upper layer and the confining box, and the nature of the material that fills the resultant gap. In the model runs, both Plasticine and rheologically softer polymer were used as fills. Where Plasticine was used, spreading was relatively limited, leading to focused, narrow-mode failure, whereas the less resistant polymer allowed greater finite extension, resulting in diffuse, wide-mode failure. The two failure modes can also be distinguished by a difference in diapir geometry when viewed in cross-section (Fig. 2).

Brittle failure of the upper layer accommodates differential stress build-up induced in the upper part of the model. The fractures that develop are oriented more or less perpendicular to the minimum principal stress and have a dominant tensile component. Fracture arrays are either distributed diffusely across the model surface or are focused within a narrow, linear central zone. Propagation and coalescence result eventually in the development of an interlinked fracture system transecting

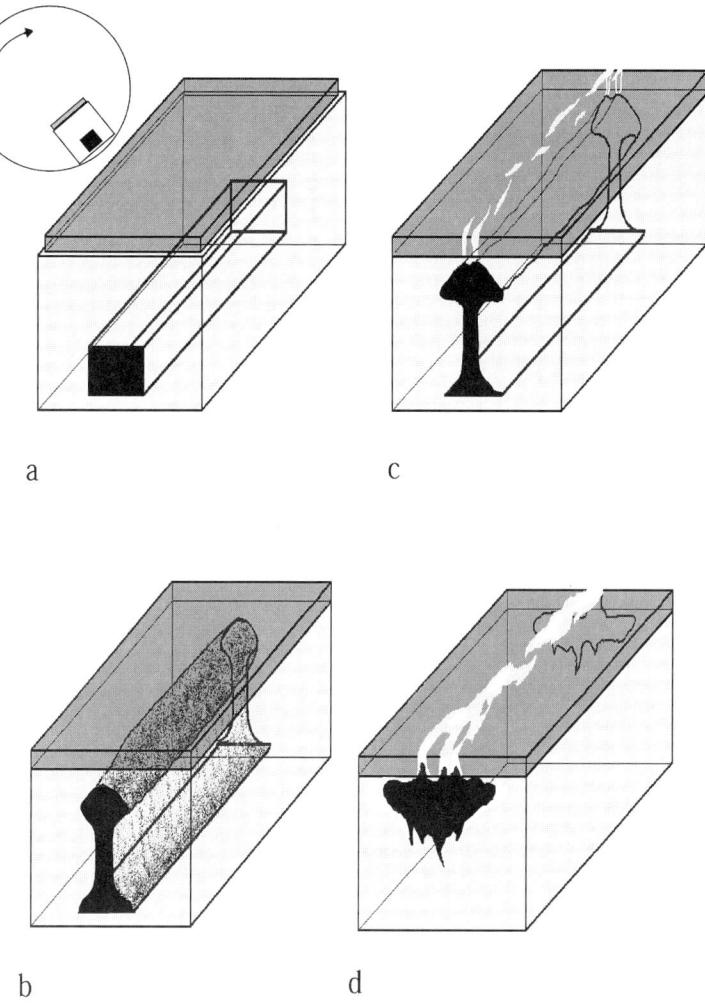

a

c

b

d

Fig. 1. Block diagrams illustrating the form of the model and successive stages of a typical model run. (**a**) Model configuration. A parallelepiped of lower density, lower viscosity silicone polymer is embedded in a matrix of slightly denser, stiffer polymer. A 2 cm thick sheet of compound, semi-brittle material with flexural rigidity is placed on the top, simulating lithosphere. (**b**) Early stage of model run. The lower density, lower viscosity material is transformed into an upwelling, linear, plume-like diapir by centrifugal acceleration. (**c**) Mid-stage. Divergent, lateral flow of denser, stiffer polymer above upwelling diapir generates extensional stress in upper layer. Stretching of latter leads to development of mode I (tensile) fractures. (**d**) Tail of diapir feeds into head as it spreads beneath upper layer. Stress concentrations lead to further fracturing and increasing dilation of existing fractures, especially above protrusions on upper surface of diapir. Upwelling polymer widens and partially fills the open fractures.

the entire upper surface, either irregular and tortuous (wide mode) or regular and close to rectilinear (narrow mode).

Each experiment was assessed by analysis of both the top, free surface and by cutting cross-sections through the model.

Results

Fracture initiation and propagation

The distributions of fractures across model surfaces are either focused (narrow-mode) or diffuse (wide-

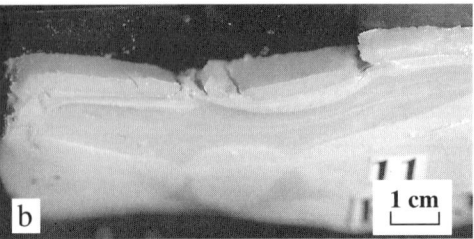

Fig. 2. Cross-sections of models after completion of model runs. (**a**) Example of narrow-mode failure, with only a single, central protrusion of the diapir associated with a focused stress concentration in upper layer. (**b**) Wide-mode failure, with zones of brittle failure in upper layer developed above upward protrusions of diapiric material. Note that the model has collapsed somewhat after removal from the centrifuge.

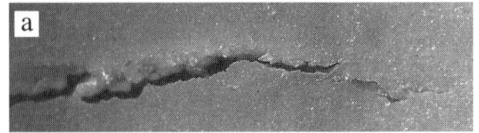

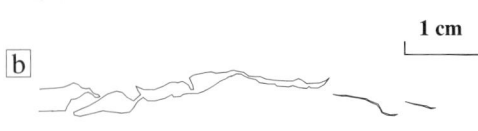

Fig. 3. Plan view of a propagating fracture set on the surface of the upper layer. Newly initiated minor cracks ahead of main tip do not show any opening between their surfaces (far right). As the main fracture propagates, it widens progressively.

mode), depending on the model set-up. In both cases, however, the dynamic and mechanical controls on fracture evolution (i.e. initiation, propagation and linkage) appear to be identical. The upper layer is as homogeneous as we could make it and is effectively isotropic. A single, isolated fracture in such a medium will tend to propagate in-plane, with both strike-parallel (lateral) and subordinate dip-parallel (vertical) components. The fracture is surrounded by a local, elliptical stress perturbation, with two zones of high stress intensity focused on the tips. The stress level is maintained more or less at a critical value as the fracture propagates through the medium. Fracture tips propagate mostly laterally, in a direction perpendicular to local σ_3, which for dilational structures must be negative (tensile).

Detailed examination of the termination of a fracture segment that has propagated reveals the presence of a damage zone around, and extending well ahead of, the tip (Fig. 3). The zone is characterized by the presence of smaller, secondary cracks (Fig. 3a) that result from the presence of tip-centred stress concentration related to fracture geometry. Some of these are consumed by subsequent migration of the tip, while the more offset

cracks that remain tend to close as the fracture propagates through the damage zone (Fig. 3b).

Narrow-mode failure In model runs where spreading of the upper layer is most restricted (by Plasticine fill), the zone of highest stress concentration is relatively narrow (Fig. 4) and is situated immediately above the apex of the upwelling diapir. The upper surface of the plume head has a broad, convex-upward form, with a single, focused further protrusion of diapiric material. The axis of the failure zone in the upper layer is aligned directly above the protrusion. The result is a laterally restricted zone of dilational fractures on the surface of the model (Fig. 4b, stage i). The segments are distributed in a relay pattern, comprising a shingled arrangement of subparallel fractures with no consistent sense of overstep, where all are subparallel to the zone in which they occur.

The limited total width of the array means that lateral offsets between adjacent structures are invariably small (Fig. 4b, stage i), thus the chances of one propagating fracture encountering the stress perturbation of a neighbouring fracture are extremely high. Under such circumstances propagation almost inevitably leads to coalescence, so that very few fractures remain isolated. Adjacent fractures only occasionally propagate tip-to-tip; most pairs overlap each other before mutual tip-to-sidewall coalescence (Fig. 4b, stage ii). The segments that result from coalescence progressively increase in their lengths and widths, while decreasing in number (Fig. 4b, stage iii). The largest segments, of stage iii, coalesce to form a through-going system in exactly the same ways that smaller scale linkages were achieved between stages i and ii.

Wide-mode failure With silicon polymer as a peripheral buffer, a greater amount of spreading of the upper layer is permitted. Diapiric impingement and

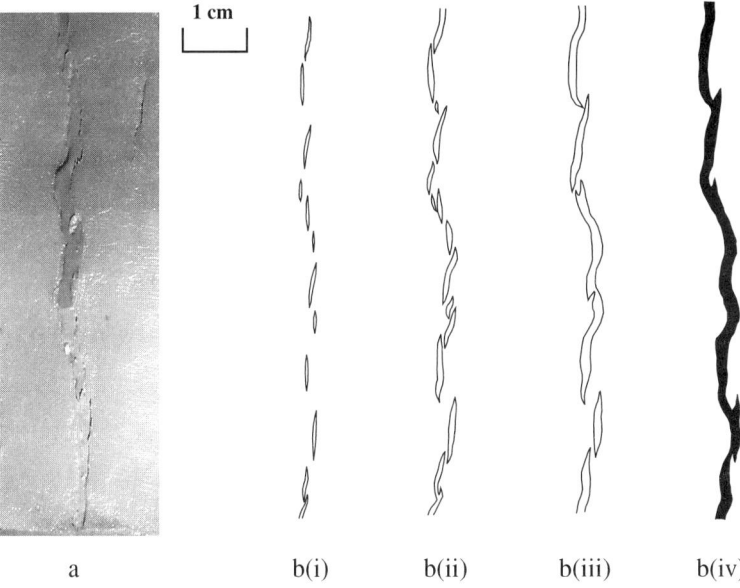

a   b(i)   b(ii)   b(iii)   b(iv)

Fig. 4. Narrow-mode failure, in plan view. (**a**) Photograph of upper layer with almost through-going, narrow-mode fracture array. (**b**) Evolution of narrow-mode failure in four stages (i-iv). Sketches depict the initiation, propagation and coalescence of fractures leading to development of through-going, narrow-mode system.

the divergent lateral flow below are initially focused, as for narrow-mode models, generating a central zone of first fractures above the apex of the plume. But as the diapiric spreading phase proceeds, local stress concentrations are induced progressively further away from the central zone (Fig. 5). Because a greater amount of finite extension is permitted in the model set-up, the finite fracture densities are much higher than those from narrow-mode failure. Brittle failure continues apace after initiation of the first fractures, and so the density increases and the distribution widens rapidly (Fig. 5a). Figure 5b illustrates a typical wide-mode run in four stages. At least 80% of the maximum number of fractures have initiated by the end of stage i. They are distributed rather diffusely across the surface of the upper layer, although there is a tendency for fractures to be larger and more densely developed in the axial zone. After stage ii, the density gradually decreases as coalescence proceeds.

As the plume head spreads laterally beneath the upper layer, further instabilities develop spontaneously on its upper surface. A few of these evolve into upward-tapering, flame-like protrusions (Figs 1d and 5a). Since the fracture density is already high by this stage, stress build-up is relieved much more by propagation and dilation of existing structures than by further fracturing (Fig. 5b, stage ii). The dilation of fractures in turn affects the shape of the diapiric body, creating space for

further local rises of buoyant material from the protrusions towards the upper surface. The largest and widest fractures are located above the diapiric protrusions and are thus most readily filled by polymer.

Although the mean orientation is more or less parallel to $\sigma 2$, fractures have a much larger strike range than any narrow-mode array. This variability, together with their wide and diffuse distribution, provides the potential for a wider variety of coalescence geometries (Fig. 5b, stage iii). From here onwards, continued growth of the largest and widest fractures relieves most of the stress build-up, so that many of the smaller examples cease to propagate. The dilations of a number of small cracks decrease between stages ii and iv, while a few examples actually close completely. This implies local compression parallel to the stretching direction, in the material between larger dilating fractures, where $\sigma 3$ is positive.

The through-going linked fracture system which results from complete coalescence is highly irregular in geometry, marked by a very wide strike variability from one end to the other, with linked fractures in a zigzag arrangement (Fig. 4b, stage iv). The system also bifurcates into two 'rift arms' over the middle third of its length. A pronounced change in strike of the system is located wherever two markedly oblique segments have coalesced.

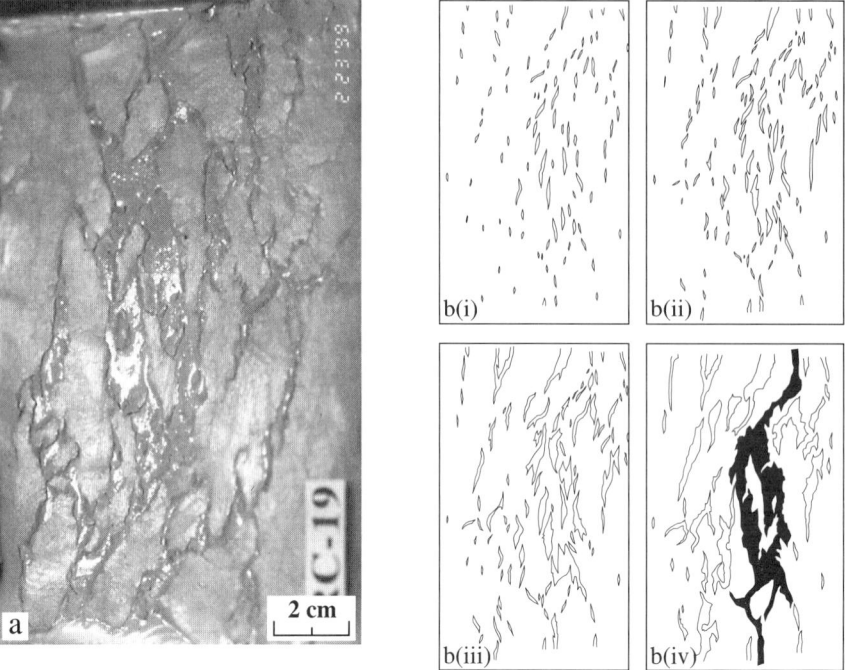

Fig. 5. Wide-mode failure, in plan view. Description corresponds to Figure 4.

Fracture coalescence

If a propagating tip encroaches into the stress perturbation of a neighbouring fracture, it tends to be deflected towards the tip zone or, more usually, the sidewall of the latter. The ensuing coalescence relieves entirely the stress concentration associated with the tip or tips involved in linkage. For a semiquantitative analysis of the fracture interactions, four particular parameters characterizing the geometries and interrelationships were measured and compared (Fig. 6). The parameters include: *offset*, the minimum separation between fractures; *overlap* of tips, measured parallel to one or both fractures; *obliquity*, the angle between the strikes of two overlapping fractures; and *propagation angle*, the angle between a propagating tip and the fracture it is directed towards.

A linked, through-going failure system represents the amalgamation of a number of discrete fracture sets, each of finite length. The geometry of the system will depend upon the variability of coalescence that, in turn, depends upon the range of individual fracture geometries and the distribution of fractures and fracture arrays. From an examination of fracture evolution on the upper surfaces of models, three types of coalescence have been distinguished between pairs of fractures (Fig. 6).

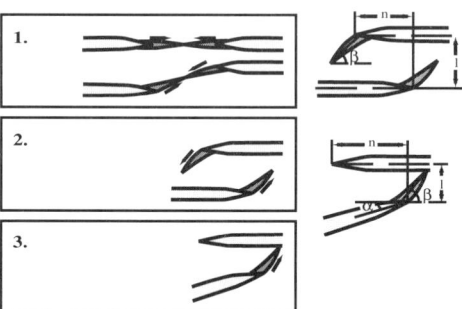

Fig. 6. Coalescence types between pairs of fractures. Two parent fractures (light shading) coalesce by propagation of one or both of their tips (dark shading). (**1**) Type 1 (tip-to-tip) coalescence. (**2**) Type 2 (mutual tip-to-sidewall) coalescence. (**3**) Type 3 (single tip-to-sidewall) coalescence. Parameters used to discriminate coalescence types are defined in the inset diagram. These include: n, overlap; l, offset; α, obliquity; β, propagation angle.

For type 1 coalescence, either a tip of one fracture propagates directly to the tip of another, or the tips of two fractures propagate and link (tip-to-tip interaction in each case). This is possible if offset is small to moderate, obliquity and propagation angle are small or zero and overlap is negative (Fig. 6.1). Types 2 and 3 both involve tip-to-sidewall interaction. In type 2 coalescence the tips of both fractures are active, but they each propagate along paths that curve in towards the sidewall of the adjacent fracture (double tip-to-sidewall) (Fig. 6.2). In this case, offset and propagation angle may be small or large, overlap positive and moderate to large, and obliquity small to moderate. Type 3 coalescence involves interaction of two non-parallel fractures. In this case, only one tip of a single fracture is activated, propagating in to the flank of the adjacent fracture (single tip-to-sidewall) (Fig. 6.3). It is associated with moderate to large values of overlap, positive overstep, obliquity and propagation angle. As a propagating tip is deflected towards a sidewall during type 2 or type 3 coalescence, the tip angle may approach 90°. At this angle the tip is most unfavourably oriented for dilational propagation, which ceases upon contact with the other fracture.

All three coalescence types in all stages of development may be seen in Figures 4 and 5. It is apparent that good examples of type 1 interaction are less common than those of types 2 or 3. One possible reason for this may lie in the fact that tip-to-tip interaction is not easily recognized in mature rift systems where coalescence obscures any preceding offset. Type 1 coalescence would leave behind a single planar crack, so that recognition of the two former segments would be difficult. This also relates to the fact that determination of type 1 coalescence is rather scale-dependent. Pairs of fractures are rarely perfectly coplanar and rarely have tips that propagate towards each other in precisely the same plane. The width of the tips of fractures is limited so that the chances of precise tip-to-tip linkage are small. It seems, therefore, that most examples of type 1 are in fact extremely small-scale examples of type 2. While examples of types 2 and 3 are commonly observed, they are in fact end members of a continuum of coalescence geometries. In Figures 3 and 4 there are examples intermediate in form between the two types.

Photographs taken during experiments illustrate in more detail the nature of each type of coalescence (Figs 7, 8 & 9). If two adjacent fractures are contained more or less on the same plane and are separated along strike (Fig. 7a), they propagate towards each other by straight paths leading to coalescence that appears as type 1 at the scale of observation (Fig. 7b). The early stage of type 2 coalescence (Fig. 8a) involves equal propagation of both segments. Ideally, the process should leave an 'island' of material upon completion, with segment bifurcation marking the zone of coalescence. More often, however (as in Fig. 8b), only a single arm remains active after coalescence, implying incomplete mutual linkage. Figure 9 shows very clearly the increase in tip angle as type 3 coalescence

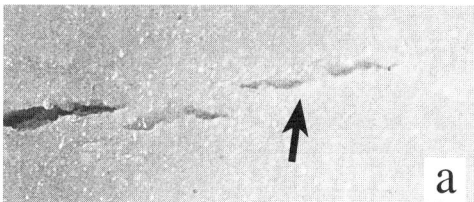

1 cm

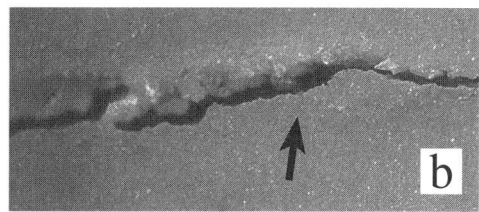

Fig. 7. An example of type 1 coalescence: (**a**) before, and (**b**) after. Location of linkage marked with arrow.

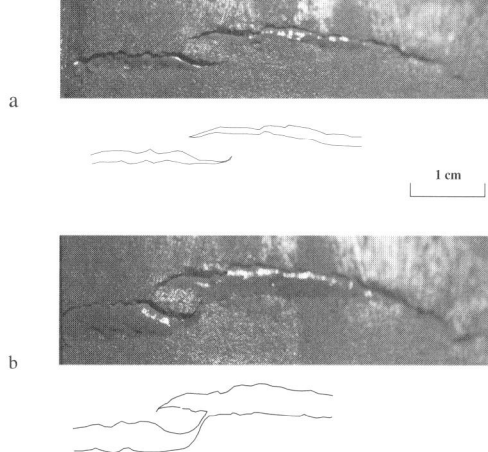

Fig. 8. An example of type 2 coalescence: (**a**) before and (**b**) after. Note that the final stage of coalescence involves only one of the tips propagating to one of the sidewalls, when both tips had been active during the earlier stage.

approaches completion, in this case varying from c. 20° as coalescence begins (Fig. 9a), to c. 80° on sidewall contact (Fig. 9b).

Precise measurements of segment parameters permit coalescence types to be discriminated graphically (Figs 10 to 13). On a graph of lateral offset versus overlap (Fig. 10), three distinct trends can be observed, each related to a specific type of

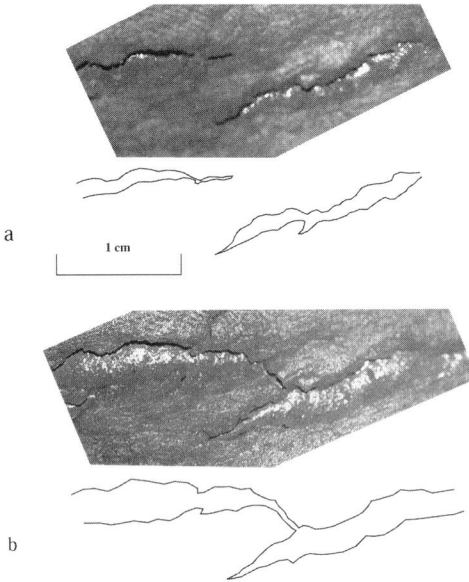

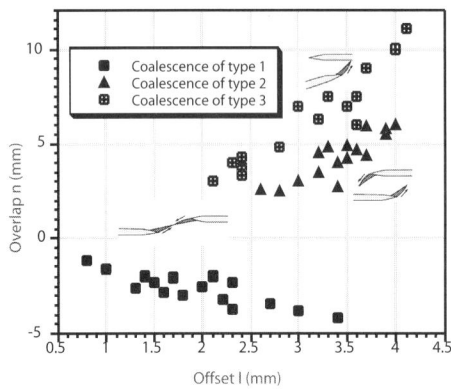

Fig. 9. An example of type 3 coalescence: (**a**) before and (**b**) after.

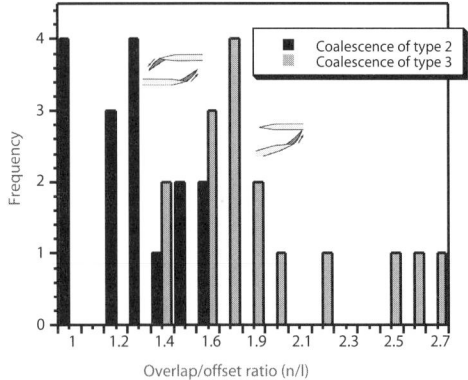

Fig. 11. Differentiation of coalescence types 2 and 3 on a histogram with overlap/offset ratio plotted against frequency.

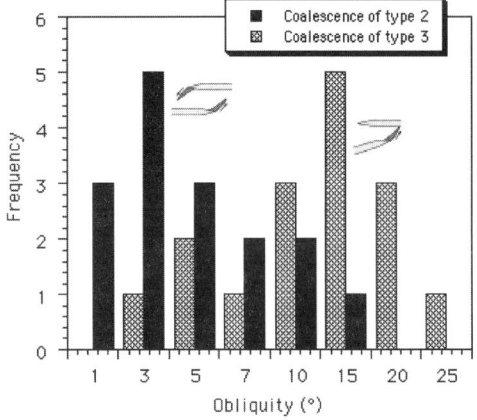

Fig. 12. Coalescence types 2 and 3 plotted on an obliquity versus frequency histogram.

Fig. 10. Graph of offset plotted against overlap, showing differentiation of all three coalescence types.

coalescence. Type 1 is clearly differentiated by negative overlap with variable but mostly small offset values. Types 2 and 3 plot in positive overlap/overstep space, but for a given value of offset, the overlap value is always lower for type 2 than it is for type 3. Thus the degree of overlap appears to be a key factor in controlling whether one or both of the tips will facilitate linkage. A graph of overlap/offset ratio versus frequency (Fig. 11) shows how well the two types of coalescence can be discriminated by means of these parameters. The angular strike difference between two fractures

provides a further control for type 2 versus type 3 coalescence. On a plot of this parameter versus frequency (Fig. 12), a prominent bimodal distribution is apparent. This is, in fact, a combination of type 2 and type 3 distributions both displaying skewed unimodal distributions with very different mean values. It may, therefore, be inferred that both tips will propagate where two fractures are subparallel, but only a single tip is likely to be active where the propagation angle exceeds about 10°. A graph of tip angle versus overlap/overstep ratio provides a further discrimination between type 2 and type 3 coalescence (Fig. 13). While plots of each type display a positive rectilinear trend, the ordinate and abscissa values both tend strongly to be lower for type 2 compared to type 3. There is, moreover, a suggestion that, while the values have a small degree of overlap, different

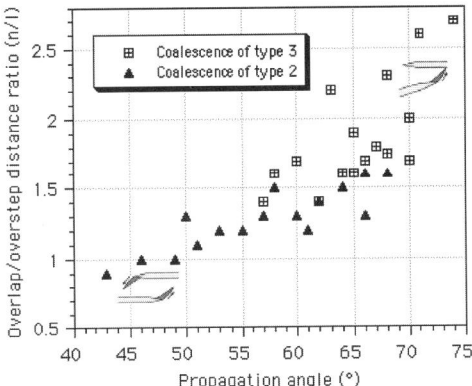

Fig. 13. Differentiation of coalescence types 2 and 3 on a graph of tip-angle plotted against overlap/offset ratio.

slopes define the best-fit lines through their trends. However, the most obvious conclusion is that type 2 coalescence is more likely where tip angles are relatively small, but beyond a tip angle of *c.* 60°, type 3 coalescence is favoured.

Discussion

Limitations of the model

Although we have taken as much care as possible to ensure an accurate scaling between the models and the natural systems, there are certain limitations to be considered. Depth-dependent strength variations are known to be important in controlling the geometry of rift zones. In particular, Benes & Davy (1996), Buck *et al.* (1999) and Brun (1999) have demonstrated that ratio of brittle strength to viscous strength, thickness and thermal state of the lithosphere are likely to be significant factors in narrow-mode versus wide-mode rifting. While none of these factors could be tested in our models, we were nevertheless able to generate both narrow-mode and wide-mode rift systems, by using two materials of different strength as buffers for the lithospheric layer. The buffers controlled the total amount that the lithospheric analogue was allowed to extend, but we see no reason why the value of finite extension should represent an important control over whether a wide or narrow rift develops. However, it may be possible that the *rate* of extension is a function of the rheology of the buffer material, so that the average strain rate in the lithospheric layer of the models is different for narrow-mode and wide-mode model runs. Since the rheology of a material is sensitive to strain-rate, this may represent an important controlling factor, and requires further study.

The modelling was designed to simulate active rather than passive rifting in nature, with the softer of the asthenospheric analogues upwelling as a plume and spreading laterally. Coupling between layers during this spreading phase means that the minimum principal stress, $\sigma 3$, is tensile in the uppermost layer. In nature, on the other hand, $\sigma 3$ is compressive during lithospheric extension, except very close to the surface in some regions. As a result of this limitation, a single tensile fracture in the model cannot be compared to a single normal fault in nature. However, a rift basin segment in nature, with an array of normal faults dipping in towards the axis, possesses a symmetry not unlike that of a single dilational fracture when scaling is taken into consideration. It is of note that significant insights have been achieved through the use of small-scale physical models where a single mode I fracture is analogous to an entire MOR rift segment (Oldenburg & Brune 1975; Pollard & Aydin 1984; Shemenda & Grocholsky 1994; Mart & Dauteuil 2000). Modelling continental lithosphere as a single layer of course represents a considerable over-simplification, but is justified by the scale of the comparison. It is of note that many numerical models of extension, including that for uniform pure shear, first quantified by McKenzie (1978), also assume the lithosphere is a single, uniform mechanical layer, and yet have remained useful for over 20 years in understanding extension on a continent-wide scale.

Comparisons with natural rift systems

Baikal rift system (BRS) The Cenozoic BRS is developed as an arcuate belt of basins on the SE flank of the Siberian craton. The system, which is over 1800 km in length, is superimposed predominantly on a Palaeozoic fold-thrust belt, and flanked to the SE by the Mesozoic Transbaikal orogenic belt (Tapponnier & Molnar 1979; Zonenshain *et al.* 1990) (Fig. 14). Strike-slip faults present at both the NE and SW limits of the rift zone imply the possibility that it is, on the largest scale, a transtensional system. Away from these bounding faults, however, a comparison of the orientations of rift basins with both horizontal principal stress directions and slip vector azimuths (Petit *et al.* 1996) reveal no significant transtensional component. Two main rifting phases have been defined: a slow phase of Oligocene to end early Pliocene age, and a recent, ongoing rapid phase. Total horizontal extension across the BRS is considered to be no more than 10 km (Artyushkov *et al.* 1990).

The Central Baikal rift zone (CBRZ) contains four major rift basins: three deep-water basins of Lake Baikal (South, Central and North Baikal basins) and the Barguzin basin to the east (Fig. 14). Sequences of smaller, variably linked rift basins are

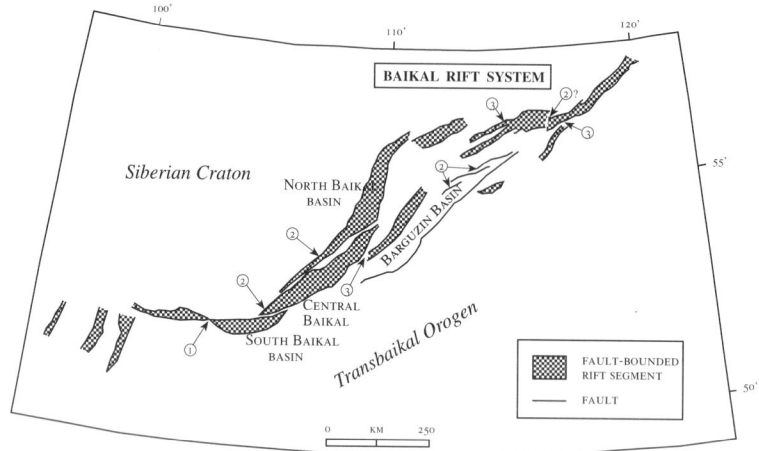

Fig. 14. The Cenozoic Baikal rift system (after Agar & Klitgord 1995) with coalescence types between rift segments marked.

present both to the NE and to the WSW of the four larger segments. In most cases, basins are flanked to the west by a major zone of normal faults, while faults on their eastern flanks tend to be less prominent and to have smaller displacements. Within the central rift zone there is a transition from focused, narrow-mode extension in the SW (the South and Central Baikal basins), to a wider, more diffuse mode to the NE, where the mean elevation is much higher (Fig. 14). Agar & Klitgord (1995) present evidence for broadening of the active Central Baikal rift zone through time, involving propagation and linkage of 10–100 km scale normal faults and the rift segments they define.

At the largest scale, we recognize some 15 discrete segments between 100 and 500 km in length. Pairs of segments display various stages of coalescence and all three types are represented. The more obvious examples of coalescence are highlighted in Figure 14. Within that part of the CBRZ defined by Lake Baikal, the south/central and central/north segment pairs are both good examples of type 2 coalescence that have proceeded almost to completion. The fourth segment, to the east of Lake Baikal, shows incipient type 3 coalescence with the Central Baikal segment. Also shown in Figure 14 are two type 2 interactions between three smaller-scale fault zones.

East African rift system (EARS) Extending from the Afar triple junction in the north to the Zambezi River in the south, the Cenozoic EARS is over 3000 km in length. The system bifurcates in its central portions, around the relatively rigid Tanzania craton that constitutes the East African plateau, with the Tanganyika and Albert rifts in the

west, and the Kenya rift in the east (Fig. 15) (Baker & Wohlenberg 1971; Baker *et al.* 1972; Ebinger 1989; Smith 1994). Both the east and west rift arms are superimposed on orogenic belts of Late Proterozoic age, which flank the craton. The earliest phase of rifting, of Pliocene to early Pleistocene age, was the result of east-west directed extension. Although rifting has been more or less continuous to the present day, after early Pleistocene times the extension direction rotated and is now oriented SE-NW (Bosworth & Strecker 1997).

The amount of horizontal crustal extension appears to vary along the system, with values obtained ranging principally between 10 and 40 km. Generally, the rifting has propagated and the rift zone widened southwards through time (Mohr 1982). With evidence of asthenospheric plume activity of Oligocene-Recent age beneath both Afar and the East African Plateau, the east-west to SE-NW extension is generally accepted to be passive. It is also achieved by narrow mode rifting, in the form of two principal, overlapping 'megasegments', each over 1000 km in length, but rarely wider than 100 km. The southern and west-central rift arms are linked to form one megasegment, while the east-central and northern rift arms are linked to form the other (Fig. 15).

At the very largest scale, the configuration of the megasegments is very like that for type 2 coalescence, although at this scale, the rigidity of the Tanzania craton compared to the flanking orogenic belts exerts the principal control on the geometry of the system. The four rift arms themselves each comprise a sequence of discrete, second-order segments, in various stages of coalescence (Fig. 15). The west-central rift arm in particular shows excel-

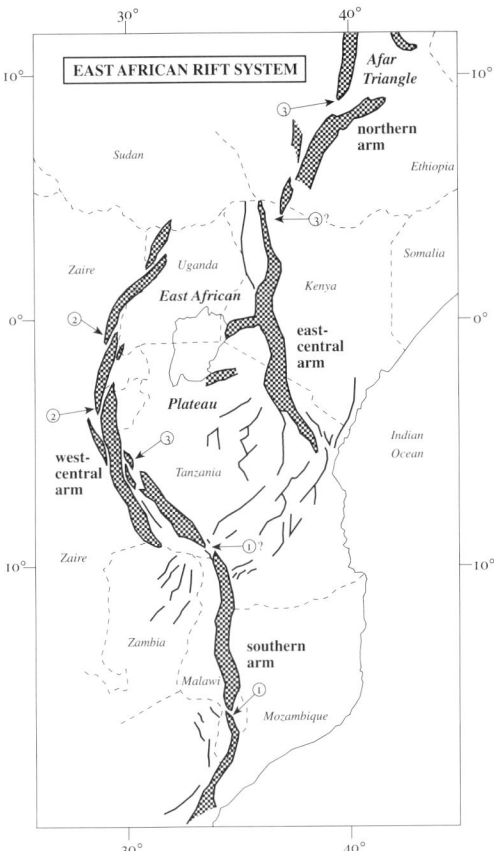

Fig. 15. The Cenozoic East African rift system (after Chapola & Kaphwiyo 1992; Tesha *et al.* 1992) with coalescence types between rift segments marked.

lent examples of type 2 coalescence. Linkages between the southern and west-central arms on the one hand, and the east-central and northern arms on the other, both appear to represent early stages of type 3 coalescence. The three segments that make up the southern arm are linked in two type 1 patterns, while the two main segments of the northern arm show type 3 interaction.

Central African rift system (CARS) Less well known and less obvious than the EARS, but developed on a similar scale, the CARS is the name given to a series of older, inactive SE-NW trending rift segments developed across Kenya and southern Sudan (Fig. 16) (Browne & Fairhead 1983; Browne *et al.* 1985; Fairhead 1988; Bosworth 1992). The system is bounded to the NW by a major NE trending system of dextral strike-slip faults with minor pull-apart basins, the Central African shear zone (Fig. 16). Rift basins are usually

flanked on both sides by segmented normal fault systems, the latter accommodating NE-SW directed horizontal extension. The main rifting phase is Mesozoic in age, with evidence of early Tertiary reactivation.

From SE to NW the system changes from narrow mode in Kenya, defined by the Anza rift, to wide mode in southern Sudan, with at least five major, parallel rift zones developed across a region up to 1500 km wide (Fig. 16). The largest of these zones, the Muglad-Abu Gabra rift, is *c.* 100 km wide and up to 1000 km long. The Anza rift is of similar length but is almost 200 km across at its widest. From their geometry, it is evident that the larger examples, including the Anza, Muglad-Abu Gabra, White Nile and Blue Nile rifts, developed by coalescence of smaller-scale segments.

Segment interactions appear to be less clearly defined and coalescence types are not as easily discriminated compared with the BRS and EARS. Transform-like zones of dextral strike-slip within the system, such as the South Sudan shear (Fig. 16), link many pairs of segments. Nevertheless, a few clear examples of direct coalescence can be distinguished of type 2 or type 3 (highlighted in Fig. 16).

Summary of observations

The natural examples of rift segments, from the three systems described above, are taken from both narrow- and wide-mode systems, and include both symmetric graben and asymmetric, detachment-controlled half-graben. The principal factors controlling the type of coalescence between natural rift segments would appear to be precisely those that govern linkages of fracture segments in the models, namely offset and overlap values, together with the difference in trend between two adjacent segments. Of course, the greater lateral restriction of rifts during focused, narrow-mode extension is such that progressive coalescence of segments, to form a through-going system, is likely to be much more readily achieved at low to moderate bulk strains, compared with wide-mode rifting. We would also expect less type 3 interaction between segments, as this would appear to demand offset values and rift-trend differences that are both rather large for narrow-mode rifting.

In two of the three natural examples, the nature of the basement exerts a notable control on the geometry of the system and the distribution of segments. In particular, the siting, orientation and size of segments at the onset of rifting are likely to be strongly controlled by the configuration of basement weaknesses. In this respect, the initial distribution of segments is crucial in determining whether linkage occurs at all and, if so, what type

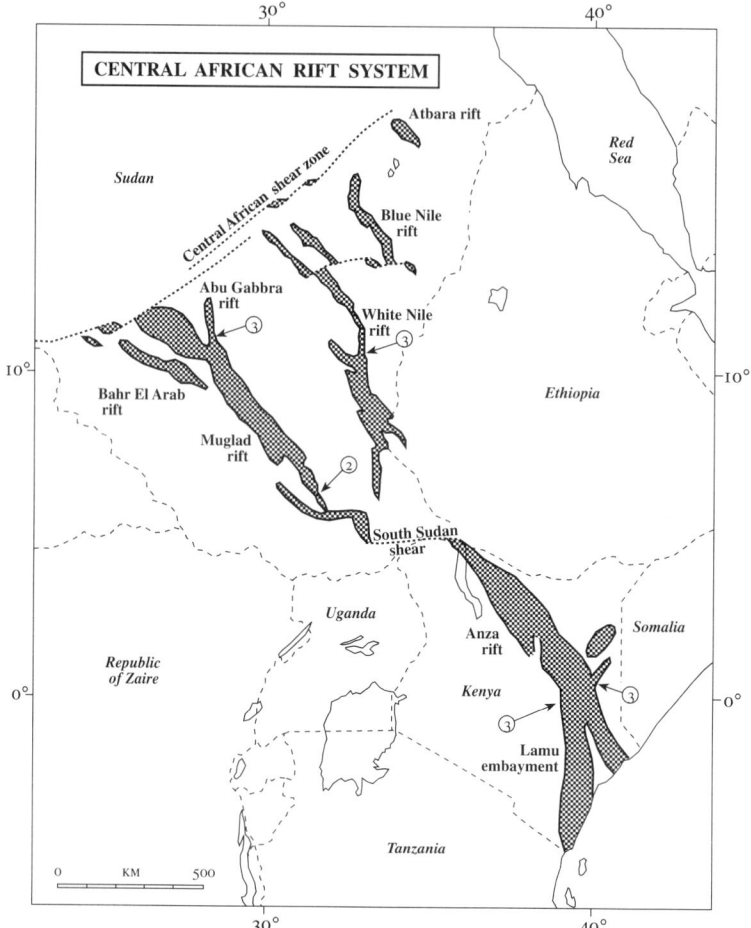

Fig. 16. The Mesozoic to early Tertiary Central African rift system (after Bosworth 1992), with coalescence types between rift segments marked.

of coalescence will take place. If rifting is not sufficiently focused when segments are developing, then the latter will be widely distributed with more chance of isolation and less potential to coalesce unless high bulk strains are achieved.

Conclusions

In the models, there are a limited number of ways in which individual fracture segments amalgamate to form a through-going system. We have outlined three distinct geometries of coalescence between pairs of fractures, involving both tip-to-tip and tip-to-sidewall interactions. These types can be clearly differentiated by comparison of key parameters, including the offset, overlap, obliquity and propagation angle between fracture pairs. In early stages of model runs, fracture initiation is more important

than propagation, but the latter dominates during later stages. The final geometry of a linked system and the range of interactions responsible depend ultimately upon the distribution and density of fractures across the model surface. The larger values of offset and obliquity, associated particularly with wide-mode failure, favour type 3 coalescence, although not exclusively so, whereas narrow-mode failure tends to be dominated more by coalescence of types 1 and 2.

Intracontinental rift systems comprise a number of discrete segments, each in the form of a rift basin bounded on one or both sides by normal faults. In order that an ocean basin can eventually develop, flanked by passive continental margins, a linked, through-going system of terrestrial rift segments must evolve. Analysis of the Baikal, East African and Central African active rift systems reveals precisely the same three types of interactions between pairs of rift segments, controlled by the same parameters as those defined for fracture segments in the models. These natural examples include elements of both narrow-mode and wide-mode rifting, with the former dominated by types 1 and 2 coalescence and the latter by type 3.

The authors would like to thank G. Mulugeta for advice on experimental design and for helpful suggestions during the modelling.

References

AGAR, S. M. & KLITGORD, K. D. 1995. Rift flank segmentation, basin initiation and propagation: a neotectonic example from Lake Baikal. *Journal of the Geological Society, London,* **152**, 849–860.

ARTYUSHKOV, E. V., LETNIKOV, F. A. & RUZHICH, V. V. 1990. The mechanism of formation of the Baikal basin. *Journal of Geodynamics,* **11**, 277–291.

ATKINSON, B. K. 1987. Introduction to Fracture Mechanics. *In:* ATKINSON, B. K. (ed.) *Fracture Mechanics of Rock.* Academic Press, London.

BAKER, B. H. & WOHLENBERG, J. 1971. Structure and evolution of the Kenya rift valley. *Nature,* **229**, 538–542.

BAKER, B. H., MOHR, P. A. & WILLIAMS, L. A. J. 1972. *Geology of the eastern rift system of Africa.* Geological Society of America, Special Papers, **136**.

BENES, V. & DAVY, P. 1996. Modes of continental lithospheric extension: experimental verification of strain localization processes. *Tectonophysics,* **254**, 69–87.

BOSWORTH, W. 1992. Mesozoic and early Tertiary rift tectonics in East Africa. *Tectonophysics,* **209**, 115–137.

BOSWORTH, W. & STRECKER, M. R. 1997. Stress field changes in the Afro-Arabian rift system during the Miocene to Recent period. *Tectonophysics,* **278**, 47–62.

BROWNE, S. E. & FAIRHEAD, J. D. 1983. Gravity study of the Central African rift system: a model of continental disruption 1. The Ngaoundere and Abu Gabra Rifts. *Tectonophysics,* **94**, 187–203.

BROWNE, S. E., FAIRHEAD, J. D. & MOHAMED, I. I. 1985. Gravity study of the White Nile Rift, Sudan and its regional tectonic setting. *Tectonophysics,* **113**, 123–137.

BRUN, J-P. 1999. Narrow rifts versus wide rifts: inferences for the mechanics of rifting from laboratory experiments. *Philosophical Transactions of the Royal Society of London,* **357**, 695–712.

BUCK, W. R. 1991. Modes of continental lithospheric extension. *Journal of Geophysical Research,* **96**, 20161–20178.

BUCK, W. R., LAVIER, L. L. & POLIAKOV, A. N. 1999. How to make a rift wide. *Philosophical Transactions of the Royal Society of London,* **357**, 671–693.

CATHLES, L. M. 1975. *The Viscosity of the Earth's Mantle.* Princeton University Press, Princeton.

CHAPOLA, L. S. & KAPHWIYO, C. E. 1992. The Malawi rift: geology, tectonics and seismisity. *Tectonophysics,* **209**, 159–164.

COWIE, P. A. & SCHOLZ, C. H. 1992. Displacement to length scaling relationships for faults: Data synthesis and discussion. *Journal of Structural Geology,* **14**, 1149–1156.

DAUTEUIL, O. & BRUN, J-P. 1993. Oblique rifting in a slow-spreading ridge. *Nature,* **361**, 145–148.

DAVIES, G. F. 1999. *Dynamic Earth.* Cambridge University Press, Cambridge.

DAVY, P. & COBBOLD, P. R. 1991. Experiments on shortening of a 4-layer model of the continental lithosphere. *Tectonophysics,* **188**, 1–25.

DIXON, J. M. & SUMMERS, L. M. 1985. Resent developments in centrifuge modelling of tectonic processes: equipment, model construction techniques and rheology of modern materials. *Journal of Structural Geology,* **7**, 83–102.

DZIEWONSKI, A. M. & ANDERSON, D. L. 1981. Preliminary reference earth model. *Physics of the Earth and Planetary Interiors,* **25**, 297–356.

EBINGER, C. J. 1989. Geometric and kinematic development of border faults and accommodation zones, Kivu-Risizi rift, Africa. *Tectonics,* **8**, 117–133.

ENGLAND, P. 1983. Constraints on the extension of continental lithosphere. *Journal of Geophysical Research,* **88**, 1145–1152.

FAIRHEAD, J. D. 1988. Late Mesozoic rifting in Africa. *In:* MANSPEIZER, W. (ed.) *Triassic-Jurassic Rifting.* Elsevier, Amsterdam, Developments in Geotectonics, **22**, 821–831.

GUPTA, A. & SCHOLZ, C. 2000. Brittle strain regime transition in the Afar depression: Implications for fault growth and seafloor spreading. *Geology,* **28**, 1087–1090.

HUBBERT, M. K. 1937. Theory of scale models as applied to the study of geologic structures. *Bulletin of the Geological Society of America,* **48**, 1459–1519.

KEEN, C. E. 1985. The dynamics of rifting: deformation of the lithosphere by active and passive driving forces. *Geophysical Journal of the Royal Astronomical Society,* **80**, 95–120.

LAMBECK, K. & JOHNSTON, P. 1998. The viscosity of the mantle: evidence from analyses of glasial rebound phenomena. *In:* JACKSON, I. N. S. (ed.) *The Earth's Mantle: Composition, Structure and Evolution.* Cambridge University Press, Cambridge, 461–502.

LISTER, G. S. & DAVIS, G. A. 1989. The origin of metamorphic core-complexes and detachment faults formed during Tertiary continental extension in the northern Colorado River region. *Journal of Structural Geology,* **11**, 65–94.

LISTER, G. S., ETHERIDGE, M. A. & SYMONDS, P. A. 1991. Detachment models for the formation of passive continental margins. *Tectonics,* **10**, 1038–1064.

LLIBOUTRY, L 1999. *Quantitative Geophysics and Geology.* Springer-Verlag, Berlin.

MACDONALD, K. C. & FOX, P. J. 1983. Overlapping spreading centres: A new kind of accretion geometry on the East Pacific Rise. *Nature,* **302**, 55–58.

MCKENZIE, D. 1978. Some remarks on the development of sedimentary basins. *Earth Planetary Science Letters,* **40**, 25–32.

MART, Y. & DAUTEUIL, O. 2000. Analogue experiments of propagation of oblique rifts. *Tectonophysics,* **316**, 121–132.

MITROVICA, J. X. 1996. Haskell [1935] revised. *Journal of Geophysical Research,* **101**, 555–569.

MOHR, P. 1982. Musings on continental rifts. *In*: PALMASON, G. (ed) *Continental and Oceanic Rifts.* American Geophysical Union, Geological Society of America, Geodynamic Series, **8**, 293–309.

MULUGETA, G. 1988. Squeeze box in a centrifuge. *Tectonophysics,* **148**, 323–335.

OLDENBURG, D. W. & BRUNE, J. N. 1975. An explanation for the orthogonality of ocean ridges and transform faults. *Journal of Geophysical Research,* **80**, 2575–2585.

PETIT, C., DEVERCHERE, J., HOUDRY, F., SANKOV, V. A., MELNIKOVA, V. I. & DELVAUX, D. 1996. Present-day stress field changes along the Baikal rift and tectonic implications. *Tectonics,* **15**, 1171–1191.

POLLARD, D. D. & AYDIN, A. 1984. Propagation and linkage of oceanic ridge segments. *Journal of Geophysical Research,* **89**, 10017–10028.

POULIMENOS, G. 2000. Scaling properties of normal fault populations in the western Corinth Graben, Greece: Implications for fault growth in large strain settings. *Journal of Structural Geology,* **22**, 307–322.

RAMBERG, H. 1981. *Gravity, Deformation and the Earth's Crust* (second edition). Academic Press, London.

SENGOR, A. M. C. & BURKE, K. C. 1978. Relative timing of rifting and volcanism on Earth and its tectonic implications. *Geophysical Research Letters,* **5**, 419–421.

SHEMENDA, A. I. & GROKHOLSKY, A. L. 1991. A formation and evolution of overlapping spreading centers (constrained on the basis of physical modeling). *Tectonophysics,* **199**, 398–404.

SHEMENDA, A. I. & GROCHOLSKY A. L. 1994. Physical modeling of slow seafloor spreading. *Journal of Geophysical Research,* **99**, 9137–9153.

SMITH, M. 1994. Stratigraphic and structural constraints on mechanisms of active rifting in the Gregory Rift, Kenya. *Tectonophysics,* **236**, 3–22.

SPYROPOULOS, C., GRIFFITH, W. J., SCHOLZ, C. H. & SHAW, B. E. 1999. Experimental evidence for different strain regimes of crack populations in a clay model. *Geophysical Research Letters,* **26**, 1081–1084.

TAPPONNIER, P. & MOLNAR, P. 1979. Active faulting and Cenozoic tectonics of the Tien Shan, Mongolia and Baikal regions. *Journal of Geophysical Research,* **84**, 3425–3459.

TESHA, A. L., EBINGER, C. J. & NYAMWERU, C. 1992. Rift-related volcanic hazards in Tanzania and their mitigation. *Tectonophysics,* **209**, 277–279.

TURCOTTE, D. L. & EMERMEN, S. H. 1983. Mechanisms of active and passive rifting. *Tectonophysics,* **94**, 39–50.

WALCOTT, R. J. 1970. Flexural rigidity, thickness, and viscosity of the lithosphere. *Journal of Geophysical Research,* **75**, 3941–3954.

Weijermars, R. & SCHMELING, H. 1986. Scaling of Newtonian and non-Newtonian fluid dinamics without inertia for quantitative modelling of rock flow due to gravity (including the concept of rheological similarity). *Physics of the Earth and Planetary Interiors,* **43**, 316–330.

WERNICKE, B. 1985. Uniform-sense normal simple shear of continental lithosphere. *Canadian Journal of Earth Science,* **22**, 108–125.

ZONENSHAIN, L. P., KUZMIN, M. I. & NATAPOV, L. M. 1990. *Geology of the USSR: A Plate-tectonic Synthesis.* American Geophysical Union, Geodynamics Series, **21**.

Hanging wall accommodation styles in ramp–flat thrust models

G. MULUGETA[1] & D. SOKOUTIS[2]

[1]*Hans Ramberg Tectonic Laboratory, Institute of Geology, University of Uppsala, Box 555, S-751 22 Uppsala, Sweden*
[2]*The Netherlands Center for Integrated Solid Earth Sciences, Faculty of Earth and Life Sciences, Vrije Universiteit, De Boelelaan 1085 HV, Amsterdam, The Netherlands*

Abstract: In this paper we study the dynamic and rheologic control of hanging wall accommodation in ramp–flat thrust models. In particular we vary the dimensionless ratio of shear strength to gravity stress to model hanging wall accommodation styles in different materials. In all models we require that the flat–ramp–flat footwall provides a surface of low frictional resistance. In viscous materials hanging wall accommodation progresses by wedge flow. In Bingham materials, wedge flow is also the preferred mode in cases where the gravity stress exceeds the yield limit of materials. Such models simulate the flow of salt or snow glaciers above ramp obstructions. At high ratios of shear strength to gravity stress the hanging wall blocks translate forward without bending and unbending to the form of the rigid footwall. In elastic–plastic strain-hardening materials ramp–flat accommodation progresses by fault-bend folding in case there is a near balance between the yield stress and gravity stress. In frictional materials hanging wall accommodation progresses by shear or kink-band nucleation above fault-bends. The shear or kink-bands which initially nucleate at the lower fault-bend change shape and reactivate by normal faulting or tensile failure at the upper fault-bend, depending on the ratio of shear strength to gravity stress. In nature, hanging wall accommodation by thrust nucleation above ramps and their subsequent reactivation may be anticipated in frictional sediments at upper crustal levels, where temperatures and pressures are low.

Material models are used to study the dynamic and rheologic control of hanging wall accommodation above undeformed footwalls, the purpose being to gain insight into how materials flow past ramps as a function of rheologic and dynamic constraints. Since the initial work of Rich (1934), it has been suggested by a number of investigators (e.g. Serra 1977; Suppe 1983; Jamison 1987; Chester *et al.* 1991; Strayer & Hudleston 1997) that many folds in layered rocks can form as they move over non-planar fault surfaces or sharp bends, by stepping up from lower to higher decollement horizons. According to the Rich model, overthrust faulting localizes in mechanically weak rocks (e.g. shales), parallel to bedding and steps upwards in mechanically strong rocks (e.g. carbonates). Rich (1934) first introduced this concept and applied it to the Pine Mountain thrust sheet in the foreland of the Appalachian mountain belt. Based on the Rich concept, Suppe (1983) formulated a kinematic fault-bend fold model built on three basic assumptions: (1) the flat–ramp–flat thrust trajectory forms first and folds form in the hanging wall by layer-parallel slip and angular kinking; (2) the footwall remains inert during deformation; and (3) the deformation conserves bed length and area. Since its inception, this geometric model has been used widely by a number of investigators as a guide for the construction of balanced cross-sections, sometimes without the critical analysis of rock rheology (Ramsay 1992). However, geometric models may not accurately represent the interrelationships between folds and faults that develop in both nature and experiments. For example, accommodation styles above undeformed footwalls may involve a wide spectrum of styles, e.g. fault-bend folding (Suppe 1983; Fig. 1a), formation of a stack of imbricate faults (Serra 1977; Fig. 1b) and viscous flow (Talbot 1981; Fig. 1c).

Here, we use a wide range of ductile and frictional materials, and vary the ratio of shear strength to gravity stress for a given material, to study different hanging wall accommodation styles in ramp–flat thrust models.

From: NIEUWLAND, D. A. (ed.) *New Insights into Structural Interpretation and Modelling*. Geological Society, London, Special Publications, **212**, 197–207. 0305-8719/03/$15
© The Geological Society of London 2003.

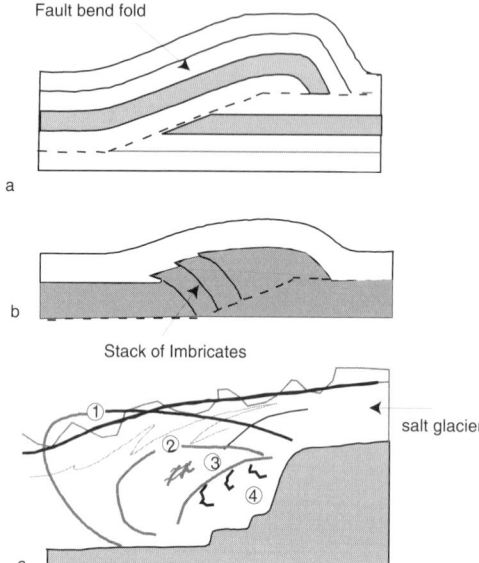

Fig. 1. Hanging wall accommodation styles in ramp regions of overthrust faults. (**a**) Fault-bend folding (Suppe 1983). (**b**) Stack of imbricates in sedimentary strata (Serra 1977). (**c**) Flow of a salt glacier over a ramp (modified from Talbot 1981): 1, flow folds; 2, crenulation; 3, slide zones; 4, static salt.

Experimental method

Set-up

Rectangular prisms of hanging wall blocks with aspect (length to thickness) ratios in the range 4 to 6 and resting on flat–ramp–flat footwalls were end-loaded in a motorized squeeze-box operating under normal gravity, or in a hydraulic squeeze-box in a centrifuge. These were the same ones as used and discussed in Mulugeta (1988). The experiments consist of sequentially shortened models with different loading conditions and layer types. At the outset, we used a ramp angle of $c.$ 30° because thrust ramps typically initiate at this angle in frictional plastic materials. We subjected all models to horizontal compression from one end at a displacement rate of $c.$ 5×10^{-2} mm s^{-1}, at a room temperature of 22.5 ± 0.5°C. The main advantage of performing models in a centrifuge is that strength of materials (scaled for gravity) can be varied within wide limits (Ramberg 1981). In some of the experiments using ductile materials, passive initially circular and square grids (printed on a vertical cross-section in the transport direction) were used to quantify hanging wall strain. A transparent fluid thoroughly lubricated the outside to protect the grids from frictional distortion against the side

walls. The circular markers deforming to finite strain ellipses allowed determination of magnitudes and orientation of principal stretch. The initial squares deforming to parallelograms permitted mapping of variations in shear strain in the different sectors of the hanging wall. In models using sand as a material analogue, we dyed the sand with different colours to visualize the internal deformation.

Dynamic scaling

Dynamic modelling of hanging wall accommodation above inert flat–ramp–flat footwalls in a normal or artificial gravity field requires scaling of strength for gravity. This is necessary because strength is a fundamental property at shallow crustal levels in the crust. For a general 2D model, the equilibrium equation of motion may be written as (e.g. Davy & Cobbold 1991):

$$\delta\sigma_{ij}/\mathrm{d}x_j + \rho g_i = \rho a_i \tag{1}$$

where a_i is acceleration of a material particle, $a_i = D^2 x_i/Dt^2$, $\delta\sigma_{ij}$ is the local stress field, i, j are Cartesian components and g_i is the gravity field. Boundary conditions are such that normal and shear stresses vanish at the surface. These have finite values at the base. Because, acceleration in terms of rate of change of velocity is negligible in geological systems (excepting earthquakes) the first term on the right-hand side of Equation 1 vanishes and it can be written in dimensionless form as ratios between model and prototype:

$$\sigma_r/x_r\rho_r g_r\,(\mathrm{d}\sigma_{ij}^*/\mathrm{d}x_j^*) + \rho^* g_i^* = 0 \tag{2}$$

where the asterisk represents non-dimensionalization, $(x_j)_n x_r = (x_j)_m$, $\rho_n\rho_r = \rho_m$, $g_n g_r = g_m$, $(\sigma_{ij})_n\sigma_r = (\sigma_{ij})_m$, and x_r, ρ_r, g_r, σ_r are model ratios of length, density, acceleration and the stress values, respectively. The subscripts n and m denote nature and model, respectively; subscript r denote ratio between model and nature.

The first term on the left-hand side of Equation 2 can be considered as a model ratio of shear stress (strength) to gravity stress, or ratio of tectonic stress to gravity stress (see appendix). The values which these dimensionless stress ratios acquire in nature or experiments determine the style of hanging wall accommodation. We vary this stress ratio in experiments to model different hanging wall accommodation styles, at different scales. We do this either by shortening models at different values of artificial gravity in a centrifuge, or by changing the shear or yield strength of materials deforming under normal gravity.

In the case that dimensionless shear stress

(strength) to gravity stress ratio $\sigma_r/x_r\rho_r g_r$ remains invariant, a model and prototype may display similar hanging wall accommodation styles at any pair of geometrically corresponding points of the model and nature:

$$\sigma_m/x_m\rho_m g_m = \sigma_n/x_n\rho_n g_n \qquad (3)$$

When the shear strength of materials exceeds the stress due to gravity, hanging wall accommodation takes place quasi-rigidly, i.e. without bending and unbending to the form of the rigid footwall. In contrast, when the shear stress or strength approaches the gravity stress, hanging wall blocks deform internally while sliding at the base. The scale ratio between a model and prototype can thus be determined as:

$$x_m/x_n = (\sigma_m/\sigma_n)\,(\rho_n/\rho_m)\,(g_n/g_m) \qquad (4)$$

If an acceleration of N times normal gravity is used in a centrifuge experiment, then $g_m = N\,g_n$. Equation 4 can then be written as:

$$x_m/x_n = (\sigma_m \cdot \sigma_n)\,(\rho_n \cdot \rho_m)\,(1/N) \qquad (5)$$

For normal gravity experiments $N = 1$, thus the strength of model materials must be reduced by the same factor as the linear dimension, i.e assuming same density of materials in nature and experiments. However, for models run in a centrifuge N can be used as a scaling parameter to study the scale effect of hanging wall accommodation (see section on experimental verification). However, to characterize the deformation for a given material the shear strength or yield stress term must be substituted by the constitutive properties of the materials used.

Rheology of model materials

We used materials with properties ranging from the ductile to the brittle field to simulate a wide spectrum of hanging wall accommodation styles, assuming that the model materials deform by many of the same mechanisms as do natural rocks.

Rhodorsil Gomme (see plot II in Fig. 2a) represented a Newtonian viscous material analogue with a density of 1.16 g cm^{-3}, and viscosity of 10^4 Pa s. Dow Corning DC-3179 (see plot I in Fig. 2a) was used as a Bingham material analogue. In viscometric tests DC-3179 exhibited complex flow behaviour (varying from approximately strain rate-softening flow with a high stress exponent $n = 7 \pm 2$ at strain rates 10^{-4} to 10^{-2} s^{-1} to Newtonian behaviour at higher strain rates (see Fig. 3a; Hailemariam & Mulugeta 1998). Dixon & Summers (1986) characterized the rheology of this material

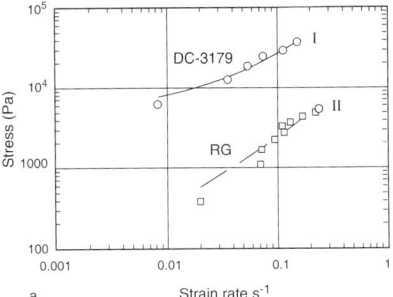

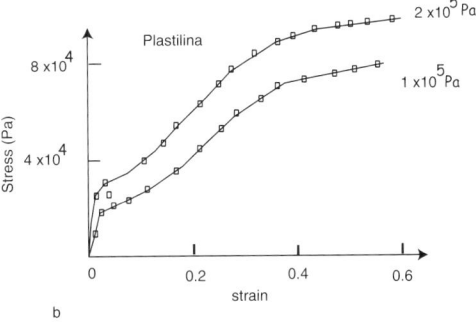

Fig. 2. (a) Log stress versus log strain rate plot for Dow Corning (DC-3179) and Rhodorsil Gomme (RG) bouncing putties. **(b)** Stress/strain behaviour of Plastilina for axial loading at confining pressures of 0.1 and 0.2 MPa.

as a Bingham fluid with a yield strength σ_o of c. 300 Pa and a viscosity of 10^4 Pa s when deformed above its yield strength. These materials simulated the viscous flow behaviour of relatively weak and ductile rocks in nature, such as salt or snow glaciers.

Plastilina simulated the elastic–plastic strain-hardening deformation behaviour of rocks. This was obtained in cylinder tests, axially compressed at a displacement rate of 5×10^{-2} mm s^{-1}, and performed within the range of confining pressures 0.1–0.5 MPa, used in the centrifuge tests (Fig. 2b). The material yielded in the stress range 10^4–10^5 Pa and strain-hardened with increasing stress to approach a plateau after 40–50% shortening (Fig. 2b). The creep viscosity varied in the range 10^7–10^8 Pa s, depending on strain rate. The stress–strain response of the competent Plastilina material closely simulated the properties of sedimentary rocks, such as sandstones and carbonates at crustal conditions (e.g. Hoshino *et al.* 1972).

We used both cohesionless and cohesive sand exhibiting Mohr–Coulomb rheology to simulate the deformation behaviour of frictional sediments. The sand material, with a mean density ρ of c. 1400 kg m^{-3} and coefficient of internal friction μ of c.

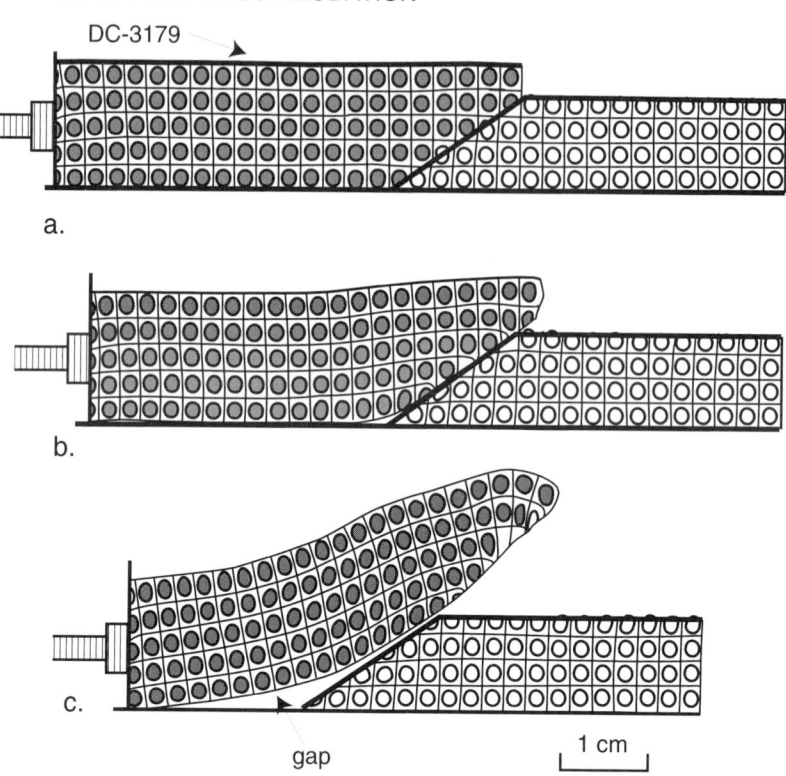

Fig. 3. Progressive stages and strain distribution in a Bingham hanging wall (DC-3179) deformed under normal gravity. (**a**) Initial undeformed stage with internal grid markers. (**b** & **c**) Deformed models after bulk shortenings of 13% and 48.4%, respectively.

0.58, consisted of pure quartz sand with particle diameters less than 0.246 mm. This was made cohesive ($\sigma_c \approx$ 200–300 Pa) by wetting with paraffin fluid.

Boundary conditions

In all models the flat–ramp–flat footwall provided a surface of low frictional drag, where movement along this surface progressed by frictional gliding with a shear stress σ_s dependent on normal stress σ_n ($\sigma_s \approx \mu_b\sigma_n$, where μ_b is coefficient of sliding friction). To induce frictional gliding with low resistance, we first lubricated the flat–ramp–flat base with paraffin oil. We then covered it with a thin film of sand. We determined the coefficient of sliding friction (in separate tests) along the base as $\mu_b \approx 0.4$, by measuring the shear stress necessary to initiate frictional sliding at a given normal stress. Thus, in the models the strength of the basal decollement was less than the strength of the deforming hanging wall. In nature, layers with reduced basal friction, zones of overpressure (e.g. Hubbert &

Rubey 1959) and ductile decollements like salt and shale (e.g Kehle 1970; Davis & Engelder 1985) can provide easy glide horizons to facilitate overthrusting.

In the sections below a number of experiments simulating hanging wall accommodation styles above rigid flat–ramp–flat footwalls are illustrated and discussed (Figs 3–8). In response to material rheology and depending on the shear or yield strength to gravity stress ratios, the hanging wall blocks moved rigidly, or thickened/thinned during migration over the rigid foot walls.

Experimental verification

Quasi-rigid accommodation

When the yield strength of materials exceeded the gravity stress ($\sigma_s/\rho gh$ >1; Fig. 3), a Bingham material, e.g. DC-3179, moved more or less rigidly above the flat–ramp–flat footwall, by supporting large voids near fault-bends. In cross-section, the strain field of quasi-rigid accommodation shows minor gradients in shear and longitudinal strain above the lower flat and ramp sectors. For example, ellipticity ($R < 1.5$) above the lower flat sector was nearly uniform with inclination of the long axes of finite strain remaining at high angles $75° < \phi < 90°$. In analogy with this experiment, hanging wall strains during quasi-rigid accommodation in nature are likely to be so uniformly low that fabrics within them would probably be poorly developed, for example, at outcrop scale c. 5 m. However, the void developed above fault-bends in Fig. 3 is geologically unrealistic in nature, at the kilometric scale. The same Bingham material (used in Fig. 3) deformed viscously when subjected to a gravity stress in excess of the yield strength, for example by deforming it in a centrifuge (Fig. 4). Scaling for this model amounted to 1 cm:11.4 km, i.e. using a yield strength of 1 MPa for the prototype, e.g. salt (Davis & Engelder 1985), and an average density for salt of 2.2 g cm^{-3}. These models illustrate the scale effect of hanging wall accommodation.

Viscous wedge flow

Figure 5 shows the strain field of viscous hanging wall accommodation above the rigid flat–ramp–flat

footwall as might occur, for example, during flow of salt glaciers above ramp obstructions. When viscous flow is combined with frictional sliding the hanging wall adjusts to the form of the rigid flat–ramp–flat footwall by accumulating pure shear strain with ellipticities, $R \approx$ 2–3 above the lower flat sector, bending strain in the ramp sector and layer-parallel extensional strain in the upper flat sector. In the ramp sector, the strain grids suffered initial layer shortening strain due to the buttressing effect of the ramp. With time, bending strains dominated the deformation in the upper sector of the ramp anticline where layer-parallel extension developed around the outer arc- and layer-parallel shortening strain developed in the inner arcs of ramp anticlines. This strain pattern is of the type known as tangential longitudinal strain (Ramsay & Huber 1983). Above the upper flat, the hanging wall was subject to extensional strain, as can be seen by the near-horizontal orientation of the long axes of finite strain ellipses exhibiting ellipticities in the range $R \approx$ 2–3.

VISCOUS WEDGE ACCOMMODATION

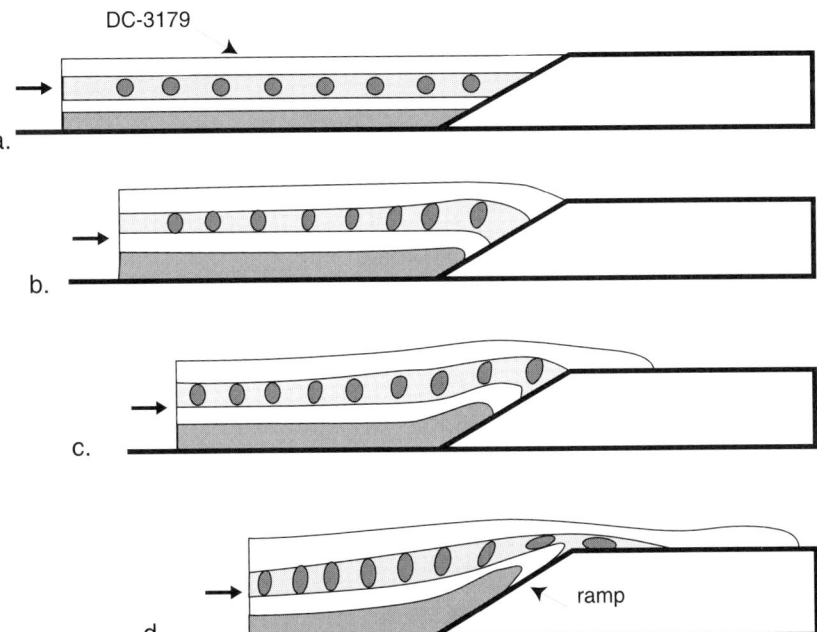

Fig. 4. Strain distribution and wedge flow in the same material (DC-3179) as Figure 3, but deformed in a centrifuge. (**a**) Initial undeformed stage with internal grid markers. (**b, c & d**) Deformed models after bulk shortenings by 16%, 30.67% and 51.3%, respectively.

VISCOUS WEDGE ACCOMMODATION

Rhodorsil Gomme

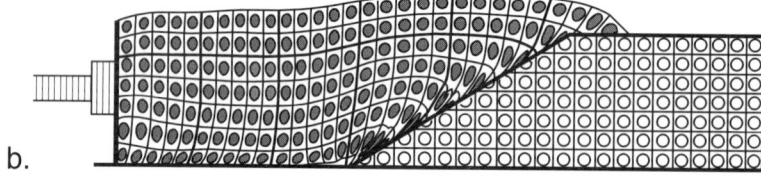

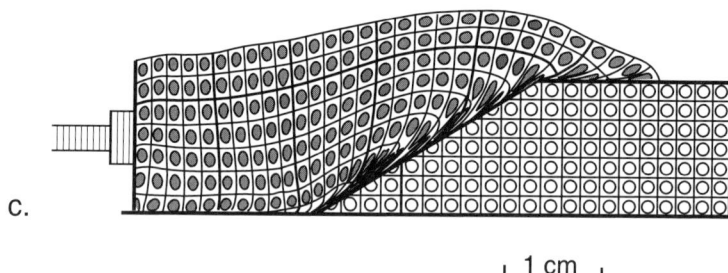

1 cm

Fig. 5. Viscous wedge flow and strain distribution in a Rhodorsil Gomme (RG) hanging wall deformed under normal gravity. (**a**) Initial undeformed stage with internal grid markers. (**b** & **c**) Deformed models after bulk shortenings of 28.6% and 54.3%, respectively.

Fault-bend folding

When there was a near balance between the yield stress and gravity stress ($\sigma_s \approx \rho gh$) the style of hanging wall accommodation closely approximated that of fault-bend folding. Figure 6 illustrates the geometry of fault-bend folding in a Plastilina hanging wall with elastic–plastic strain-hardening rheology. In this model fold shapes were smooth and became broader and flat-topped upwards in the stratigraphy. During progressive hanging wall accommodation, the ramp anticlines changed amplitude and form in accordance with the dip variations and the amount of movement on the underlying flat–ramp–flat footwall. However, in this particular model and at the last stage of shortening the deformation accumulated layer shortening due to enhanced basal drag, as reflected by increase in limb dip and decrease in interlimb angles of ramp folds with progressive shortening

(Fig. 6c). In any case, the ductile materials used above are not appropriate to simulate the faulting and fracturing in ramp regions of overthrust faults so commonly observed in low-temperature sedimentary environments. We study these accommodation styles in frictional materials, as discussed below.

Kink-band nucleation and migration

In frictional materials hanging wall accommodation progressed by nucleation of a series of kink- or shear-bands at the lower fault-bend (Fig. 7). These changed shape, sheared and localized to thrusts as deformation progressed (as discussed by Mulugeta & Koyi 1992; Colletta *et al.* 1991; Merle & Abidi 1995; Bonini *et al.* 2000). As in the previous models, the shear strength to gravity stress ratio was instrumental in controlling the style of frictional hanging wall accommodation. For

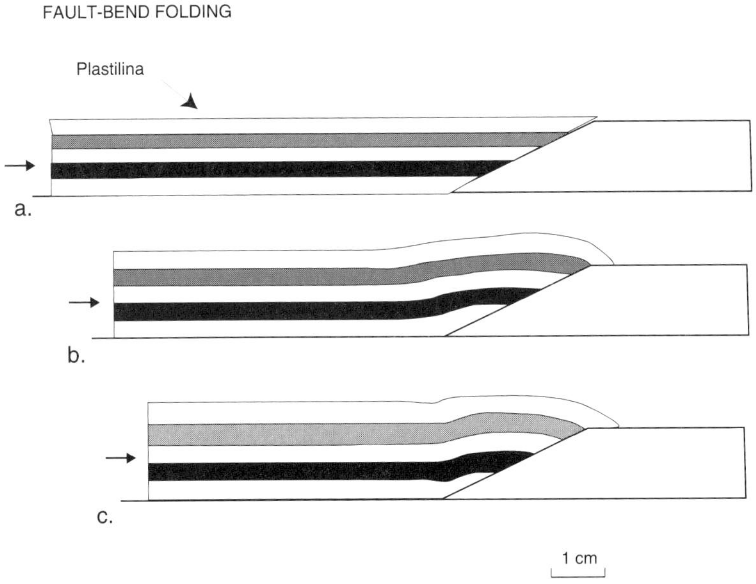

FAULT-BEND FOLDING

Fig. 6. Progressive stages in fault-bend folding of a Plastilina hanging wall deformed in a centrifuge. (**a**) Initial undeformed stage. (**b** & **c**) Deformed models after bulk shortening by 13.8% and 23.6%, respectively.

example, in models with low shear strength to gravity stress ratio ($\sigma_s \cdot \rho gh \leq 1$; Fig. 7), flat-topped ramp anticlines, dissected by imbricates, developed at the early stages of convergence (Fig. 7a). With time, these became more rounded. Foreland-dipping normal faults started to appear at 40% bulk shortening (Fig. 7c), partly reactivating older thrusts during extensional collapse of the hanging wall above the upper flat. The kink interlimb angles increased when these were re-utilized as normal faults (cf. Mulugeta & Koyi 1992).

By comparison, with increase in the shear strength to gravity stress ratio ($\sigma_s / \rho gh \approx 2$; Fig. 8), a single shear-band nucleated above the lower fault-bend and was reactivated later by tensile failure, at the upper fault-bend. However, in the experimental models the mechanics of kink-band migration is quite different from that suggested by some kinematic models (e.g. Suppe 1983). In Suppe's kinematic model the synformal axis of a kink-band remains spatially fixed at the lower fault-bend while the anticlinal axial surface migrates along the ramp until it reaches the top of the ramp upon which its position becomes fixed relative to the fault surface. The kink model specifically predicts that layers only change shape as a result of deformation and strain localization by simple shear displacements in the kinked sectors of the structures. By comparison, in material models using cohesionless sand (e.g. Merle & Abidi 1995; Bonini *et al.* 2000; and this paper), both axial surfaces

in a kink-band migrate towards the upper flat and change shape in the process.

Discussion

Applications

The experimental models discussed above considered the dynamic and rheologic control of hanging wall accommodation above pre-defined ramp–flat footwalls. In nature, the quasi-rigid hanging wall accommodation above a fault-bend, exhibiting both a void space and low strains (or poorly developed fabrics), may be envisioned in stiff members at outcrop scale, or at scales less than 50 m (Fig. 3). Such voids are also present at early stages of deformation in unscaled models of Chester *et al.* (1991) and numerical models of Strayer & Hudleston (1997). In nature, such gaps (which provide areas of low pressure) may act as sinks for mobile fluids; however, their formation is not likely at the kilometric scale.

Hanging wall accommodation by viscous wedge flow may be relevant to deformation of salt and ice glaciers. For example, Talbot (1981) has discussed flow of salt glaciers upstream of ramp obstructions, such as scarp faces of bedrock ridges (Fig. 1c), where these surmounted such obstructions by developing passive asymmetric folds as well as discrete slide zones. Others (e.g. Cobbold & Quinquis 1980; Hudleston 1983; Brun & Merle 1988) have

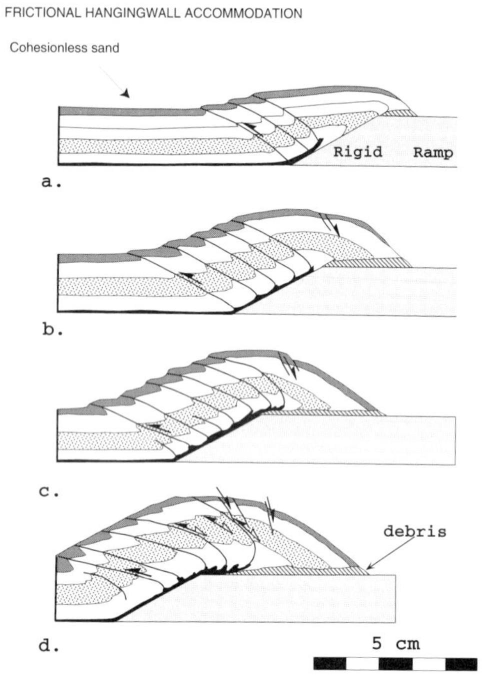

FRICTIONAL HANGINGWALL ACCOMMODATION

Cohesionless sand

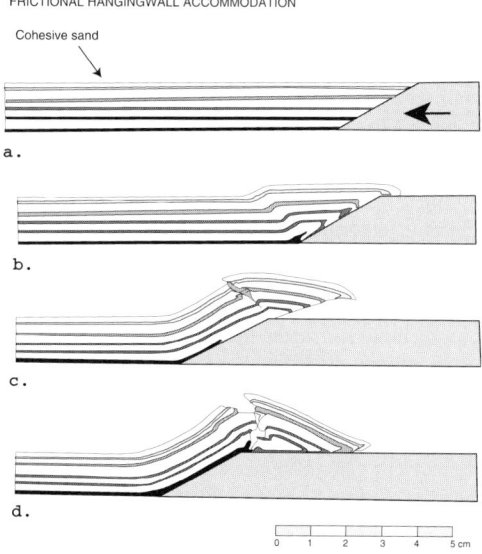

a.

b.

c.

d.   5 cm

Fig. 7. Frictional hanging wall accommodation in cohesionless sand showing serial nucleation of serial kink-bands at the lower fault-bend which later became reactivated by normal faulting at the upper fault-bend. (**a**) Initial undeformed stage. (**b, c** and **d**) Deformed models after bulk shortenings of 40%, 60% and 80%, respectively (from Bonini *et al.* 2000).

FRICTIONAL HANGINGWALL ACCOMMODATION

Cohesive sand

a.

b.

c.

d.   0   1   2   3   4   5 cm

Fig. 8. Same as in Figure 7 but for cohesive sand, showing reactivation by tensile failure at the upper fault-bend.

discussed development of flow perturbations that can arise above topographic irregularities, either by inhomogeneous strain acting on initially planar layers, or by passive amplification of layers during unstable shear flow. In agreement with previous experiments using viscous media (e.g. Berger & Johnson 1980, 1982; Erickson & Jamison 1995) our models show contributions by layer-parallel shortening, shear and extension in the different sectors of the hanging wall.

The special geometric features of fault-bend folding in the experiment with elastic–plastic strain-hardening rheology (Fig. 6) arose from the necessity of making geometric adjustments of the hanging wall to conform to the shape of the underlying flat–ramp–flat footwall. This required a shear strength of the hanging wall blocks in near balance with the gravity stress. This model shows the smooth forms of ramp folds which broaden upwards in the stratigraphy (Fig. 6).

In materials with Mohr–Coulomb rheology, hanging wall accommodation took place by shear- or kink-band nucleation above fault-bends and migration of these as the hanging walls translated forwards above the rigid footwall. The kink-folds which nucleated above the lower fault-bend subsequently sheared and changed shape during forward translation up the ramp. The pattern of faulting in the frictional hanging walls (e.g. Fig. 7) can be compared to the curved reverse faults observed in ramp regions of overthrust faults (e.g. Serra 1977; see Fig. 1b), or in experiments using rock analogues (Morse 1977). This geometry is presumably strongly controlled by friction along the ramp, shear strength and homogeneity of the hanging wall. Moreover, the results of experiments using Mohr–Coulomb materials, migrating over a rigid ramp–flat footwall, show that extensional reactivation of early shear bands may vary depending on the ratio of shear strength to gravity stress(cf. Figs 7& 8). Thus, this dimensionless parameter needs to be taken into consideration in any analysis of ramp–flat thrust accommodation in both nature and experiments.

Limitations

The above models considered the dynamic and rheologic control of hanging wall accommodation above pre-defined ramp–flat footwalls with the restriction that the initial footwall and basal thrust trajectory remained unchanged during deformation. By comparison, in nature there can be significant footwall deformation in combination with changing fault geometries during slip accommodation (e.g. Evans & Neves 1992; Apperson & Goff 1991; Apperson 1993).

In nature, hanging wall accommodation may

also involve effects of material anisotropy which the models discussed above have not addressed. However, previous experiments (e.g. Chester *et al.* 1991; Lan & Hudleston 1995; Strayer & Hudleston 1997) have shown that mechanical anisotropy can play a big role in controlling the kink geometry of ramp folds.

Moreover, external geological variables, such as erosion and sedimentation cycles, can impose a significant control and hence complicate the geometry of hanging wall accommodation (see discussion in Merle & Abidi 1995). Another parameter which affects ramp–flat accommodation pertains to changes in material and decollement response which can induce structural modifications in the hanging wall; for example, if the thrust transport is affected by changes in material rheology or by slip-hardeing on faults (e.g. Chester *et al.* 1991). Such changes make it difficult to analyse the time history of hanging wall accommodation, based on the final deformation geometry alone.

Finally, due to the limitations discussed above, we do not suggest that the models discussed here can be used directly as replicas of specific ramp–flat thrust systems in nature; rather, they may be used to provide some insight into the dynamic and rheologic control of hanging wall accommodation in single thrust sheets against which natural hanging wall accommodation styles may be assessed.

Conclusions

From the discussion above, it is clear that geometric analysis alone, without consideration of dynamic and rheologic constraints, is insufficient to characterize fold-thrust styles. Based on the results of the experiments we draw the following conclusions.

(1) The experimental models illustrate the scale effect of hanging wall accommodation, by shortening models at different *g*-values in a centrifuge.
(2) At high ratios of yield stress to gravity stress, hanging wall blocks translated forward rigidly, i.e without bending and unbending to the form of the rigid footwall.
(3) In viscous materials hanging wall accommodation progressed by wedge flow. In Bingham materials this required a gravity stress in excess of the yield stress.
(4) Of all the models discussed above, hanging wall accommodation in elastic–plastic strain-hardening materials closely matched the geometric form of fault-bend folding, when there was a near balance between the yield limit of materials and the gravity stress.
(5) In both nature and experiments, changes in

decollement response and material strength can induce modifications to hanging wall accommodation, which in turn can make kinematic restoration to the initial template difficult.

We would like to thank J. P. Burg, M. Leroy and D. A. Nieuwland for reviews and for helpful suggestions which improved the manuscript. D. Sokoutis kindly acknowledges financial support from the Netherlands Center for Integrated Solid Earth Sciences (ISES) and the Netherlands Organization for Scientific Research (NWO).

Appendix

Estimates of tectonic stress to gravity stress ratio

We consider balance of forces in the horizontal direction (Equation A1) to estimate the ratio of tectonic stress to gravity stress necessary to produce sliding along the flat–ramp–flat surface. The tectonic driving force F_D per unit width is resisted by the shear force F_S per unit width acting on the flat base and the resistance force per unit width due to the presence of the ramp F_R. The force balance in the horizontal direction can be written as:

$$F_D = F_S + F_R \tag{A1}$$

Here, we consider frictional sliding along the base where we approximate the shear stress by Amonton's law.

Let us first estimate the force F_R necessary to initiate sliding up the ramp, by separating the hanging wall block into two regions, a rectangular region and a triangular ramp region (Fig. A1). To determine forces acting in the rectangular region, we need to determine the normal and shear stresses acting on the ramp (Equations A2, A3; Fig. A1).

Resolving forces in terms of the normal N and shear forces T acting on the ramp, where W is weight per unit width of the triangular wedge:

$$N = W\cos \beta + F_R \sin \beta \tag{A2}$$

$$T = F_R \cos \beta - W \sin\beta \tag{A3}$$

Substituting the value for W which is $W = 1/2\rho gh^2\cot\beta$

$$T = F_R \cos\beta - 1/2\rho gh^2 \cot\beta \sin\beta \tag{A4}$$

$$N = F_R \sin\beta + 1/2\rho gh^2 \cot\beta \cos\beta \tag{A5}$$

The condition for sliding along the ramp from Amonton's law is $T = \mu_b N$, where μ_b is coefficient of sliding friction, ρ is density and g is gravity.

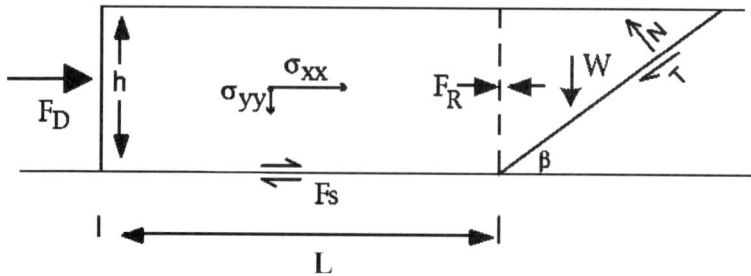

Fig. A1. Force balance in a hanging wall block.

$$T = \mu_b N \tag{A6}$$

F_R can be calculated by substituting the values for T and N from Equations A4 and A5 in Equation A6, which is:

$$F_R \cos\beta - 1/2\rho gh^2 \cot\beta \sin\beta$$
$$= \mu_b(F_R \sin\beta + 1/2\rho gh^2 \cot\beta \cos\beta)$$

$$F_R = \frac{\rho gh^2 \cot\beta}{2}\left[\frac{\mu_b + \tan\beta}{1 - \mu_b \tan\beta}\right] \tag{A7}$$

The shear force F_S acting on the flat sector is:

$$F_s = \sigma_s L = \mu_b\sigma_{yy} = \rho ghL \tag{A8}$$

i.e assuming that σ_{yy} has the lithostatic value ρgh; L is the horizontal length of the flat sector of the hanging wall.

The total driving force per unit width F_D can be written as:

$$F_D = \mu_b\rho ghL + \frac{\rho gh^2 \cot\beta}{2}\left[\frac{\mu_b + \tan\beta}{1 - \mu_b \tan\beta}\right] \tag{A9}$$

The net horizontal force per unit width driving the hanging wall block and acting at the rear end is obtained by integrating the horizontal normal stress σ_{xx}. This is made up of a lithostatic contribution ρgy and a constant tectonic contribution $\Delta\sigma_{xx}$:

$$F_D = \int_0^h \sigma_{xx}d_y$$
$$= \int_0^h (\Delta\sigma_{xx} + \rho gy)\, dy \tag{A10}$$
$$= \Delta\sigma_{xx}h + \frac{\rho gh^2}{2}$$

Thus, the compressional deviatoric stress $\Delta\sigma_{xx}$ can be calculated as:

$$\Delta\sigma_{xx} = \mu_b\rho gL + \frac{\rho gh}{2} \tag{A11}$$
$$\left[\frac{1 + \mu_b \cot\beta}{1 - \mu_b \tan\beta} - 1\right]$$

The tectonic stress to gravity stress ratio can be found from Equation A11 by dividing both sides by ρgh.

$$\frac{\Delta\sigma_{xx}}{\rho gh} = \mu_b\frac{L}{h} + \frac{1}{2}\left[\frac{1 + \mu_b \cot\beta}{1 - \mu_b \tan\beta} - 1\right] \tag{A12}$$

Taking, $\mu_b \approx 0.4$, $L/h \approx 5$, $\beta = 30°$, we find from Equation A12 that $\Delta\sigma_{xx}/\rho gh \approx 3$, i.e a tectonic stress about three times the lithostatic stress is necessary to emplace the hanging wall block. However, as discussed in the text, the tectonic stress has to be replaced by the constitutive material properties of single layer blocks to assess the effect of rheology in hanging wall accommodation.

References

APPERSON, K. D. 1993. Structural controls on early layer parallel shortening. *Eos*, **74**, 301.
APPERSON, K. D. & GOFF, D. F. 1991. Deformation of thrust ramps and footwalls observed in numerical models. *Eos*, **72**, 514–515.
BERGER, P. & JOHNSON, A. M. 1980. First order analysis of deformation of a thrust sheet moving over a ramp. *Tectonophysics*, **70**, T9–T24.
BERGER, P. & JOHNSON, A. M. 1982. Folding of passive layers and forms of minor structures near terminations of blind thrust systems-application to the central Appalachian blind thrust. *Journal of Structural Geology*, **4**, 343–353.
BONINI, M., SOKOUTIS, D., MULUGETA, G. & KATRIVANOS, E. 2000. Modelling hanging wall accommodation above rigid thrust ramps. *Journal of Structural Geology*, **22**, 1165–1179.
BRUN, J. P. & MERLE, O. 1988. Experiments on folding in spreading-gliding nappes. *Tectonophysics*, **145**, 129–139.
CHESTER, J. S., LOGAN, J. M. & SPANG, J. H. 1991.

Influence of layering and boundary conditions on fault-bend and fault propagation folding. *Geological Society of America Bulletin,* **103**, 1059–1072.

COBBOLD, P. R. & QUINQUIS, H. 1980. Development of sheat folds in shear zones. *Jounal of Structural Geology,* **2**(1/2), 119–126.

COLLETTA, B., LETOUZEY, J., PINDEO, R., BALLARD, J. F. & BALÉ, P. 1991. Computerized X-ray tomography analysis of sandbox models: Examples of thin-skinned thrust systems. *Geology,* **19**, 1063–1067.

DAVIS, D. M. & ENGELDER, T. 1985. The role of salt in fold thrust belts. *Tectonophysics,* **119**, 67–88.

DAVY, PH. & COBBOLD, P. R. 1991. Experiments on shortening of a 4-layer model of the continental lithosphere. *Tectonophysics,* **188**, 1–25.

DIXON, J. M. & SUMMERS, J. M. 1986. Another word on the rheology of silicone putty: Bingham. *Jounal of Structural Geology,* **8**, 593–595.

ERICKSON, S. G. & JAMISON, W. R. 1995. Viscous-plastic finite element models of fault-bend folds. *Journal of Structural Geology,* **17**, 561–573.

EVANS, J. P. & NEVES, D. S. 1992. Footwall deformation along Willard thrust, Sevier orogenic belt: implications for mechanisms, timing and kinematics. *Geological Society of America Bulletin,* **104**, 516–527.

HAILEMARIAM, H. & MULUGETA, G. 1998. Temperature-dependent rheology of some bouncing putties used as rock analogs. *Tectonophysics,* **294**, 131–141.

HOSHINO, K., KOIDE, H., INAMI, K., IWAMURA, S. & MITSUI, S. 1972. *Mechanical properties of Japanese Tertiary sedimentary rocks under high confining pressures.* Geological Survey of Japan Report, **244**.

HUBBERT, M. K. & RUBEY, W. W. 1959. Role of fluid pressure in mechanics of overthrust faulting. *Geological Society of America Bulletin,* **70**, 115–166.

HUDLESTON, P. J. 1983. Strain patterns in ice cap and implications for strain variations in shear zones. *Journal of Structural Geology,* **5**(3/4), 455–463.

JAMISON, W. R. 1987. Geometric analysis of fold development in overthrust terranes. *Journal of Structural Geology,* **9**, 207–219.

KEHLE, R. 1970. Analysis of gravity sliding and orogenic translation. *Geological Society of America Bulletin,* **81**, 1641–1664.

LAN, L. & HUDLESTON, P. J. 1995. Angular folds developed in single layers and rheological implications. *Geological Society of America Abstracts with Programs,* **27**, 123.

MERLE, O. & ABIDI, N. 1995. Approche expérimentale du functionnementdes rampes émergentes. *Bulletin de la Société Géologique de France,* **166**, 439–450.

MORSE, J. D. 1977. Deformation in the ramp regions of overthrust faults; experiments with small scale rock models. *Twenty-Ninth Annual Field Conference. Wyoming Geological Association Guidebook,* 457–470.

MULUGETA, G. 1988. Squeeze box in a centrifuge. *Tectonophysics,* **148**, 323–335.

MULUGETA, G. & KOYI, H. 1992. Episodic accretion and strain partitioning in a model sand wedge. *Tectonophysics,* **202**, 319–333.

RAMBERG, H. 1981. *Gravity, Deformation and the Earth's Crust.* Academic Press, New York.

RAMSAY, J. G. 1992. Some geometric problems of ramp-flat thrust models. *In:* McClay, K. R. (ed.) *Thrust Tectonics.* Chapman & Hall, London, 191–200.

RAMSAY, J. G. & HUBER, M. I. 1983. *The Techniques of Modern Structural Geology.* Vol. 1. Academic Press, New York.

RICH, J. L. 1934. Mechanics of low-angle overthrust faulting as illustrated by Cumberland thrust block, Virginia, Kentucky & Tennessee. *AAPG Bulletin,* **118**, 1584–1596.

SERRA, S. 1977. Styles of deformation in the ramp regions of overthrust faults. *Twenty-Ninth Annual Field Conference. Wyoming Geological Association Guidebook,* 487–498.

STRAYER, L. M & HUDLESTON, P. J. 1997. Numerical modelling of fold initiation at thrust ramps. *Jounal of Structural Geology,* **19**, 551–566.

SUPPE, J. 1983. Geometry and kinematics of fault-bend folding. *American Journal of Science,* **283**, 684–771.

TALBOT, C. J. 1981. Sliding and other deformation mechanisms in a glacier of salt, S. Iran. *In:* McCLAY, K. R. & PRICE, N. J. (eds) *Thrust and Nappe Tectonics.* Geological Society, London, Special Publications, **9**, 173–183.

Erosional forcing of basin dynamics: new aspects of syn- and post-rift evolution

E. BUROV[1] & A. POLIAKOV[2]

[1]University of Pierre et Marie Curie, Paris, France
[2]CNRS/University Montpellier II, Montpellier, France

Abstract: We revisit a number of important topics associated with the problem of interactions between surface and subsurface processes during syn- and post-rift evolution. To demonstrate the importance of these interactions and to verify a number of earlier ideas on rift evolution, we use a fully coupled three-fold mechanical behaviour, surface processes, heat transport numerical model, which combines brittle-elastic-ductile rheology, fault localization, erosion and sedimentation mechanisms. The model simulates fault formation causing brittle strain localization. Fault distribution and evolution are thus outputs of the model, allowing for new, geologically sensible constraints on the results. The numerical algorithm accounts for 'true' surface erosion/sedimentation, that is, the numerical elements are eliminated (eroded) and created (sedimented) with respective changes in properties. The results show that sedimentation in the basin and erosion on the rift flanks strongly control the mode of extension. In particular, active erosion/sedimentation on the synrift phase results in more pronounced thinning and widening of the basin, so that the apparent coefficients of extension increase by a factor of 1.5–2. Surface loading/unloading results in lithospheric flexure. Flexural stresses in places of maximum bending exceed lithospheric strength and create zones of localized weakening that partly or completely compensate strengthening due to cooling in the post-rift phase, when the subsidence rates also accelerate. Erosional unloading on the rift shoulders has the opposite effect, producing local strengthening and flexural rebound. Pressure gradients induced by subsidence/rebound result in lower crustal flow that controls 20–30% of subsidence rates, stability of the rift shoulders and drives some post-rift extension or compression. By taking account for the intermediate and lower crustal rheology, new explanations for some synrift phase effects such as polyphase subsidence of the basement provoked by crustal flow and 'switching' of the level of necking from one competent lithospheric level to another are suggested. Syn- and post-rift stagnation, upward and downward accelerations find a natural explanation within our model without the necessity to invoke external mechanisms.

The majority of present-day models of continental rifting do not associate mechanisms of rift necking and subsidence with syn- and post-rift surface processes (sedimentation, surface transport and erosion). Some workers demonstrate the importance of a number of secondary effects of sedimentation such as retarded cooling due to heat screening by sediments with low thermal conductivity (Stephenson *et al.* 1989; England & Richardson 1980). Others account for simplified elastic flexural response to surface loading/unloading by erosion and sedimentation (Kooi & Beaumont 1994; Balen *et al.* 1995). Finally, Burov & Cloetingh (1997) considered the influence of surface processes on basin evolution using a rheologically realistic semi-analytical model, yet limited to consideration of the post-rifting phase and neglecting faulting. Poliakov *et al.* (in press) conducted a complementary study accounting for synrift erosion, but also using a con-

tinuous (no faulting) analytical model with highly simplified rheology and erosion laws.

Formation and maintenance of rift flank escarpments was explained by different mechanisms: lateral heat transport and associated thermoelastic effects, variation in horizontal and vertical forces associated with the far-field regime and lithosphere-asthenosphere interactions, phase changes, flexural forces, small-scale convection, underplating (e.g. Cloetingh *et al.* 1982; Braun & Beaumont 1989; Beaumont *et al.* 1992; Chéry *et al.* 1992; Kusznir & Karner 1985; White & McKenzie 1988; Kusznir 1991; Watts & Torne 1992; Bassi 1995; Cloetingh & Burov 1996). However, previous studies do not account for the fact that for sufficiently deep basins, evolution of the rift must be influenced by production and lateral redistribution of important amounts of normal surface load (Burov & Cloetingh 1997). The sedimen-

From: NIEUWLAND, D. A. (ed.) *New Insights into Structural Interpretation and Modelling.* Geological Society, London, Special Publications, **212**, 209–223. 0305-8719/03/$15

tation and surface denudation result in hundreds of megapascals of temporal and lateral variations of over burden stress and thus are as important as any other tectonic forces acting on the rift system. Erosion is faster on elevated topography and steep slopes (gravity-driven surface processes, wind and physical weathering) and on newly created topography (chemical and all types of climate weathering: van der Beek *et al.* 1995; van Balen *et al.* 1995; Burov & Cloetingh 1997). Gravity drives short-range land-sliding and long-range fluvial transport that carry eroded material from high relief to low relief areas. Consequently, the uplift on the flanks and subsidence in the basin are interlinked not only via tectonic mechanisms but also via surface processes. Erosion on the rift flanks controls sedimentary load accumulated in the basin. In most continental rift zones, loading and unloading of the crust by erosion is equivalent to the surface load created by mountain ranges (5–10 km of sediments, e.g. Pannonian basin, Albert rift, Baikal rift). This allows us to suggest that erosional unloading and sedimentary loading are as important as the other mechanisms controlling basin dynamics. Surface process-related loads may weaken the lithosphere via: (1) faulting in the brittle part; (2) flow in the ductile parts (Fig. 1b; see also Lobkovsky & Kerchman 1992; Hopper & Buck 1996); (3) erosional thinning of the uppermost strong brittle crustal layers followed by their isostatic replacement by weak ductile material coming from beneath (Fig. 1a). Burov & Cloetingh (1997) already suggested that this should result in more localized loading in the central parts of the basin, enhanced extension and maintenance of high strength closer to the flanks. This idea is supported by a number of recent field studies (e.g. Ebinger *et al.* 1989, 1999). As known from rock mechanics studies, the lithosphere behaves as a *de facto* elastic-plastic-ductile medium (e.g. Ranalli 1995). Its rheology is far more complex than the linear elastic or viscous rheology assumed in most previous basin models. One of the main properties of the lithospheric rheology, which differs from commonly inferred linear models, is a possibility to change the mechanical behaviour from elastic to brittle or ductile when the deformational stress exceeds depth-specific pressure- and temperature-controlled yield stress limits. This results in strong mechanical weakening of the material in zones affected by changes in the mechanical behaviour. The strongly non-linear properties of the stress-strain relationships characterizing the lithospheric rheology should influence the subsidence rates and subsidence phases known from tectonic geomorphology and fission-track age/length patterns (Rohrman *et al.* 1995).

Previous studies of interactions between surface and subsurface processes (Burov & Cloetingh 1997; Poliakov *et al.* in press) were focused on basin evolution-erosion interactions using continuous semi-analytic lithospheric models. In the present study, we numerically investigate the influence of surface processes on both syn- and post-rift evolution with primary emphasis on the interplay with faulting and ductile flow during the synrift phase. For this purpose we use a specially modified numerical code, Paravoz, originating from the algorithm used in the FLAC method (Cundall 1989), which allows for brittle and viscous strain localization and accounts for 'true' surface erosion. That is, the numerical elements are eliminated ('eroded') and created ('sedimented') with appropriate changes in physical properties.

Erosion and basin evolution: conceptual background

Erosion is a highly selective process. Inside the same system, the erosion rate may vary by several orders of magnitude. It is lowest on flat topography and fastest on steep, rough topography, such as forming rift flanks, and on crests of tilted blocks (Fig. 1). In isostatically balanced systems, unloading (erosion) of newly created topography results in compensatory rock uplift. At the surface, this uplift occurs at the expense of strong brittle upper crustal material removed by erosion and compensated by the ascent of the weak, ductile crustal material from below. Consequently, the mechanically competent part of the system vanishes without additional tectonic extension (Fig. 1). Such non-extensional thinning of the upper crust facilitates further regional extension. In the vicinity of the rift flanks, the crustal up-flow is predictably most intensive, because the ductile crust is squeezed from the centre to the flanks of the rift due to the increasing normal load (sediments) in the centre. This increase results from two mechanisms: (1) lithospheric necking, due to which the ductile crustal channel thins much faster in the centre than at the sides; (2) preferential growth of the surface load in the rift centre due to accumulation of sedimentary material. Consequently both tectonic and surface processes are linked by the isostatic response and lower crustal flow.

Basin subsidence is caused by three major factors: (1) crustal thinning on the synrift phase; (2) thermal cooling; and (3) sediment loading. Rifting results in thinning of the low viscosity lower crust, which may gradually restore its initial thickness in the post-rift phase. As expected, dynamic pressure differences created by erosion/sediment loading in the lower crust (10–50 MPa; Burov & Cloetingh 1997) can compensate other contributions to the non-hydrostatic pressure (Fig. 1). The pressure dif-

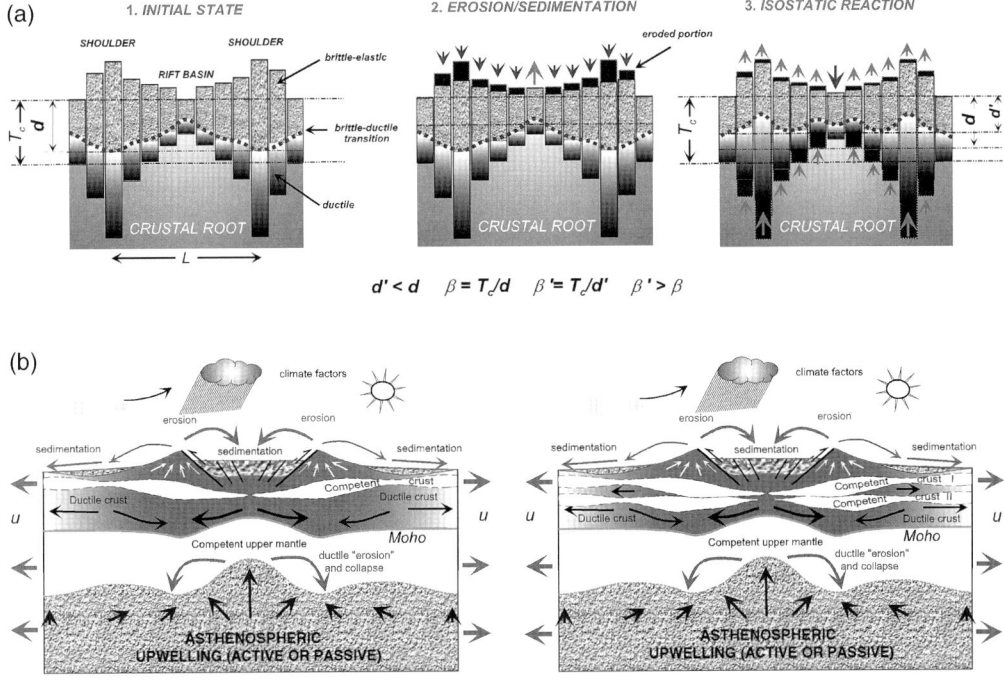

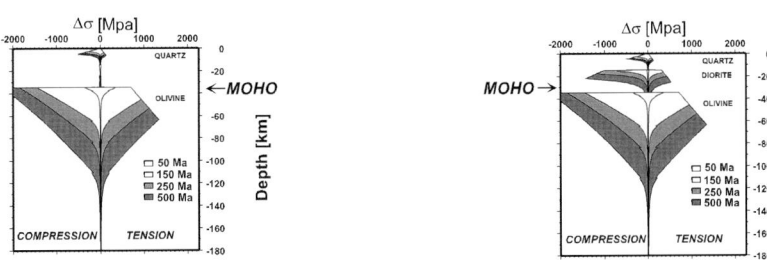

Fig. 1. (a) Simplified cartoon explaining weakening of the extended crust due to erosional removal and reworking of strong brittle parts with isostatic replacement by weaker ductile parts in the ideal case of local isostasy. L is the initial reference length, T_c is the normal crustal thickness, d is the thickness of the thinned crustal layer before erosion, d' is the thickness of this layer after erosion, β is the stretching factor before erosion, β' is the stretching factor after erosion. Note reduction of the thickness of the strong brittle portion of the upper crustal layer. **(b)** Set-up of the numerical problem. Sediments derived from erosion on slopes of rift shoulders result in increase of load in the basin. Strong parts of the crust and of the mantle lithosphere bend and weaken by flexural yielding (local reduction of the thickness of the competent cores). As a result, the integrated strength drops beneath the basin and shoulders (bottom) and becomes even lower than that immediately after extension. The lower crustal material flows from the centre of the basin towards the shoulders facilitating their uplift. The basic lithological structures are shown, one (left) with ductile intermediate and lower crust and another (right) with competent intermediate or lower crust resulting in appearance of a second competent and ductile crustal layer. The corresponding rheological yield-stress envelopes (yield stress as a function of the depth and thermotectonic age) are shown in the bottom of the figure. u is horizontal extension rate.

ference resulting from crustal flow is equal to $2\mu * \dot{\epsilon}$ where μ is the effective viscosity and $\dot{\epsilon}$ the strain rate. The flow stress must be equal to the ductile yield strength, which is 10 to 50 MPa in the lower crust (Burov & Cloetingh 1997). Let us assume a

simple case of local isostasy considered in McKenzie (1978). Let us also assume some commonly inferred parameters: the initial total crustal thickness $h_c = 30$–40 km, subsidence rate $\Delta \nu = 1$ mm a^{-1} ($\sim 3 \times 10^{-11}$ m s^{-1}), and coefficient of extension $\beta = h_c/(h_c - \Delta w) = 2$. Here Δw is the value of crustal thinning. Based on these parameters, the depth-averaged vertical strain rate $\dot{\epsilon}_y = dv/dy$ will be on the order of 10^{-15} s^{-1} (dv/dy ranges from $(\Delta v + \Delta v \rho_m/\Delta \rho)/h_c$ to $(\Delta v + \Delta v \rho_m/\Delta \rho)/(h_c - \Delta w)$, where ρ_m is the density of the mantle and $\Delta \rho$ is the difference between the overlying material and that of the mantle; $\rho_m/\Delta \rho$ ranges from 3 to 5). Continuity and mass conservation require equivalent horizontal and vertical strain rates. Thus the associated horizontal strain rate in the ductile crust should be on the same order as the vertical strain rate. This conclusion could be a trivial matter if the subsidence was entirely caused by extension and thermal cooling of the whole lithosphere. Yet the presence of a weak ductile crust, which cannot always support extra sedimentary loading, may result in downward deflection of the crustal brittle-ductile boundary. This boundary will subside into the ductile crustal channel, which will force horizontal crustal outflow and thinning of the channel (Fig. 1b). The horizontal strain rate in the ductile crust must be the same as the vertical strain rate, with both rates depending on the viscosity of the lower crust. For example, strong lower crust means zero differential subsidence, whereas weak lower crust (i.e. viscosity of 10^{19} Pa s) means initiation of crustal flow already after deposition of several hundred metres of sediments.

In rheologically stratified lithosphere consisting of alternating weak and strong layers, the latter may be allowed to slip with respect to one another. The strain rates thus can vary with depth. The lower or intermediate crust is remarkably weaker than the upper crust and mantle, and the ductile crust thus can be thinned faster than the other layers (e.g. Royden & Keen 1980). For this reason, during the synrift phase basin subsidence can be primarily accommodated in the lower or intermediate crust (shallow level of necking), whereas the other lithospheric layers just bend down or upward preserving their initial thickness. Assuming that, we can conclude that the mean crustal β factor of 2 may associated with much higher lower crustal β factor (from 10 to 1000) and, consequently, with strain rates 5–500 times higher (10^{-14} to 10^{-13} s^{-1}) than the depth-averaged value for the whole crust. The associated flow stresses may exceed 50–100 MPa. The contribution of the ductile crustal flow in the observed subsidence rates thus should be as important as that of the 'global' forces associated with rifting and thermal subsidence.

Numerical model: fully coupled approach

Loads acting on the lithosphere can be subdivided into (Fig. 1): (1) positive loads (topography and deposited eroded material); (2) 'negative' loads due to erosion in the uplifted areas; (3) various sub-surface loads including asthenospheric instability, thermal forces, crustal flow, regional forces; (4) restoring isostatic loads tending to compensate the loads 1–3. The lithospheric response depends on the mechanical properties of the lithosphere, which are partly controlled by its transient thermal state. Consequently, modelling of basin evolution requires a fully coupled approach accounting for surface processes, mechanical and thermal evolution of the system. All these factors are accounted for in the three-fold (surface processes, mechanical behaviour, heat transfer) numerical code Paravoz (Poliakov et al. 1993) based on the FLAC algorithm (Cundall 1989). Paravoz is a mixed finite-element/differences fully explicit time-marching Lagrangian scheme, a detailed explanation of which can be found in Cundall (1989). The Paravoz code can handle rheologically complex behaviours with large strains, including localization and propagation of non-predefined brittle shear bands, which simulate faults, power law creep and various kinds of strain softening and work hardening materials (examples of large-scale geodynamic applications can be found in: Burov & Molnar 1998; Gerbault et al. 1999; Burov & Guillou-Frottier 1999). For the purposes of the given study, we additionally modified the basic version of the code (Poliakov et al. 1993) to include: (a) the erosion/sedimentation model from (Burov & Cloetingh 1997); (b) the rheological and lithological model from Burov & Cloetingh (1997); (c) the heat advection and diffusion processes and initial temperature distribution model from Burov & Diament (1995).

Surface processes

The evolution of surface loads is described by transport and erosion laws derived from geomorphologic, stratigraphic and hydrologic observations (e.g. Carson & Kirkby 1972). The most commonly inferred models of surface processes include short-range erosion laws (parabolic diffusion equation) and fluvial transport laws (channel flow model) (Carson & Kirkby 1972; Gossman 1976; Kirkby 1986; Leeder 1991; Willgoose et al. 1991; Beaumont et al. 1992):

$$dh/dt = \nabla(k^*(x,y,h, \nabla h) \nabla h) \quad (1)$$
$$\text{(slope erosion by diffusion)}$$
$$q_{fe} = -K_r q_r dh/dl \text{ (fluvial transport)} \quad (2)$$

where h is topography, t is time, x is horizontal coordinate, k^* is an experimentally adopted scale-dependent coefficient of erosion, which can be a function of coordinates x,y and of the local slope ∇h, q_r is river discharge, dh/dl is the slope in the direction of the river drainage, K_r is a non-dimensional transport coefficient and l is the distance along the transporting channel. The use of the diffusion law for slope erosion is based on the observation that the erosion rates are not spatially constant but are strongly dependent on the local topography slopes because steeper slopes are more affected by gravity sliding, mechanical and climatic (e.g. wind) weathering. For 2D mechanical models described below, a 1D diffusion law is used:

$$dh/dt = k(x) \, (\partial h/\partial x)^n \, \partial^2 h/\partial x^2 = k^* \partial^2 h/\partial x^2 \qquad (3)$$

where $k(x)$ is a scale-dependent linear coefficient of erosion related to a generalized coefficient of erosion k^* as $k^* = k(x)(\partial h/\partial x)^n$. The parameter n can be equal to 0, 1, 2, 3. The case $n = 0$ is referred to as simple, linear, or zero-order erosion. The other cases are referred to as non-linear n-order erosion. The use of non-linear erosion laws is justified by the fact that not only the local erosion rates but also surface erodibility may be strongly slope dependent (e.g. steep slopes are more influenced by chemical alteration). Consequently, simple linear dependence of the erosion rate on the local slope assumed in linear diffusion equation may be not sufficient in the case of, for example, arid climatic environments, and stronger non-linear relations are needed. In particular, first-order non-linear erosion tends to form much less smoothed, sharpened topography features.

Rheology

Rheology and lithological structure are defined through yield-stress envelopes derived from rock mechanics data (e.g. Kirby & Kronenberg 1987; Kohlstedt *et al.* 1995). We use a non-linear brittle-elastic-ductile rheology for a granite-dominated upper crust, quartz-diorite or quartz-controlled lower crust, and olivine-dominated mantle. The ductile part obeys power law stress (σ) and exponential temperature (T) – stress/strain rate ($\dot{\epsilon}$) dependence:

$$\dot{\epsilon} = A^* \exp\left(-H^*/RT\right) (\sigma_1 - \sigma_3)^n, \qquad (4)$$

where σ_1 and σ_2 are the principal stresses, A^*, H^*, R, and n are the material constants explained in Table 1.

The brittle part follows Byerlee's law (Ranalli 1995), which is approximated by Mohr-Coulomb plasticity with friction angle 30° and cohesion of 20 MPa.

The elastic part is defined for commonly inferred values of Young's modulus and Poisson's ratio (Table 1; Turcotte & Schubert 1982).

Boundary and initial conditions

We used a constant extensional velocity as lateral boundary conditions at the both sides of the model, free surface as the upper boundary condition, and pliable Winkler (i.e. isostatic) basement as the bottom boundary condition. The rectangular numerical mesh is composed of quadruple elements (40 000–125 000), each constructed of two couples of overlapped triangular elements (Cundall 1989). The initial thermal structure is defined from the thermal age of the lithosphere at the time of rifting, calculated from the half-space cooling model (Burov & Diament 1995). A very small (100°C) Gaussian shape temperature anomaly at the base of the model is used for the initial perturbation needed to initiate rifting in the passive mode (e.g. Chery *et al.* 1992). The initial thickness of the upper and lower crustal layer is 20 km and 20 km, respectively (see Table 1). The total vertical and horizontal dimensions of the model vary, depending on the initial thermal structure, from 80 km $\times$ 20 km to 200 km $\times$ 150 km.

Experiments and results

We have conducted three sets of experiments employing major initial lithospheric structures conditioned by possible combinations of weak and strong rheological layers: the mechanical mantle layer can be (1) weaker than its crustal counterpart, or (2) stronger than it, or (3) equally competent. The ratio of the competence of the thickness of the mechanical mantle layer to that of the competent crustal layers is largely controlled by the geotherm (thermotectonic age, *ta*). Thus it is convenient to consider various lithospheric structures as a function of the lithospheric age. Three most representative cases may be envisaged: (1) a very young hot lithosphere with mechanically weak mantle part (thermotectonic age 50 Ma) - in this situation the thickness of the mechanical mantle does not exceed 10 km and crust plays a major role in the overall mechanical response of the lithosphere; (2) intermediate 'Jurassic' lithosphere (thermotectonic age 175 Ma) where the crust and mantle have approximately equal integrated rigidities; (3) cold lithosphere (thermotectonic age of 400 Ma) with thick mechanical mantle, which dominates the mechanical response.

The second scenario is of special interest for our study, since in this case weak ductile portions of the lower and intermediate crust can form flow channels delimited by rigid but highly flexible crustal and mantle layers.

For each set of experiments, we studied four basic situations: (1) slow extension (5 mm a^{-1}), no erosion; (2) slow extension, rapid erosion compara-

Table 1. *Values of parameters used*

Variable/parameter	Values and units	Comments
crustal thickness	40 km	continental crust
upper crustal thickness	20 km	continental crust
lower crustal thickness	20 km	continental crust
extension velocity	5, 10, 25 mm a^{-1}	background velocity on both sides
thermotectonic age, t_a	50, 250, 400 Ma	'young', 'intermediate', 'old' plate
coefficient of erosion k	0, 500, 1000 m^2 a^{-1}	zero, 'intermediate', 'rapid' erosion
background strain rate $\dot{\varepsilon}$	s^{-1}, variable, 10^{-17} to 10^{-13} s^{-1}	obtained from calculations
Young's modulus E	80 GPa	all rocks
Poisson's ratio ν	0.25	all rocks
universal gas constant R	8.314 J mol^{-1} K^{-1}	used in power law
power law constant A_{c1}^*	5×10^{-12} Pa^{-n} s^{-1}	wet granite (upper crust)
power law constant n_{c1}	3	wet granite (upper crust)
creep activation energy H^*_{c1}	190 kJ mol^{-1}	wet granite (upper crust)
power law constant A_{c2}^*	5.01×10^{-15} Pa^{-n} s^{-1}	dry diorite (lower crust)
power law constant n_{c2}	2.4	dry diorite (lower crust)
creep activation energy H^*_{c2}	212 kJ mol^{-1}	dry diorite (lower crust)
power law constant A_m^*	7×10^{-14} Pa^{-n} s^{-1}	olivine
power law constant n_m	3	olivine
creep activation energy H^*_m	520 kJ mol^{-1}	olivine
density ρ_s	2300 kg m^{-3}	uncompacted sediment
density ρ_{c1}	2650 kg m^{-3}	upper crust
density ρ_{c2}	2900 kg m^{-3}	lower crust
density ρ_m	3330 kg m^{-3}	lithospheric mantle
density ρ_a	3250 kg m^{-3}	asthenosphere
gravity constant g	9.8 m s^{-1}	
initial lithospheric thickness a_t	250 km	used to compute initial geotherms
temperature at the base of the lithosphere T_m	1330°C	used to compute initial geotherms
thermal diffusivity χ_{c1}	8.3×10^{-7} m^2 s^{-1}	upper crust
thermal diffusivity χ_{c2}	6.7×10^{-7} m^2 s^{-1}	lower crust
thermal diffusivity χ_m	8.75×10^{-7} m^2 s^{-1}	mantle lithosphere
thermal conductivity k_s	1.6 W m^{-1} K^{-1}	uncompacted sediment
thermal conductivity k_c1	2.5 W m^{-1} K^{-1}	upper crust
thermal conductivity k_{c2}	2 W m^{-1} K^{-1}	lower crust
thermal conductivity k_m	3.5 W m^{-1} K^{-1}	mantle lithosphere
radiogenic decay depth h_r	10 km	upper crust
surface heat production H_s	9.5×10^{-10} W kg^{-1}	upper crust

ble with the rock uplift/subsidence rate (that is $k = $ 200–1000 m^2 a^{-1} for our experiments); (3) rapid extension (25 mm a^{-1}), no erosion; (4) rapid extension (25 mm a^{-1}), rapid erosion (500 m^2 a^{-1}).

The faults on the sides of the rift are not pre×defined but are initialized by the numerical code as a result of deformation. Hence, the fault distribution is one of the important outputs of the model, allowing for better constraints on the results of the experiments than in commonly inferred approaches. In particular, we studied the influence of the extension scenario and surface processes on fault localization, distribution and activity.

Case 1: young lithosphere, mantle weaker than crust

In this scenario, corresponding to $ta = 50$ Ma, the upper crustal layer is the only layer which can stay cold enough to preserve important strength. Consequently, this layer controls the strength and the mechanical response of the lithosphere. The necking level can be only very shallow in this case. The experiments with rapid extension (25 mm a^{-1} on both sides), in which the lithosphere was extended without and with concurrent erosion, produced quite different results (Fig. 2). In the case with no erosion the model-generated topography and faulted structures closely reproduce those of oceanic slow spreading zones (Buck & Poliakov 1998). This is expected, since there is little erosion at the sea bottom, and thus it is natural that the predicted structures are similar to those observed in the oceans. In the case of fast synrift erosion, the topography is highly different from the no-erosion case, not because it is smoothed by surface processes, but, importantly, the thickness and the entire deep structure of the rift also significantly

(a)
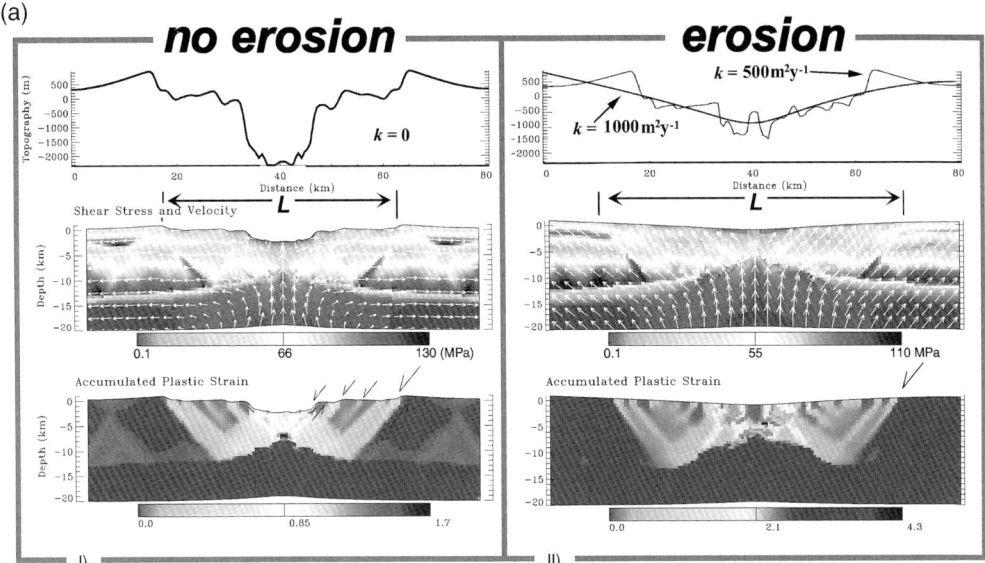

(b)
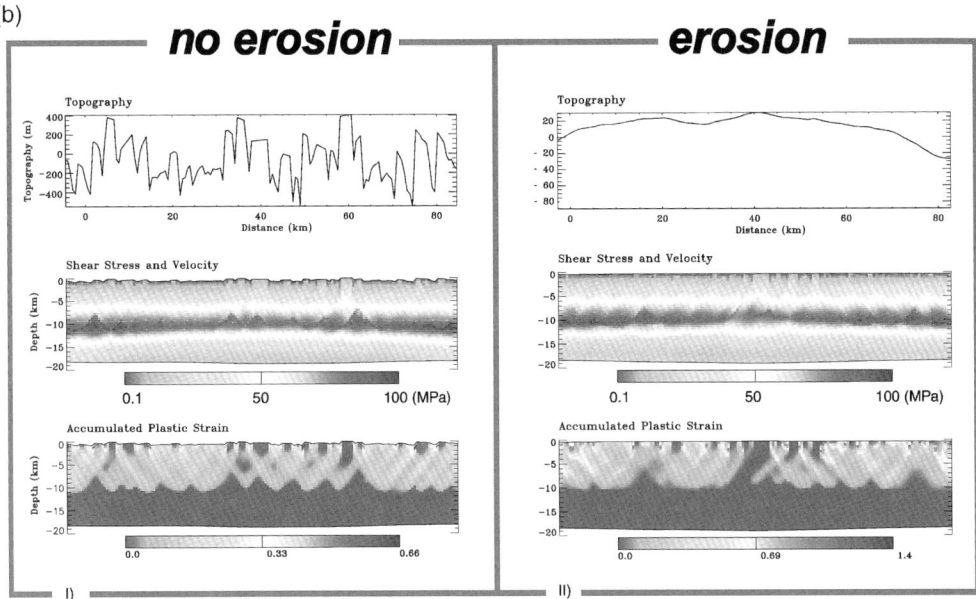

Fig. 2. Numerical modelling of synrift extension of a young lithosphere (age 50 Ma, upper crust dominates in the mechanical response): computed surface topography (top); shear stress and velocity vectors (middle); accumulated plastic strain (bottom). (**a**) The experiments shown in I and II are identical except that in case I there is no erosion, whereas in case II there is rapid erosion ($k = 1000$ m^2 a^{-1}). Also shown in (a) is surface topography for intermediate coefficient of erosion ($k = 500$ m^2 a^{-1}). One can see that erosion results in stronger thinning of the rift and produces a larger basin (according to the mechanism of Fig. 1) than in the case without erosion. (**b**) The same experiment as in (a) is represented, but for fast spreading lithosphere (50 mm a^{-1} on both sides). I shows the case without erosion, and II shows the case with rapid erosion ($k = 1000$ m^2 a^{-1}).

differ from the oceanic-like case: for the same boundary conditions, the rift became almost two times thinner at the centre and 1.5 times wider. Consequently, the β coefficient determined from subsidence curves using McKenzie's approach or from estimates of crustal thinning, will also be two times larger than in the first case. Yet, the amount of tectonic extension is the same in both cases. This effect of erosion is explained in Figure 1a: the erosion destroys the uppermost cold and strong crustal layer, which creates space for uplift of a weak ductile material and, consequently, results in faster localized thinning in the eroded areas. This effect would be amplified in case of the stress boundary conditions.

To test the model (compare our case with oceanic fast spreading zones), we have also conducted a fast extension experiment (50 mm a^{-1} on both sides), presented in Figure 2b. The structures resulting from the experiments without erosion strongly resemble those of oceanic zones of fast spreading. In the case with erosion, two times higher vertical acceleration of the crustal blocks can be also seen (compare the maximum strain rates in two cases).

Case 2: intermediate age lithosphere, mantle as strong as crust

This group of experiments relates to middle-aged lithospheric structure for which the mechanical thickness of the crust and mantle are approximately the same, and the lower crust is also strong enough to play a significant mechanical role. The upper, lower crust and competent mantle can be mechanically decoupled from each other by low strength ductile layers resulting from differences in creep activation temperatures specific for different crustal lithologies. Consequently, the mechanical behaviour of different layers becomes partly independent, and the equivalent elastic thickness of the system is much smaller than just the sum of the elastic thickness of each layer (Burov & Diament 1995), resulting in very 'weak' behaviour of the rift. This is certainly one of the most delicate cases since rift necking may occur simultaneously on different levels (Fig. 3), and the location of the maximum strain zone may switch from one depth level to another. Depending on the role of the surface processes as well as at which moment extension has ceased, the level of necking may be quite different, from very shallow to very deep. Erosion and sedimentation in the synrift stage may invert the direction of vertical crustal movement. Indeed, rapid erosional unloading, together with the lower crustal flow which it provokes, may cause a temporary uplift in the middle of the basin followed by a slow (with respect to thermal) subsidence. In the first case (no erosion), lower crustal flow also allows for relative uplift and subsidence of the upper crustal, lower crustal and mantle parts. Consequently, in this case the surface process may

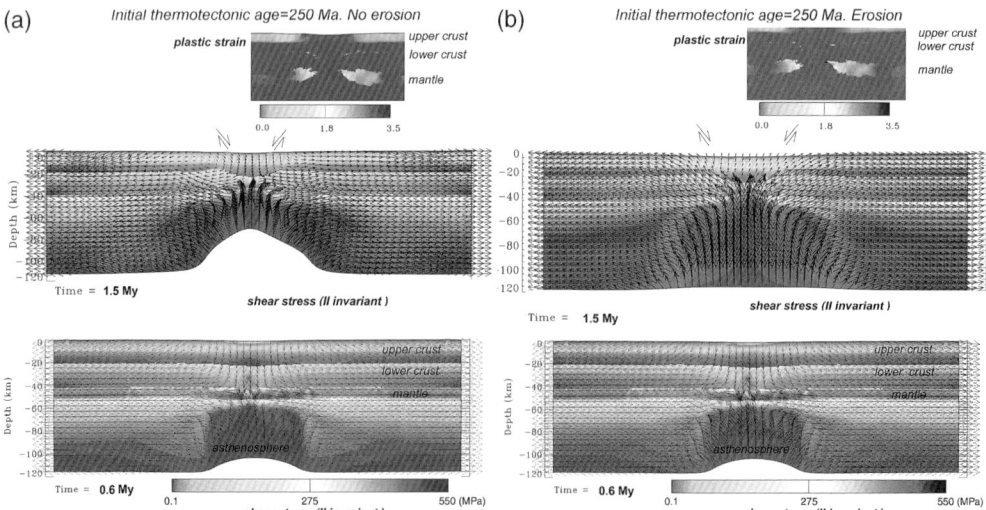

Fig. 3. Numerical model of synrift extension and erosion of a middle-aged lithosphere (250 Ma) with competent middle crustal and mantle layers (shear stress and velocity vectors). Subsidence and uplift are caused by interplay between sedimentary loading and mechanical response. (**a**) Experiments without erosion. (**b**) Experiments with intermediate erosion ($k = 500$ m^2 a^{-1}). For the developed stage, accumulated plastic strain ($\times 100\%$) in the central part of the rift is also shown in the top insert. The lower figure corresponds to 0.6 Ma since onset of rifting; the top figure corresponds to a developed rifting stage (1.5 Ma since onset of rifting).

not only accelerate or retard extension as in case 1, but also affect the level of necking resulting in different geometries of the rift basement and subsidence patterns.

During extension, some or all of the crustal and mantle rigid cores vanish in the centre of the rift being replaced by weak ductile or brittle zones. The rupture of the lower (intermediate) crustal core reduces its resistance to vertical uplift of the mantle layer. The latter rapidly ascends resulting in slower subsidence. In other cases, extension does not end in rupture, but in joining of the strong cores of the upper, lower crustal and mantle layers, resulting in flexural strengthening and in outward lateral expulsion of the ductile crust. This expulsion may lead to uplift on the rift flanks and crustal thickening outside of the basin. When the strong layers join each other under the basin forming a single 'neck', a mechanical coupling occurs, which leads to a step-like increase in the flexural thickness of the system (about two times; Burov & Diament 1995). The subsidence of the basin is thus instantly slowed down and is possibly followed by its lateral enlargement.

Mechanical coupling of competent layers below the basin is preceded by layer 'welding' under the rift flanks (Fig. 3), which locally doubles the elastic thickness in the flank area (Burov & Cloetingh 1997). The inelastic flexural yielding is also lowest under the flanks. This localized flexural border strengthening results in flexural uplift of the rift flanks and helps to maintain the flanks through the time. Indeed, gravity and apatite fission track studies of, for example, the East African rift system, indicate that continental lithosphere undergoing extension maintains considerable strength during the synrift stage, leading to long-lived rift flank uplift (Ebinger *et al.* 1989, 1991; Bechtel *et al.* 1990; van der Beek *et al.* 1995).

Since the mechanical strength and extension of the different strong layers may be quite different, at some stages one of the layers may deflect much more strongly than the others, creating additional space under the basin. This space may be filled by the ductile crust from outside of the basin, delaying subsidence or even uplifting the centre of the basin. As was pointed out by Kaufman & Royden (1994), the crustal and mantle lithosphere may have highly different β coefficients. Here we show that necking may occur on different levels in the crust and in the mantle, so that the internal crustal levels may also exhibit quite different coefficients of extension and even deflect in opposite directions. For example, the upper crust may subside while the lower or intermediate crust moves upward. Since the competent core of the upper crustal layer may be quite thin (a few kilometres), it follows that upper crustal subsidence or uplift may be significantly affected by normal loads generated by surface processes.

Finally, in the post-rift stages, our experiments confirm the semi-analytical model of Burov & Cloetingh (1997) who have shown that the surface processes and induced crustal flow may enhance basin subsidence. In addition to their model, another effect is revealed by the numerical experiments: the presence of a low viscosity lower or intermediate crustal layer may result in detachment from the mantle lithosphere, and also lead to slowed subsidence at the surface, since subsidence (due to cooling) of the deep mantle layers may have no immediate effect at the surface. Thus the mantle may subside separately, with the mantle-crust 'gap' being filled by rapid lower crustal flow.

Case 3: old lithosphere, mantle stronger than crust

In this part of the experiments the strong mantle layer was considerably thicker than the strong crustal layers, conditioned by a cold (400 Ma) geotherm. In this case (Fig. 4), there is no significant crust-mantle decoupling and flow, and crustal deformation is controlled by the mantle lithosphere. Yet, even in the case of very strong lithosphere, the surface processes remain important. Erosion and sedimentation can provoke quiescence and even uplift periods during the synrift phase, and favour development of asymmetric extension patterns by amplifying local anomalies in surface uplift rate.

Discussion: coupling between surface and subsurface processes

Evolution of topography and crust

The topography produced by our coupled model is geologically realistic compared to common geological and gydynamic hypotheses, e.g. rift structures and fault distributions (Salvenson 1978). Major stages of rift development and fault evolution were reproduced (Fig. 5), such as half-graben, graben, continental extensional basins, and oceanic basins.

Influence of the erosion law

The isostatic uplift of the rift shoulders in response to erosion is an important mechanism for maintaining a relatively high rate of material flux from hillslopes. However, erosion of the drainage divides on the top of the rift shoulders results in their retreat from the centre of the basin. At the same time the sedimentary wedge migrates towards the

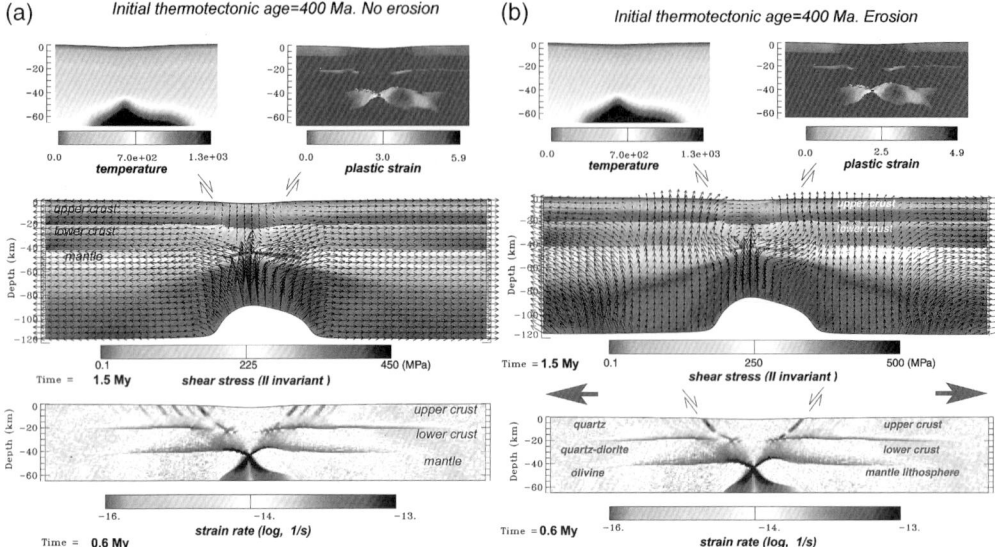

Fig. 4. Extension of an old (400 Ma) lithosphere. (**a**) No erosion. (**b**) Intermediate erosion ($k = 500$ m^2 a^{-1}). All other notations as for Figure 3.

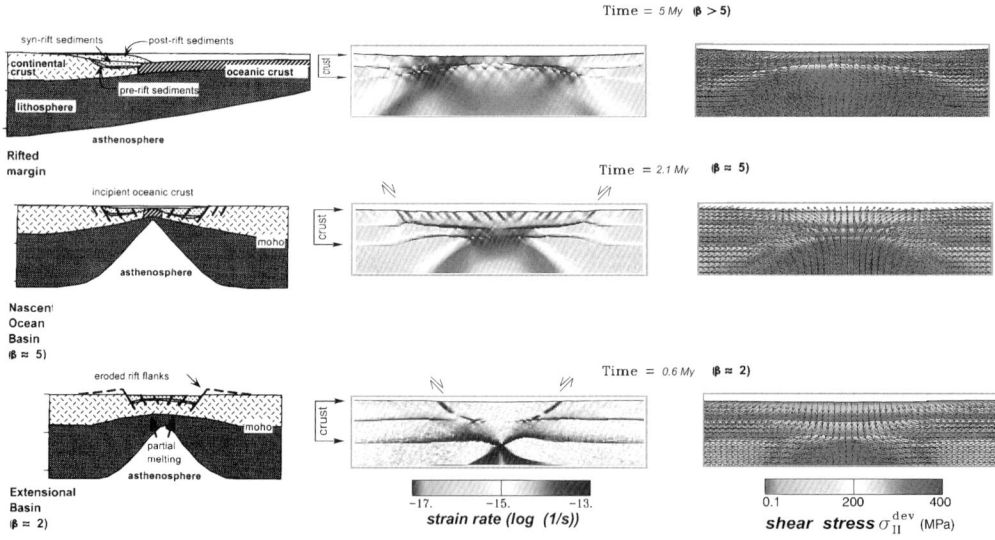

Fig. 5. Various stages of rift evolution produced by the model at different stages of extension (increasing β factor) and compared to the common geotectonic models (adapted from Salvenson 1978; see also Buck 1991). Numerical set-up corresponds to that of Figure 3b.

centre of the basin causing stress variations and characteristic stratigraphic onlap patterns that can be matched with observations. The geometry of rift shoulders and stratigraphic patterns is also highly dependent on the assumed erosion law. We used a zero-order short range diffusion (Equation 2) for the short-range erosion and assumed flat deposition

as a response to long-range fluvial transport surface processes. The transition from short-range to long-range processes was introduced for simplicity by the assumption that flat-deposition 'switches on' at highs below 100 m from the outer side of the shoulders. We also tested first-order non-linear erosion to check the importance of the erosion law.

The major effect is that it tends to keep the hillslopes steeper than in the linear erosion case (Fig. 4). Thus it favours more localization of the rift shoulders than the conventional erosion law. The relief produced by the non-linear erosion is also more realistic than the relief produced by linear erosion.

When erosion is low, basin subsidence is retarded. The related pressure gradient might not be sufficient to counteract the pressure gradients due to density contrasts between the crust and mantle beneath the rift shoulders. The net flux in the lower crust can be reversed in this case (Figs 3 & 4). It will thus retard subsidence of the basin and accelerate collapse of the rift shoulders (Fig. 6).

Post-rift extension and compression

The accumulation of the eroded matter also requires an adequate increase of the basin volume in time. This can occur both in the vertical and horizontal directions, by an increase of the depth of the basin due to subsidence but also by progressive horizontal spreading and onlap of sedimentary deposits. A logical effect of the latter process is widening of the basin resulting in additional extension. This gravity-driven extension is facilitated by secondary post-rift extension due to diverging flow in the lower crust.

Erosion cannot respond immediately to changes in surface uplift, particularly because it is conditioned by a number of independent factors such as climate and surface erodibility. Also, rock viscosity limits the rate of the response of the lower crust and asthenosphere to alternations in surface load. This naturally introduces some delay in the feedback between the surface and subsurface processes. As is known from general studies of feedback systems (e.g. cybernetics, theory of automation, theory of oscillators, operations research theory, etc.), positive feedback with no or in-phase time shift with respect to the input signal results in amplification of the system reply, and the system can be even made to resonate. Out-of-phase (e.g. delayed positive feedback) may result in various oscillations on the output of the system, especially in cases of rapid changes on the input or in the behaviour of the feedback itself. For this reason, one can expect extensional and compressional oscillations, as well as oscillations in the rate of subsidence caused by transient imbalances between the forces of the gravity collapse, lower crustal flow and erosion. Such oscillations in the rate of basin subsidence are indeed observed is several cases (e.g. Dnieper-Donez basin). Though they can probably be explained by eustatic changes or deep mantle processes, the 'feedback' nature is not excluded. Such oscillations were also demonstrated in previous semi-analytical orogenic models (Avouac & Burov 1996), as well as in the numerical experiments of the present study. Even though

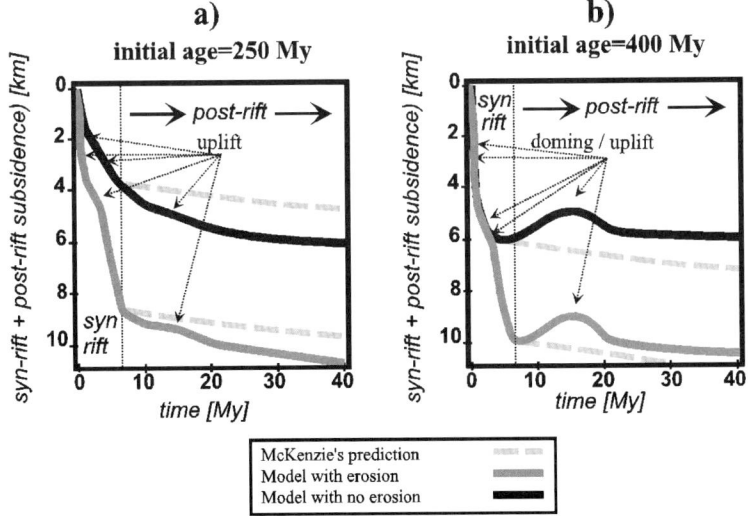

Fig. 6. Subsidence curves corresponding to the case of Figure 3 compared to the classic McKenzie (1978) model. (**a**) Subsidence curves for thermotectonic age 250 Ma. (**b**) Subsidence curves for thermotectonic age 400 Ma. Grey dashed lines show predictions for respective McKenzie models. Note periods of relative syn- and post-rifting uplifts related to structural changes (disappearance of certain mechanical layers) and misbalances between the surface processes and tectonic reaction.

we are confident that these oscillations are not numerical artifacts (their period ranges from 10 ka to several hundred thousand years, which is much longer than the period of possible artificial oscillations due to propagation of numerical waves (<0.1 ka), this topic itself definitely requires a separate detailed study.

Conclusions

Erosion and sedimentation enhance rift thinning and stabilize uplift on the rift shoulders, even during the synrift phase, which is the result of the present study, and largely during the post-rift phase (here the numerical results confirm the analytical predictions made by Burov & Cloetingh 1997). The role of surface processes in the synrift phase is especially important for rifts with young and middle-age thermal structure, where it can result in increase by a factor of 2 or more in extension than without synrift erosion. This has very important consequences for most common stratigraphic evolution models based upon McKenzie's (1978) or Royden & Keen's (1980) method of estimation of the coefficients of extension.

This study shows that the syn- and post-rift evolution of a rift is to a large extent a result of coupling between surface processes (erosion and sedimentation) and response of the lithosphere that includes both rebound effects of localized weakening due to load, and flow in the lower crust. In confirmation of the previous simplified studies (e.g. Burov & Cloetingh 1997), we conclude that flow in the lower crust facilitates both subsidence and crustal thinning, uplift of the rift shoulders and variation in the width of the basin (secondary extension). If either erosion or subsidence terminates for some reason, the lower crustal flow will facilitate collapse of rift shoulders and thickening of the crust and uplift of the basin. Erosion and sedimentation in the post-rift phase can be responsible for delays and accelerations of subsidence with respect to common kinematic/thermal models. It is evident that the carrying capacity and preferential direction of the fluvial network largely control timing of sedimentary filling of the basin (e.g. Kooi & Beaumont 1994). The tectonic reaction to morphological loading and unloading depends on the viscosity and other thermomechanical parameters of the system. Consequently, important delays may appear between the tectonic action (subsidence) and morphological reaction (surface processes). In our case this results in uplift and subsidence events without apparent reason.

We have demonstrated that synrift and post-rift subsidence may occur at a slower rate than inferred from common models (20–25%), and can be characterized by sufficiently long periods of stag-

nation or uplift resulting from interplay between different rheological layers, lower or intermediate crustal flow and surface processes. Rupture of the quasi-elastic core of the intermediate crustal level may result in rapid uplift of the mantle layer. Joining ('welding together') of the rigid layers followed by expulsion of the ductile crust results in temporal flexural strengthening, stagnation of subsidence and widening of the basin. In the post-rift phase, multiple periods of stagnation result from interplay between surface processes and the mechanical response of the lithosphere. This model explains most of the actively discussed deviations from thermal subsidence models such as slow or rapid subsidence, periodical stagnations, and uplifts.

The model explains basin evolution patterns using only relatively well-constrained internal crustal structure, and observable surface processes as boundary/initial conditions. It does not need to invoke external, poorly constrained phenomena such as phase transitions or the inversion of tectonic stresses.

Surface processes and interplay between various mechanical layers in the lithosphere result in different levels of necking for the same initial structure and boundary conditions. In different stages of extension, the level of necking may switch from deep to shallow and vice versa. Thus, depending on the duration of the tectonic extension, the basin starts its post-rift subsidence from a shallow or deep level of necking. Consequently, the traditionally estimated level of necking cannot be directly used to infer the initial lithospheric structure or subsidence.

Evolution of the surface load in time due to sedimentation in the hinterland and erosion in higher flank areas continuously changes the strength of the underlying lithosphere. Because flexure and inelastic effects significantly change the geometry of the crust and Moho, and especially the level of necking, it appears that it is difficult to trust in estimates of the β-factors made on the basis of observations of crustal geometry or backstripping reconstructions. We have demonstrated that the presence of low viscosity lower crust of laterally variable thickness may introduce an important time-dependent contribution to the mechanical response of the lithosphere. This effect must be taken into account not only in basin modelling, but also in models of post-glacial rebound of the lithosphere used to determine the effective viscosity of the asthenosphere. One can predict some extensional and compressional oscillations, as well as oscillations in the rate of subsidence caused by transient imbalances between the forces of the gravity collapse, lower crustal flow and erosion.

In contrast to previous studies (e.g. England

1983; Dunbar & Sawyer 1988), and in confirmation of some qualitative analytical results obtained in (Burov & Cloetingh 1997; Burov & Poliakov 2001), we show that post-rift strengthening results in greater integrated strength in the middle of the rift basin than immediately outside its flanks. The rift flank areas stay weak because they are permanently locally weakened by flexure and by enhanced conductive (in the lithosphere) and also convective (in the asthenosphere) heat transport on the borders of the extended area. If extension continues, these areas nucleate new rifting (not to be mixed with the results of Lavier *et al.* (2000) on short distance normal fault jumping in extending ideal brittle layer). This secondary rifting does not require any new far-field extension episode, because self-stressing due to the gravity spreading and induced flows in the ductile crust and the asthenosphere is already sufficient. Interplays between strengthening and extension at the end of a long (e.g. longer than 5 Ma) synrift phase or at the beginning of a post-rift phase may explain so-called abandoned rifts.

We greatly benefited from the highly instructive and detailed review provided by the first anonymous reviewer and T. den Bezemer. A co-creator of Paravoz, Y. Podladchikov, is deeply thanked for generously sharing his knowledge and experience. We also benefited from discussions with C. Ebinger, S. Cloetingh, M.-P. Doin, P. Van der Beek and J. Chery. This study is supported by IT Program of INSU (CNRS).

Appendix

Numerical model and rheological assumptions

For numerical experiments (Fig. 1b) we adopted the code Paravoz (Poliakov *et al.* 1993), which allows for mixed brittle, elastic, viscous and non-Newtonian temperature, stress and strain rate-dependent power-law rheology and complex geometrical structures. This code belongs to the so-called Fast Lagrangian Analysis of Continua family (FLAC) (Cundall 1989) of large strain, fully explicit time-marching numerical algorithms based on a Lagrangian 'moving grid' method. It is not clear whether FLAC codes should belong to finite difference or to finite element methods, since they use a mixed formulation, which includes an explicit method to solve algebraic finite-difference equations, but implicit, matrix-oriented solution schemes used in finite element methods. As in finite element methods, the FLAC method uses arbitrarily shaped numerical elements. To allow small strain solutions to work for large strains, FLAC codes rely on very small time steps. They

solve for near incremental strains and then explicitly numerically integrate them, because the result of the model is by default the cumulative effect of all the small strains over all time steps.

The major advantage of the FLAC method is its capacity to model initialization and evolution of non-predefined faults, which is crucial for this study. This algorithm and its application to caldera modelling were described in detail in Burov & Guillou-Frottier (1999).

As with other FLAC codes, Paravoz solves Newton's equations of motion in continuum mechanics formulation:

$$\rho \partial \nu_i / \partial t - \partial \sigma_{ij} / \partial x_j - \rho g_i = 0 \qquad (A1)$$

where ν is velocity, g is the acceleration due to gravity and ρ is the density. The numerical mesh moves with the material, and at each time step the new positions of the mesh grid nodes are calculated from the current velocity field (Cundall 1989). As mentioned, the code is explicit and uses very short time steps and very small elements. Since no global stiffness matrix is needed, it is a trivial matter to update coordinates at each time step in large-strain mode. Equation (A1) is solved in the local evolving coordinates, thus the strain can be small with respect to moving Lagrangian coordinates, but large with respect to a fixed Cartesian grid. The local area rotation at large strain is accounted for by adjusting appropriate tensor components, e.g. stress tensor components σ_{ij} are adjusted as $\sigma_{ij} = \sigma_{ij} + (\omega_{ik}\sigma_{kj} - \sigma_{ik}\omega_{kj})$ where the finite rotation ω is given by $\omega_{ij} = \frac{1}{2}(\partial u_i / \partial x_j - \partial u_j / \partial x_i)$. Solution for velocities at mesh points is used to calculate element strains ϵ_{ij}. These strains are employed in the constitutive relations yielding element stresses σ_{ij} and equivalent forces $\rho \partial \nu_i / \partial t$, which provide input for the next calculation cycle. For elastic and brittle materials the constitutive relations have a linear form:

$$\epsilon_{ij} = A\sigma_{ij} + A_0 \qquad (A2)$$

where A, A_0 are constitutive parameters (Table 1). For a ductile rheology these relations become:

$$\dot{\epsilon}_{ij} = A\sigma^{n-1}\sigma_{ij} \qquad (A3)$$

where $\dot{\epsilon}_{ij}$ is the strain rate and $\sigma = (\frac{1}{2}\sigma_{ij}\sigma_{ij})^{\frac{1}{2}}$ is the effective stress (second invariant of the deviatoric stress). The variables n (the effective stress exponent) and A (constitutive parameter) describe the properties of a specific material (Table 1). For ductile materials, n usually equals 2–4 and A is depth and temperature dependent ($A = A_0 \exp(-H/RT)$). For the brittle and elastic materials A is usually only depth dependent. Yet, A and A_0

can be functions of strain or stress for softening or hardening materials. To allow for explicit solution of the governing equations, the FLAC method employs a dynamic relaxation technique based on introduction of artificial inertial masses in the dynamic system. The adaptive remeshing technique allows strain localizations resulting in the formation of faults. The method does not use inherent rheology assumptions, in contrast with common finite-element techniques.

For the elastic rheology, we use the following constitutive parameters: E (Young's) modulus = 80 GPa and ν (Poisson's ratio) = 0.25. The brittle behaviour is presented by Mohr-Coulomb plasticity with cohesion softening (friction angle 30°, cohesion decreases from 20 MPa at zero strain to 0 at 1% strain; Gerbault et al. 1998).

Since the ductile rheology is temperature dependent, the mechanical balance equations are coupled with the heat transport equations:

$$div(\mathbf{k}\nabla T) - \rho C_p \partial T/\partial t + H_r = \mathbf{v}\nabla T \qquad (A4)$$

where $\mathbf{v}$ is the velocity tensor, C_p is the specific heat, $\mathbf{k}$ is the thermal conductivity tensor, H is radiogenic heat production per unit volume (here we use the commonly inferred values adopted, e.g. in Burov & Diament (1995). The solution of the right-hand side (diffusive) and left-hand side (advective) part of Equation A4 is separated: the latter is calculated automatically when solving the equations of motion, whereas the former is computed using a separate procedure.

The size of the mesh elements was between 50 m × 50 m and 250 m × 250 m.

References

AVOUAC, J. P. & BUROV, E. B. 1996. Erosion as a driving mechanism of intracontinental mountain growth. *Journal of Geophysical Research,* **101**, 17747–17769.

BASSI, G. 1995. Relative importance of strain rate and rheology for the mode of continental extension. *Geophysical Journal International,* **122**, 195–210.

BEAUMONT, C., FULLSACK, P. & HAMILTON, J. 1992. Erosional control of active compressional orogens. *In:* MCCLAY, K. R. (ed.) *Thrust Tectonics.* Chapman & Hall, London, 1–31.

BECHTEL, D., FORSYTH, D. W., SHARPTON, V. L. & GRIEVE, R. A. F. 1990. Variations in effective elastic thickness of the North American lithosphere. *Nature,* **343**, 636–638.

BRAUN, J. & BEAUMONT, C. 1989. A physical explanation of the relation between flank uplifts and the breakup unconformity at rifted continental margins. *Geology,* **17**, 760–765.

BUCK, W. R. 1991. Modes of continental lithospheric extension. *Journal of Geophysical Research,* **96**, 20161–20178.

BUCK, W. R. & POLIAKOV, A. N. B. 1998. Abyssal hills formed by stretching oceanic lithosphere. *Nature,* **392**, 272–275.

BUROV, E. B. & CLOETINGH, S. 1997. Erosion and rift dynamics: new thermomechanical aspects of post-rift evolution of extensional basins. *Earth and Planetary Science Letters,* **150**, 7–26.

BUROV, E. B. & DIAMENT, M. 1995. The effective elastic thickness (Te) of continental lithosphere: What does it really mean? *Journal of Geophysical Research,* **100**, 3905–3927.

BUROV, E. B. & GUILLOU-FROTTIER, L. 1999. Thermomechanical behavior of large ash-flow calderas, *Journal of Geophysical Research,* **104(10)**, 23081–23109.

BUROV E. B. & MOLNAR, P. 1998. Gravity anomalies over the Ferghana Valley (central Asia) and intracontinental Deformation. *Journal of Geophysical Research,* **103**, 18137–18152.

BUROV, E. B. & POLIAKOV, A. N. B. 2001. Erosion and rheology controls on synrift and postrift evolution: Verifying old and new ideas using a fully coupled numerical model. *Journal of Geophysical Research,* **106**, 16461–16481.

CARSON, M. A. & KIRKBY, M. J. 1972. *Hillslope Form and Processes.* Cambridge University Press, Cambridge.

CHÉRY, J., LUCAZEAU, F., DAIGNIERES, M. & VILOTTE, J.-P. 1992. Large uplift of rift flanks: A genetic link with lithospheric rigidity? *Earth and Planetary Science Letters,* **112**, 195–211.

CLOETINGH, S. & BUROV, E. B. 1996. Thermomechanical structure of European lithosphere: constraints from rheological profiles and EET estimates. *Geophysical Journal International,* **124**, 695–723.

CLOETINGH, S. A. P. L., WORTEL, M. J. R. & VLAAR, N. J. 1982. Evolution of passive continental margins and initiation of subduction zones. *Nature,* **297**, 139–142.

CUNDALL, P. A. 1989. Numerical experiments on localization in frictional materials. *Ingenieur-Archiv,* **59**, 148–159.

DUNBAR, J. A. & SAWYER, D. S. 1988. Continental rifting at pre-existing lithospheric weakness. *Nature,* **333**, 450–452.

EBINGER, C. J., BECHTEL, T. D., FORSYTH, D. W. & BOWIN, C. O. 1989. Effective elastic plate thickness beneath the East African and Afar Plateaux and dynamic compensation of the uplifts. *Journal of Geophysical Research,* **94**, 2883–2901.

EBINGER, C. J., KARNER, G. D. & WEISSEL, G. D. 1991. Mechanical strength of extended continental lithosphere: constraints from the western rift system, Africa. *Tectonics,* **10**, 1239–1256.

EBINGER, C., JACKSON, J., FOSTER, A. & HAYWARD, N. 1999. Extensional basin geometry and the elastic lithosphere. *Philosophical Transactions of the Royal Society of London,* **A357**, 741–762.

ENGLAND, P. 1983. Constraints on extension of the continental lithosphere. *Journal of Geophysical Research,* **88**, 1145–1152.

ENGLAND, P. & RICHARDSON, S. W. 1980. Erosion and the age dependence of the continental heat flow. *Geophysical Journal of the Royal Astronomical Society,* **62**, 421–437.

GERBAULT, M., POLIAKOV, A. N. B. & DAIGNIERES, M.

1998. Prediction of faulting from the theories of elasticity and plasticity; what are the limits? *Journal of Structural Geology,* **20**, 301–320.

GERBAULT, M., BUROV, E. B., POLIAKOV A. & DAGNIERES, M. 1999. Do faults trigger folding in the lithosphere? *Geophysical Research Letters,* **26(2)**, 271–274.

GOSSMAN, H. 1976. Slope modelling with changing boundary conditions – effects of climate and lithology. *Zeitschrift fr Geomorphologie N.F.,* Suppl. Bd. **25**, 72–88.

HOPPER, J. R. & BUCK, W. R. 1996. The effect of lower crustal flow on continental extension and passive margin formation. *Journal of Geophysical Research,* **101**, 20175–20194.

KAUFMAN, P. S. & ROYDEN, L. H. 1994. Lower crustal flow in an extensional setting: Constraints from the Halloran Hills region, eastern Mojave Desert, California. *Journal of Geophysical Research,* **99**, 15723–15739.

KIRBY, S. H. & KRONENBERG, A. K. 1987. Rheology of the lithosphere: selected topics. *Review of Geophysics,* **25**, 1219–1244.

KIRKBY, M. J. 1986. A two-dimensional model for slope and stream evolution. *In*: ABRAHAMS, A. D. (ed.) *Hillslope Processes.* Allen and Unwin, Boston, 203–224.

KOHLSTEDT, D. L., EVANS, B. & MACKWELL, S. J. 1995. Strength of the lithosphere: Constraints imposed by laboratory experiments. *Journal of Geophysical Research,* **100**, 17587–17602.

KOOI, H. & BEAUMONT, C. 1994. Escarpment evolution on high-elevation rifted margins: insights derived from a surface processes model that combines diffusion, advection and reaction. *Journal of Geophysical Research,* **99**, 12191–12210.

KUSZNIR, N. & KARNER, G. 1985. Dependence of the flexural rigidity of the continental lithosphere on rheology and temperature. *Nature,* **316**, 138–142.

KUSZNIR, N. J. 1991. The distribution of stress with depth in the lithosphere: thermo-rheological and geodynamic constraints, *Philosophical Transactions of the Royal Society of London,* **A337**, 95–110.

LAVIER, L. L., BUCK, W. R. & POLIAKOV, A. N. B. 2000. Factors controlling normal fault offset in ideal brittle layer. *Journal of Geophysical Research,* **105**, 23431–23442.

LEEDER, M. R. 1991. Denudation, vertical crustal movements and sedimentary basin infill. *Geologische Rundschau,* **80(2)**, 441–458.

LOBKOVSKY, L. I. & KERCHMAN, V. I. 1992. A two-level concept of plate tectonics: application to geodynamics. *Tectonophysics,* **199**, 343–374.

MCKENZIE, D. 1978. Some remarks on the development of sedimentary basins. *Earth and Planetary Science Letters,* **40**, 25–32.

POLIAKOV A. N. B., PODLADCHIKOV, Y. & TALBOT, C. 1993. Initiation of salt diapirs with frictional overburden: numerical experiments. *Tectonophysics,* **228**, 199–210.

POLIAKOV, A. N. B., PODLADCHIKOV, Y. Y. & YUEN, D. A. (in press). A model of sedimentary basin formation with phase-transition and erosion: explanation of syn-rift uplift and stratigraphic onlap. *Tectonophysics.*

RANALLI, G. 1995. *Rheology of the Earth* (second edition). Chapman & Hall, London.

ROHRMAN, M., VAN DER BEEK, P., ANDRIESSEN, P. & CLOETINGH, S. 1995. Meso-Cenozoic morphotectonic evolution of southern Norway: Neogene domal uplift inferred from apatite fission track thermochronology. *Tectonics,* **14**, 704–718.

ROYDEN, L. & KEEN, C. E. 1980. Rifting process and thermal evolution of the continental margin of Eastern Canada determined from subsidence curves, *Earth and Planetary Science Letters,* **51**, 343–361.

SALVENSON, J. O. 1978. Variations in the geology of rift basins; a tectonic model. *Conference Proceedings Los Alamos Scientific Laboratory,* **7487**, 82–86.

STEPHENSON, R. A., NAKIBOGLU, S. M. & KELLY, M. A. 1989. Effects of astenosphere melting, regional thermoisostasy, and sediment loading on the thermomechanical subsidence of extensional sedimentary basins. *In*: PRICE, R. A. (ed.) *Origin and Evolution of Sedimentary Basins and their Energy and Mineral Resources.* American Geophysical Union, Washington, Geophysical Monographs, **48**, 17–27.

TURCOTTE, D. L. & SCHUBERT, G. 1982. *Geodynamics. Applications of Continuum Physics to Geological Problems.* Wiley, New York.

VAN BALEN, R., VAN DER BEEK, P. A. & CLOETINGH, S. A. P. L. 1995. The effect of rift shoulder erosion on stratal patterns at passive margins: Implications for sequence stratigraphy. *Earth and Planetary Science Letters,* **134**, 527–544.

VAN DER BEEK, P., ANDRIESSEN, P. & CLOETINGH, S. 1995. Morphotectonic evolution of rifted continental margins; inferences from a coupled tectonic-surface processes model and fission-track thermochronology. *Tectonics,* **14**, 406–421.

WATTS, A. B. & TORNE, M. 1992. Crustal structure and the mechanical properties of extended continental lithosphere in the Valencia through (western Mediterranean). *Journal of the Geological Society, London,* **149**, 813–827.

WHITE, N. & MCKENZIE, D. P. 1988. Formation of the "Steer's Head" geometry of sedimentary basins by differential stretching of the crust and mantle. *Geology,* **16**, 250–253.

WILLGOOSE, G., BRAS, R. L. & RODRIGUES-ITURBE, I. 1991. A coupled channel network growth and hillslope evolution model 1. Theory. *Water Research,* **27**, 1671–1684.

Vertical movements of the Paris Basin (Triassic–Pleistocene): from 3D stratigraphic database to numerical models

C. ROBIN[1], P. ALLEMAND[2], E. BUROV[3], M. P. DOIN[4], F. GUILLOCHEAU[5], G. DROMART[6] & J.-P. GARCIA[7]

[1]*Département de Géologie Sédimentaire, FRE 2400, Université Pierre et Marie Curie (Paris 6), Case 116, 75252 Paris Cedex 05, France (e-mail: robin@ccr.jussieu.fr)*
[2]*Centre des Sciences de la Terre, UMR 5570, Université Claude Bernard (Lyon 1), 69622 Villeurbanne cedex, France*
[3]*Laboratoire de Tectonique, UMR 7072, Universite Pierre et Marie Curie, 4 Place Jussieu, 75252 Paris cedex 05, France*
[4]*Laboratoire de géologie, Ecole Normale Supérieure, UMR 8538, 24 rue Lhomond, 75231 Paris cedex 05, France*
[5]*Géosciences-Rennes, UMR 6118, Université de Rennes I, 35042 Rennes cedex, France*
[6]*Centre des Sciences de la Terre, UMR 5125 'Paléoenvironnements et Paléobiosphère', Université Claude Bernard (Lyon 1), 69622 Villeurbanne cedex, France*
[7]*Biogéosciences-Dijon, UMR 5561, Université de Bourgogne, France*

Abstract: A 3D stratigraphic database has been constructed from the inspection of 1100 wells and outcrops in the Paris basin. The database contains 88 surfaces correlated at high temporal resolution using sequence stratigraphy. For each well and each surface, the present-day depth, the depositional environment and the lithology between two layers are available. This database provides a key to quantify the tectonics associated with this intracratonic basin and to model the thermal and mechanical processes at the origin of the tectonics.

Three types of numerical modelling have been carried out in order (1) to better constrain the long-term thermal subsidence and its cause, (2) to characterize the spatial and temporal evolution of the crustal tectonics during the 'extensional' period and (3) to test a lithospheric folding origin during the end-Cretaceous to present-day compressional period. The philosophy of these three models are different.

The Chablis model for the lithospheric thermal evolution is used to predict the long-term subsidence of the Paris Basin. The thermal evolution of the lithosphere is computed, taking account of a constant temperature or heat flow at the base of the lithosphere, temperature- and pressure-dependent thermal characteristics, metamorphism in the crust, top-crustal erosion and phase transition in the mantle. The long-term subsidence of the Paris basin results from the decay of a thermal anomaly initiated during late Variscan times. The subsidence data can be explained by short- (Stephano-Autunian) as well as long- (Stephano-Triassic) lasting extension. These hypotheses both implicitly refer to extensional collapse of the Variscan belt.

The characterization of the spatial and temporal evolution of the crustal tectonics during the thermal relaxation period has been need to quantify the local effect of the sediment load on vertical crust movements. From sedimentary thickness and bathymetric data, maps of relative tectonics have been drawn at a time scale around 500 ka. These maps show two different tectonic behaviours: (1) narrow regions with a high horizontal gradient of tectonics (faults), and (2) domains with a diffuse subsidence correlated with topographic domes and high rates of sedimentation. The geometrical and temporal characteristics of the regions of diffuse subsidence are compatible with a model of flow of the lower crust if the thickness of the flowing channel is at least 20 km with a viscosity of 10^{20} Pas.

The Tertiary characteristics of the Paris Basin could be the record of large-scale lithospheric folding. The numerical experiments demonstrate that extremely low (0.2 mm a^{-1}) shortening

From: NIEUWLAND, D. A. (ed.) *New Insights into Structural Interpretation and Modelling.* Geological Society, London, Special Publications, **212**, 225–250. 0305-8719/03/$15

rates are largely sufficient to induce large-scale low-amplitude folding under low maximum values of tectonic stresses (*c.* ~50 MPa). These values suggest that alpine compression is largely sufficient to activate this deformation.

From the data collected in this database and from the models described here, the evolution of the Paris Basin is better understood. The Paris Basin Meso-Cenozoic evolution can be described as a long-term thermal subsidence, inherited from the Permian extension and perturbed by intraplate deformations in reaction to the geodynamic events occurring in western Europe, i.e. the Ligurian Tethys opening and closure, and the Atlantic opening. Those tectonic events modify in space and time both subsidence and facies distributions. The Paris Basin was initially an 'extensional' basin which progressively evolved into a compressional one, temporarily (lower Berriasian and late Aptian) and then permanently (late Turonian to present day). The present-day geometry of the Paris Basin is the consequence of lithospheric folding occurring mainly during the Tertiary. In consequence, (1) the Paris Basin is not still a subsiding basin but an uplifted area, and (2) during the Jurassic and part of the Cretaceous, the surrounding present-day outcropping basement massifs were subsiding areas flooded by the sea.

The Paris Basin is certainly among the most extensively studied sedimentary basins since the middle part of the eighteenth century. With the development of plate tectonic concepts and the renewal of sedimentary basin studies, the Paris Basin has been reinvestigated (Brunet 1981; Brunet & Le Pichon 1982; Perrodon & Zabek 1990). It is now considered as an intracratonic basin in a regime of decreased thermal subsidence, coming into existence during a period of rifting in Permo-Triassic times (Perrodon & Zabek 1990). High resolution subsidence measurements in the 1990s suggest a more complex mechanical behaviour of the basin, which can record intraplate force variations (Guillocheau 1991; Loup & Wildi 1994).

In order to better understand the relationships between the different tectonic wavelength controls and the sedimentary fill, a stratigraphic database has been constructed, based on 1100 wells and 88 surfaces (Guillocheau *et al.* 2000). The purpose of this paper is to use this database to test or to develop different thermomechanical models in order to better constrain the vertical movements of the lithosphere beneath the Paris Basin.

Three types of numerical modelling have been carried out.

1. The long-term thermal subsidence and its cause is predicted using the Chablis model (Doin & Fleitout 1996).
2. The knowledge of spatial and temporal evolution of the crustal tectonics during the 'extensional' period is based on the qualification of the vertical effect of the sediment load on the vertical crust movements.
3. The end-Cretaceous/Cenozoic lithospheric folding of the Paris Basin is tested using the Paravoz code (Poliakov *et al.* 1993; Burov & Molnar 1998; Gerbault *et al.* 1999).

The 3D stratigraphic database: a sequence stratigraphic correlation of well-logs

The 3D stratigraphic database is based on well-log correlations. Around 1100 wells have been correlated using the principles of high-resolution sequence stratigraphy, i.e. the stacking pattern of parasequences (Van Wagoner *et al.* 1988, 1990; Homewood *et al.* 1992, 2000) which have been applied to the Paris Basin earlier (Bessereau *et al.* 1995; Bourquin *et al.* 1996, 1998). The principle of correlation is to identify the transgressive–regressive cycles of shortest duration (20 000 years to 400 000 years) which can be identified on well-logs and which are called parasequences (Van Wagoner *et al.* 1988, 1990) or genetic units (Homewood *et al.* 1992, 2000). This requires a calibration of the well-logs in term of sedimentary environment which is achieved by comparison with cores and/or outcrops. The vertical stacking of the parasequences in lower order composite transgressive–regressive units of longer duration provides the means for correlation to the next well. Two lower orders of transgressive–regressive cycles have been recognized: 10–40 Ma (major cycles) and 1–15 Ma (minor cycles). The time lines resulting from correlations of the stacking pattern are calibrated by biostratigraphy (Guillocheau *et al.* 2000).

This stratigraphic database contains surfaces of stratigraphic cycles (maximum flooding surface, flooding surface and unconformities) of 10–40 and 1–15 Ma duration (Fig. 2). Eighty-eight surfaces are available with information on the present-day depth, the depositional environment and the lithology between two surfaces. Isopach maps were compiled for time intervals between 3 and 20 Ma (Guillocheau *et al.* 2000).

The accommodation space, defined as the sum of tectonic and eustatic variations (Jervey 1988), has been systematically measured (Robin *et al.* 1996, 1998; Robin 1997; Prijac *et al.* 2000).

Geological framework and crustal structure of the Paris Basin

Today, four Cadomian/Variscan basement massifs surround the Paris Basin: the Armorican Massif in the west, the Massif Central in the south, the Vosges in the east and the Ardennes in the NE. Eastwards, the Rhine and Bresse Grabens bound the Vosges basement and the Paris Basin. Both the Paris Basin and the Cadomian/Variscan basements are incised by the present-day river network (Loire, Seine, Meuse and Moselle). Cretaceous and Tertiary deformation and erosion have exhumed Mesozoic sediments and underlying basement. This process was particularly effective in the eastern part of the basin, exposing the large 'rings' of Mesozoic sediments.

The structure of the continental crust beneath the Paris Basin is poorly known. One deep seismic line has been shot in the northern part of the basin, between Evreux in the SW and Valenciennes in the NE (ECORS project: Cazes & Toreilles 1987). The upper crust is characterized by SW dipping oblique reflectors and the lower layered crust thins westward toward the Armorican block and disappears in the NE, below the London–Brabant block (Fig. 1). The only fault, which cuts across both the upper crust and the Moho, is the Bray Fault. The mean depth of the Moho (Autran *et al.* 1994; Chantraine *et al.* 1996) is around 35 km.

The nature of the upper part of the Paris Basin basement has been established from deep drilling, gravimetric and magnetic data (Autran *et al.* in Mégnien 1980; Autran *et al.* 1994). It is now established that the basement belongs to four different Variscan domains (Chantraine *et al.* 1996; Fig. 1): the Armorican domain (central-Armorican zone and Cadomian block) in the west, the internal domain (Liguro-Arverne zone and Morvan-Vosges zone) in the south and SE, the Saxo-Thuringian zone which pinches out to the west in the central part of the basin along the Bray Fault, and the Rheno-Hercynian zone in the north.

The extension, the stratigraphy and the geodynamic significance of the Permian sediments below the Meso-Cenozoic deposits are still poorly known (Autran *et al.* in Mégnien 1980; Mascle 1990; Perrodon & Zabek 1990). The only point of agreement is the discontinuous and isolated nature of these small basins, a consequence of non-deposition and/or post-depositional subsequent erosion. The most important area of Permian sedimentation is the Saar-Nahe Basin, extending from Germany in the east to the eastern part of the present-day Paris Basin (Korsch & Schäfer 1995; Schäfer & Korsch

1998). The relationships with the other French Permian basins, now interpreted as syn- to postorogenic extensional gravitational collapse basins (Burg *et al.* 1994*a,b*), have not been fully discussed. But because of crustal thickening during the Variscan continental collision below the Paris Basin south of the Variscan front, such an origin may be assumed for the Permian basins located below the present-day Paris Basin.

Major stratigraphic cycles and the geological history of the Paris Basin

The Meso-Cenozoic evolution of the Paris Basin can be subdivided into five main steps recognized by thickness and facies variations through time (Guillocheau *et al.* 2000) (Fig. 2).

The *Scythian to Toarcian* is characterized by an arcuate subsiding area along NE–SW and east–west to ENE–SSW directions, by general onlaps in the north and west, and by the creation of accommodation space (5–30 m Ma^{-1}) (Fig. 3).

At end Triassic (Rhaetien), the northwestern part of the basin became slightly subsident along a NW–SE trend.

The intra-Norian unconformity (base 'Grès à roseaux' Formation) records the beginning of subsidence in the central part of the present-day Paris Basin.

The Hettangian to Pliensbachian time is characterized by short and medium wavelength deformation which contrasts with the medium wavelength flexural controls of the Triassic and the Toarcian. At that time, the Paris Basin recorded the different steps of opening of the Ligurian Tethys.

The spatial subsidence pattern is consistent with both east–west and north–south extension during this time interval, but no tilted blocks were encountered. Extension was directed east–west to NW–SE during Hettangian to Pliensbachian time.

The *Aalenian to Tithonian* is characterized by NW–SE medium wavelength flexural control, beginning with some short wavelength controls during the Aalenian/Lower Bajocian (and locally during the Bathonian) and changing to a large wavelength flexural pattern during the Kimmeridgian/Tithonian times (Fig. 4). The kinematics of the Dogger deformation is poorly known.

The large wavelength control was coeval with the highest rate of accommodation space creation (20–40 m Ma^{-1}), and with the development of progradational (Callovo-Oxfordian) and aggradational (Kimmeridgian–Tithonian) carbonate platforms.

These changes in both tectonic wavelength and carbonate platforms during the Callovian are con-

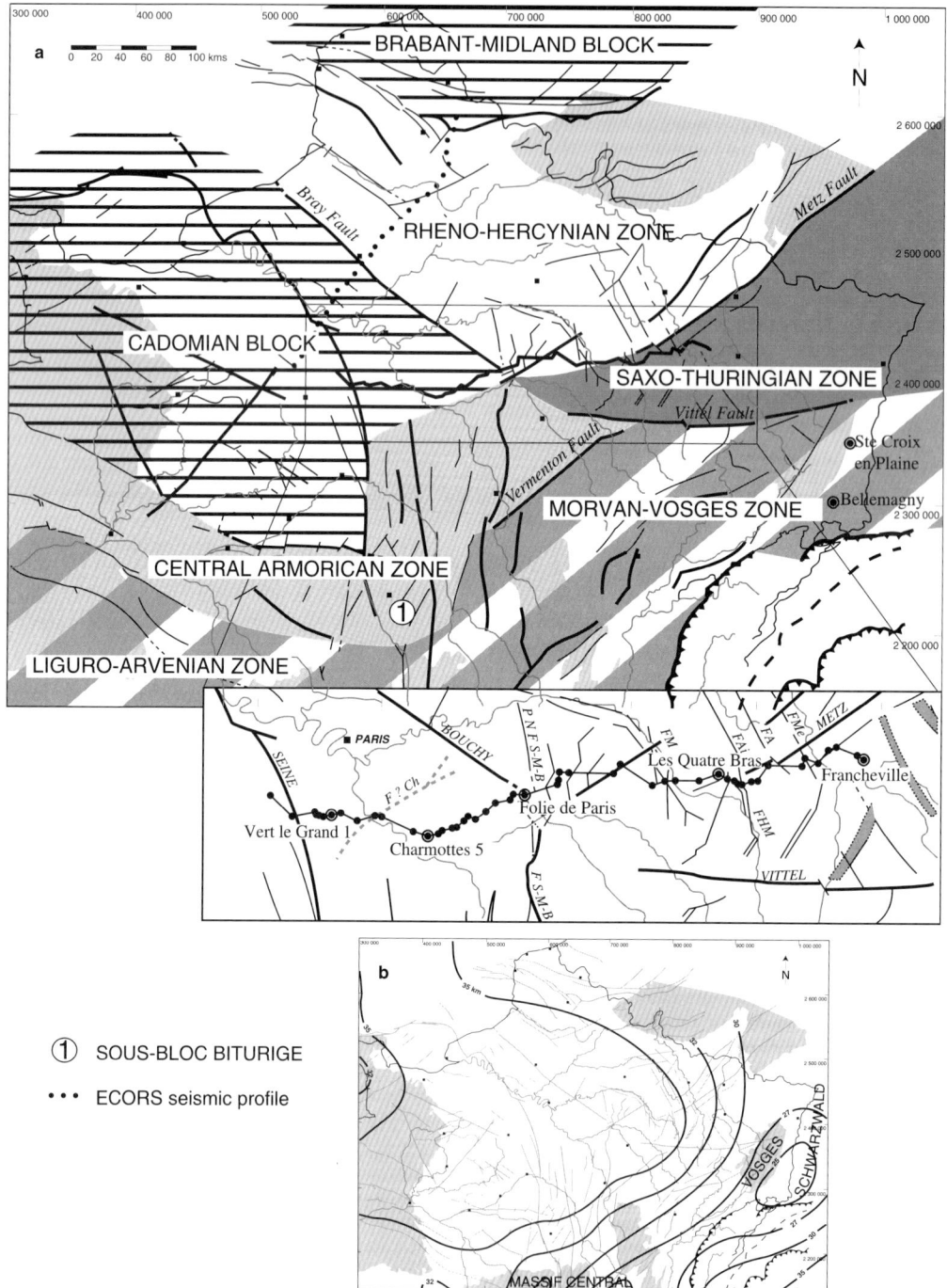

Fig. 1. (**a**) Main structural domains of the Cadomian–Variscan basement below the Paris Basin and location of the Nancy–Rambouillet transect. (**b**) Moho isobaths in kilometres.

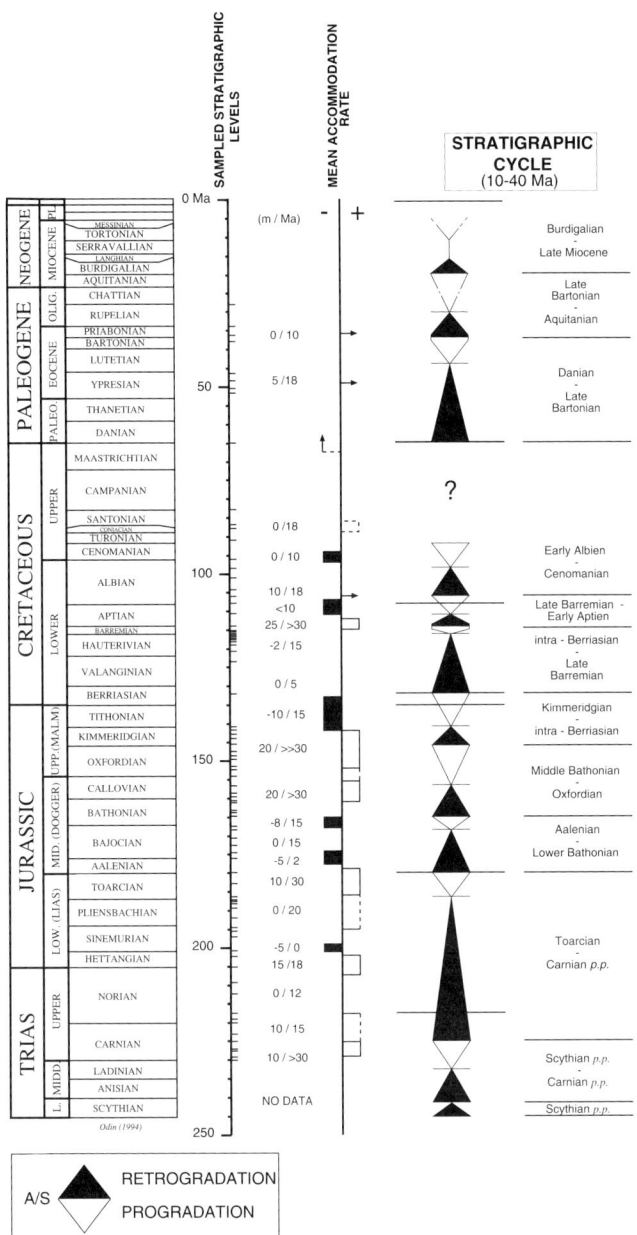

Fig. 2. Major 10–40 Ma duration stratigraphic cycles, unconformities and associated accommodation rate variations (mean values), recognized from sequence stratigraphic studies.

temporaneous with oceanic accretion in the Ligurian Tethys and the transformation of the Subalpine Basin into a passive margin.

The Aalenian unconformity, a major basin discontinuity, seems to reflect short wavelength east–west to ENE–WSW compression, contemporaneous with the doming of the North Sea.

The *Berriasian to late Aptian* is characterized by a NW–SE medium wavelength flexural control bounded by two major unconformities (late-Cimmerian: Jurassic/Cretaceous boundary, Lower/Upper Berriasian boundary; 'Austrian': late Aptian) contemporary with a sharp decrease of the subsidence rates (0–10 m Ma^{-1}) and a change in

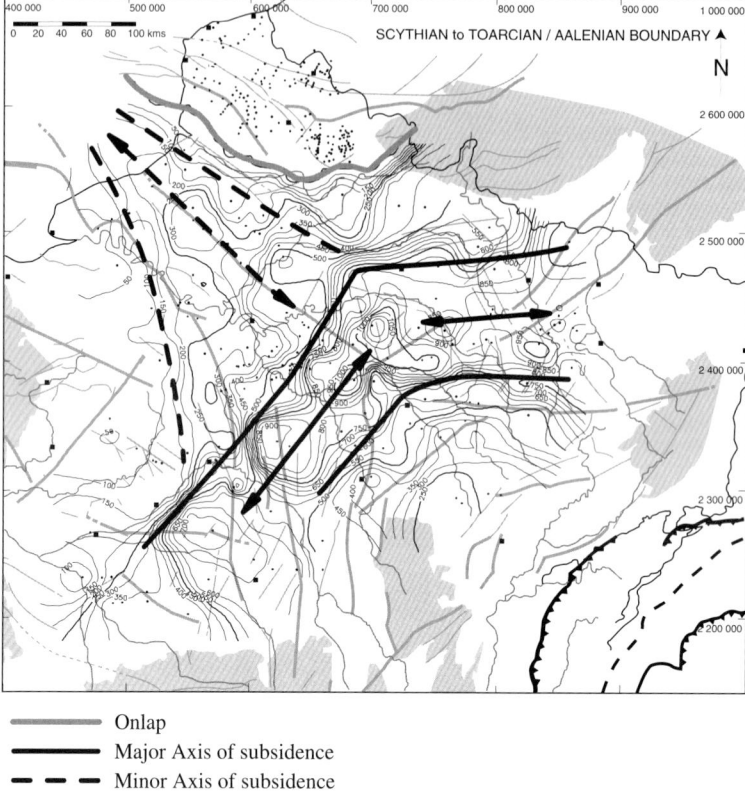

Fig. 3. Isopach map from the base Scythian (top of the basement) to the Toarcian–Aalenian boundary (Aalenian unconformity). Contours in metres.

the sedimentary system with the development of the 'Wealden' siliciclastic deltas (Fig. 5).

Both unconformities result from NE–SW to east–west compression along the eastern border of the Armorican domain and along the southern border of the Dinant synclinorium (Artois/Ardennes). They are contemporaneous with the early stages of extension in the Bay of Biscay (rift: late Cimmerian) and of oceanic accretion (passive margin: Late Aptian). This time was also the first period of lithosphere buckling in the Paris Basin (Guillocheau *et al.* 2000).

The *Late Aptian to Turonian* is characterized by a NW–SE medium wavelength flexural control with an increase of the subsidence rate and a change from a siliciclastic system (tidal-dominated Greensands) to carbonate platforms (Cenomano-Turonian chalks) with post-unconformity onlaps toward the west and the Northeast (Fig. 6).

The *Turonian to Recent* is characterized by a decrease of the subsidence rate, sediment by-pass and finally uplift and erosion, in a generally com-

pressional setting. Three main periods of compression can be defined: (1) the Late Cretaceous (short and medium wavelengths, 'Subhercynian' and 'Laramide') paroxysm of Senonian deformation, (2) from the Lutetian to the Lower Oligocene (short wavelength, 'Pyrenean') and (3) in the Late Miocene (medium wavelength). Subsidence rates sharply decreased in Late Cretaceous time, falling from 10–20 m Ma^{-1} to very low rates (0–10 m Ma^{-1}) in the Lower Oligocene. Sediment bypass occurred from the Upper Oligocene to the Pliocene with local deposition along the present-day Loire River. Uplift started in the late Lower/early Middle Pleistocene.

Late Cretaceous and Late Miocene middle wavelength deformations reflect lithosphere buckling in response to north–south and NW–SE compression (Guillocheau *et al.* 2000). Late Eocene/Lower Oligocene extension occurred east of the basin, whereas compression was located in the western part. Minor compression was recorded in the earliest Burdigalian.

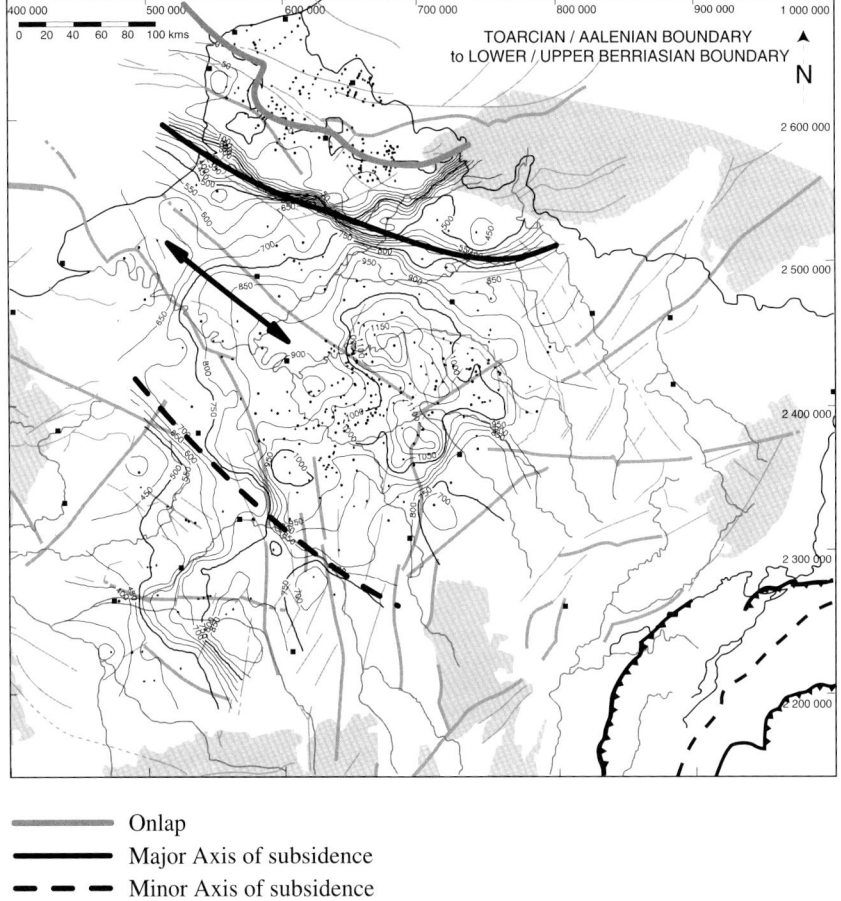

```
          Onlap
          Major Axis of subsidence
  —  —  — Minor Axis of subsidence
```

Fig. 4. Isopach map from the Toarcian–Aalenian boundary (Aalenian unconformity) to the Lower/Upper Berriasian unconformity (Neo-Cimmerian unconformity). Contours in metres.

The present-day shape of the Paris Basin is the result of compression during the Tertiary and to a lesser degree in the Early Cretaceous (Fig. 7).

Long term (230 Ma) thermal subsidence history

Thermal evolution of the lithosphere

The long-term tectonic subsidence of the Paris Basin reflects the exponential thermal decay of a hot lithosphere towards its cooler thermal equilibrium state (Brunet & Le Pichon 1982; Robin 1997; Robin *et al.* 1998, 2000; Prijac *et al.* 2000; Guillocheau *et al.* 2000) (Fig. 8). We here define the model that enables the simulation of lithospheric cooling, choose the parameters that yield a best fit between modelled and measured tectonic subsidence curves, and then compare the simulated initial

thermal evolution of the lithosphere with the palaeogeography, sedimentary geometries, magmatism, and metamorphism concerning the late Variscan evolution in the Paris Basin area. Finally the Neogene uplift will be quantified by comparison of the predicted present-day basement depth to its actual elevation.

Simulations of lithospheric thermal cooling

The idea of the existence of an equilibrium thermal state for the lithosphere has almost always been included in attempts to model long-term thermal subsidence of continental basins (e.g. Royden & Keen 1980; Sclater & Christie 1980; Brunet & Le Pichon 1982). This equilibrium is easily described by a 'mantle heat flow' q_m. If this mantle heat flow is zero, the equilibrium thermal thickness of the lithosphere tends to infinity. A mantle heat flow of

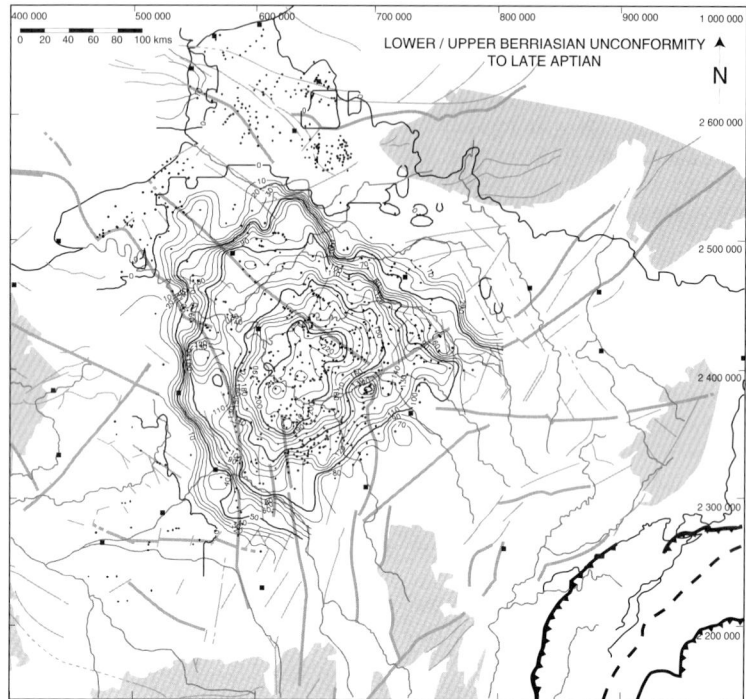

Fig. 5. Isopach map from the Lower/Upper Berriasian unconformity (Neo-Cimmerian unconformity) to the late Aptian unconformity (Austrian unconformity). Contours in metres.

about 50 to 30 mW m^{-2} results in equilibrium thermal lithospheric thicknesses of between 90 to 150 km, the exact value of which depends on the crustal heat production. Such an amount of heat can only be transferred by a strong convective flow beneath the lithosphere. The heat transfer beneath a rigid conductive lid has been extensively studied (Davaille & Jaupart 1994; Doin *et al.* 1997; Dumoulin *et al.* 1999; Solomatov & Moresi 2000) and is shown to result from the detachment of small and cold downwelling drips from the base of the lithosphere. It only depends on the viscosity of the mantle just below the lithosphere and on the temperature and pressure dependence of the viscosity. The evolution through time of this basal heat transfer q_m when a lithosphere returns to its thermal equilibrium state, has long been debated (Parsons & McKenzie 1978; Yuen & Fleitout 1984; Davaille & Jaupart 1994; Dumoulin *et al.* 2000). However, numerical simulations show that, as long as one does not artificially fix perfectly flat isotherms in the initially extended lithosphere, the heat transfer at the base of the lithosphere remains controlled by the asthenospheric viscosity during the whole cooling period (Dumoulin 2000). We will therefore assume that this heat transfer q_m is constant (Chablis model: Doin & Fleitout 1996)

and refer to Prijac *et al.* (2000) for a comparison between the Plate and Chablis models.

We postulate that the long-term tectonic subsidence registered by marine sedimentation from 230 Ma to 30 Ma results from the Stephanian–Autunian extensional collapse of the Variscan belt, from about 300 Ma ago until 280 Ma ago (Burg *et al.* 1994*a*, *b*; Lefort & Jaffal 1994; Rey *et al.* 1997). As the Moho depth is presently at about 35 km beneath the Paris Basin (Cazes *et al.* 1985; Autran *et al.* 1994), the initial lithospheric structure defined at 300 Ma includes a crust thickened by a β factor. As we consider extension within an orogen, we have to choose an initial thermal state representative of an orogen. We assume that, at the end of the Variscan orogeny, the mantle lithosphere was not thickened and that thermal relaxation in the thickened crustal root had taken place. The initial geotherm is thus the steady-state solution of the equation of heat conduction with a heat transfer q_m applied at the mantle temperature T_m. The assumed lithospheric evolution is summarized in Figure 9.

To compute the thermal evolution of the lithosphere we solve the one-dimensional heat equation in the lithosphere taking into account the upward flow due to erosion and to finite extension rate as

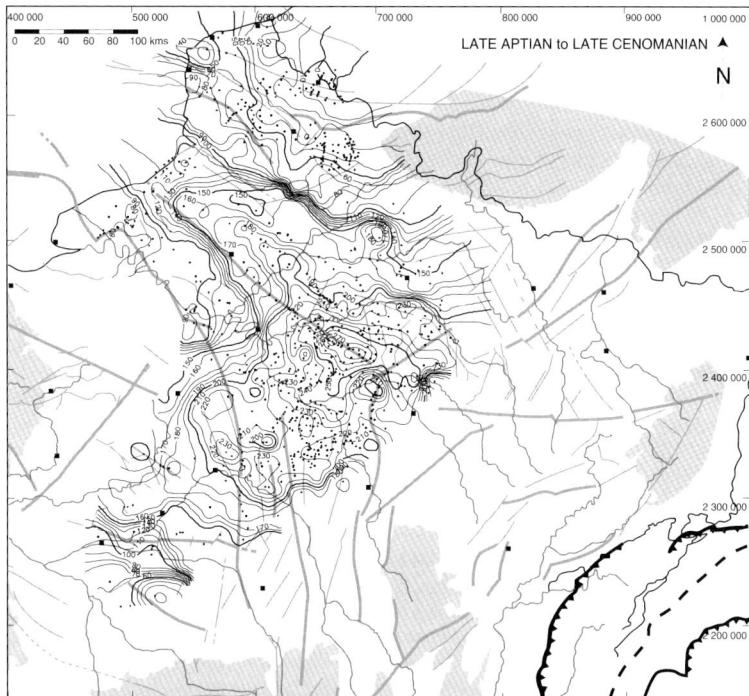

Fig. 6. Isopach map from the late Aptian unconformity (Austrian unconformity) to the late Cenomanian (*Actinocamax plenus* Marls). Contours in metres.

in Jarvis & McKenzie (1980). The conductivity and heat capacity dependences on temperature are included as well as the radioactive heat production in crustal layers. The basal boundary condition is a constant heat transfer q_m applied at the mantle temperature T_m. The thermal subsidence is computed, assuming local isostasy, from the temperature field with a thermal expansion coefficient depending both on temperature and pressure. In addition to thermal contraction, we include the effects on subsidence of the change in depth of the spinel/garnet phase transition in the mantle and of metamorphic changes of density in the crust (assuming a granodioritic upper crust and a lower crust of intermediate composition, whose densities are given by Bousquet *et al.* 1997). Modelling constants are defined in Table 1.

Fit of the tectonic subsidence curves

Accommodation curves have been constructed from the dating (according to the time-scale of Gradstein *et al.* 1994) and correlation of 88 markers from the Triassic to the Tertiary. The oldest sediments included in these curves correspond to the first marine transgression in the Triassic. We thus record the progressive onlap of German domain from west to east (Bourquin *et al.* 1995; Bourquin & Guillocheau 1996; Goggin *et al.* 1997). The tectonic subsidence curves have been corrected for sediment loading and compaction (Sclater & Christie 1980) and first-order absolute sea level variations (Haq *et al.* 1987). They correspond to 51 wells located on an east–west profile between Nancy and Rambouillet and to two wells located in the Rhinegraben (Fig. 1). Cenozoic erosion affected 30 Ma and older sediments, their age increasing gradually towards the eastern side of the basin. On the easternmost side of the profile and on the Rhinegraben wells, the subsidence curve is limited to the late Triassic to Middle Jurassic.

Fitting tests have been performed on the well 'La Folie de Paris' which is located in the deep central part of the basin and was only slightly affected by Tertiary tectonics (Fig. 1). The model was adjusted to the tectonic subsidence curve between 230 Ma and 30 Ma. We tuned the equilibrium lithospheric thickness Z_L^E (which is equivalent to tuning the mantle heat flow q_m) and the stretching factor β. A value of 120 km for Z_L^E and of 1.58 for β, together with the parameters set defined in Table 1, yielded a good fit between the modelled and constructed tectonic subsidence curves (Fig. 10a). The root mean square difference between

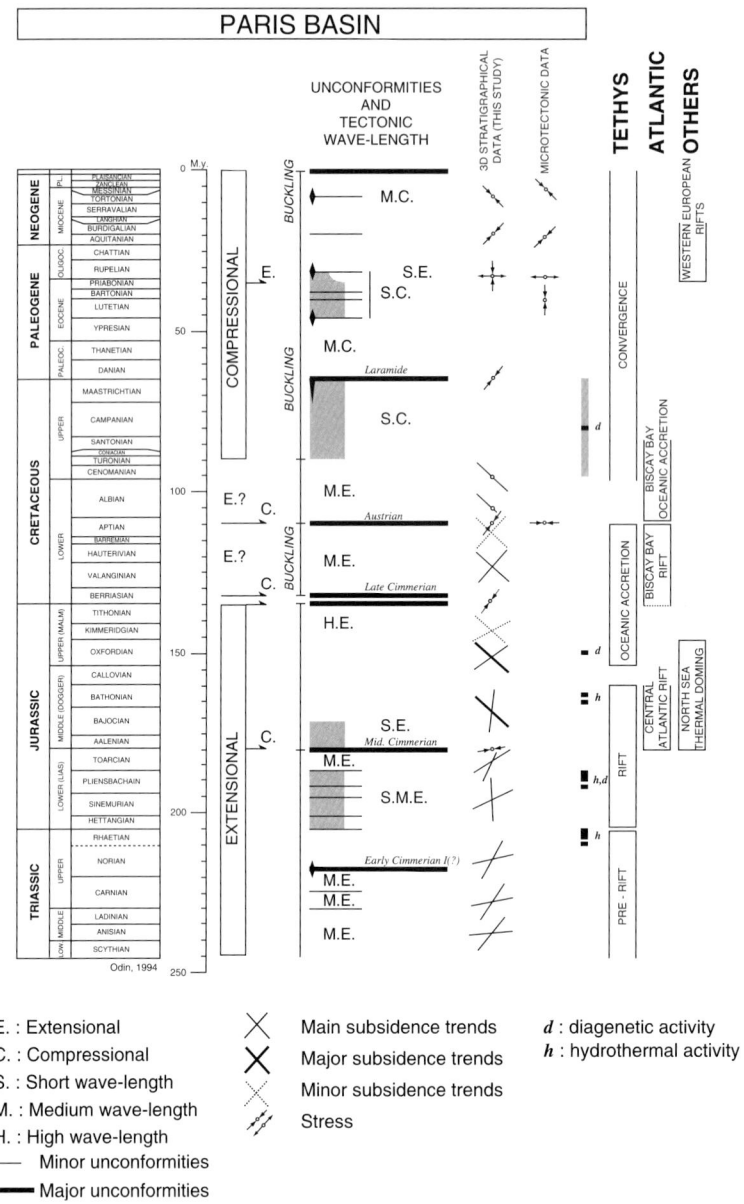

Fig. 7. Geodynamic evolution of the Paris Basin.

both curves is 49 m. The inferred present-day thermal lithospheric thickness of about 110 km and the present-day heat flow of 74 mW m^{-2} agrees with geophysical observations (e.g. Suhadolc *et al.* 1990; Lucazeau & Vasseur 1989). The predicted thermal lithospheric evolution is displayed in Figure 10b.

To fit the tectonic subsidence data of other wells, we assumed that the equilibrium lithospheric thickness ($Z_L^E = 120$ km) remains constant across the Paris Basin, and adjusted the value of β. The root mean square deviation between constructed and modelled curves is bracketed between 34 and 56 m. β is maximum in the centre of the basin (about 1.6) and decreases towards the sides (Fig. 11a). If we had assumed a longer extension duration, the obtained stretching factors would have been smaller (see Prijac *et al.* 2000). Modelled pseudo-

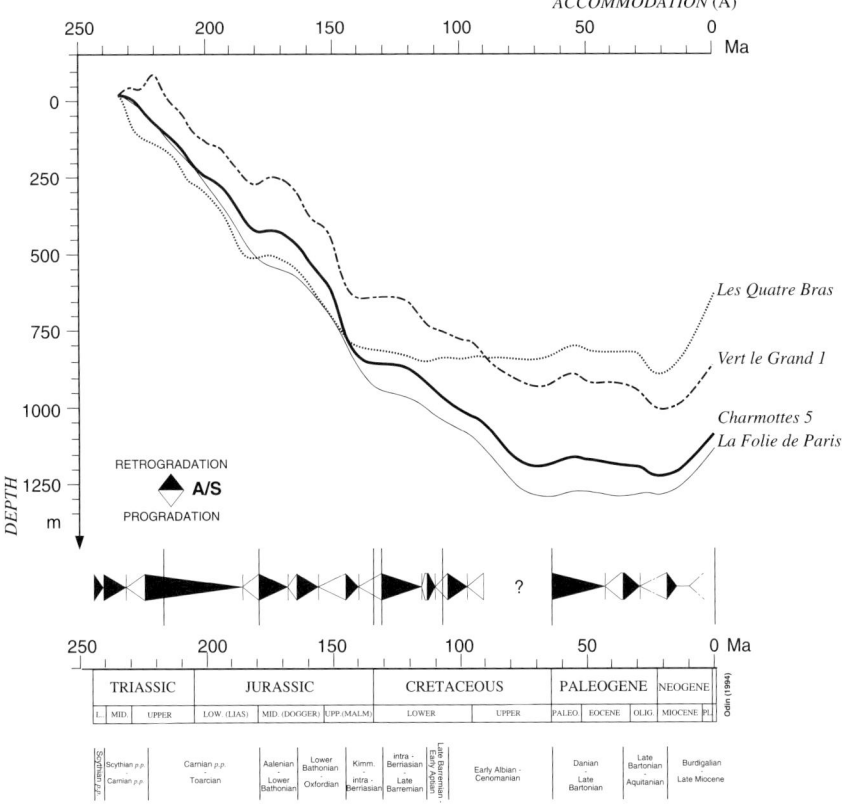

Fig. 8. Accommodation curves calculated for the wells 'Les Quatre Bras', 'Vert-le-Grand 1', 'Charmottes 5' and 'La Folie de Paris': correlation with the 10–40 Ma duration stratigraphic cycles.

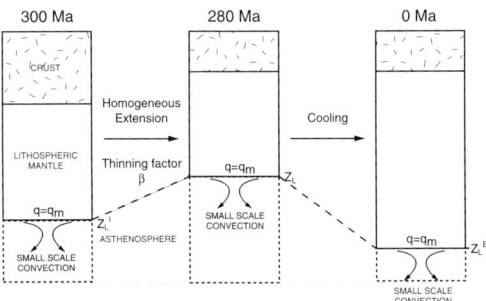

Fig. 9. Sketch describing the evolution of the lithosphere (Chablis model). The bottom boundary condition is a constant applied heat flow q_m on the isotherm defining the base of the lithosphere (dashed line).

2D temperature and melt cross-sections across the basin are shown in Figure 11b and c at 300 Ma and 280 Ma. The high thermal regime from 300 to 280 Ma implies an important crustal melting. Asthenospheric melting is inferred to have been weak at 300 Ma, but to have increased significantly towards the end of extension (at about 280 Ma).

The model predicts a high mountain range at 300 Ma (end of Variscan orogeny), evolving at the end of extension towards a relatively flat plateau at about 1000 m altitude in the basin centre (Fig. 12). Synrift intramontane sedimentation in fault-bounded depressions and relief erosion are expected for this period. From 280 Ma onwards

Table 1. *Modelling constants*

Description	Value
Radiogenic heat production, upper crust*	1.3×10^{-6} W m^{-3}
Radiogenic heat production, lower crust*	0.3×10^{-6} W m^{-3}
Radiogenic heat production, lithospheric mantle	0
Final thickness of the upper crust	22 km
Final total thickness of the crust	35 km
Erosion rate	3.5×10^{-5} m a^{-1}
Erosion duration	55 Ma
Stretching duration	20 Ma
Temperature at the base of the lithosphere	1350°C
Temperature at the surface	0°C
Average thermal conductivity, lithospheric mantle†	3.1 W km^{-1}
Average thermal conductivity, upper crust†	3.0 W km^{-1}
Average thermal conductivity, lower crust†	2.5 W km^{-1}
Average heat capacity†	1.24×10^3 J gK^{-1}
Equivalent thermal expansion coefficient†	3.75×10^{-5} K^{-1}
Density, lithospheric mantle (at T_M)	3250 kg m^{-3}
Density contrast garnet–spinel phases	50 kg m^{-3}
Density, seawater	1030 kg m^{-3}

*Average heat production value as calculated by Rudnick & Fountain (1995) for areas covered by palaeozoic orogens in western Europe and western United States.
†The expressions of the conductivity, heat capacity and thermal expansion coefficient as a function of temperature are given in Doin & Fleitout (1996).

we model a slow general subsidence without relief creation (Fig. 10a). At about 230 Ma the basement passes below sea level (Fig. 12).

Discussion

The Stephanian to Triassic tectonostratigraphic, magmatic and metamorphic evolutions implied by our simulation compare well with available observations from the Paris Basin itself or surrounding areas (Prijac *et al.* 2000). The modelled high geothermal gradients within the crust account for the intense granitoid generation (e.g. Carron *et al.* 1994; Ploquin *et al.* 1994; Schaltegger *et al.* 1996; Schaltegger 1997), and the widespread high T–low P metamorphism (e.g. Vielzeuf & Pin 1989; Gardien *et al.* 1997) recorded by late Variscan orogenic units cropping out around the Paris Basin. The simulation also reproduces the transition from magmatism of mainly crustal origin to bimodal

magmatism, described by Turner *et al.* (1992) as characterizing orogenesis followed by lithospheric thinning. Bimodal volcanism present in Stephanian and Autunian sediments peaked at the Early Saxonian (Henk 1993; Korsch & Schäfer 1995). Sedimentation occurred within fault-controlled isolated intramontane troughs during Stephanian and Autunian (Vallé *et al.* 1988; Faure & Becq Giraudon 1993; Doré 1994; Pelhate 1994; Burg *et al.* 1994*a*, *b*; Rey *et al.* 1997). At the end of the Autunian we may assume that the tectonic setting was similar to the Basin and Range, with numerous isolated basins (Mascle 1990; Bachmann *et al.* 1987) and metamorphic core complexes (e.g. Malavieille 1993). Sedimentation areas widen and flatten progressively from Saxonian to Triassic until first lakes and then the sea progressively flooded the subsiding area (Faure 1995; Carasco 1987; Van der Driessche & Brun 1989).

One-dimensional solutions to the heat conduction equation are reasonable in the case of the Paris Basin. The present-day shape of the Paris Basin is mostly the result of Cenozoic deformation (large-scale folding and uplift) and erosion (Guillocheau *et al.* 2000). Permian, Triassic and Jurassic sediments are still preserved within the Rhinegraben, Bresse and Burgundy troughs whereas they have been partly eroded on the Rhenish shield (Anderle 1987) and on the Vosges and Black Forest Massifs. Jurassic sediments have probably been deposited on the Massif Central and later eroded (Enay *et al.* 1980; Guillocheau *et al.* 2000; Barbarand *et al.* 2001). During the Triassic and Jurassic the Paris Basin was part of a large subsiding area including the Iberian Basin (Van Wees *et al.* 1998), the Aquitaine and Provence Basins, the Brittany, Western Approaches (Ziegler 1987*a*), the Celtic Sea, Channel and Wessex Basins (Lake & Karner 1987; Underhill & Stoneley 1998), the external Alps (Loup 1992), and the Southern German Basin (Bachmann *et al.* 1987; Schröder 1987), all of them being located on the former Hercynian fold belt (Ziegler 1982, 1987*b*). These regions experienced continental sedimentation in intramontane depressions during Carboniferous and Permian, followed by progressive marine transgression during Triassic and Early Jurassic and by continued Jurassic subsidence. A complex rift system with localized extensive deformation, linked with the opening of Tethys and associated with volcanism, developed during Late Permian and Triassic and affected mostly the basins south of the Massif Central (Ziegler 1982). Rifting pulses also occurred during Middle/Upper Jurassic and Lower Cretaceous (Berriasian–Aptian) in basins located on the northern (North Sea) and the western (Bay of Biscay) sides of the Paris Basin, under the influence of the Atlantic opening. They are contempor-

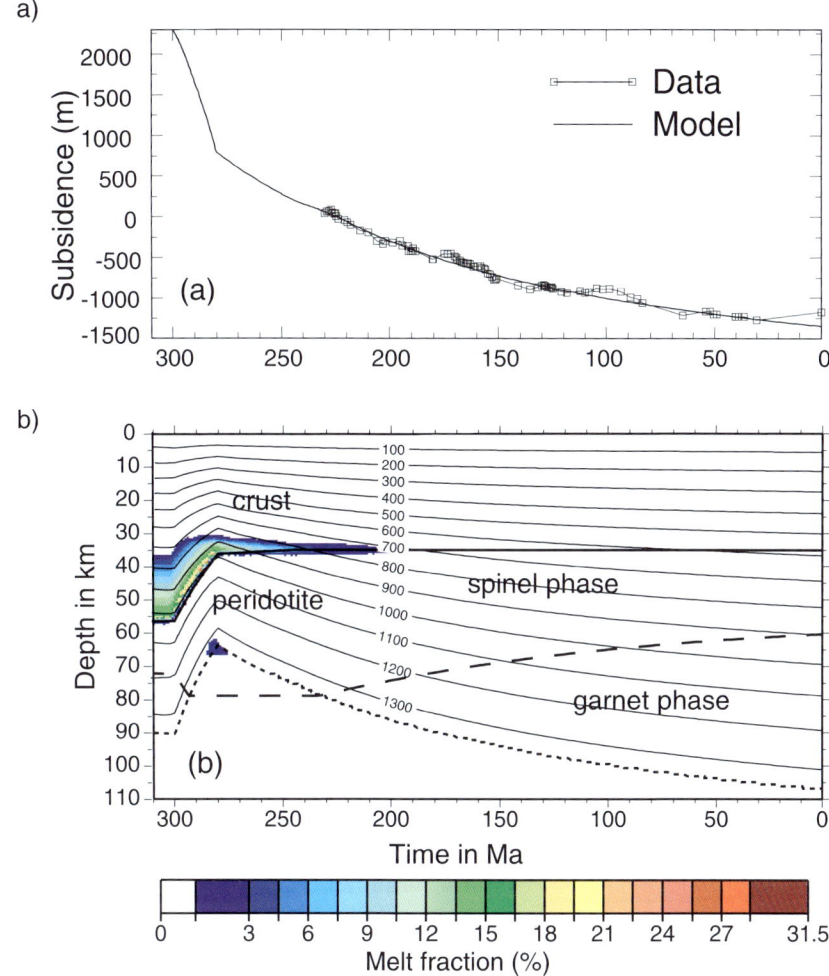

Fig. 10. (a) Fit of the tectonic subsidence curve of the well 'la Folie de Paris' with the thermal model. (b) Simulated thermal evolution of the lithosphere (thin black lines are isotherms) and estimated melt fraction in the crust and mantle. The thick solid, thick dotted, and thick dashed lines represent the depth of Moho, the base of the lithosphere, and the spinel–garnet phase transition, respectively.

aneous with large-scale deformations in the Paris Basin (Guillocheau *et al.* 2000). Therefore, the Paris Basin thermal subsidence cannot be considered as the consequence of a local thermal rejuvenation of a lithosphere surrounded by much cooler lithosphere, as was for example inferred for the Michigan Basin (Kaminski & Jaupart 2000). Such an assumption would imply quicker cooling of the lithosphere, which cannot apply to the Paris Basin area. Lateral thermal conduction effects coming from the Paris Basin borders are difficult to assess without modelling the thermal evolution of the lithosphere beneath the surrounding basins. However, as wavelengths characterizing the long-term Paris Basin subsidence are larger than 800

km, lateral gradients in temperature are inferred to be much smaller than vertical gradients. This justifies the one-dimensional approximation in the heat equation. Furthermore, departure from isostasy due to elastic flexure is expected to be less than a few per cent and does not affect our conclusions.

Quantifying Cenozoic uplift

The long-term thermal subsidence trend can be used to estimate the thickness of eroded or non-deposited sediments during the Tertiary. We will here define the uplift as the present-day height difference *h* between the modelled long-term tectonic subsidence and the reported tectonic subsidence

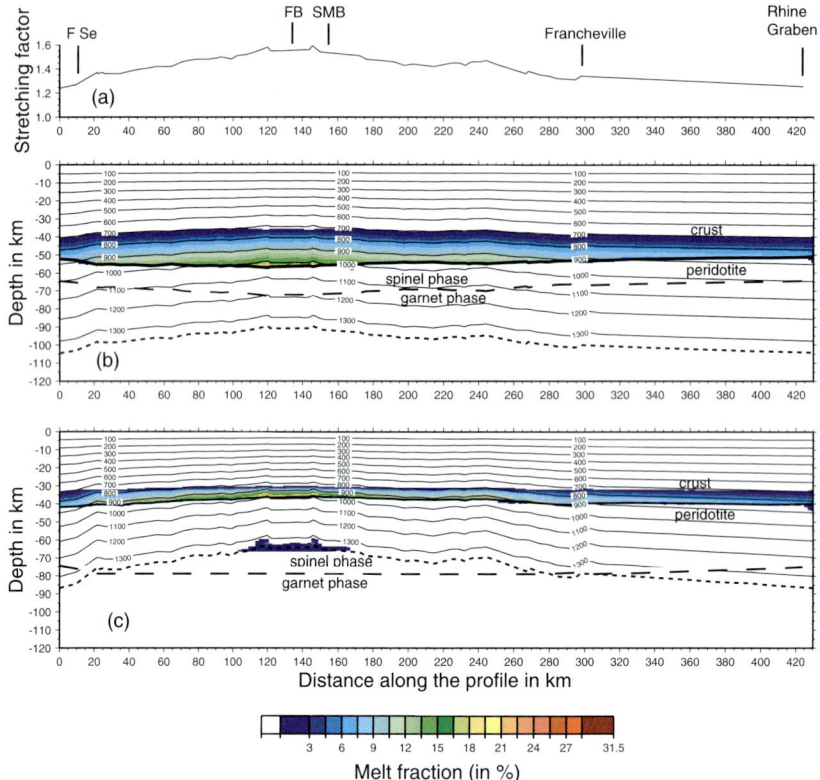

Fig. 11. (**a**) Variation of the stretching factor along the profile defined on Figure 1. (**b**) Simulated lithospheric structure at 300 Ma, showing isotherms, the Moho (thick solid line), the base of the lithosphere (dotted line), and the spinel–garnet phase transition (dashed line). (**c**) Same as (**b**) but at the end of extension (280 Ma). F Se, Seine Fault; FB, Bray Fault; SMB, Saint Martin de Bossenay Fault. Models have been interpolated between Francheville and the Rhinegraben.

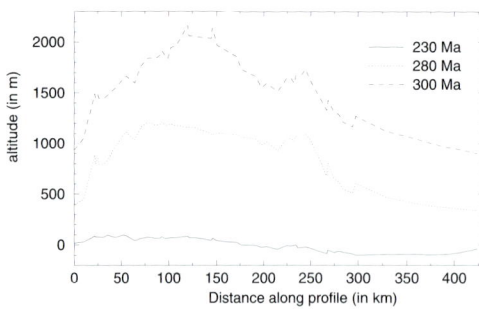

Fig. 12. Predicted topography at 300 Ma, 280 Ma and 230 Ma along the profile shown on Figure 1.

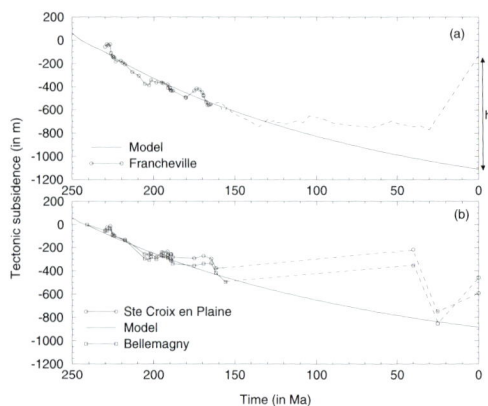

(Fig. 13a). The modelled curve represents the depth below sea level that would have been reached if the basement had not been uplifted. Multiplying h by the density difference between mantle (ρ_m) and water (ρ_w) gives the excess weight due to uplift and

Fig. 13. Tectonic subsidence curves for (**a**) Francheville well, (**b**) the two wells located within the Rhinegraben, together with the best fitting modelled curve.

erosion. The resulting gravity anomaly could then roughly be estimated by the Bouguer formula $2\pi G(\rho_m - \rho_w)h$. This mass excess may or may not be compensated by asthenospheric upwelling and lithospheric heating. We cannot compute the uplift for wells located within the Rhinegraben as we did not model the effect on subsidence of rifting. For points located on the profile between Nancy and the Rhinegraben, where most sediments have been eroded, we built 'synthetic' subsidence curves defined by a streching factor that is a linear combination of those derived for 'Francheville' and for the Rhinegraben wells (Fig. 13). The present-day altitude of the basement at each of these points was then used to calculate the equivalent position below sea level of the uplifted basement (using backstripping technique). The uplift profile shows a continuous increase towards the eastern side of the basin (0 to 2700 m for the Vosges basement) (Fig. 14). The uplift of the Vosges Massif would have been underestimated by about one-third if the thickness of the eroded or non-deposited sediments had not been taken into account.

This uplift has been related to different phenomena: flexure of rift flanks (Weissel & Karner 1989), forebulge of the subducting lithosphere beneath the Alps (Gutscher 1995), compressional tectonics (Cloetingh & Burov 1996), or doming of the lithosphere in answer to asthenospheric upwelling (Lyon-Caen & Molnar 1989; Ziegler 1992). The uplift origin depends on its timing.

1. A first uplift is evident from the distribution of pre-Tertiary sediments within the Rhinegraben (Anderle 1987). This uplift decreased in amplitude towards the south of the Rhinegraben where Late Jurassic sediments have been preserved (Sittler 1985). This event, taking place between Late Jurassic and Palaeocene, cannot be dated accurately. However, the preservation of Cretaceous sediments until late Campanian to the west (Paris Basin), south (Helvetic realm: Loup 1992), SE (Molasse Basin: Bachmann *et al.* 1987), and NE (Lower Saxony Basin: Betz *et al.* 1987) of the Rhinegraben,

suggests that Cretaceous sediments may have been deposited and later eroded in the Rhinegraben area.

2. After the Rhine rift initiation during the Eocene, the Rhenish shield became a peneplain partly covered by a shallow Oligocene sea. The massif was uplifted from the Miocene until today. The present-day altitude of Oligocene sandstone yields an uplift of about 500 m since the Miocene (Demoulin 1989), whereas a 240 m uplift has been inferred for the Quaternary from river terraces (Anderle 1987). The Vosges and Black Forest massifs are believed to have undergone updoming since the Middle Miocene (Laubscher 1987; Ziegler 1992), while the south Rhinegraben ceased to subside.

More generally in Western Europe, Ziegler (1987*c*) recognized three main phases of compressional deformation. The first inversion phase, which culminated from Santonian to Campanian is restricted to the basins located to the NE of the Paris Basin (from the Bohemian Massif to the Netherlands and the Sole Pit Basin: Shröder 1987; Betz *et al.* 1987). The second phase, at the end of Cretaceous and Palaeocene, triggered general compressive deformation in the Alpine foreland. It initiated inversion in basins on the western side of the Paris Basin (Western Approaches, Wessex, Celtic Sea, Channel Basins). In the latter basins, this phase was first followed by subsidence (Hillis 1991), and later by the main inversion and uplift phase, during the Oligocene/Miocene. This uplift, during which 0.3 to 3 km of sediments were eroded, extended beyond the inverted basins (Van Hoorn 1987; Lake & Karner 1987; Hillis 1991; Bray *et al.* 1998). The Pays de Bray fault records transpressional deformation during the second and third phase.

Temporal and spatial scales of crustal tectonics in the Paris Basin during Middle Jurassic (Dogger)

Measurement of the relative vertical displacement from stratigraphic data

The characterization of the spatial and temporal evolution of the crustal tectonics during the thermal relaxation period of the Paris Basin allow us to quantify the local effect of the sediment load on vertical crust movements. The measurements are based on a database containing bathymetry of time-lines, and thickness and lithology between two successive time-lines. As shown on Figure 8, the accommodations measured on four wells of the Paris Basin are globally similar. However, accommodations vary in details from well to well at a

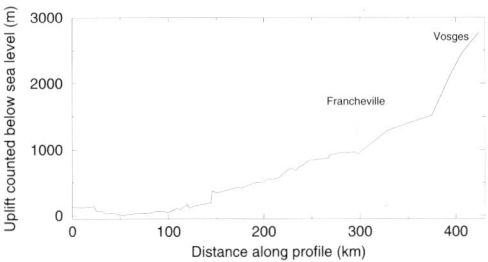

Fig. 14. Estimated uplift *h* along the profile.

scale of less than some million years. Phases of acceleration or deceleration of accommodation occur non-simultaneously with an amplitude of some tenths of metres per million years. As the pulses of accommodation are not synchronous, the origin of this signal has to be searched for in tectonics rather than in eustasy. To quantify this, the eustatic part of the signal has to be subtracted from the accommodation signal. The absolute vertical displacement of a point located in the basement of the basin under the sedimentary layer submitted to compaction is given by:

$$T_A = (Hc_{A2} + Es_{A2} + B_{A2}) \\ - (\Delta + B_{A1} + Hc_{A1}) \tag{1}$$

with T_A is the absolute vertical tectonics at the well site A relative to the Earth centre, Hc_{A1} (resp. Hc_{A2}) is the thickness of the sedimentary layer submitted to compaction at the well site A at time 1 (resp. time 2), B_{A1} (resp. B_{A2}) is the bathymetry at time 1 (resp. time 2), Es_{A2} is the thickness of the sedimentary layer deposited at point 1 between times 1 and 2, Δ is the eustatic variation between times 1 and 2.

There is no agreement on a chart of eustatic variation at time steps lower than a million years for Jurassic time (e.g. Haq *et al.* 1988, Robin *et al.* 1996, 1998, 2000; Robin 1997). The parameter Δ is difficult to estimate and will induce large uncertainty in the measure of the tectonics. However, the vertical displacement can be estimated relative to a reference well site B on which the same calculation is done. The relative vertical displacement that we call 'Relative Tectonics' is equal to the difference between the two displacements (Fig. 15):

$$Tr = T_A - T_B \\ = (Hc_{A2} - Hc_{A1}) + (B_{A2} - B_{A1}) \\ + Es_{A2} - (Hc_{B2} - Hc_{B1}) \tag{2} \\ - (B_{B2} + B_{B1}) - Es_{B2}$$

The eustatic variation cancels out in Equation 2 (Fig. 15). This method has been used by Dromart *et al.* (1998) to study the displacement of normal fault at high temporal resolution. The absolute value of the relative tectonics has no meaning as it depends on the reference point. Only spatial gradients of relative tectonics are indicative of the relative vertical displacement between the well sites. If a tectonic component exists at a larger scale than the area of study, it is not measured by this method. This small-scale component of tectonics, as the thermal subsidence, is taken into account in the global eustatic part and thus removed.

Application to the Middle Jurassic (Dogger) of the Paris Basin

A set of 164 sites distributed over a surface of around 83 000 km^2 (380 per 220 km) has been processed for studying the relative tectonics during the Bathonian–Callovian interval. Site distribution and spacing are fairly regular even though the southwestern basin is marked by a scarcity of boreholes due to the lack of subsurface hydrocarbon reservoirs. A cumulated length of about 2000 m of cores has been inspected in the Bathonian–Callovian of the Paris Basin, representing 6% of the total logging. Eventually, a total number of 24 time-lines could have been traced on the scale of the Paris Basin, through the 100–300 m thick sedimentary pile of the Bathonian–Callovian carbonate platform, over a duration of about 10 Ma. Final timelines tie distinct depositional environments across the sedimentary sequences, and delineate packages of rocks whose thickness and composition vary as a function of their vertical and lateral location. Each layer has been decompacted according to the principles developed by Van Hinte (1978). As it is difficult to evaluate precisely the burial of each well site, a maximum constant burial depth of 1500 m has been supposed for all the well sites. For the decompaction process, the existence of a uniform and compactable 1000 m thick layer has been supposed under the Bathonian–Callovian units. The velocity of relative tectonics was obtained by dividing the relative tectonics by the duration of the intervals, which is 500 ka on average.

The time evolution of the maximum difference of relative tectonics is shown in Figure 16. For a given time-line, this value is computed from the difference of the mean of the 30 maximum and

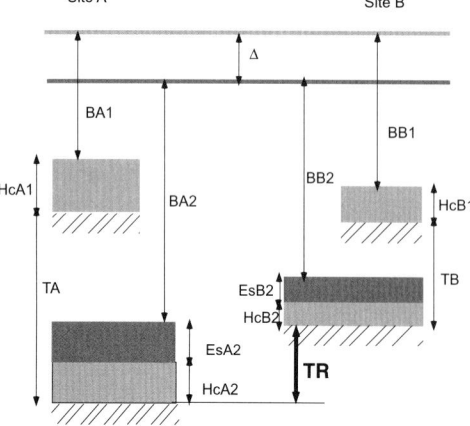

Fig. 15. Principles of the calculation of relative tectonics. See text for details.

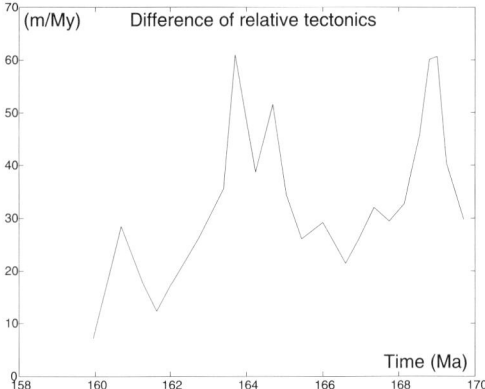

Fig. 16. Difference between the maximum and the minimum of relative tectonics as a function of time during the Bathonian and Callovian.

minimum values of relative tectonics in order to reduce the effect of extreme non-significant values. This parameter describes the intensity of tectonics. No obvious temporal trend is visible during the studied period. The average velocity of relative tectonics is around 40 m Ma^{-1} with three peaks of velocities larger than 60 m Ma^{-1}. These velocities are low compared to velocities measured on passive margins on which faults slip at rates larger than 100 m Ma^{-1}. The cumulative sum in time of relative tectonics for a well site represents the progressive vertical displacement of this point relative to the reference site. The mean of this parameter at a given time estimated on all the well sites is equal to the mean of the relative tectonics. This parameter is reported relative to time on Figure 17 and is between -35 and $+20$ m and does not show any particular trend with time. At the scale of Bathonian–Callovian, the Paris Basin is submitted to a homogeneous subsidence. This demonstrates that the subsidence is not localized on particular zones. The difference of relative tectonics between two wells has been reported according to the dis-

tance between the wells (Fig. 18). As there are 164 site wells, this represent around 12 000 differences. Each point of the curve represents the sliding mean of 500 differences of relative tectonics. Three different behaviours can be observed. In the first one, the difference of velocity of relative tectonics is independent of the distance and is around 10 m Ma^{-1}. The tectonics of the Paris Basin occurred on blocks smaller than the resolution of the method, which is about 20 km. In the second kind of behaviour, the difference of relative tectonics increases with distance to 120 km. There is no correlation between velocity and distance for larger distances. The tectonics occurred here on blocks, which have a size of 120 km. This is the average distance, which limits the main faults of the Paris Basin (Fig. 1). For other periods, the difference of relative tectonics increases in proportion to the distance. This can be related to a tilt of the Paris Basin, which can reach values of 5 to 10 m Ma^{-1} per 100 km.

The values of relative tectonics measured on wells sites have been linearly interpolated to obtain maps of velocities of relative tectonics. Figure 19 shows the map of the period between 165.5 and 164.7 Ma. During this period, an average of 35 m of sediments was deposited. The map of water depth shows a platform located around 20 m depth in the centre and in the eastern part of the Paris Basin. The bathymetry of the western part is more variable and locally reached 60 m. The rocks which have been deposited on the platform are oolithic calcarenites characteristic of shoreface or lagoon. Two kinds of tectonic behaviour are visible on the map of relative tectonics: (1) zones of strong gradient of vertical velocity which are correlated with the main faults and which are independent of the velocity of sedimentation and marked mainly by water depth variation; and (2) zones of diffuse sub-

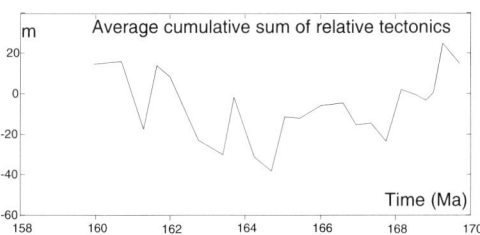

Fig. 17. Average of the relative tectonics as a function of time during the Bathonian and Callovian.

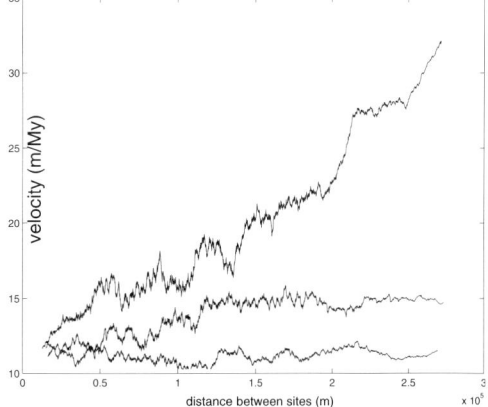

Fig. 18. Three characteristic behaviours of the velocity of relative tectonics as a function of distance.

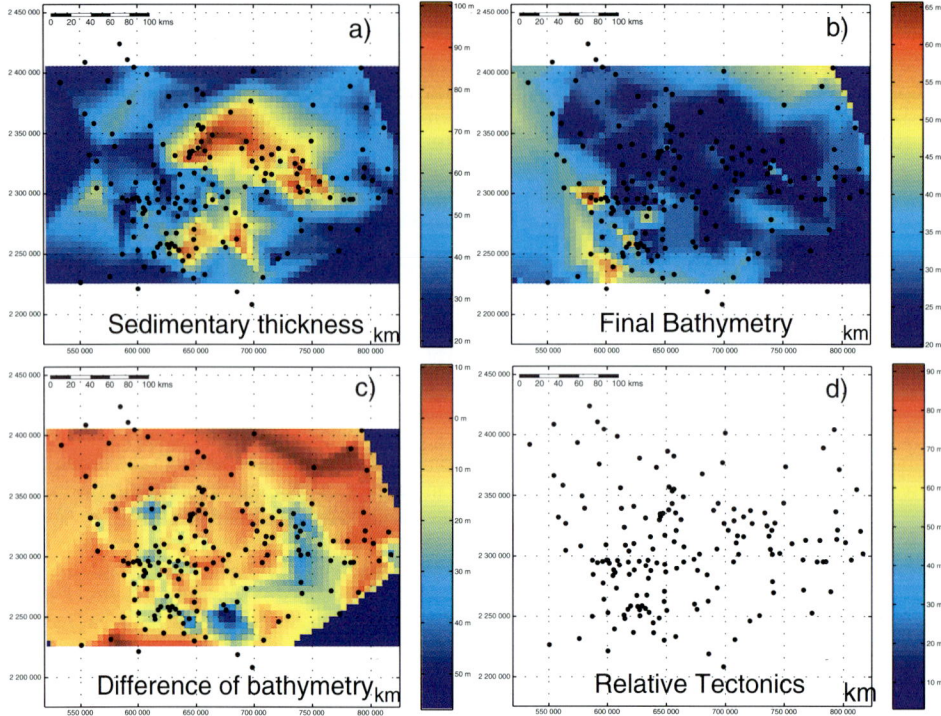

Fig. 19. (*a*) Map of sedimentary thickness; (**b**) map of bathymetry; (**c**) map of change in bathymetry; (**d**) map of relative tectonics in the Paris Basin during a period of 0.8 Ma in the Bathonian. Each dot represents a well site.

sidence with a 30 km radius lens shape which are correlated with strong velocities of sedimentation. In the second kind of zone, the lack of faults is not certain. If faults exist, they are spaced out from less than 5 km and associated with conjugated structures working at the same velocity. The difference of subsidence between the basin border and the centre of such domains is around 20 m Ma^{-1}. The sediments that were deposited in the selected zones have been mainly produced by algae and other micro-organisms and have been deposited *in situ*. The high of these oolitic shoals above the platform was 10 m at a maximum. During the selected periods, the variations of bathymetry are low or nil on these platforms. These shoals, which have a thickness of 30 to 40 m, subside during their construction.

Discussion

During the Bathonian–Callovian period, the Paris Basin was submitted to two types of tectonic behaviour: (1) a localized deformation which is not correlated with sedimentation; and (2) a diffuse deformation correlated with a flat or positive topography and with high velocity of sedimentation. The first behaviour is observed along the major faults of the Paris Basin. This classical tectonic

activity was generated by strain and stress applied at the boundary of the basin. Guillocheau (1991) has demonstrated that the fault activities are correlated with major changes occurring at the limit of the lithospheric plate that supports the Paris Basin. The second behaviour seems to be associated with sedimentation. The tectonic velocity is in the same range as the velocity of bioconstruction. As the effect of compaction is removed from the calculation, the downward motion has an origin located in the basement of the basin. It has been demonstrated that flow of the lower crust can have induced topographic modification at the scale of some tenths of metres (Kruse *et al.* 1991) and that topography and sedimentation can be the cause of the flow of the lower crust (Burov & Cloetingh 1997). The data described here suggest that the same mechanism is also possible for the sedimentary bodies of the Paris Basin. The topography generated by the sedimentary construction could have been dampened permanently by the flow in a viscous channel composed by the crust or a part of it. For a Newtonian viscosity of the crust the time constant of the phenomenon is (Lliboutry 1976):

$$\tau = \frac{4\eta}{\rho g H^3 k^2} \qquad (3)$$

where ρ is density, g gravity, H thickness of the crustal channel, k wave number, and η viscosity of the channel. For a mean viscosity of the crust of 10^{20} Pa s, a density of 2700 kg m^{-3}, a thickness of the crustal channel of 15 km and a sedimentary body with a wave number of 3×10^{-5} m^{-1} (wavelength of 100 km), the time constant of the system is around 100 ka. This constant is in the same range as the duration of the sedimentary construction (shoals, build-up etc.). This mechanism is thus possible for classical parameters of the crust and could explain the differential spatial accommodation observed in the basin.

The end-Cretaceous/Cenozoic lithospheric folding

Lithospheric folding as a possible mechanism of deformation in the Paris Basin

As was previously suggested in a number of studies (e.g. Stephenson & Cloetingh 1991; Burov *et al.* 1993; Cloetingh *et al.* 1999), lithospheric folding is a common primary response to onset of tectonic compression.

Folding is associated with unstable periodic deformation of layered structures caused by stress/strain continuity unconformities at the interfaces between rheological layers of different mechanical resistance (e.g. strong brittle and weak ductile crust) (Biot 1961). The folding theory suggests that a simple relationship exists between the observed folding wavelength and the thickness of the folded competent layers (folding wavelength λ is roughly proportional to four to six thicknesses of the competent lithospheric layer; e.g. Cloetingh *et al.* 1999). Yet, the geological and geophysical recognition of folding requires a multidisciplinary approach: with the exception of the ocean lithosphere, folding is rarely reflected directly in the topography. Because of morphogenic activity modifying the surface topography and the possibility of crust–mantle decoupling, folding is more often better reflected in the sedimentary records, differential subsidence and uplift data, fault spacing, seismic refraction and reflection profiles, and in the data on Moho topography (known from seismic reflection, gravity and other data) (Cloetingh *et al.* 1999). Mantle deformation has normally very long wavelength characteristics requiring long geophysical and geological cross-sections obtained using deep penetration techniques. Deep seismic cross-sections are unfortunately unavailable for the Paris Basin, and the Bouguer gravity data thus present the major data source on geometry of the Moho (Lefort & Agarwal 1996), which reveal large-scale

undulations with a period of several hundred kilometres.

During the Cretaceous–Palaeocene (55–65 Ma), Lutetian–Oligocene (35–40 Ma) and Miocene periods (10 Ma), compressive signatures were registered in the sedimentary cover of the Paris Basin (Guillocheau *et al.* 2000; this paper), and one can question how the lithosphere could respond to this compression. The presence of periodical Bouguer gravity anomalies with spatial period of more than 400 km (Lefort & Argarwal 1996), quite typical for known cases of lithospheric folding for similar age, gives rise to the idea of development of a compressional instability.

Mechanical model of folding in the Paris Basin

We conducted a set of numerical experiments allowing us to investigate the development of folding instabilities in 'realistic' conditions close to that of the Paris Basin. Our numerical model accounts for brittle–elastic–ductile rheology explicitly taken from the laboratory data, temperature, large strains, variable strain rates, and the possibility of lithospheric faulting during deformation.

Model set-up and numerical experiments on regular folding in realistic conditions

Figure 20 presents a basic setup for the numerical model of folding. As was previously shown (e.g., Vilotte *et al.* 1993; Burov *et al.* 1993) brittle–elastic–ductile lithospheric rheology and lateral discontinuities strongly affect the dynamics of the lithospheric deformation. To investigate the development of folding instabilities in brittle–elastic–ductile multilayers, we adopted the Fast Lagrangian Analysis of Continua (FLAC)-based (Cundall 1989) finite element code Paravoz (Poliakov *et al.* 1993), which has a good record in modelling of folding and buckling instabilities in non-linear media (Burov & Molnar 1998; Gerbault *et al.* 1999; Cloetingh *et al.* 1999; see these publications for details of the numerical method). The numerical box consisted of up to 200×100 elements. Free surface and Winkler forces were used for conditions on horizontal boundaries, and horizontal velocities were used as the lateral boundary conditions. Since the available data on Alpine compression and pre-Alpine history do not allow us to constrain the latter with sufficient precision, anything between 2 and 0.1 mm a^{-1} is possible. We thus varied the lateral rates within these limits and considered different durations of shortening, from 300 Ma (age of the whole rigion) to 50 Ma (age of Alpine collision).

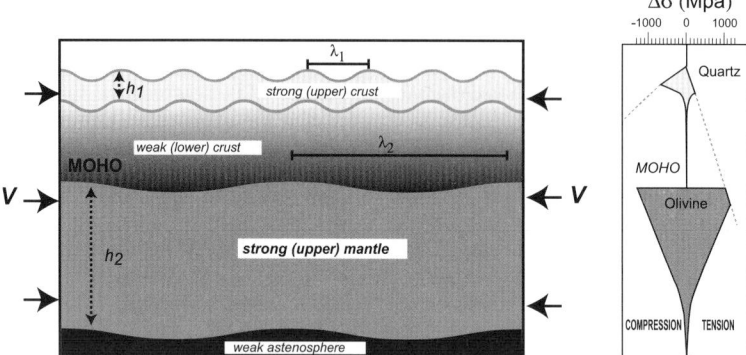

Fig. 20. Set-up for the numerical model of folding. v is horizontal velocity; upper crust (competent core thickness h_1), lower crust and mantle layer (thickness of the competent core h_2) are defined via corresponding rheologies (yield stress envelope on the right, where $\Delta\sigma$ is the maximal differential stress) and physical properties, see text and Table 2. If competent mantle and crust are mechanically decoupled, two characteristic wavelengths of folding can be observed (λ_1 and λ_2).

The finite amplitude of deformation in an unstable compressive regime can be influenced by many often-negligible factors. Therefore our primary goal was only to obtain reasonable amplitude of deformation applying reasonable forces (i.e. not exceeding mechanical strength of the lithosphere).

We tested a significant number of situations encompassing most of the possible scenarios starting from hot continental lithosphere (initial temperature distribution corresponds to a thermotectonic age less than 50 Ma (time of basin formation) and ending with a much colder thermal structure (thermotectonic age 250 Ma related to the onset of the Alpine compression). Figure 21 presents results of 'slow' experiments with non linear brittle–elastic–ductile rheology derived from rock mechanics data (Table 2) assuming initially 50 Ma old quartz-controlled crust and olivine-controlled mantle and slow (0.2 mm a^{-1}) shortening rate during 300 Ma. Figure 22 demonstrates the results that explored the possibility of folding related to onset of the Alpine compression (initial thermotectonic age 250 Ma), at 'faster' shortening rate (0.5 mm a^{-1}). Both figures demonstrate the development of pronounced folding over a NW transect encompassing the Paris Basin and the Armorican Massif, with the difference that faster folding promotes a shorter wavelength. This difference in the wavelength results from different stress levels associated with different strain rates and different rates of rheological hardening due to the thermal cooling of the lithosphere. The uncertainties in the initial thermal structure and rheology leave sufficient freedom for about 50% wavelength variations; that is about 50% shorter wavelength could be obtained for the case of Figure 21 and 50%

longer one for the case of Figure 22. The quartz–olivine rheology used in the experiments is the same as in Cloetingh *et al.* (1999) (Table 2) and infers a weak lower crust. Folding of a lithosphere with such a mechanically weak lower crust (of which there are a number of examples in central Asia; e.g. Burov & Molnar 1998) usually can be characterized by two characteristic wavelengths. The shorter one would correspond to crustal folding (30–60 km), and the larger one (200–400 km) could correspond to mantle folding (Cloetingh *et al.* 1999). Yet at small shortening rates, the associated crustal stresses (50 MPa) are insufficient for strong crustal–mantle decoupling. Consequently, for slow deformation, whole lithospheric folding may be expected. Indeed, Figures 21 and 22, presenting results for the initial thermotectonic ages of 50 and 250 Ma, demonstrate development of a single large wavelength of 200–400 km, yet with shorter wavelength components associated with localized zones of crust–mantle decoupling. This experiment demonstrates good correspondence with the Biot (1961) theory. Surface processes like erosion and sedimentation significantly prolong the lifetime of folding, since they decrease the effect of gravity by (1) filling the downward flexed basins and thus reducing the restoring force, and (2) cutting the upward flexed basement and thus unloading the lithosphere in the uplifted areas.

Discussion

The numerical experiments show the possibility of development of large-scale lithospheric folding in the Paris Basin, either initiated at early stages of basin formation 300 Ma ago or associated with

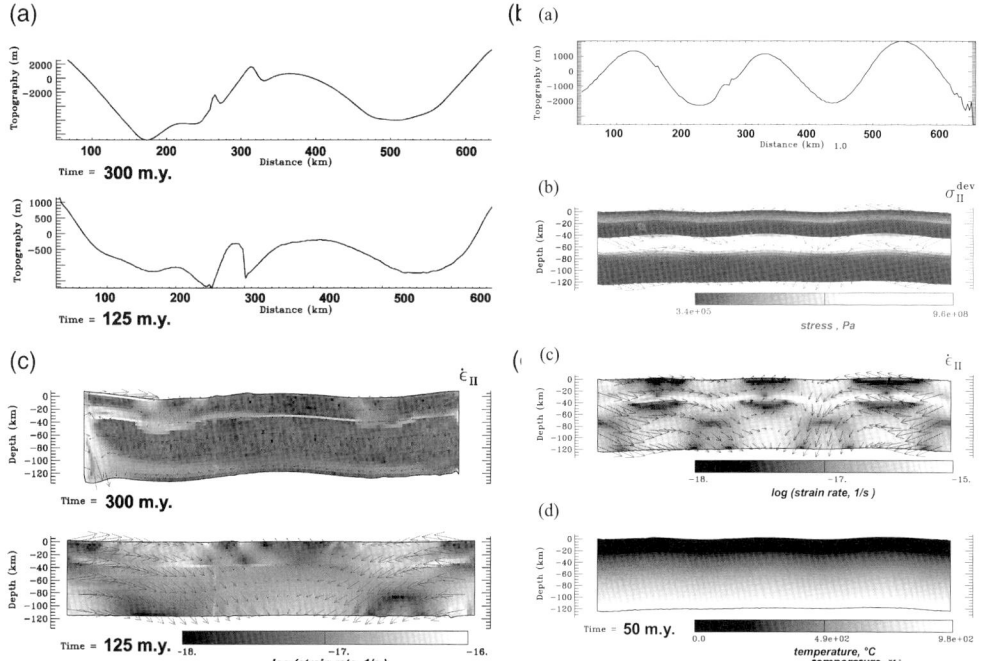

Fig. 21. Experiments on very slow but long symmetric shortening of initially 50 Ma old lithosphere (shortening rate 0.2 mm a^{-1} on each side of the box). Scale-dependent coefficient of erosion 200 m^2 a^{-1}. Note characteristic wavelength of coupled crust–mantle folding and zones of partial crust–mantle decoupling leading to a shorter wavelength component (see text). (**a**) Topography at 125 Ma and 300 Ma after onset of shortening. (**b**) Shear stress invariant and velocity field at 125 Ma and 300 Ma after onset of shortening. (**c**) Strain rate field and velocity field at 125 Ma and 300 Ma after onset of shortening. (**d**) Temperature field at 125 Ma and 300 Ma after onset of shortening.

Fig. 22. Experiments on 'fast' symmetric shortening of a 250 Ma old lithosphere (shortening rate 0.5 mm a^{-1} on each side of the box). Scale-dependent coefficient of erosion 200 m^2 a^{-1}. (**a**) Topography; (**b**) shear stress and velocity field; (**c**) strain rate and velocity field; (**d**) temperature field. Note characteristic wavelength of coupled crust–mantle folding and zones of partial crust–mantle decoupling leading to a shorter wavelength components in some places (see text).

Table 2. *Rheological parameters adopted for the numerical model*

Mineral/rock	A (Pa^{-n} s^{-1})	H (kJ mol^{-1})	n
Quartzite (dry)	5×10^{-12}	190	3
Diorite (dry)	5.01×10^{-15}	212	2.4
Diabase (dry)	6.31×10^{-20}	276	3.05
Olivine/dunite (dry)	7×10^{-14}	520	3

The ductile creep and elastic parameters are taken from Cloetingh *et al.* (1999). The elastic moduli used for all materials throughout this paper are E (Young's) modulus = 0.8 GPa and ν (Poisson's ratio) = 0.25. The brittle properties are represented by Mohr–Coulomb plasticity with friction angle 30° and cohesion 20 MPa (Gerbault *et al.* 1999).

Alpine collision about 50 Ma ago. In all cases the shortening rate used in the experiments (from 0.2 to 0.5 mm a^{-1}) was far below the accuracy of all possible field estimates (1 mm a^{-1}). The experiments demonstrate that extremely low (0.2 mm a^{-1}) shortening rates are largely sufficient to induce large scale/low amplitude lithospheric folding over 50–300 Ma time spans. Yet due to the impossibility of obtaining better constraints on the shortening rate as well as because the amplitude of the vertical deflection observed in the region is quite small, we admit that various mechanisms of basin subsidence are compatible with the data, including simple flexure due to the load by the Alpine system. Nevertheless, the periodic deformation at the same wavelength is more or less well expressed to the northwest of the area, which would not be the case if the northeastern part of the basin was simply flexed down by the load of the Alps. We have demonstrated that under condition of NW–SE Alpine compression, whole lithospheric folding should develop on the scale of the Paris Basin, since the extremely slow shortening rates used in the experiments are sufficient to initiate an unstable compressional deformation at a scale of the basin. The maximal values of tectonic stresses needed to activate this folding are low (depth averaged stress 50 MPa, magnified for about 10–20 times in localized zones; Figure 21), which suggests that Alpine compression is largely sufficient to activate this deformation. The amplitude of deformation during the Alpine compression period is, however, too small to preclude the possibility of other mechanisms. Thus it is not precluded that the same amplitude could have been achieved using a more common model of thermal subsidence. Consequently, only the periodicity of the deformation observed in gravity and stratigraphy data that demonstrated the clear physical possibility of folding matching the observed wavelengths, can be used as an argument in favour of the folding scenario. A more developed model to be produced in future could use additional information on fault geometry and distributions. Such a model could give a more decisive response to the problem of evolution of the Paris Basin.

Conclusion

The use of a stratigraphic database, constructed from the inspection of 1100 wells, correlated according to the principles of sequence stratigraphy, was the basis of three different types of numerical modelling in order to better constrain the thermomechanical evolution an the vertical motion of the intracratonic Paris Basin (Triassic to present day).

Both stratigraphic data and models revisited the Paris Basin history. The Meso-Cenozoic evolution of the Paris Basin can be described as a long-term thermal subsidence, inherited from the Permian extension and perturbed by intraplate deformation in reaction to the geodynamic events occurring in western Europe, i.e. opening and closing of the Ligurian Tethys and opening of the Atlantic. These tectonic events modified both subsidence and facies distribution in space and time. The Paris Basin is first an 'extensional' basin (Triassic–Jurassic) which progressively evolved to a compressional one, first only temporarily (lower Berriasian and late Aptian) and then permanently (late Turonian/present day).

The simulation of the lithospheric thermal cooling using the Chablis code (Doin & Fleitout 1996) and the fit with the tectonic subsidence curves suggest a high thermal regime from 300 to 200 Ma, with extension and topographic change going from mountain range (300 Ma) to flat plateau (280 Ma). This is consistent with the few geological data available for the Permian basins beneath the Paris Basin. This type of model was helpful for estimating Cenozoic uplift on the eastern part of the basin (toward the Rhinegraben), which could be around 2700 m (Vosges basement).

The measure of the relative tectonic displacement from stratigraphic data, and its application to the Middle Jurassic carbonate platforms, show two types of tectonic behaviour: (1) a localized deformation which is not correlated with sedimentation; and (2) a diffuse deformation correlated with a flat or positive topography and with high velocity of sedimentation. The first one is related to stress and strain applied at the boundary of the basin (intraplate deformations). The second one is due to the sediment load.

The use of mechanical models such as the Paravoz code (Poliakov et al. 1993) suggest a lithospheric folding origin for the Paris Basin from end Cretaceous to present day. The numerical experiments demonstrate that extremely low (0.2 mm a^{-1}) shortening rates are largely sufficient to induce large-scale low amplitude folding under low maximum values of tectonic stresses (c. 50 MPa). These values suggest that Alpine compression is largely sufficient to activate this deformation.

S. Cloetingh and H. Doust are thanked for their useful reviews of the manuscript.

References

ANDERLE H. -J. 1987. The evolution of the South Hunsrück and Taunus Borderzone. *Tectonophysics*, **137**, 101–114.

AUTRAN, A., LEFORT, J. P., DEBEGLIA, N., EDEL, J. B. & VIGNERESSE, J. L. 1994. Gravity and magnetic expression of terranes in France and their correlation beneath overstep sequences. *In*: KEPPIE J. D. (ed.), *Pre-Mesozoic Geology in France and Related Area*. Springer Verlag, Berlin, 49–72.

BACHMANN G. H., MÜLLER M. & WEGGEN K. 1987. Evolution of the Molasse Basin (Germany, Switzerland). *Tectonophysics*, **137**, 77–92.

BARBARAND, J., LUCAZEAU, F., PAGEL, M. & SERANNE, M. 2001. Burial and exhumation history of the southeastern Massif Central (France) constrained by an apatite fission-tract thermochronology. *Tectonophysics*, **335**, 275–290.

BENEK, R., KRAMER, W., MCCANN, T. et al. 1996. Permo-Carboniferous magmatism of the Northeast German Basin. *Tectonophysics*, **266**, 379–404.

BESSEREAU G., GUILLOCHEAU F. & HUC A. Y. 1995. Source rock occurrence in a sequence stratigraphic framework: the example of the Lias of the Paris Basin. *In*: HUC, A. Y. (ed.) *Studies in Geology*. American Association of Petroleum Geologists, 273–301.

BETZ D. , FUHRER F., GREINER G. & PLEIN E. 1987. Evolution of the Lower Saxony Basin, *Tectonophysics*, **137**, 127–170.

BIOT, M. A. 1961. Theory of folding of stratified viscoelastic media and its implications in tectonics and orogenesis. *Geological Society of America Bulletin,* **72**, 1595–1620.

BOURQUIN S. & GUILLOCHEAU F. 1996. Keuper stratigraphic cycles in the Paris Basin and comparison with cycles in other Peritethyan basins (German Basin and Bresse-Jura Basin). *Sedimentary Geology,* **105**, 159–182.

BOURQUIN S., GUILLOCHEAU F. & FRIEDENBERG R. 1995. High-resolution stratigraphy in the Triassic series of the Paris Basin: geodynamic implications. *Cuadernos de Geologia Iberica,* **19**, 337–362.

BOURQUIN, S., RIGOLLET, C. & BOURGES, P. 1998. High-resolution sequence stratigraphy of an alluvial fan-delta environment: stratigraphic and geodynamic implications – An example from the Keuper Chaunoy Sandstones, Paris Basin. *Sedimentary Geology,* **121**, 207–237.

BOUSQUET, R., GOFFÉ, B., HENRY, P., LE PICHON, X. & CHOPIN, C. 1997. Kinematic, thermal and petrological model of the Central Alps; Lepontine metamorphism in the upper crust and eclogitisation of the lower crust. *Tectonophysics,* **273**, 105–127.

BRAY, R. J., DUDDY, I. R. & GREEN, P. F. 1998. Multiple heating episodes in the Wessex Basin: implications for geological evolution and hydrocarbon generation. *In:* UNDERHILL, J. R. (ed.) *Development, Evolution and Petroleum Geology of the Wessex Basin.* Geological Society, London, Special Publications, **133**, 199–213.

BRUNET, M. F. 1981. *Etude quantitative de la subsidence du bassin de Paris.* Thesis, Université Pierre et Marie Curie, Paris VI.

BRUNET, M. -F. & LE PICHON, X. 1982. Subsidence of the Paris Basin. *Journal of Geophysical Research,* B., **87**, 8547–8560.

BURG J. P., VAN DEN DRIESSCHE J. & BRUN J. P. 1994*a*. Syn- to post-thickening extension: mode and consequences. *Comptes-Rendus de l'Académie des Sciences Paris,* Série II, **319**, 1019–1032.

BURG, J. P., VAN DEN DRIESSCHE, J. & BRUN, J. P. 1994*b*. Syn- to post-thickening extension in the Variscan Belt of Western Europe: Mode and structural consequences. *Géologie de la France,* **3**, 33–51.

BUROV E. B. & MOLNAR P. 1998. Gravity anomalies over the Ferghana valley (central Asia) and intracontinental deformation, *Journal of Geophysical Research,* **103**, 18137–18152.

BUROV, E.B., LOBKOVSKY, L.I., CLOETINGH, S. & NIKISHIN, A.M. 1993. Continental lithosphere folding in Central Asia (part 2), constraints from gravity and topography. *Tectonophysics,* **226**, 73–87.

BUROV, G. & CLOETINGH, S. 1997. Erosion and rift dynamics: new thermomechanical aspect of post-rift evolution of extensional basin. *Earth and Planetary Sciences Letters,* **150**, 7–26.

CARASCO, B. 1987. *Les grabens Stéphano-permiens de l'est de la France: évolution tectonosédimentaire, développement des faciès lacustres carbonatés et sapropéliques.* Thesis, Université Louis Pasteur, Strasbourg.

CARRON, J. P., LE GUEN DE KERNEIZON, M. & NACHIT, H. 1994. Variscan granites from Brittany. *In:* KEPPIE, J. D. (ed.) *Pre-Mesozoic Geology in France and Related Areas.* Springer-Verlag, Berlin, 231–239.

CAZES, M. & TORREILLES, G. 1987. *Etude de la crote terrestre profonde; profil Nord de la France; structures hercyniennes.* Technip, Paris.

CAZES, M., TOREILLES, G., BOIS, C. *et al.* 1985. Structure of the Hercynian crust of Northern France; first results of the ECORS profile. *Bulletin de la Société géologique de France,* **6**, 925–941.

CHANTRAINE J., AUTRAN A. & CAVELIER C. 1996. *Carte géologique de la France à 1/1 000 000.* Bureau de Recherches Géologiques et Minières, Service Géologique National, Orléans.

CLOETINGH, S. & BUROV, E. 1996. Thermomechanical structure of European continental lithosphere: constraints from rheological profile and EET estimates. *Geophysical Journal International,* **124**, 695–723.

CLOETINGH, S., BUROV, E. & POLIAKOV A. 1999. Lithosphere folding: primary response to compression? (from central Asia to Paris Basin). *Tectonics,* **18**, 1064–1083.

CUNDALL, P. A. 1989. Numerical experiments on localization in frictional materials. *Ingenieur-Archiv,* **59**, 148–159.

DAVAILLE, A. & JAUPART, C. 1994. Onset of thermal convection in fluids with temperature-dependent viscosity: application to the oceanic mantle. *Journal of Geophysical Research,* **99**, 19853–19866.

DEMOULIN, A. 1989. Les transgressions oligocènes sur le Massif Ardenne-Eifel. *Annales de la Société Géologique de Belgique,* **112**, 215–224.

DOIN, M. P. & FLEITOUT, L. 1996. Thermal evolution of the oceanic lithosphere; an alternative view. *Earth and Planetary Sciences Letters,* **142**, 121–136.

DOIN, M. -P., FLEITOUT, L. & CHRISTENSEN, U. 1997. Mantle convection and stability of depleted and undepleted continental lithosphere. *Journal of Geophysical Research,* B, **102**, 2771–2787.

DORÉ, F. 1994. The Variscan Orogeny in the Armorican Massif; Permian of the Armorican Massif. *In:* KEPPIE, J. D. (ed.) *Pre-Mesozoic Geology in France and Related Areas.* Springer-Verlag, Berlin, 169–171.

DUMOULIN, C. 2000. *Convection mantellique et structure de la lithosphère.* Thesis, Université Paris Sud-Orsay.

DUMOULIN, C., DOIN, M. -P. & FLEITOUT, L. 1999. Heat transport in stagnant lid convection with temperature- and pressure-dependent Newtonian or non-Newtonian rheology. *Journal of Geophysical Research,* **104**, 12759–12778.

DUMOULIN, C., DOIN, M. -P. & FLEITOUT L. 2001. Numerical simulations of the cooling of an oceanic lithosphere above a convective mantle. *Physics of the Earth and Planetary Interior,* 125, 45–65.

ENAY, R., MANGOLD, C., CARIOU, E. *et al.* 1980. *Synthèse paléogéographique du Jurassique franNais.* Documents des Laboratoires de Géologie de Lyon, Hors Série, **5**.

FAURE, M. 1995. Late orogenic Carboniferous extensions in the Variscan French Massif Central. *Tectonics,* **14**, 132–153.

FAURE, M. & BECQ-GIRAUDON, J. F. 1993. The succession of extensional episodes during the Carboniferous crustal thinning of the French Central Massif. *Comptes-Rendus de l'Académie des Sciences Paris,* Série II, **316**, 967–973.

GARDIEN, V., LARDEAUX, J.-M., LEDRU, P., ALLEMAND, P. & GUILLOT, S. 1997. Metamorphism during late orogenic extension; insights from the French Variscan belt. *Bulletin de la Société géologique de France,* **168,** 271–286.

GERBAULT, M., BUROV, E. B., POLIAKOV A. & DAGNIERES, M. 1999. Do faults trigger folding in the lithosphere? *Geophysical Research Letters,* **26,** 271–274.

GOGGIN V., JACQUIN T. & GAULIER J. -M. 1997. Three dimensional accomodation analysis of the Triassic in the Paris Basin: a new approach in unravelling the basin evolution with time. *Tectonophysics,* **282,** 205–222.

GRADSTEIN, F. M., AGTERBERG, F. P., OGG, J. G., HARDENBOL, J., VAN VEEN, P., THIERRY, J. & HUANG, Z. 1994. A Mesozoic time scale. *Journal of Geophysical Research,* B, **99,** 24051–24074.

GUILLOCHEAU, F. 1991. Mesozoic large-scale transgressive-regressive cycles of tectonic origin in the Paris Basin. *Comptes-Rendus de l'Académie des Sciences Paris,* Série II, **312,** 1587–1593.

GUILLOCHEAU, F., ROBIN, C., ALLEMAND, P. *et al.* 2000. Meso-cenozoic geodynamic evolution of the Paris Basin: 3D stratigraphic constraints. *Geodinamica Acta,* **13,** 189–246.

GUTSCHER, M. -A. 1995. Crustal structure and dynamics in the Rhine graben and the Alpine foreland. *Geophysical Journal International,* **122,** 617–636.

HAQ, B. U., HARDENBOL, J. & VAIL, P. R. 1987. Chronology of fluctuating sea levels since the Triassic. *Science,* **235,** 1156–1167.

HAQ, B. U., HARDENBOL, J. & VAIL, P. R. 1988. Mesozoic and Cenozoic chronostratigraphy and cycles of sea level change. *In:* WILGUS, C. K. *et al.* (eds) *Sea-level Changes: an Integrated Approach.* Society of Economic Paleontologists and Mineralogists, Boulder, Special Publications, **42,** 71–108.

HENK, A. 1993. Late orogenic basin evolution in the Variscan Internides; the Saar-Nahe Basin, Southwest Germany. *Tectonophysics,* **223,** 273–290.

HILLIS, R. R. 1991. Chalk porosity and tertiary uplift, Western Approaches trough, SW UK and NW French continental shelves. *Journal of the Geological Society, London,* **148,** 669–679.

HOMEWOOD, P., GUILLOCHEAU, F., ESCHARD, R. & CROSS, T. A. 1992. Corrélations haute résolution et stratigraphie génétique : une démarche intégrée. *Bulletin des Centres de Recherches Exploration-Production Elf-Aquitaine,* **16,** 357–381.

HOMEWOOD, P., MAURIAUD, P. & LAFONT, F. 1999. Best practices in Sequence Stratigraphy for explorationists and reservoir engineers. *Bulletin des Centres de Recherches Exploration-Production Elf-Aquitaine, Mémoires,* **25.**

JARVIS, J. T. & MCKENZIE, D. P. 1980. Sedimentary basin formation with finite extension rates. *Earth and Planetary Sciences Letters,* **48,** 42–52.

JERVEY, M. T. 1988. Quantitative geological modeling of siliclastic rock sequences and their seismic expressions. *In:* WILGUS, C. K., HASTINGS, B. S., KENDALL, C. G., POSAMENTIER, H. W., ROSS, C. A. & VAN WAGONER, J. C. (eds) *Sea-level Change: an Integrated Approach.* Society of Economic Paleontologist and Mineralogists, Boulder, Special Publications, **42,** 47–69.

KAMINSKI E. & JAUPART C. 2000. Lithosphere structure beneath the Phanerozoic intracratonic basins of North

America. *Earth and Planetary Sciences Letters,* **178,** 139–149.

KORSCH, R. J. & SCHÄFER, A. 1995. The Permo-Carboniferous Saar-Nahe Basin, South-West Germany and North-East France; basin formation and deformation in a strike-slip regime. *Geologische Rundschau,* **84,** 293–318.

KRUSE, S., MCNUTT, M., PHIPPS-MORGAN, J., ROYDEN, L. & WERNICKE, B. 1991. Lithospheric extension near lake Mead, Nevada: a model for ductile flow in the lower crust. *Journal of Geophysical Research,* **96,** 4435–4456.

LAKE S. D. & KARNER G. 1987. The structure and evolution of the Wessex Basin, southern England: an example of inversion tectonics. *Tectonophysics,* **137,** 347–378.

LAUBSCHER, H. P. 1987. Die tektonische Entwicklung der Nordschweiz. *Eclogae Geologicae Helviticae,* **80,** 287–303.

LEFORT, J. P. & AGARWAL, B. N. P. 1996. Gravity evidence for an Alpine buckling of the crust beneath the Paris Basin. *Tectonophysics,* **258,** 1–14.

LEFORT, J. -P. & JAFFAL, M. 1994. Les témoins de l'extension post-hercynienne dans la crote inférieure litée de la Manche occidentale. *Bulletin des Centres de Recherches Exploration-Production Elf-Aquitaine,* **18,** 421–436.

LLIBOUTRY, L. 1976. Isostasie, propriétés rhéologiques du manteau supérieur. *In:* COULOMB, J. & JOBERT, G. (eds) *Traité de Géophysique interne.* Masson, Paris, 473–505.

LOUP, B. 1992. Mesozoic subsidence and stretching models of the lithosphere in Switzerland (Jura, Swiss Plateau and Helvetic realm). *Eclogae Geologicae Helviticae,* **85,** 541–572.

LOUP, B. & WILDI, W. 1994. Subsidence analysis in the Paris Basin: a key to Northwest European intracontinental basins? *Basin Research,* **6,** 159–177.

LUCAZEAU, F. & VASSEUR, G. 1989. Heat flow density data from France and surrounding margins. *Tectonophysics,* **164,** 251–258.

LYON-CAEN H. & MOLNAR P. 1989. Constraints on the deep structure and dynamic processes beneath the Alps and adjacent regions from an analysis of gravity anomalies. *Geophysical Journal International,* **99,** 19–32.

MCKENZIE, D. 1978. Some remarks on the development of sedimentary basins. *Earth and Planetary Sciences Letters,* **40,** 25–32.

MALAVIEILLE, J. 1993. Late orogenic extension in mountain belts; insights from the Basin and Range and the late Paleozoic Variscan belt. *Tectonics,* **12,** 1115–1130.

MASCLE, A. 1990. Géologie pétrolière des bassins permiens franNais. Comparaison avec les bassins permiens du Nord de l'Europe. *Chroniques de Recherches Minières,* **499,** 69–86.

MEGNIEN C. (ed.) 1980. Synthèse géologique du Bassin de Paris, I, stratigraphie et paléogéographie. *Mémoires du Bureau de Recherches Géologiques et Minières,* **101.**

PARSONS, B. & MCKENZIE, D. P. 1978. Mantle convection and the thermal structure of the plates. *Journal of Geophysical Research,* A, **83,** 4485–4496.

PELHATE, A. 1994. The Variscan Orogeny in the Armorican Massif; Carboniferous of the Armorican Massif.

In: KEPPIE, J. D. (ed.) *Pre-Mesozoic Geology in France and Related Areas*. Springer-Verlag, Berlin, 162–168.

PERRODON, A. & ZABEK, J. 1990. Paris Basin. *In*: LEIGHTON, M. W., KOLATA, D. R., OLTZ, D. F. & EIDEL, J. J. (eds) *Interior Cratonic Basin*. American Association of Petroleum Geologists, Memoirs, **51**, 633–678.

PLOQUIN, A., BRIAND, B., DUBUISSON, G. *et al.* 1994. The Massif Central; igneous activity; Caledono-Hercynian magmatism in the French Massif Central. *In*: KEPPIE, J. D. (ed.) *Pre-Mesozoic Geology in France and Related Areas*. Springer-Verlag, Berlin, 341–378.

POLIAKOV, A. N. B., PODLADCHIKOV, YU. & TALBOT, C. 1993. Initiation of salt diapirs with frictional overburden: numerical experiments. *Tectonophysics*, **228**, 199–210.

PRIJAC, C., DOIN, M. -P., GAULIER, J. -M. & GUILLOCHEAU, F. 2000. Subsidence of the Paris Basin and its bearing on the late Variscan lithosphere evolution: a comparison between Plate and Chablis models. *Tectonophysics*, **323**, 1–38.

REY, P., BURG, J. P. & CASEY, M. 1997. The Scandinavian Caledonides and their relationship to the Variscan Belt. *In*: BURG, J. -P. & FORD, M. (eds) *Orogeny Through Time*. Geological Society, London, Special Publications, **121**, 179–200.

ROBIN, C. 1997. *Mesure stratigraphique de la déformation : Application à l'évolution jurassique du Bassin de Paris*. Mémoires Géosciences-Rennes, **77**, 293 pp. and Thèse, Université de Rennes 1 (1995).

ROBIN, C., GUILLOCHEAU, F. & GAULIER, J. M. 1996. Measurement of tectonic and eustatic signals from stratigraphic information within an intracontinental basin; application to the Lower Jurassic of the Paris Basin. *Comptes-Rendus de l'Académie des Sciences Paris*, Série II, **322**, 1079–1086.

ROBIN, C., GUILLOCHEAU, F. & GAULIER, J. M. 1998. Discriminating between tectonic and eustatic controls on stratigraphic record in the Paris Basin. *Terra Nova*, **10**, 323–329.

ROBIN, C., GUILLOCHEAU, F., ALLEMAND, P., BOURQUIN, S., DROMART, G., GAULIER, J. -M. & PRIJEAC, C. 2000. Echelles de temps et d'espace du contrôle tectonique d'un bassin flexural intracratonique: le bassin de Paris. *Bulletin de la Société géologique de France*, **171**, 181–196.

ROYDEN, L. & KEEN, C. E. 1980. Rifting process and thermal evolution of the continental margin of Eastern Canada determined from subsidence curves. *Earth and Planetary Sciences Letters*, **51**, 343–361.

RUDNICK, R. L. & FOUNTAIN, D. M. 1995. Nature and composition of the continental crust: a lower crustal perspective. *Reviews of Geophysics*, **33**, 267–309.

SCHÄFER, A. & KORSCH, R. J. 1998. Formation and sediment fill of the Saar-Nahe Basin (Permo-Carboniferous, Germany). *Zeitschrift der deutschen geologischen Gesellschaft*, **149**, 233–269.

SCHALTEGGER, U. 1997. Magma pulses in Central Variscan Belt; episodic melt generation and emplacement during lithospheric thinning. *Terra Nova*, **9**, 242–245.

SCHALTEGGER, U., SCHNEIDER, J. -L., MAURIN, J. -C. & CORFU, F. 1996. Precise U-Pb chronometry of 345–340 Ma old magmatism related to syn-convergence extension in the Southern Vosges (Central Variscan

Belt). *Earth and Planetary Sciences Letters*, **144**, 403–419.

SCHRODER, B. 1987. Inversion tectonics along the western margin of the Bohemian Massif. *Tectonophysics*, **137**, 93–100.

SCLATER, J. G. & CHRISTIE, P. A. F. 1980. Continental stretching; an explanation of the post-Mid-Cretaceous subsidence of the Central North Sea basin. *Journal of Geophysical Research*, **85**, 3711–3739.

SITTLER C. 1985. Les hydrocarbures d'Alsace dans le contexte historique et géodynamique du fossé rhénan. *Bulletin des Centres de Recherches Exploration-Production Elf-Aquitaine*, **9**, 335–371.

SOLOMATOV, V. S. & MORESI, L. -N. 2000. Scaling of time-dependent stagnant lid convection: Application to small-scale convection on Earth and other terrestrial planets. *Journal of Geophysical Research*, **105**, 21795–21817.

STEPHENSON, R. A. & CLOETINGH, S. 1991. Some examples and mechanical aspects of continental lithospheric folding. *Tectonophysics*, **188**, 27–37.

SUHADOLC, P., PANZA, G. F. & MUELLER, S. 1990. Physical properties of the lithosphere-asthenosphere system in Europe. *Tectonophysics*, **176**, 123–135.

TURNER, S., SANDIFORD, M. & FODEN, J. 1992. Some geodynamic and compositional constraints on "post-orogenic" magmatism. *Geology*, **20**, 931–934.

UNDERHILL, J. R. & STONELEY, R. 1998. Introduction to the development, evolution and petroleum geology of the Wessex Basin. *In*: UNDERHILL, J. R. (ed.) *Development, Evolution and Petroleum Geology of the Wessex Basin*. Geological Society, London, Special Publications, **133**, 1–18.

VALLÉ, B., COUREL, L. & GELARD, J. P. 1988. Synsedimentary and syndiagenetic tectonic markers in the shear regime of the Blanzy-Montceau Stephanian basin, Central Massif, France. *Bulletin de la Société géologique de France*, 8ème Série, **4**, 529–540.

VAN DEN DRIESSCHE, J. & BRUN, J. -P. 1989. Un modèle de l'extension paléozoïque supérieur dans le Sud du Massif Central. *Comptes-Rendus de l'Académie des Sciences Paris*, Série II, **309**, 1607–1613.

VAN HINTE, J. E. 1978. Geohistory analysis – application of micropaleontology in exploration geology. *AAPG Bulletin*, **62**, 201–222.

VAN HOORN, B. 1987. The South Celtic Sea / Bristol Channel Basin: origin, deformation and inversion history. *Tectonophysics*, **137**, 309–334.

VAN WAGONER, J. C., POSAMENTIER, H. W., MITCHUM, R. M., VAIL, P. R., SARG, J. F., LOUTIT, T. S. & HARDENBOL, J. 1988. An overview of the fundamentals of sequence stratigraphy and key definitions. *In*: WILGUS, C. K., HASTINGS, B. S., KENDALL, C. G. ST. C., POSAMENTIER, H. W., ROSS, C. A. & VAN WAGONER, J. C. (eds) *Sea-level Change: an Integrated Approach*. Society of Economic Paleontologists and Mineralogists, Special Publications, **42**, 39–45.

VAN WAGONER, J. C., MITCHUM, R. M. JR, CAMPION, K. M. & RAHMANIAN, V. D. 1990. Siliciclastic sequence stratigraphy in well logs, cores, and outcrops: concepts for high resolution correlation of time and facies. American Association of Petroleum Geologists, *Methods in Exploration Series*, **7**.

VAN WEES, J. D., ARCHE, A., BEIJDORFF, C.,G., LÓPEZ-

GÓMEZ, J. & CLOETINGH, P. L. 1998. Temporal and spatial variations in tectonic subsidence in the Iberian Basin (eastern Spain): inferences from automated forward modelling of high-resolution stratigraphy (Permian-Mesozoic). *Tectonophysics,* **300**, 285–310.

VIELZEUF, D. & PIN, C. 1989. Geodynamic implications of granulitic rocks in the Hercynian Belt. *In*: DALY, J. S., CLIFF, R. A. & YARDLEY, B. W. D. (eds) *Evolution of Metamorphic Belts*. Geological Society, London, Special Publications, **20**, 343–348.

VILOTTE, J. P., MELOSH, J., SASSI, W. & RANALLI, G. 1993. Lithosphere rheology and sedimentary basins. *Tectonophysics,* **226**, 89–95.

WEBER, C., HIRN, A., PHILIP, H. & ROCHE, A. 1980. Image géophysique de la France; Geophysics of France. *In*: Géologie de la France, *Bureau de Recherches Géologiques et Minières, 26ème Congrès de Géologie Internationale*, Colloque **C7**, 25–50.

WEISSEL, J. K. & KARNER, G. D. 1989. Flexural uplift of rift flanks due to mechanical unloading of the lithosphere during extension. *Journal of Geophysical Research,* **94**, 13919–13950.

YUEN, D. A. & FLEITOUT, L. 1984. Stability of the oceanic lithosphere with variable viscosity: an initial value approach. *Physics of the Earth and Planetary Interior,* **343**, 173–185.

ZIEGLER, P. A. 1982. Triassic rifts and facies patterns in Western and Central Europe. *Geologische Rundschau,* **71**, 747–772.

ZIEGLER, P. A. 1987a. Celtic Sea-Western Approaches area: an overview. *Tectonophysics,* **137**, 285–289.

ZIEGLER, P. A. 1987b. Evolution of the Western Approaches Trough. *Tectonophysics,* **137**, 341–346.

ZIEGLER, P. A. 1987c. Late Cretaceous and Cenozoic intra-plate compressional deformations in the Alpine foreland – a geodynamic model. *Tectonophysics,* **137**, 389–420.

ZIEGLER, P. A. 1992a. European Cenozoic rift system. *Tectonophysics,* **208**, 92–111.

ZIEGLER, P. A. 1992b. Geodynamics of rifting and implications for hydrocarbon habitat. *Tectonophysics,* **215**, 221–253.

Modelling and processing of 3D seismic data collected over the overlapping spreading centre on the East Pacific Rise at 9° 03′ N

R. HOBBS, C. H. TONG & J. PYE

Bullard Laboratories, Department of Earth Sciences, Madingley Road, Cambridge, CB3 0EZ, UK

Abstract: Oceanic ridges are important as the locus for the generation of new crust associated with the movement of plates on the Earth's surface. The geometry of the axial magma chamber (AMC) under these ridges is poorly constrained. This is especially true for the AMC under an overlapping spreading centre (OSC) where the ridge undergoes a small lateral offset, typically less than 8 km. One such OSC exists on the East Pacific Rise at 9° 03′ N. Both 2D and more recently 3D seismic reflection surveys have been conducted over the area in an attempt to establish the relationship of the AMC with the two overlapping limbs of the axial rise. A major problem for imaging the AMC is caused by the seafloor, which has a large velocity contrast and has rapid changes in topography. This presents a major challenge for the processing of data. This paper aims to show that the understanding the effects of acquisition and processing can be facilitated by accurate 3D modelling of the seismic wavefield. In this paper we use a modelling method based on complex-screens, which is both robust and fast for large 3D models with deep water and diffractive interfaces. We compare synthetics with real data and explore the sensitivity of processing 3D data from shot gather to final 3D depth-migrated image. We show that the combination of severe seabed topography and 3D velocity variation within the oceanic crust causes distortion and complex focusing of the reflected energy. Also, for data with only sparse reflectivity it is not possible to construct an adequate velocity model for depth imaging from the reflection data alone. This problem may be overcome if there is an alternative method to determine velocity, e.g. refraction data, but care must be exercised during interpretation given the different resolutions of the two datasets. Lessons from the modelling exercise can be used to aid the interpretation of the real data and examples are shown for comparison; however, the geological implications of the results are outside the scope of this paper.

In 1985 a two-ship experiment shot a series of 2D profiles over the East Pacific Rise between 8° 50′ N and 13° 30′ N (Derick *et al.* 1987). This comprehensive 2D program included reflection profiles both along and across strike (Kent *et al.* 1993). More recently, the overlapping spreading centre (OSC) at 9° 03′ N has been the subject of a 3D reflection and refraction survey (Singh *et al.* 1999; Kent *et al.* 2000) which included the collection of detailed swath bathymetry over a 23 × 20 km box centred over the OSC (Fig. 1). The data and results obtained from this survey (Tong 2000) form the basis of this paper.

The ridge between major fracture zones is segmented by a series of overlapping spreading centres such as the OSC described here. The classic model for melt supply to a ridge segment is in the centre of a segment with attenuated supply to the segment ends (Macdonald *et al.* 1988). An alterna-

tive view by Lonsdale (1983) suggests that the whole OSC may be underlain by a single large magma body. A review of the current understanding of the OSC can be found in Kent *et al.* (2000). They recognized that the axial magma chamber (AMC) at 9° 03′ N on the East Pacific Rise was apparently skewed with an event extending from the northeastern limb westwards beneath the central basin of the OSC; the extension of the AMC into the OSC basin is not seen for the southwestern limb. They explained this asymmetry by a model where in the north the melt is being extracted from a large zone that extends beneath the OSC whereas to the south the melt is being extracted from a narrower zone directly beneath the axial rise. Recent work by Collier & Singh (1997) shows that the magma body may only be about 100 m thick even in sections of the AMC with particularly strong reflections. To date, interpretations have been larg-

From: NIEUWLAND, D. A. (ed.) *New Insights into Structural Interpretation and Modelling.* Geological Society, London, Special Publications, **212**, 251–259. 0305-8719/03/$15

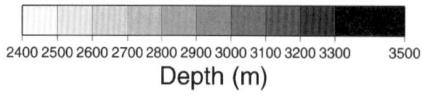

2400 2500 2600 2700 2800 2900 3000 3100 3200 3300   3500
Depth (m)

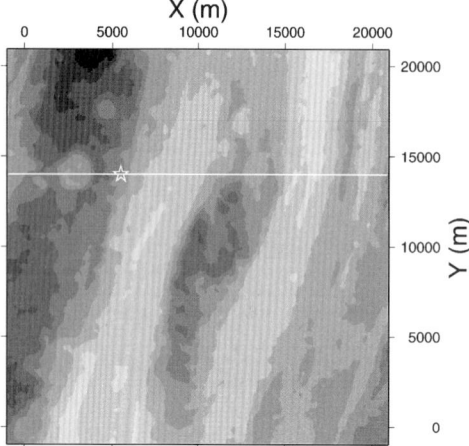

Fig. 1. Bathymetry map of the area showing the two axial ridges of the overlapping spreading centre on either side of the overlap basin. The white line shows the location of the seismic sections shown in subsequent figures. The white star shows the location of the shot point for the modelled shot gathers shown in Figure 5 and the location of the CMP gathers shown in Figure 6.

ely based on time-migrated sections and velocity models derived from 2D refraction surveys with widely spaced 2D ocean bottom seismometers (OBS). We will show that knowledge of the 3D velocity structure is crucial to advancing our understanding of the interaction of the AMC with the axial rise and we will test the sensitivity of the imaging process with 2D and 3D synthetic data.

The real data

The 3D data were shot in 1997 (Singh *et al.* 1999) using the academic research vessel *RV Maurice Ewing*. The survey was acquired using a single source with a single 3 km hydrophone cable over a 20 km × 23 km area with a cross-line spacing of 100 m and an in-line shot spacing of 37.5 m; the 3.1 km receiver had a group length of 25 m. In the in-line direction the CMP bin spacing was 12.5 m with the wavefield fully sampled up to frequencies of 60 Hz for propagation velocity of 1500 m s^{-1}, whereas the CMP spacing of 100 m in the cross-line direction means that frequencies over 7.5 Hz are aliased. A major issue was how to define the stacking velocity field. The CMP gathers were

dominated by incoherent scattered energy with only the occasional pickable event from the AMC. Standard semblance-based velocity analysis could not be used. The chosen strategy used constant velocity stacks (CVS) which enabled the picking of stacking velocity through the whole data volume where primary events could be recognized. These were then smoothed and interpolated to provide a velocity model for the whole volume. This method could only give a crude velocity model, but the depth of target meant that error in the stacking velocity field would not have a significant effect on the result (McBride *et al.* 1993). Examples of time migrated images from the data volume can be seen in Kent *et al.* (2000).

Whilst shooting the 3D normal incidence reflection data, wide-angle and refraction data were recorded on a grid of 19 OBS placed within the survey area. These were inverted independently by two groups (Bazin *et al.* 2001; Tong 2002) using different techniques to obtain a p-wave velocity model. The results are consistent and show that the velocity structure beneath the seafloor is variable, though there is some broad correlation with bathymetry (Fig. 2). This OBS velocity model is far more robust than the one derived from the stacking velocity determined from the reflection data given the problems mentioned above. The vertical resolution of the OBS velocity model is 200 m (Tong 2000). The horizontal model is smoothed over 2000 m within the inversion method to control short wavelength structure that could not be constrained by the recorded data. These effects combined with realistic picking errors equate to a possible error of about 10% in the sub-seafloor velocity model. Refraction data cannot constrain the velocity or thickness of low velocity layers in the subsurface but if such zones are present and have a thickness greater than the vertical resolution then these will be observed in the data as a offset dependent discontinuities in the recorded data. This effect was not observed on the OBS data hence we believe the model derived by Tong (2000) is reliable and represents the long wavelength velocity structure of this region.

Modelling

As the correct result for the real data is unknown, the method of analysing modelled data was chosen to investigate the potential sensitivity of the final 3D reflection image. The modelling program used is based on the Complex Elastic Screen (Wu 1994; Wild & Hudson 1998). In principle, the model is represented by a series of thin slabs with constant p- and s-wave velocities. For a heterogeneous volume this assumption is wrong but provided the error is small, i.e. the wave has only locally under-

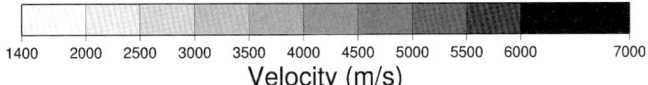

Velocity (m/s)

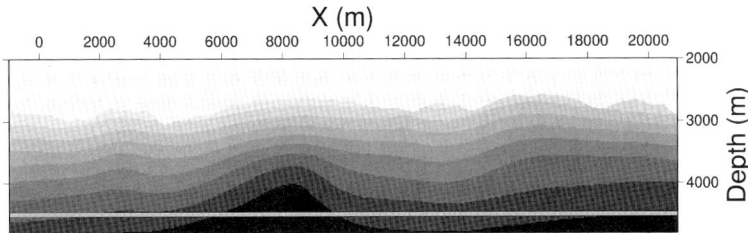

Fig. 2. A vertical cross-section of the velocity cube along the line marked on Figure 1. The model is constructed from swath bathymetry measurements combined with the 3D velocity model derived from ocean bottom seismometers. The low velocity layer at 4500 m has been added to the model to represent the axial magma chamber. Here this body extends across the model but on the real data it will be restricted to the regions below the axial ridges and possibly the central basin.

gone a single scattering event, this error can be minimized by the application of appropriate corrections on the output face of the slab. These corrections advance or delay the wavefield depending on whether the actual local velocity is higher or lower than the constant value.

So for a homogeneous layer, like the sea water column, a single slab will suffice, though it is beneficial to have a small number of intermediate steps to control the effects of Fourier domain wrap-around caused by transforms to and from the wave-number domain. However for the seabed, layers can be set at very close intervals to accurately describe the 3D surface. The optimum spacing is a function of complexity and maximum frequency to be contained in the synthetic seismograms. Inherent in the modelling method is the concentration of computational effort in the regions of most complexity. The source wavefield is moved through the model by applying the appropriate phase shift for any given frequency in the complex 2D wavenumber domain. Local corrections, described above, to wavefield are applied in the spatial domain on the output face of the slab. Also at this stage a small-angle approximation to the Zeoppritz equations are applied to any location where the elastic properties between the input and output faces of the slab have changed. The reflected component is stored whilst the transmitted component is moved onwards through the model. On completion of the forward transmission, the program works from the base of the model collecting the stored reflected energy at each screen and summing it into the backscattered wavefield as it is propagated back to the surface. The program

completes the simulation by outputting the total backscattered wavefield at predefined receivers. The critical part of the computer code is the efficiency of the 2D Fourier transforms used to flip the wavefield between wave-number and spatial domains. We have found a self-optimizing FFT routine (Frigo & Johnson 1997) gives excellent results. The code includes all p- and s-wave components and the conversions between them up to an angle that is 80% of the local critical angle. The code cannot model turning rays. So the event labelled as layer 2a by Kent *et al.* (2000) which is caused by refraction in the high velocity gradient in the near surface, is missing. The code also includes a 1D local approximation for attenuation.

The model

As described above the model is broken up into a series of slabs. Four slabs span the water column down to the highest point of the seabed, which is 2.4 km below the sea surface. The seabed topography was obtained from the swath bathymetry and the sub-seafloor p-wave velocity model from Tong (2000). At a depth of 4.5 km a low velocity zone 100 m thick was imposed over the whole model with a velocity of 3 km s^{-1}, corresponding to an AMC (Collier & Singh 1997) (Fig. 2). S-wave velocities were calculated from the p-wave velocities using the formula, $v_s = 0.86v_p$, except in the water layer where the s-wave velocity was set to 0.0 km s^{-1}. A simple model was chosen for the density: the density was fixed at 1000 kg m^{-3} for the water layer and for the subsurface $\rho = 0.2v_p + 1300.0$ kg m^{-3}. For attenuation, the p-wave velocities were grouped into four ranges: for velocities below 1.5 km s^{-1} (water) $Q_p = \infty$; for velocities in the range 1.5 to 5.0 km s^{-1} (upper crust and AMC) $Q_p = 75$; for velocities in the range 5.0 to 7.0 km s^{-1} (mid-crust) $Q_p = 100$; and for velocities greater than 7.0 km s^{-1} $Q_p = 500$. Q_s values where set to $Q_p/3$.

This 1D model of the AMC extending across the whole volume is unrealistic as it is expected that the AMC does not extend laterally outside the immediate vicinity of the axial rise and OSC. However, the purpose of this exercise is to test the imaging of an AMC rather than generate an exact model to fit the data. A vertical slice through the final model is shown in Figure 2, and Figure 3 shows a number of horizontal slices through the model showing the complexity of the seafloor and the p-wave velocity model.

The modelling program was run in two modes. In the first mode, the code was set to simulate the collection of reflection data along the profile marked on Figure 1, with the same acquisition parameters as for the real data. Three runs were made: one using the full 3D model; a second using a 2D model based on the velocity structure directly beneath profile; and a third with a 1D model based on the velocity structure directly beneath the source-point. In the second mode, the code was set to perform exploding reflector simulations, once for the whole 3D volume and then for the single 2D profile used above for the shot simulation. For the whole 3D simulation the receiver spacing was set to simulate the CMP geometry of the real data with a trace spacing of 12.5 m in-line and 100 m cross-line. For the 2D line the traces were spaced at 12.5 m. The source function for all runs was a zero-phase Ormsby wavelet, with a bandwidth of 4.5 to 25 Hz.

Results: real data

Figure 4a shows an unmigrated stacked profile taken from the 3D volume; the AMC reflector can be seen as the low frequency event starting under the OSC basin (X = 13 000 m, time 4.7 s) and apparently rising up beneath the eastern limb of the axial ending (X = 17 000 m, time 4.3 s). The signal-to-noise ratio of these data is poor and the event is emergent out of the noise at both ends. Inspection of the profile shows the AMC reflection disappears beneath the area of most rapid bathymetry change (X = 14 000–15 500 m). Figure 4b shows the same after 2D post-stack time migration using the velocity model from the OBS data. Rather than resolving the geometry, the time migration has exaggerated the distortion caused by the bathymetry change making the event narrow and disjointed. Figure 4c shows the 2D post-stack depth migrated

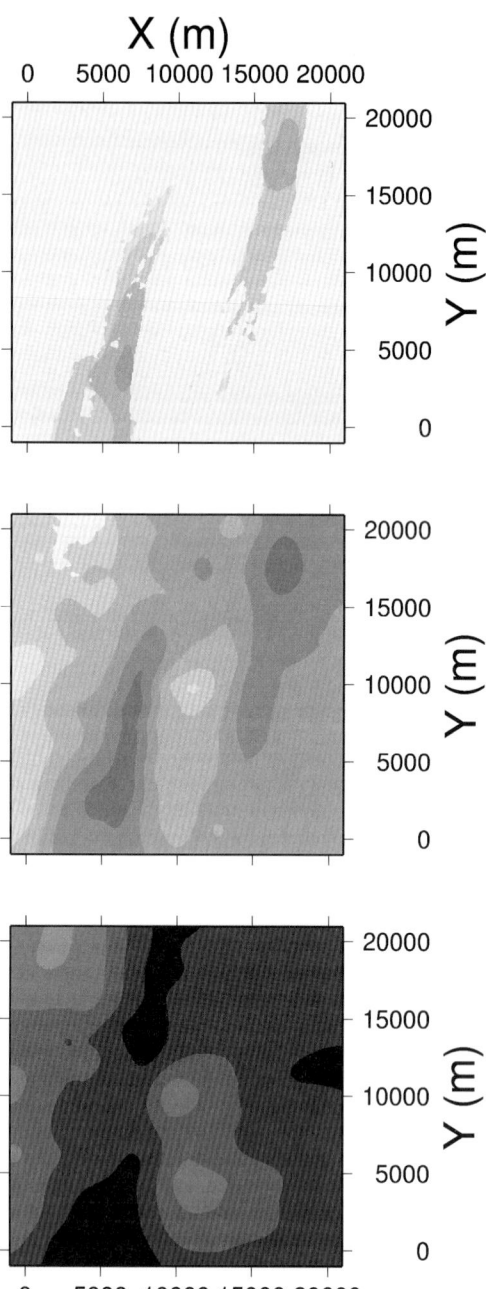

Fig. 3. Three slices through the model cube at the depths of 2.64, 3.2 and 4.4 km, showing the relationship between velocity and bathymetry. Velocity scale as in Fig. 2.

image, again using the velocity model from the OBS data. In this image the AMC centred around 4.5 km depth, looks flatter and extends further west than suggested by the time migration. There is still

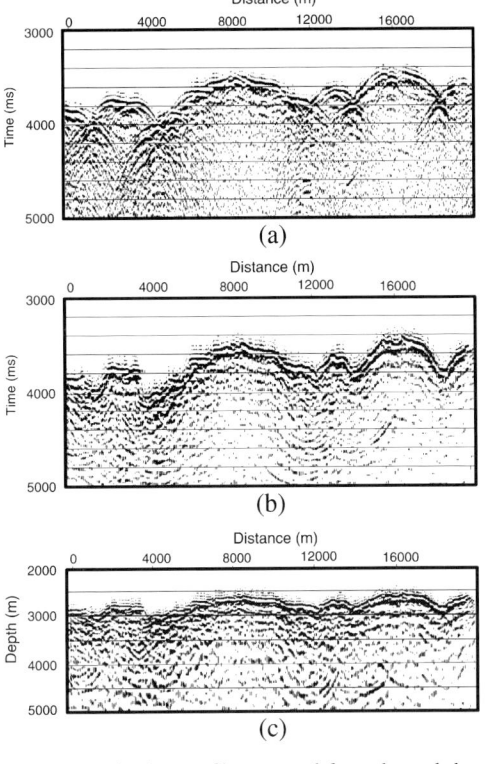

Fig. 4. Seismic data profile extracted from the real data volume along the line marked in Figure 1: (**a**) DMO stacked data; (**b**) with time migration using the velocity field derived from the OBS; (**c**) with depth migration using the same velocity field.

a sharp crest of about 300 m between the two segments. 'Smiles' can be seen on both migrations. The area under the OSC basin is contaminated by high amplitude disjoint reflections (Fig. 4a) possibly caused by interference with out-of-plane energy. The 'smiles' on the AMC under the axial ridge event suggest that the velocities are too high but as shown below in the synthetic model it is possible that the velocities in this area are actually too low. A local increase in velocity in the crust above the AMC would tend to depress the sharp crest and make the event more laterally continuous. Another cause for these 'smiles' is the loss of signal from the AMC at the edge of the ridge. The migration operator expects a diffraction where there is a truncation of energy. When this diffraction is missing, due to attenuation, the migration operator will generate a 'smile'. The geometry of the AMC is one of the key issues that motivated the project so it is critical to determine whether this apparent structure is real or an artefact within the likely errors of the experiment.

Results: synthetic model

Pre-stack data

A single gather with the same source point was selected from each of the three shot simulations (Fig. 5); the location is marked in the map (Fig. 1). The 1D approximation yields the expected result with two hyperbolic events, one from the seafloor and the other from the low velocity 'AMC' reflector. Measurements of moveout give the expected stacking velocities as those derived from the input model. The loss of amplitude in the seafloor reflection at the larger offsets is an artefact of the modelling code that is starting to suppress energy that is beyond 80% of critical angle. The 2D simulation produces the most complex gather. The overall seafloor event is non-hyperbolic due to a regional slope in the vicinity of the gather and is locally distorted by a step of about 200 m (260 ms) between -500 to -1500 m offset. This event also generates a diffraction tail which dips back towards the nearer offsets at 4250 ms. The 'AMC' reflection is severely distorted in both travel time and amplitude, again principally caused by the regional and local perturbations in the seafloor. In general, the 3D simulation shows the same character as the 2D simulation but the diffraction event is not as clear and the overall frequency content is lower. These phenomena are explained by the fact that the 3D wavefront additionally samples the reflection surface out of the plane of the profile so the effects of local roughness are averaged out. The corresponding CMP gathers (Fig. 6) correct some of the distortion effects. The seafloor on the 2D and 3D simulations can now be modelled by a hyperbola. The 'AMC' event is still severely broken up in the 2D and 3D simulations and has a distorted amplitude versus offset characteristic when compared to the 1D simulation. The strong diffraction on seen in the 2D shot gather now appears as a hyperbolic event.

The 3D simulation data were sorted into the CMP domain and passed through the same CVS analysis as the real data. Using the criteria of maximum amplitude, the optimum stacking velocity field was picked and interpolated onto a regular grid (Fig. 7). When compared with the input model (Fig. 2) there is little similarity; this observation shows that for areas with a complex velocity structure stacking velocities are not suitable for imaging. It is interesting to note that the velocity fluctuation in the model is, in general, anti-correlated to the bathymetry. This is because of the 1D approximation used in velocity analysis. For example, consider a CMP with its centre point located over a bathymetric low: as one moves to larger offsets, a reflection from a sub-seafloor

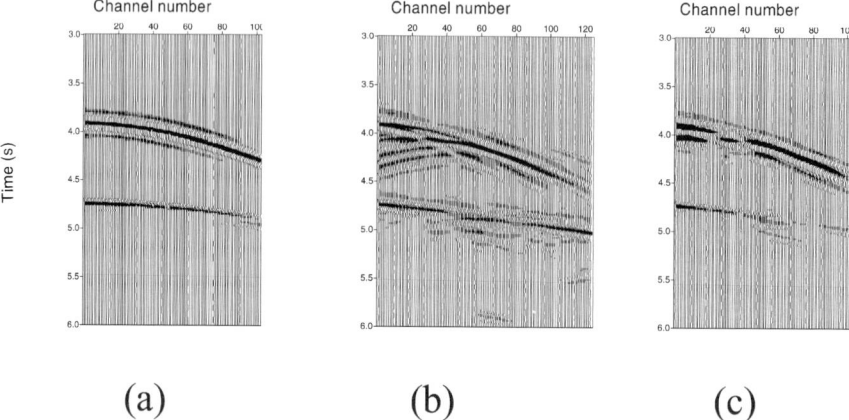

Fig. 5. Three simulations of shot data using the complex-screen modelling code. The shot point is marked on Figure 1: (**a**) result assuming a 1D velocity structure taken from directly under the shot-point; (**b**) result assuming a 2D velocity structure in the the plane of the shot gather; (**c**) result using the full 3D model including out-of-plane scattered energy.

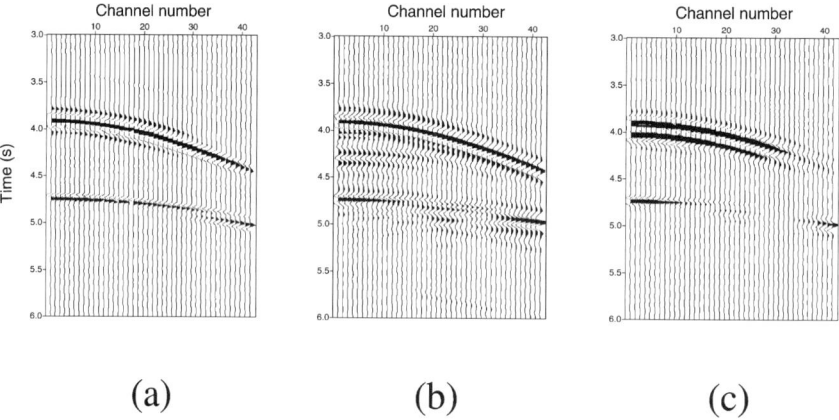

Fig. 6. CMP gathers at the location marked on Figure 1: (**a**) result assuming a 1D velocity structure taken from directly under the CMP; (**b**) result assuming a 2D velocity structure in the plane of the profile; (**c**) result using the full 3D model. The seafloor is hyperbolic in all three simulation but note the decreasing resolution with higher dimension. Also note the distortion at the axial magma chamber reflection in both amplitude and travel time caused by the propagation path; these effects are particularly marked in the 2D simulation including a spurious event at 4.3 s which originates from a diffraction seen on the 2D shot gather shown in Figure 5b.

reflector will travel through higher velocity material than predicted by the 1D assumption made in velocity analysis, so the moveout is reduced and the apparent stacking velocity is increased. The converse is true for the bathymetric highs.

Post-stack data

The stacked synthetic data (Fig. 8a) can be compared to the real data (Fig. 4a). The seafloor is similar as are the geometry and amplitude behaviour of the 'AMC' reflector. Except on the real data the AMC reflector is restricted to the eastern limb

and occurs about 100 ms earlier beneath the axial rise. Making the synthetics agree more closely to the real data requires either the 'AMC' reflector under the axial rise to be shallower by about 200 m or that there is a local increase in velocity by about 10% between the seafloor and the AMC reflector under the axial rise.

Time migration is used where there is no robust velocity model (Yilmaz 1987). Though it collapses diffraction hyperbolae and moves events up-dip, time migration does not accurately reconstruct the true geometry of the reflectors except for the special cases where the velocity is constant or the

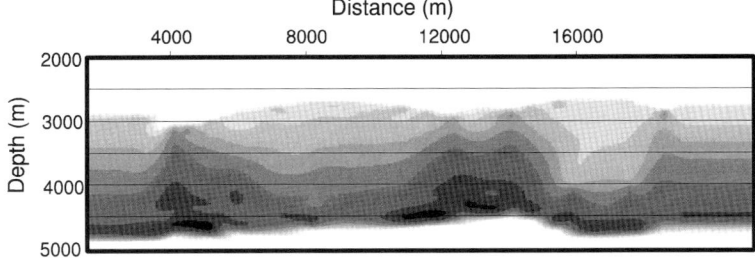

Fig. 7. Interval velocity versus depth derived from the stacking velocity field. When compared to Figure 2 the velocities shown here are lower and are anti-correlated with bathymetry whereas the velocities derived from the OBS data are correlated with bathymetry. This is an expected outcome that arises from the 1D approximation made in the normal moveout correction. Velocity scale as in Fig. 2.

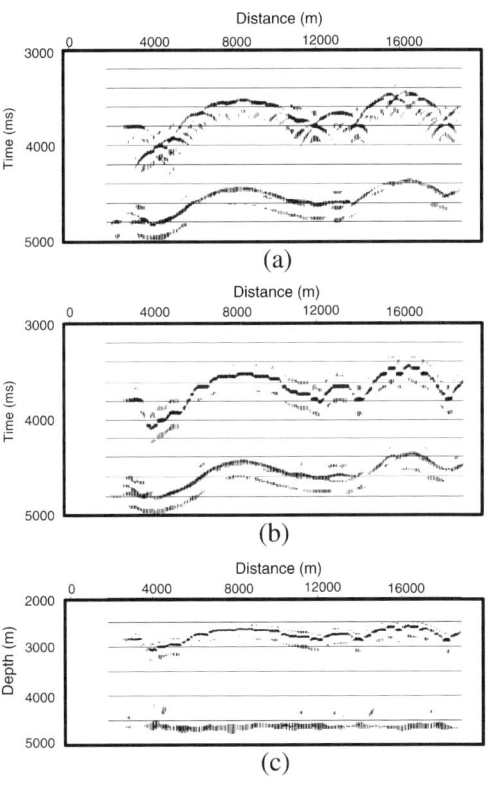

Fig. 8. DMO stacked modelled data: (**a**) raw stack; (**b**) with time migration using the velocity field derived from the OBS; (**c**) with depth migration using the same velocity field. The depth migration has correctly removed the travel-time distortion from the axial magma chamber reflection at 4500 m. However, the migration has not recovered the correct amplitudes leaving significant variations particularly in regions where the seafloor is rough or steeply sloping as can be seen at X = 14 000 m.

reflectors are horizontal in a 1D velocity medium. This error in the reconstruction gets worse with increasing dip of the structures and increasing lateral velocity perturbation. The results presented here are produced using a 2D finite difference code and the velocity model from Tong (2000). For the OSC data time migration for the seafloor yields an acceptable result because the seawater can be considered as a constant velocity layer (Fig. 8b). However, the ability to image the AMC is severely compromised by the 3D nature of the seafloor and subsurface. The 2D post-stack depth migrated image (Fig. 8c) restores the correct geometry though the amplitude variations on the AMC are not removed. These variations are at their most severe where the seafloor is roughest or where there are abrupt changes in bathymetry.

Using these stack data, sensitivity trails were carried out to assess the effects of possible errors in the OBS velocity model to try to quantify the significance of the apparent structure seen on the depth migrated image of the real data (Fig. 4c). For the vertical direction we would expect to be able to resolve structures down to 50 m in the migrated reflection image whereas the OBS velocity model has a resolution of 200 m. For the horizontal direction the ideal migrated reflection image should resolve structure at 100 m whereas the OBS velocity model is smoothed over 2000 m. So there is an issue of the reliability of the depth-migrated reflection image given that the resolution of the OBS velocity model is about an order of magnitude lower than the expected resolution of the reflection data. The first test (Fig. 9a) examined the effect of velocity error. The result shows that the principal effect is to displace the AMC by about 300 m in a vertical direction that depends on whether the sub-seafloor velocities are increased or decreased by 10%. There is a secondary effect in the continuity of the reflector which deteriorates as as the velocity

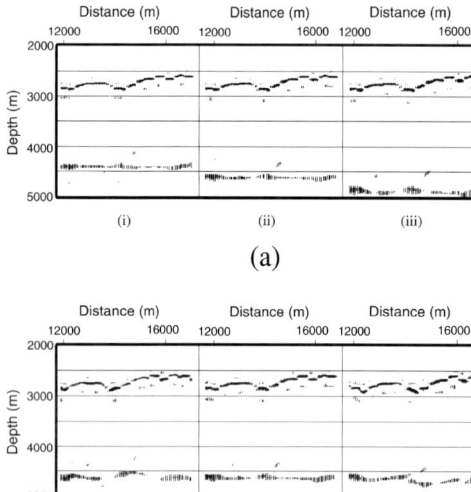

Fig. 9. Sensitivity tests on the synthetic data. (**a**) Testing the effect of possible error in the subsurface velocity field: (i) the velocity is 10% too low; (ii) the exact result; (iii) the velocity is 10% too high, the errors cause bulk shifts of the AMC reflector by ±200 m and, as the velocity increases, causes the reflection to break up. (**b**) Testing the effect of poor registration between the velocity field and the reflection data: (i) the velocity field is shifted 200 m towards lower X values relative to the reflection data prior to migration; (ii) exact result; (iii) as (i) but shifted 200 m towards higher X values, though the effect is not as marked as that shown in (a); this error can produce sharp offsets in the target horizon that could lead to a misinterpretation of the data.

increases. The second test examined the effect of registration of the OBS velocity model with the reflection data. As the two datasets have been processed independently there is a potential for misalignment. Figure 9b shows three test panels where the OBS velocity model has been displaced by ±200 m horizontally along the line of the profile relative to the exact position. The 'AMC' remains at the correct depth but false structure is introduced with apparent offsets of over 100 m.

The results show that a velocity model derived from OBS data can be used to image seismic reflection data where there is poor signal-to-noise ratio or where there is little primary reflectivity. However, caution must be exercised in the interpretation of the final reflection depth image as it is possible that small-scale velocity heterogeneities that are below the resolution of the OBS data will introduce artificial structure. A possible way to improve the result would be to use pre-stack imaging using the OBS velocity model as a starting

point, though the poor signal-to-noise ratio and lack of primary reflectivity would severely limit any analysis of the image gathers to improve the shorter wavelength velocity perturbations.

Conclusions

The 3D dataset from the East Pacific Rise presents a number of challenges for data processing. The principal issue is the recovery of the sub-seafloor geometry of the AMC reflector as this has major implications as to the mechanism of seafloor spreading which is a cornerstone of plate tectonics. The AMC target is buried beneath a rough seafloor with rapid changes in bathymetry and a varying sub-seafloor velocity structure. The picking of the stacking velocity field is highly subjective as the influence of velocity structure and bathymetry cannot be resolved within the 1D approximation implicit in the method. This is further complicated by the paucity of reflectors in the subsurface. Hence, the 3D velocity model obtained from the stacking velocities cannot be used as a basis for depth imaging. However, a second velocity field obtained from 3D wide-angle OBS data may be substituted for post-stack depth imaging. Depth migration is shown to be a robust procedure for inverting the seismic data, provided the velocity field does not vary too much. Sensitivity tests show the final image to be robust to likely errors. There is a caution that needs to be observed in the interpretation of the depth image due to the significantly different resolution scales in the OBS velocity model and the reflection image. One must be sensitive to interpret the gross structure which can be resolved in the velocity model and not to infer conclusions from short wavelength structure that could be caused by local velocity perturbations that may not have been resolved.

So given the tests on the synthetic data what can be believed on the migrated image of the real data shown in Figure 4c?

1. There is an axial magma chamber beneath the eastern limb of the overlapping spreading centre that extends westwards under the basin.
2. The depth to the axial magma chamber is 4.5 km ± 300 m from the sea-surface or 1.9 km ± 300 m from the seafloor.
3. The detailed geometry of the axial magma chamber has to be treated with caution, particularly the crest under the axial rise as a velocity perturbation that could cause this effect cannot be resolved in the OBS velocity model. This crest could be caused by the velocities being too low directly under the axial rise by about 10%. This is consistent with the error seen in the travel time to the axial magma

chamber reflectors on the unstacked real and synthetic data, with the reflection being about 100 ms late compared to the real data.

4. The poor signal-to-noise ratio of the real data limits the interpretation to areas where the reflection from the axial magma chamber is strong and modelling shows that amplitude perturbations along the magma chamber reflection may be caused by local imaging problems and may not be caused by lateral variations in the size or composition. The low signal-to-noise ratio and limited distribution of reflections in the real data will limit the use of pre-stack depth imaging tools to improve the lateral resolution of the velocity model derived from the OBS.

The use of synthetic data are demonstrated as an aid to understanding the complex interaction between data acquisition geometry and processing, which is particularly critical for this type of data where reflectivity is poor and the more conventional techniques to estimate the velocity structure for depth imaging fail. This has direct relevance to the hydrocarbons industry in its attempts to image structure beneath salt and basalt. In both these cases the overburden contains major lateral and vertical velocity perturbations which will have similar effects to the examples shown in this paper.

The authors wish to acknowledge the NERC BIRPS programme, the BIRPS Industrial Associate Scheme members, the NSF RIDGE programme for funding the ARAD project that collected and processed the 3D data discussed in this paper. Cambridge Earth Science contribution number ES7062.

References

BAZIN, S., HARDING, A. J., KENT, G. M. *et al.* 2001. Three-dimensional shallow crustal emplacement at the 9° 03′ N overlapping spreading center on the East Pacific Rise: Correlations between magnetization and tomographic images. *Journal of Geophysical Research,* **106**, 16 101–16 118.

COLLIER, J. S. & SINGH, S. C. 1997. Detailed structure of the top of the melt body beneath the East Pacific Rise at 9° 40′ N from waveform inversion of seismic data. *Journal of Geophysical Research,* **102**, 20 287–20 304.

DERICK, R. S., BUHL, P., VERA, E. E., MUTTER, J. C., MADSEN, J. A. & BROCKER, T. M. 1987. Multichannel seismic imaging of a crustal magma chamber along the East Pacific Rise. *Nature,* **326**, 35–41.

FRIGO, M. & JOHNSON, S. G., 1997. *The fastest Fourier Transform in the west.* Technical report of the MIT Laboratory for Computer Science, MIT-LCS-TR-728.

KENT, G. M., HARDING, A. J. & ORCUTT, J. A. 1993. Distribution of melt beneath the East Pacific Rise between the Clipperton transform and the 9° 17′ Deval from forward modelling of common depth point data. *Journal of Geophysical Research,* **98**, 13 945–13 970.

KENT, G. M., SINGH, S. C., HARDING, A. J. *et al.* 2000. Evidence from three-dimensional seismic reflectivity images for enhanced melt supply beneath mid-ocean-ridge discontinuities. *Nature,* **406**, 614–618.

LONSDALE, P. 1983. Overlapping rift zone at the 5.5 deg S offset of the East Pacific Rise. *Journal of Geophysical Research,* **88**, 9393–9406.

MACDONALD, K. C., FOX, P. J., PERRAM, L. J. *et al.* 1988. A new view of the mid-ocean ridge from the behaviour of ridge axis discontinuities. *Nature,* **335**, 217–225.

MCBRIDE, J. H., LINDSEY, G., SNYDER, D. B., HOBBS, R. W. & TOTTERDELL, I. J. 1993. Some problems in velocity analysis for marine deep seismic profiles. *First Break,* **11**, 345–356.

SINGH, S. C., SINHA, M. C., HARDING, A. J. *et al.* 1999. Preliminary results are in from mid-ocean ridge three-dimensional seismic reflection survey. *EOS,* **80**, 181–185.

TONG, C. H. 2000. Three-dimensional tomography study of 9° 03′ N overlapping spreading centre on the East Pacific Rise. PhD Thesis, University of Cambridge.

WILD, J. & HUDSON, J. A. 1998. A geometrical approach to the elastic complex screen. *Journal of Geophysical Research,* **103**, 707–726.

WU, R. -S. 1994. Wide-angle elastic one-way propagation in heterogeneous media and an elastic wave complex-screen method. *Journal of Geophysical Research,* **99**, 751–765.

YILMAZ, O. 1987. *Seismic Data Processing.* Society of Exploration Geophysicists, Tulsa.

3D finite element model of major tectonic processes in the Eastern Mediterranean

O. HEIDBACH* & H. DREWES

Deutsches Geodätisches Forschungsinstitut (DGFI), Marstallplatz 8, D-80539 München, Germany
Present address: Geophysical Institute, Karlsruhe University, Hertzstrasse 16, D-76187 Karlsruhe, Germany (e-mail: oliver.heidbach@gpi.uni-karlsruhe.de)

Abstract: A three-dimensional finite element model of the Eastern Mediterranean was developed in order to investigate the tectonic processes and model parameters which are mainly responsible for the observed surface deformation. The main faults are modelled as surfaces with Coulomb friction. Boundary conditions are slab pull at the Hellenic arc and the displacement for the African and Arabian plate taken from the rigid plate model NUVEL-1A. The rheology is elasto-visco-plastic. By varying the model properties a set of parameters can be determined which leads to a minimal mean deviation comparing the velocity field of the models with the space geodetic observations at 42 sites from global positioning system (GPS) and satellite laser ranging (SLR). The best result was attained with a low friction coefficient between 0.2 and 0.45 along the subduction zones and the main faults, a slab pull stress of 150 MPa and different cohesive strengths C in areas with compressional ($C = 460$ MPa) and extensional tectonic regimes ($C = 40$ MPa). The velocity field of the models shows the overall pattern of a westward escape of Anatolia and the SSW movement of the Aegean region.

The convergence between the African, Arabian and Eurasian plates in the Central and Eastern Mediterranean is leading to a complex tectonic situation (Kahle & Mueller 1998; Mueller & Kahle 1993). Broad deformation belts have developed along the border of the plates (Drewes 1993, 1998). Subduction zones and continental collision alternate along the northern border of the African and Arabian plate (Fig. 1). Recent observations from space geodesy and new geophysical data offer new insights into deformation processes and the deep structure of the Eastern Mediterranean area. As a first approximation the relative movements can be described through rigid plates (DeMets *et al.* 1994). Additionally, McKenzie (1972) defined several rigid blocks which move passively between the three main plates, Africa, Arabia and Eurasia. Passively in this context means that these blocks have no direct driving force in terms of boundary edge forces (ridge push or slab pull). The most pronounced block in this area is the Anatolian-Aegean block, which is framed by the North and the South Anatolian fault system and the subduction zones along the Hellenic and the Cypriotic arc (Fig. 1). However, the description of the plate and the block motion with a rigid body model is not able to explain the observed internal deformation. There-

fore we apply in this approach a model with a continuous deformation.

The existing models of the Eastern Mediterranean are 2D approaches using elastic rheology with dynamic (e.g. Meijer & Wortel 1996, 1997) or kinematic boundary conditions (Lundgren *et al.* 1998) or models using power-law rheology with mixed boundary conditions (e.g. Cianetti *et al.* 1997). The presented model extends to 3D in order to avoid a horizontal parameterization of the slab pull stress, to take into account the internal forces due to lateral density variations and to investigate the effect of rheological layering on the surface deformation. The velocity field of the model is compared with global positioning system (GPS) and satellite laser ranging (SLR) observations at 42 sites on the Aegean-Anatolian block from Reilinger *et al.* (1997), Kaniuth *et al.* (1999) and Noomen *et al.* (1996).

Tectonic development of the Eastern Mediterranean

According to the rigid body model NUVEL-1A (DeMets *et al.* 1994) the African plate moves in its eastern part at approximately 0.9 cm a^{-1} in a northerly direction in a fixed Eurasian plate reference

From: NIEUWLAND, D. A. (ed.) *New Insights into Structural Interpretation and Modelling.* Geological Society, London, Special Publications, **212**, 261–274. 0305-8719/03/$15

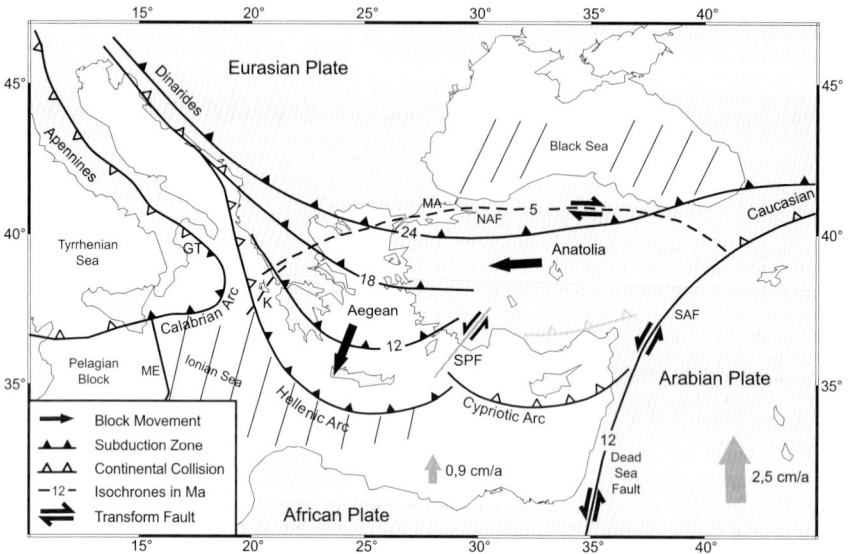

Fig. 1. Tectonic map of the eastern Mediterranean. The hatched area marks the Ionian Sea which is floored by 90 Ma old oceanic crust, a remainder of the Neotethys. The grey and black arrows give the movement direction of the Arabian and the African plate and of the Anatolian-Aegean block relative to a fixed Eurasian plate. GT, Gulf of Tarent; ME, Malta escarpment; K, Kephalonia fault; MA, Marmara Sea; NAF, North Anatolian fault; SAF, South Anatolian fault; SPF, Strabo-Plini fault. The isochrons show the southward migration of the Hellenic arc subduction zone and the onset time of the major faults. Isochrons are deduced from Ziegler (1988), Meulenkamp et al. (1988) and Dewey et al. (1989).

system. Due to the opening of the Gulf of Aden the Arabian plate separated from the African plate along the Dead Sea fault system in the middle Miocene and started its own independent northward movement with an average velocity of about 2.5 cm a^{-1} (DeMets et al. 1994; Westaway 1994; Robertson & Grasso 1995). At the end of the Miocene, when continental collision began at the northern edge of the Arabian plate, westward escape of the Anatolian block started along the North and the South Anatolian fault, probably triggered and maintained by the ongoing SSW retreat of the Hellenic arc subduction zone in the southern Aegean (Fig. 1). At the same time the subduction process of the African plate under the European plate terminated in several areas of the Mediterranean. In late Miocene and Pliocene the subduction zones in the Mediterranean have been drastically narrowed. The oceanic crust is already completely consumed by subduction west of the Malta escarpment and probably east of Crete while a substantial part south of the Hellenic arc, the Ionian sea, is still present (Fig. 1) (Ben-Avraham et al. 1988; Reuther et al. 1993). Since continental collision started, the already subducted slab is still pulling downwards. Due to the increasing tensional stresses detachment of the subducted slab might occur. This is proposed by several investigations

using seismic tomography data (Spakman 1986, 1990, 1991; Wortel et al. 1990; Wortel & Spakman 1992; Spakman et al. 1993; de Jonge et al. 1994; Papazachos et al. 1995; Kissling & Spakman 1996). Slab detachment could have happened west of the Malta escarpment, under the Apennines north of the Gulf of Tarent, under the Dinarides north of the Kephalonia fault and under the Cypriotic arc east of the island Crete. Figure 2 presents a three-dimensional sketch looking from the NE towards the SW onto the Hellenic arc subduction zone showing slab detachment under the Dinarides. It is still under debate whether and how subduction under the Cypriotic arc continues (Ben-Avraham et al. 1988; Ben-Avraham & Ginzburg 1990). In this area the resolution of seismic tomography is low and does not give a clear picture of the deep structure.

Today the last remainder of an approximately 90 Ma old oceanic crust is observed only in the Ionian Sea where two subduction zones exist (Mueller & Kahle 1993). The Calabrian arc subduction zone is 300–400 km long and sharply bent (Fig. 1). The Hellenic arc subduction zone starts south of the Kephalonia fault and terminates SE of Crete near the left lateral Strabo-Plini strike-slip fault system. Due to the roll-back of the hanging slab in the upper mantle both subduction zones are retreating,

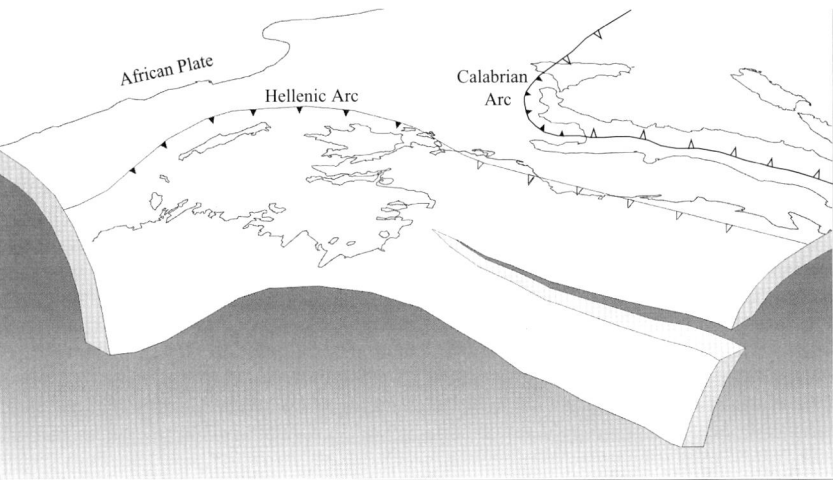

Fig. 2. Three-dimensional sketch of the subducted slab beneath the Hellenic arc in the eastern Mediterranean (modified after Meijer & Wortel 1996). The view direction is from NE towards SW. The slab geometry is based on the results of seismic tomography (e.g. Spakman *et al.* 1993).

probably since the beginning of the Miocene. The retreat velocity of the Calabrian and the Hellenic arc subduction zone is approximately 3 cm a^{-1}. The direction of these migrations is SE for the Calabrian arc and SSW for the Hellenic arc. Both subduction zones therefore migrate more or less in the opposite direction compared to the north-south directed convergence of the African plate with the Eurasian plate. This process formed the back-arc spreading center in the Tyrrhenian Sea and the back-arc basin in the Aegean.

Geophysical, geological and geodetic observations

Observations from space geodetic methods, such as satellite laser ranging (SLR) and global positioning system (GPS), provide accurate information about the movement of discrete points (Hager *et al.* 1991). Figure 3 displays the results of Kaniuth *et al.* (1999), Reilinger *et al.* (1997) and Noomen *et al.* (1996) for the Eastern Mediterranean. The expected westward directed movement of Anatolian and the SSW movement of the Aegean can clearly be seen.

The Aegean area is characterized by high seismicity within the broad deformation belts along its borders and low seismicity in the Central Aegean Sea (Jackson & McKenzie 1988). However, the Anatolian part of the block has a significant low seismicity in its central area. From the seismological strain rate (Jackson & McKenzie 1988; Jackson *et al.* 1992) and from geological indicators (Mercier *et al.* 1987; Meulenkamp *et al.* 1988) it

is known that the Anatolian part of the block moves mainly rigidly with only little internal deformation. The Aegean area is deformed by extension which is expressed through a thinned crust and a higher surface heat flow (Hurtig 1995). The crustal thickness in the western part of Anatolia is 30–35 km compared to the central Aegean Sea where it is decreases to 20–25 km (Makris & Stobbe 1984; Geiss 1987*b*). This variation of crustal thickness induces a horizontal pressure gradient whose resulting body forces support the westward drift of western Anatolia. This process is often called the gravitational collapse (Le Pichon *et al.* 1995; Meijer & Wortel 1996). The fault plane solutions of the seismic events in the northern part of the Aegean Sea reflect extension and strike-slip. Along the North and the South Anatolian fault mainly strike-slip occurs and near the Hellenic and the Cypriotic subduction zone thrusting dominates (Taymaz *et al.* 1990, 1991*a*, *b*; Ben-Avraham *et al.* 1995). These tectonics regimes are indicated in the World Stress Map database (Müller *et al.* 2000).

Description of the finite element model

The simulation procedure for a numerical model involves five consecutive steps: abstraction of the observed phenomena into a simple physical model, construction of the model geometry, discretization of this model geometry, definition of the boundary conditions model rheology and application of the numerical algorithm.

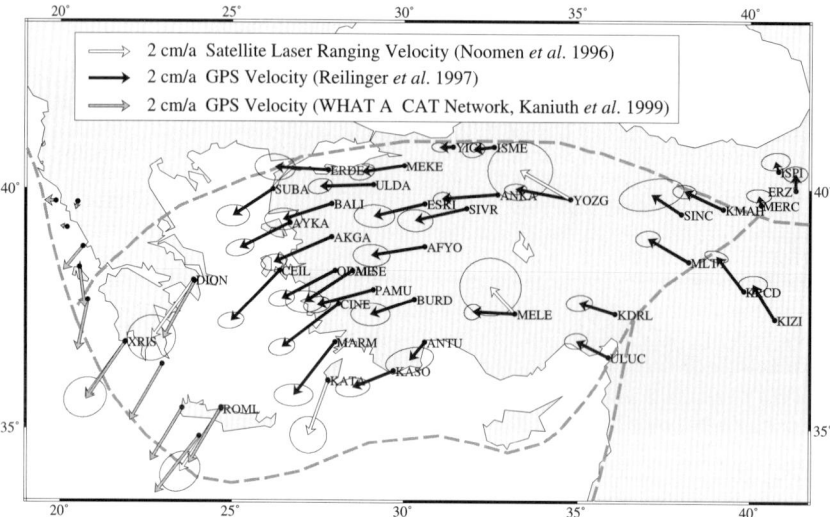

Fig. 3. Results from geodetic observation networks with global positioning system (GPS) and satellite laser ranging (SLR). The velocities are shown relative to a fixed Eurasian plate reference system. The ellipses show the 3σ error of the observations.

Model geometry and its discretization

Assuming that the indentation of the Arabian plate and the ongoing subduction at the Hellenic arc are the main driving forces for the observed surface deformation and the movement of the Anatolian-Aegean block, the model geometry has to allow relative movement between the main blocks. These movements take place at the Hellenic arc, the Cypriotic arc, the Dead Sea fault, the North and the South Anatolian fault system and the border between the Arabian and the European plate. Small-scale tectonic features have to be neglected since the spatial resolution of the discretization cannot resolve such features. Also the model has to reflect the curvature of the Earth. Due to the size of the investigated area a flat approximation might lead to deviations (Antonioli *et al.* 1998). Since the variation of the crustal and lithospheric thickness induces internal forces due to lateral density variations, the model should reflect these density variations (Fig. 4). The data for the Moho variations and lithospheric thickness are taken from Geiss (1987*a*) and Heidbach (2000). The topography is also included. The border of the model is chosen at a distance of at least 500 km from the boundaries of the Anatolian-Aegean block, in order to reduce boundary effects (the influence of the model margins on the investigated area).

In order to meet these requirements the geometry of the model is divided into four modules: the Arabian plate, the African plate including the Adriatic block, the Eurasian platform and the

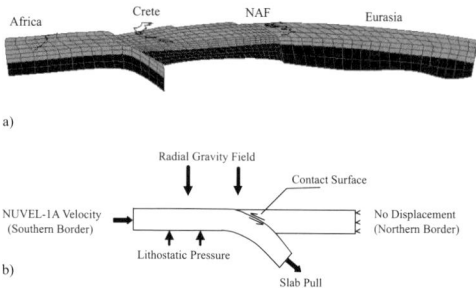

Fig. 4. (**a**) 3D cross-section through the finite element model in NE direction. The grey colour indicates the crust and the black colour the lithospheric part of the upper mantle. Both layers show thickness variations from south towards north. (**b**) 2D sketch with the applied boundary conditions used in the models.

Anatolian-Aegean block. These blocks are interconnected with so-called three-dimensional contact surfaces (MARC 1999). Along these contact surfaces Coulomb friction is defined and relative movement of the elements is allowed. The model consists of 15 000 brick elements with eight nodes each. The size of the elements is highly variable, that is, where large deformations are expected the element size is reduced (e.g. Anatolian-Aegean block) and vice versa, where small deformations might occur (e.g. Eurasian platform) the discretization is coarse.

Assumptions and boundary conditions

The whole model including the boundary conditions from NUVEL-1A, the GPS and SLR velocities and the model results are considered in fixed Eurasian plate reference frame. The three models presented in this paper are based on the following assumptions and boundary conditions (Fig. 4).

(1) Accelerations are neglected, since gravity effects are much higher than accelerations of the tectonic processes.
(2) Incompressibility is assumed (Boussinesq approximation). In the case of steady-state creep, which is reached in the calculated model time, the change of the rock volume is rather small. The effect of the elastic proportion of the total deformation (elastic + plastic + creep) is relatively small. Therefore the volume change due to elastic compression in the model is negligible.
(3) Temporal constant thermal properties.
(4) Radial Earth, i.e. no membrane stresses due to north or south movement.
(5) No phase transitions or other chemical processes.
(6) Radial gravity field is implemented.
(7) Lithostatic pressure at the bottom of the model. No shear forces are applied at the bottom of the model following the idea that the viscosity contrast between the lithosphere and the asthenosphere is too high in order to produce a reasonable coupling.
(8) Kinematic constraints from NUVEL-1A at the models southern border for the African and the Arabian plate.
(9) Fixed northern border of the model and free slip along the model's eastern and western borders in north-south direction, i.e. movement in east-west direction at the model's sides is suppressed.
(10) Coulomb friction along the contact surfaces of the blocks.
(11) Slab pull stress in the direction of the subducting slab of the Hellenic arc. The results from seismic tomography give information as to where and at which angle the slab pull stress have to be applied. At the Cypriotic arc and north of the Kephalonia fault no slab pull stress is assumed since slab detachment has already occurred and continental collision processes are taking place at present.

In a previous investigation it was shown that the influence of the subduction along the Calabrian arc on the strain rate and the velocity field pattern of the Anatolian-Aegean block is very small (Heidbach 2000). Therefore the slab pull stresses

of the Calabrian arc are neglected in the presented models.

Software and solving procedure

The discretization of the model has been implemented with the mesh generator PATRAN Version 6.2 (1996). The solver used for the numerical problem is the commercial software package MARC/MENTAT which also includes postprocessing with a graphical user interface. With a user subroutine the results of the models were exported in a generic mapping tool (GMT) input format. All of the result files for this paper were produced with GMT provided by Wessel & Smith (1991; http://www.soest.hawaii.edu/gmt/).

The calculation time is 0.5 Ma and the model is solved in 50 time steps with an implicit algorithm. The solving routine needs about 1 gigabyte workspace on a SP2/RS6000 and up to 48 hours CPU time.

Element properties and rheology

The complete model behaves as an extended Bingham body with Maxwell rheology (Fig. 5). The rheology of the upper crust is described with an elastic-plastic body. Its yield strength σ_y is computed using the Mohr-Coulomb criterion with the cohesive strength C.

$$\sigma_y = 0.6 \, \sigma_n + C$$

where σ_n is the normal pressure and the given value for the internal friction μ' is 0.6. The rheology of the lower crust and the lithosphere is described with a visco-elastic body.

In the upper crust, in order to simulate a plastic behaviour, stresses above the yield strength are not relaxed with an instantaneous ideal plastic process, but with a retardation according to a relatively low viscosity of 10^{18} Pa s. In the upper crust the cohesive strength C stress is 100 MPa. In the first model with linear Newton viscosity the lower crust has a

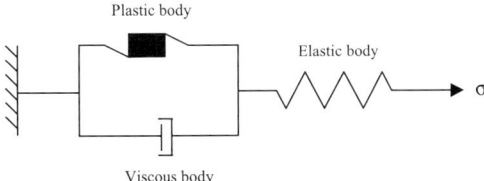

Fig. 5. The combination of the plastic St Venant body parallel to a viscous Newtonian body and in series to an elastic Hooke body is called the extended Bingham body. The viscosity of the Newtonian body is either linear (model 1) or non-linear (models 2 and 3).

viscosity of 10^{21} Pa s and the upper mantle 10^{22} Pa s. The lower crust and the upper mantle have no yield strength. Viscous deformation occurs at any state of stress.

The expression for the deviatoric shear stress tensor σ'_{ij} of the extended Bingham body is:

$$\sigma'_{ij} = \begin{cases} 2G\epsilon'_{ij} & \text{for } \sigma \leq \sigma_y \\ \sigma_y + 2\eta\dot{\epsilon}'_{ij} & \text{for } \sigma > \sigma_y \end{cases}$$

where G is the shear modulus, ϵ'_{ij} the deviatoric strain, σ_y the yield strength, η the viscosity and the time derivative of ϵ'_{ij} is the deviatoric strain rate. The viscosity can also be described with the effective viscosity η_{eff} determined from a power law, i.e. in our model approach a non-linear temperature dependent dislocation creep. As a function of stress the effective viscosity can be expressed as

$$\eta_{\text{eff}} = \frac{1}{2A} e^{(Q/RT)} (\sigma_1 - \sigma_3)^{(1-n)}$$

or as a function of deformation rate

$$\eta_{\text{eff}} = \frac{1}{2} A^{(-1/n)} e^{(Q/nRT)} \dot{\epsilon}_s^{\left(\frac{1}{n}-1\right)}$$

where A is a material constant, n the stress exponent, Q the activation energy, R the gas constant, T the temperature in Kelvin, and σ_1, σ_3 the largest and the smallest principal stress respectively. The values taken for the power law are listed in Table 1. The density of the upper crust is 2750 kg m^{-3}, 2950 kg m^{-3} for the lower crust and 3150 kg m^{-3} for the upper mantle and the oceanic crust.

Results of the finite element model

Three different models will be presented. The first one uses linear Newtonian viscosity and focuses on the investigation of the influence of a variation of the slab pull stress and different friction coef-

ficients along the modelled faults and the subduction zones. In the second and third models non-linear viscosity is applied. The third model also considers the influence of different cohesive strengths in an extensional or compressional regime.

Model 1 with linear Newtonian viscosity

In comparison with the geodetic observations, a value of 150 MPa for the slab pull stress (equal to 1.5×10^{13} N m^{-1} of a 100 km thick lithosphere) gives the best result. In order to get reasonable displacement rates the friction coefficient has to be reduced to values of 0.2 for the North and South Anatolian fault and the Dead Sea fault and rift system, 0.35 for the Hellenic arc and 0.45 for the Cypriotic arc, the suture zone in the Dinarides and the Apennines. These low values might be caused by a high fluid pressure and fine grained structure of the rocks within the faults or by frictional heating (van den Beukel & Wortel 1988; Bird & Kong 1994; Michel & Janssen 1996). Other model configurations with higher friction coefficients (0.6–0.8) and/or higher slab pull stresses (200–300 MPa) lead to subsidence rates of more than 0.8 cm a^{-1} in the Aegean Sea and a strong coupling along the North Anatolian fault which is not observed in geological or geodetic data. The best fitting results are shown in Figure 6 where the strain rate and the velocity field are presented. The strike-slip movements along the North Anatolian fault system are obvious, but the absolute values are smaller by a factor of two compared to geological and geodetic observations (Fig. 7). The modelled velocities are between 0.6 and 1.2 cm a^{-1} while geodesy observes up to 3.0 cm a^{-1} (Reilinger et al. 1997) and geology up to 2.5 cm a^{-1} for Anatolia along the eastern part of the North Anatolian fault (Barka & Hancock 1984; Barka & Kadinsky-Cade 1988; Westaway 1994). However, the lateral escape due to the indentation of the Arabian plate and the SSW movement of the Aegean is reflected

Table 1. *Parameters for temperature controlled dislocation creep*

Rock type, mineral (location)	A (Pa^{-n} s^{-1})	n	Q (kJ mol^{-1})
(upper crust)	10^{-18}	1	0
Granite (dry, lower crust)	3.16×10^{-26}	3.3	186.3
Granite (wet, lower crust)	7.94×10^{-16}	1.9	140.6
Diabase (dry, lower crust)	6.31×10^{-20}	3.05	276.0
Diabase (wet, lower crust)	1.26×10^{-16}	2.4	212.0
Olivine (dry, upper mantle)	7.00×10^{-14}	3.0	510.0
Olivine (wet, upper mantle)	3.98×10^{-25}	4.5	498.0

Data are taken from Carter & Tsenn (1987).
A = material constant; n = stress exponent; Q = activation energy.

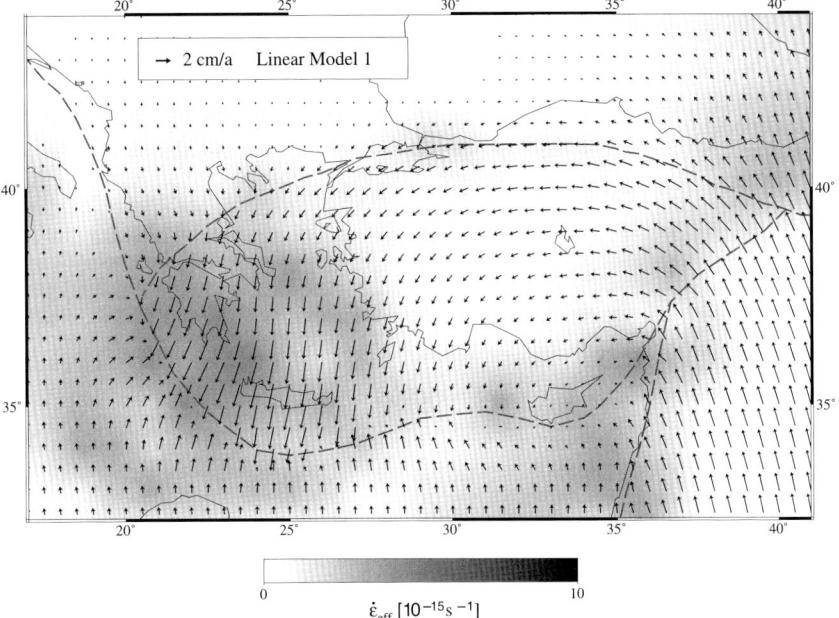

Fig. 6. Surface strain rate and velocity field for model 1 with linear Newtonian viscosity and homogeneous cohesive strength of 100 MPa. Further parameters are: 150 MPa slab pull stress, friction coefficients between 0.2 and 0.45 along the modelled faults as described in the text.

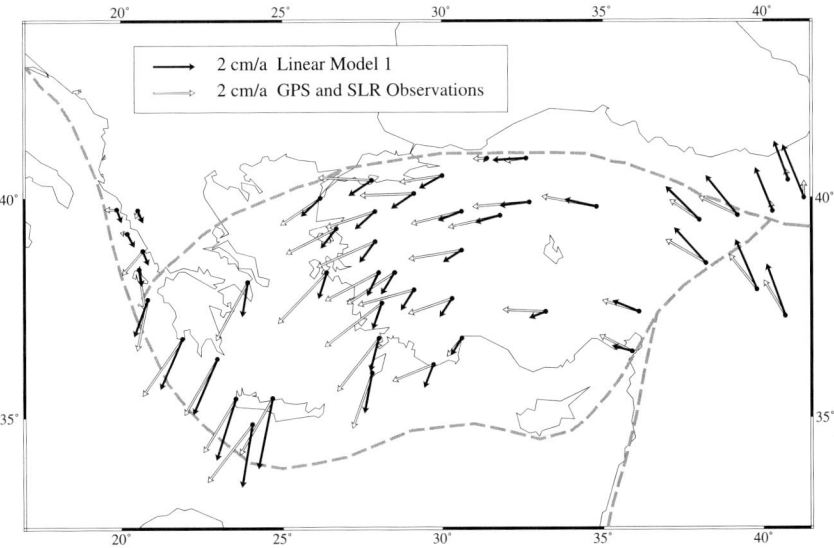

Fig. 7. Comparison of the modelled velocity field with the velocities from geodetic observations. Model 1 with linear Newtonian viscosity and homogeneous cohesive strength of 100 MPa. Further parameters are: 150 MPa slab pull stress, friction coefficients between 0.2 and 0.45 along the modelled faults as described in the text.

in the model. Within the Anatolian-Aegean block, areas with different strain rates can be identified. The eastern Anatolian part has low strain rates and thus more rigid body movements and the western Aegean part has higher strain rates indicating extension, i.e. deformation. The modelled strain rate in the Aegean Sea (3×10^{-15} to 5×10^{-15} s^{-1}) is in good agreement with the seismologically determined deformation rates from Jackson & McKenzie (1988).

Different models had been set up to test the influence of different viscosities in the lower crust and the upper mantle. A reduction of the viscosity in the lower crust and the upper mantle by one order in the Aegean due to a higher surface heat flow did neither affect the velocity nor the strain rate field. Also the impact of a higher density contrast between the two crustal layers and the upper mantle (increase of the density contrast from 200 kg m^{-3} to 250 kg m^{-3}) on the surface deformation pattern is also relatively small. But neglecting the density contrast, i.e. applying a uniform density of 3000 kg m^{-3} in the model, gives inferior results in western Anatolia.

Models 2 and 3 with non-linear viscosity

In models 2 and 3 the temperature-controlled dislocation creep, described above, is applied for the viscosity. The constant temperature field as an input parameter was calculated from measured surface heatflow data (European Thermal Atlas; Hurtig 1995) and from data for heat production and thermal conductivity taken from publications of Čermák & Bodri (1995) and Seipold (1995). The results of this model show that the non-linear viscosity leads to an amplification of strain rates (6×10^{-15} to 8×10^{-15} s^{-1}) and velocities in the Aegean and to a lower velocity field in the Anatolian region (Figs 8 & 9). Again, a variation of the effective viscosity through a different temperature field and application of wet or dry rheology parameters (Table 1) for the power law have only a small effect on the surface strain rate and velocity field.

The indentation of the Arabian plate into Eurasia in the model still does not lead to a rigid body movement of the Anatolian block. The kinematic energy from the indentation process is partly consumed by compressive deformation processes. One possible explanation for this could be that the strength of the rock under compression is underestimated. Since the velocity of the Arabian indenter cannot be raised and the friction coefficient is already very low, the only model parameter which is left in order to increase the westward escape velocity of the Anatolian block are the plasticity parameters of the Mohr-Coulomb cri-

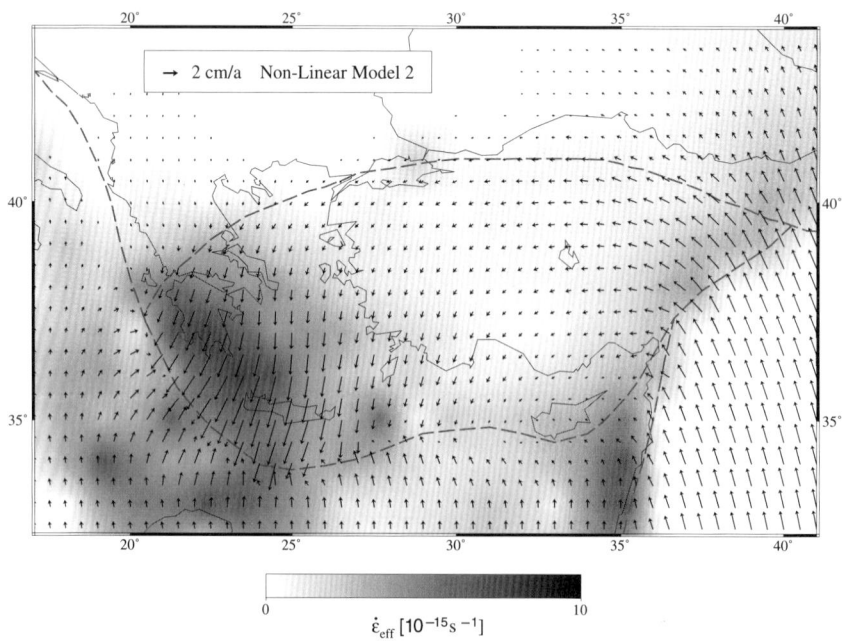

Fig. 8. Surface strain rate and velocity field for model 2 with non-linear viscosity and homogeneous cohesion stress of 100 MPa. Further parameters are: 150 MPa slab pull stress, friction coefficients between 0.2 and 0.45 along the modelled faults.

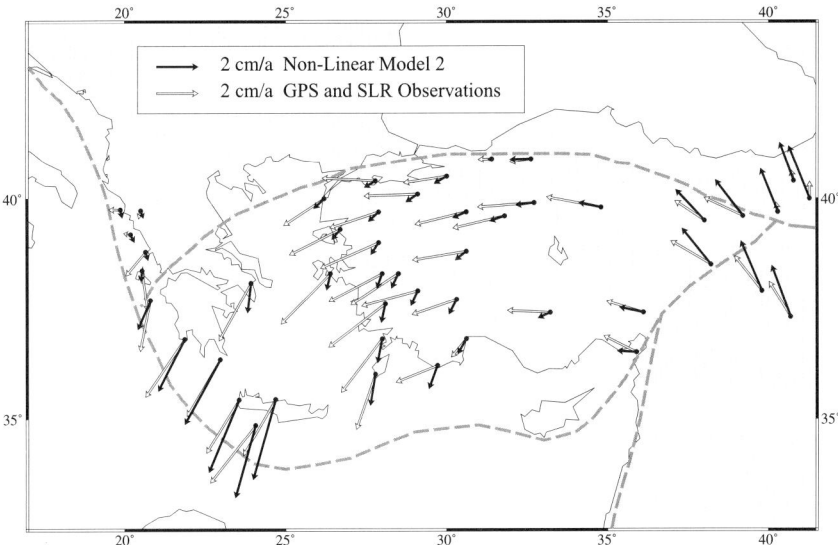

Fig. 9. Comparison of the velocity field with the observed velocities from geodetic observations. Model 2 with non-linear viscosity and homogeneous cohesion stress of 100 MPa. Further parameters are: 150 MPa slab pull stress, friction coefficients between 0.2 and 0.45 along the modelled faults.

terion. A higher value for the internal friction of $\mu' = 1$ did not give satisfactory model results. Therefore in model 3 the effect of a variable the cohesive strength C is investigated. In the previous models C was homogeneous all over the model (100 MPa). After Brace (1964) the compressive strength of granite under extension is up to 11 to 12 times higher in comparison to its tensile strength. Therefore in the Anatolian region and in the Caucasian mountains, which are, as a first approximation, under compression (Müller *et al.* 1992; Rebaï *et al.* 1992), a cohesive strength of 460 MPa is assumed. For the Aegean which is mainly affected by an extensional stress field, C is set to 40 MPa. The appropriate yield stress envelopes representing the brittle and ductile crustal strength in the Aegean and Anatolian area is displayed in Figure 10.

As a result, the velocity in the Anatolian region is increasing and values around 2.0 cm a^{-1} are reached. The strain rates are decreasing slightly (Fig. 11). As a first approximation the Anatolian block behaves as a rigid unit with little internal deformation. In the Aegean block the effect of the lowered cohesive strength is smaller and can only be observed in the western and northwestern parts. Most of the changes are due to the higher yield strength of the Anatolian block. A comparison of the modelled velocity field with the geodetic observation on the Anatolian-Aegean block is shown Figure 12. Calculating the deviation between the two vector sets results in a mean deviation of $\pm 12°$

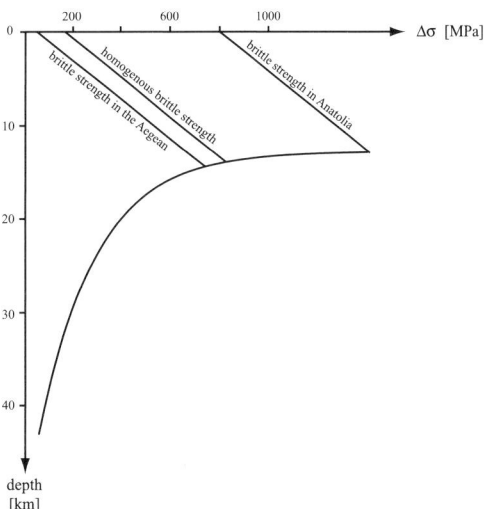

Fig. 10. Representative yield stress envelopes for the crust of model 2 with homogeneous cohesive strength and model 3 with variation of the cohesive strength for the Aegean and Anatolia. The given strain rate is 10^{-15} s^{-1} for the dislocation creep of dry granite according to the values of Table 1. The differential stress $\Delta\sigma$ is defined as $\Delta\sigma = 0.5(\sigma_1 - \sigma_3)$.

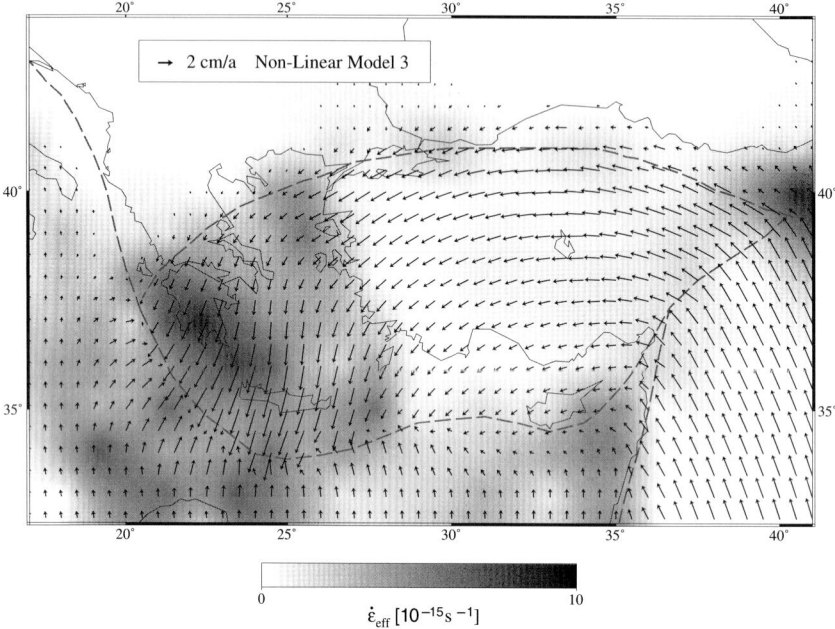

Fig. 11. Surface strain rate and velocity field of model 3 with non-linear viscosity and variable cohesive strength (40 MPa in the Aegean block, 460 MPa in the Anatolian block and the Caucasian region). Further parameters are: 150 MPa slab pull stress, friction coefficients between 0.2 and 0.45 along the modelled faults.

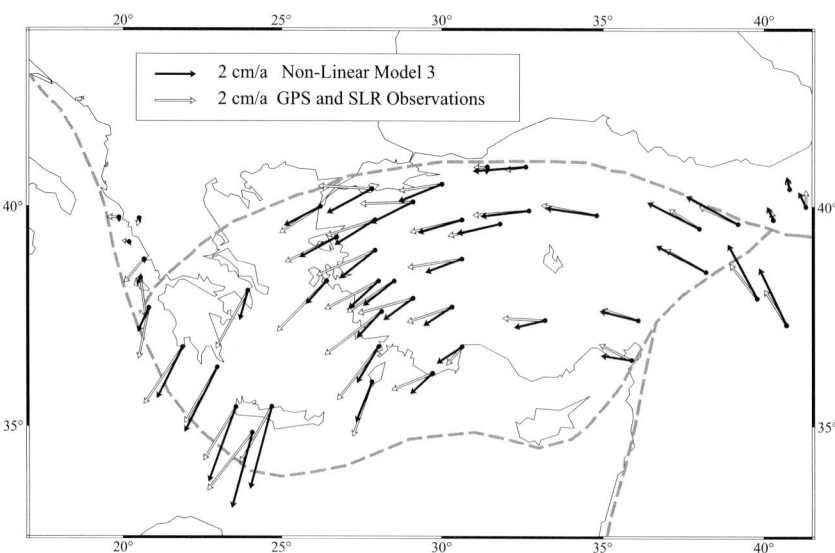

Fig. 12. Comparison of the velocity field with the observed velocities from geodetic observations. Model 3 with non-linear viscosity and variable cohesive strength (40 MPa in the Aegean block, 460 MPa in the Anatolian block and the Caucasian region). Further parameters are: 150 MPa slab pull stress, friction coefficients between 0.2 and 0.45 along the modelled faults.

for the azimuth and ±24% for the magnitudes. Large deviations occur in western Anatolia near the coastline of the Aegean Sea and south of the Marmara Sea. In the Anatolian block and near the Hellenic arc deviations are low.

Discussion and conclusions

The presented models have four results in common.

(1) The viscosity of the lower crust and the upper mantle has only a small effect on the surface deformation processes and the main pattern of the velocity field. Application of different viscosities, the variation of the temperature field and the application of wet or dry parameters for the temperature-controlled dislocation creep show this insensitivity in the presented model results.

(2) The velocities are mainly controlled by the yield strength in the upper crust, the friction coefficient applied to the modelled faults and the values taken for the slab pull stresses.

(3) The models support the hypothesis that the subduction process north of the Kephalonia fault and east of Crete towards the Cyprus arc has been terminated.

(4) Indentation of the Arabian plate into Eurasia and the retreat of the Hellenic arc subduction zone induced by the slab pull are the main processes responsible for the observed deformation pattern and for the observed velocity field in the Anatolian-Aegean block. Both deformation processes, the indentation and the subduction zone retreat, are needed in order to model the observed velocity field pattern. This escape tectonics pattern always needs a pushing part and another process which provides the space for the escape. These two processes are interwoven. Their influence radii are intersecting each other in western Anatolia. Neglecting or underestimating the retreat of the Hellenic arc subduction zone or the indentation of the Arabian plate into Eurasia would lead either to locally restricted back-arc spreading in the Aegean region or to mountain building processes in eastern Anatolia.

A mantle drag as an additional boundary condition is therefore not necessary as a rule. This does not imply that an additional driving force does not exist, but from our model results it is shown that it is probably of small importance for the surface processes in the investigated region. In all models the Anatolian block moves almost as a rigid unit which is expressed by the very low strain rate. In contrast, the Aegean block is strongly deformed

through extensional tectonics. This process is induced by the retreat of the Hellenic arc subduction zone. Inside the Anatolian-Aegean block model 3 matches the geodetic observations best, but in the transition zone between western Anatolia and the east Aegean coast line deviations in magnitude and azimuth are significant. These deviations might be explained by different effects.

(a) The Strabo-Plini fault system which has not been introduced as a contact surface into the model geometry. Sinistral movement along this fault system could probably have an important influence on the velocity field east of Crete. But this effect could not account for the large deviations further north.

(b) The effect of the varying Moho depth might be underestimated. The data on which the Moho depth estimation is based are quite sparse in that area. In the case of a larger difference of the crustal thickness between the western part of the Anatolian block and the central Aegean, additional body forces would occur and might lead to a stronger deformation.

(c) A basal drag could deliver an additional driving force.

(d) Episodic behaviour of the upper crustal deformation process. From the relatively short period of geodetic observations the average value of the velocity field must not reflect the long-term behaviour which is represented in the model. Assuming a relatively quiet decade of seismicity, the geodetic observations could have observed an episodic effect which is not comparable with the average value of the model. But since the main pattern of the observed velocity field is comparable with the results from geology and kinematic models it can be assumed that the deviation induced by seismic events will have only local effects, but will not change the overall pattern of the observed velocity field.

An unexpected result of the modelling is the need for extreme values for the frictional sliding at the main faults and the subduction zones and the very high value for the cohesive strength under compression for the upper crust. The value for frictional sliding along the North Anatolian fault and Hellenic arc is 0.2 which is one-third of the proposed value from literature (Kohlstedt et al. 1995). The value for the cohesive strength under compression has to be four times higher than the maximum value given from laboratory investigations (Brace 1964). In order to reach the magnitude of the GPS and SLR observed displacements in the Anatolian-Aegean block these values are needed. This result corresponds with the finite element model from Bird & Kong (1994) where a very low friction coefficient of 0.17 was also favoured in order to meet the geologically observed stain rates in the San Andreas fault system in California. Cianetti *et al.* (1997) even assumed the North Anatolian fault to be frictionless in their 2D approach in order to model the observed velocity field.

We investigated the large-scale tectonic processes which are responsible for the observed crustal deformation behaviour and the main pattern of the velocity field with a three-dimensional finite element model. The merit of this three-dimensional approach is that the lithological and rheological layering is accounted for. Also the slab pull effect can be introduced directly and not through a horizontal parameterization in terms of the suction forces induced by retreat of the Hellenic arc subduction zone. Even though rheological and lithological layering have only a small effect on the surface deformation pattern, the cumulative impact onto the model results can be seen. The modelling cannot resolve the local effects caused, for example, by small-scale fault systems, but it provides valuable boundary conditions which can be used for small-scale models.

The authors are grateful to the Deutsche Forschungsgemeinschaft (DFG) who financed the project (Project number Dr 143/5–2). Thanks also to the software company McNeal Schwendler for the support and advice with their Finite Element package MENTAT/MARC. We are also thankful to the two reviewers J. -D. van Wees and D. Nieuwland for their helpful and constructive reviews.

References

ANTONIOLI, A., PIERSANTI, A. & SPADA, G. 1998. Stress diffusion following large strike-slip earthquakes: a comparison between spherical and flat-earth models. *Geophysical Journal International*, **133**, 85–90.

BARKA, A. A. & HANCOCK, P. I. 1984. Neotectonic deformation patterns in the convex northward arc of the North Anatolian fault. *In*: DIXON, J. E. & ROBERTSON, A. H. F. (eds) *The Geological Evolution of the Eastern Mediterranean*. Geological Society, London, Special Publications, **17**, 763–777.

BARKA, A. A. & KADINSKY-CADE, C. 1988. Strike-slip fault geometry in Turkey and its influence on earthquake activity. *Tectonics*, **7**, 663–684.

BEN-AVRAHAM, Z. & GINZBURG, A. 1990. Displaced terranes and crustal evolution of the Levant and the Eastern Mediterranean. *Tectonics*, **9**, 613–622.

BEN-AVRAHAM, Z., KEMPLER, D. & GINZBURG, A. 1988. Plate convergence in the Cyprean Arc. *Tectonophysics*, **146**, 231–240.

BEN-AVRAHAM, Z., LYAKHOVSKY, V. & GRASSO, M. 1995. Simulation of collision zone segmentation in the central Mediterranean. *Tectonophysics*, **243**, 57–68.

BIRD, P. & KONG, X. 1994. Computer simulations of California tectonics confirm very low strength of major

faults. *Geological Society of America Bulletin*, **106**, 159–174.

BRACE, W. F. 1964. Brittle fracture of rocks. *In*: JUDD, W. R. (ed.) *State of Stress in the Earth's Crust*. Elsevier, New York.

CARTER, N. J. & TSENN, M. C. 1987. Flow properties of continental lithosphere. *Tectonophysics*, **136**, 27–63.

ČERMÁK, V. & BODRI, L. 1995. Three-dimensional deep temperature modelling along the European geotraverse. *Tectonophysics*, **244**, 1–11.

CIANETTI, S., GASPERINI, P., BOCCALETTI, M. & GIUNCHI, C. 1997. Reproducing the velocity and stress fields in the Aegean region. *Geophysical Research Letters*, **24**, 2087–2090.

DE JONGE, M., WORTEL, M. J. R. & SPAKMAN, W. 1994. Regional scale tectonic evolution and the seismic velocity structure of the lithosphere and upper mantle: The Mediterranean region. *Journal of Geophysical Research*, **99**, 12091–12108.

DEMETS, C., GORDON, R. G., ARGUS, D. F. & STEIN, S. 1994. Effect of recent revisions to the geomagnetic reversal time scale on estimates of current plate motions. *Geophysical Research Letters*, **21/20**, 2191–2194.

DEWEY, J. F., HELMAN, M. L., TURCO, E., HUTTON, D. H. W. & KNOTT, S. D. 1989. Kinematics of the western Mediterranean. *In*: COWARD, H. *et al.* (eds) *Alpine Tectonics*, Blackwell Scientific Publications, Oxford, 265–283.

DREWES, H. 1993. A deformation model of the mediterranean from space geodetic observations and geophysical predictions. *IAG Symposium*, **112**, 373–378.

DREWES, H. 1998. Combination of VLBI, SLR and GPS determined velocities for actual plate kinematic and crustal deformation models. *IAG Symposium*, **119**, 377–382.

GEISS, E. 1987a. A new compilation of crustal thickness data for the Mediterranean area. *Annals of Geophyics*, **5B**(6), 623–630.

GEISS, E. 1987b. Die Lithosphäre im mediterranen Raum. Ein Beitrag zu Struktur, Schwerefeld und Deformation. PhD thesis, *Deutsche Geodätische Kommission, Serie C: Dissertation*, No. 332, München.

HAGER, B. H., KING, R. W. & MURRAY, M. H. 1991. Measurement of crustal deformation using the global positioning system. *Annual Review of Earth and Planetary Science*, **19**, 351–387.

HEIDBACH, O. 2000. Der Mittelmeerraum – Numerische Modellierung der Lithosphären-dynamik im Vergleich mit Ergebnissen aus der Satellitengeodäsie. PhD thesis, *Deutsche Geodätische Kommission, Serie C: Dissertation*, No. 525, München.

HURTIG, E. 1995. Temperature and heat-flow density along European transcontinental profiles. *Tectonophysics*, **244**, 75–83.

JACKSON, J. & MCKENZIE, D. 1988. The relationship between plate motions and seismic moment tensors, and the rates of active deformation in the Mediterranean and Middle East. *Geophysical Journal*, **93**, 45–73.

JACKSON, J., HAINES, J. & HOLT, W. 1992. The horizontal velocity field in the deforming Aegean Sea region determined from the moment tensors of earthquakes. *Journal of Geophysical Research*, **97**, 17657–17684.

KAHLE, H. -G. & MUELLER, S. 1998. Structure and dynamics of the Eurasian-African/Arabian plate boundary system: Objectives, tasks and resources of the WEGENER Group. *Journal of Geodynamics,* **25**, 303–326.

KANIUTH, K., DREWES, H., STUBER, K. *et al.* 1999. Crustal deformations in the central Mediterranean derived from the WHAT A CAT GPS project. *Proceedings of the 13th Working Meeting on European VLBI for Geodesy and Astronomy,* Wettzell, 192–197.

KISSLING, E. & SPAKMAN, W. 1996. Interpretation of Tomographic Images of Uppermost Mantle Structure: Examples from the Western and Central Alps. *Journal of Geodynamics,* **21(1)**, 97–111.

KOHLSTEDT, D. L., EVANS, B. & MACKWELL, S. J. 1995. Strength of the lithosphere: Constraints imposed by laboratory experiments. *Journal of Geophysical Research,* **100**, 17587–17602.

LE PICHON, X., CHAMOT-ROOKE, A., LALLMANT, S., NOOMEN., R. & VEIS, G. 1995. Geodetic determination of the kinematics of central Greece with respect to Europe: Implications for eastern Mediterranean tectonics. *Journal of Geophysical Research,* **100**, 12675–12690.

LUNDGREN, P., GIARDINI, D. & RUSSO, R. M. 1998. A geodynamic framework for eastern Mediterranean kinematics. *Geophysical Research Letters,* **25**, 4007–4010.

MAKRIS, J. & STOBBE, C. 1984. Physical properties and state of the crust and upper mantle of the Eastern Mediterranean sea deduced from geophysical data. *Marine Geology,* **55**, 347–363.

MCKENZIE, D. 1972. Active tectonics of the Mediterranean region. *Geophysical Journal of the Royal Astronomical Society,* **30**, 109–185.

MARC 1999. *User Manuals A-F, Version 7.3.* MARC Research Analysis Corporation, Palo Alto.

MEIJER, P. T. & WORTEL, M. J. R. 1996. Temporal variation in the stress field of the Aegean region. *Geophysical Research Letters,* **23(5)**, 439–442.

MEIJER, P. T. & WORTEL, M. J. R. 1997. Present-day dynamics of the Aegean region: A model analysis of the horizontal pattern of stress and deformation. *Tectonics,* **16**, 879–895.

MERCIER, J. L., SOREL, D. & SIMEAKIS, K. 1987. Changes in the state of stress in the overriding plate of a subduction zone: the Aegean Arc from the Pliocene to the Present. *Annales Tectonicae,* **1(1)**, 20–39.

MEULENKAMP, J. E., WORTEL, M. J. R., VAN WAMEL, W. A., SPAKMAN, W. & HOOGERDUYN STRATING, E. 1988. On the Hellenic subduction zone and the geodynamic evolution of Crete since the late Middle Miocene. *Tectonophysics,* **146**, 203–215.

MICHEL, G. W. & JANSSEN, C. 1996. Deformation along an apparent seismic barrier: a paleoseismological study along the North Anatolian Fault. *Annali de Geophysica,* **36**, 647–662.

MÜLLER, B., ZOBACK, M. L., FUCHS, K. *et al.* 1992. Regional Patterns of Tectonic Stress in Europe. *Journal of Geophysical Research,* **97**, 11783–11803.

MÜLLER, B., REINECKER, J., HEIDBACH, O. & FUCHS, K. 2000. The 2000 release of the World Stress Map. WWW address: http://www.world-stress-map.org.

MUELLER, S. & KAHLE, H.-G. 1993. Crust-mantle evolution, structure and dynamics of the Mediterranean-Alpine region. *Contribution of Space Geodesy to Geodynamics. Crustal Dynamics, Geodynamics Series,* **23**, AGU, Washington, 249–298.

NOOMEN, R., SPRINGER, T. A., AMBROSIUS, B. A. C. *et al.* 1996. Crustal deformations in the Mediterranean area computed from SLR and GPS observations. *Journal of Geodynamics,* **21(1)**, 73–96.

PAPAZACHOS, C. B., HATZIDIMITRIOU, P. M., PANAGIOTOPOULOS, D. G. & TSOKAS, N. 1995. Tomography of the crust and upper mantle in southeast Europe. *Journal of Geophysical Research,* **100**, 12405–12422.

PATRAN 1996. *PATRAN User Manuals, Version 6.2.* McNeal Schwendler, USA

REBAÏ, S., PHILIP, H. & TABOADA, A. 1992. Modern tectonic stress field in the Mediterranean region: evidence for variation in stress directions at different scales. *Geophysical Journal International,* **110**, 106–140.

REILINGER, R. E., MCCLUSKY, S. C., ORAL, M. B. *et al.* 1997. Global Positioning System measurements of present-day crustal movements in the Arabia-Africa-Eurasia plate collision zone. *Journal of Geophysical Research,* **102**, 9983–9999.

REUTHER, C. -D., BEN-AVRAHAM, Z. & GRASSO, M. 1993. Origin and role of major strike-slip transfers during plate collision in the central Mediterranean. *Terra Nova,* **5**, 249–257.

ROBERTSON, A. H. F. & GRASSO, M. 1995. Overview of the Late Tertiary-Recent tectonic and paleo-environmental development of the Mediterranean region. *Terra Nova,* **7**, 114–127.

SEIPOLD, U. 1995. The variation of thermal transport properties in the Earth's crust. *Journal of Geodynamics,* **20**, 145–154.

SPAKMAN, W. 1986. Subduction beneath Eurasia in connection with the Mesozoic Tethys. *Geologie en Mijnbouw,* **65**, 145–153.

SPAKMAN, W. 1990. Tomographic images of the upper mantle below central Europe and the Mediterranean. *Terra Nova,* **2**, 542–553.

SPAKMAN, W. 1991. Delay-time tomography of the upper mantle below Europe, the Mediterranean, and Asia Minor. *Geophysical Journal International,* **107**, 309–332.

SPAKMAN, W., VAN DER LEE, S. & VAN DER HILST, R. 1993. Travel-time tomography of the European-Mediterranean mantle down to 1400 km. *Physics of the Earth and Planetary Interiors,* **79**, 3–74.

TAYMAZ, T., JACKSON, J. & WESTAWAY, R. 1990. Earthquake mechanisms in the Hellenic Trench near Crete. *Geophysical Journal International,* **102**, 695–731.

TAYMAZ, T., EYIDOGAN, H. & McKENZIE, D. 1991*a*. Source Parameter of large earthquakes in the East Anatolian fault zone. *Geophysical Journal International,* **106**, 537–550.

TAYMAZ, T., JACKSON, J. & McKENZIE, D. 1991*b*. Active tectonics of the north and central Aegean Sea. *Geophysical Journal International,* **106**, 433–490.

VAN DEN BEUKEL, J. & WORTEL, R. 1988. Thermo-mechanical modelling of arc-trench regions, *Tectonophysics,* **154**, 177–193.

WESSEL, P. & SMITH, W. H. F. 1991. Free software helps map and display data. *EOS Transactions, American Geophysical Union,* **72**, 441.

WESSEL, P. & SMITH, W. H. F. 1999. *The GMT-SYSTEM. Technical Reference and Cookbook (Version 3.1).* University of Hawaii, University of California, Hawaii, San Diego.

WESTAWAY, R. 1994. Present-day kinematics of the Middle East and eastern Mediterranean. *Journal of Geophysical Research,* **99**, 12071–12090.

WORTEL, M. J. R. & SPAKMAN, W. 1992. Structure and dynamics of subducted lithosphere in the Mediterranean region. *Proceedings of the Koninklijke*

Nederlandse Akademie Van Wetenschapen, **95(3)**, 325–347.

WORTEL, M. J. R., GOES, S. D. B. & SPAKMAN, W. 1990. Structural and seismicity of the Aegean subduction zone. *Terra Nova,* **2**, 554–562.

ZIEGLER, P. A. 1988. Evolution of the Arctic-North Atlantic and the Western Tethys. American Society of Petroleum Geologists, Tulsa, Memoirs, **43**.

3D discrete kinematic modelling of sedimentary basin deformation

T. CORNU[1], F. SCHNEIDER[1] & J.-P. GRATIER[2]

[1]IFP, 1–4 avenue de Bois-Préau, 92500 Rueil-Malmaison, France (e-mail:
tristan.cornu@ifp.fr; frederic.schneider@ifp.fr)
Present address: Faculty of Earth and Life Sciences, Tectonics, Vrije Universiteit, De
Boelelaan 1085, 1081 HV Amsterdam, The Netherlands (e-mail: tristan.cornu@falw.vu.nl)
[2]LGIT, Observatoire de l'université de Grenoble, IRIGM, BP 53, 38041 Grenoble, France
(e-mail: Jean-Pierre.Gratier@obs.ujf-grenoble.fr)

Abstract: 3D coupled backward and forward deformation of geological layers is a new step in
basin modelling. This problem can be treated with a mechanical or a kinematic approach.
Because of the difficulties met with the mechanical approach, the kinematic approach is more
often used. The kinematic model described here allows a geologically acceptable path to be
built, which takes into account an incremental evolution of time. To obtain a better description
of 3D geometries, the model uses a full hexaedric discretization. The discrete neutral surface
of each layer is used to perform the flexural slip deformation.

The tectonic deformation of sedimentary rocks is
one of the main problems in basin modelling
(Schneider *et al.* 1996). In order to integrate tectonic deformation, including folds and faults, two
approaches can be used: a mechanical approach,
based on mechanical laws, or a kinematic
approach, based on geometrical assumptions. The
mechanical approach has given satisfying results
for two-dimensional studies, and has been compared to experiments in a sandbox (Barnichon
1998) or applied to field cases (Hassani 1994; Niño
et al. 1998). Nevertheless, a 3D mechanical model,
which integrates all the relevant geological parameters, has not yet been proposed. The complexity
of the phenomenon at the geological time and
space scales, the lack of an adapted rheological law
(Ramsay & Huber 1987; Lamoureux-Var 1997),
and the difficulty of finding the right boundary conditions may explain this. The mechanical problem
is a problem of large deformations and large displacements, so even a model like UDEC (Cundall
1988; Hart *et al.* 1988) has restrictive assumptions
for applications to natural processes.

Because of the difficulties with the mechanical
approach, an alternative is the kinematic approach.
Such an approach is sufficiently representative of
the natural processes. It can serve for future computations of transfer problems (thermic, fluids),
evolution of rock attributes (porosity, permeability,
thermal conductivity), and development of natural

processes (sedimentation, erosion, compaction). To
obtain a better description of the 3D geometries,
the model proposed here is patterned after the 2D
discrete approach developed by Diviès (1997).
Such a discrete kinematic model allows one to
build an acceptable geological path for backward
and forward modelling. Unlike models of
unfolding like UNFOLD (Gratier *et al.* 1991; Gratier & Guillier 1993) or PATCHWORK (Bennis *et
al.* 1991), which unfold layers instantaneously, the
new model uses an incremental evolution with
time. Until now, most work on forward kinematic
models has been performed in two dimensions
(Suppe 1983; De Paor 1988*a,b*; Zoetemeijer 1992;
Waltham 1989; Contreras 1990) based on the
assumption of area conservation proposed by
Dahlstrom (1969). However, some 2.5D models
have been proposed from these 2D models
(Wilkerson & Medwedeff 1991; Shaw *et al.* 1994;
Egan *et al.* 1998). They extend the area conservation to volume conservation, but are limited to
cylindrical cases, built from topologically equivalent 2D sections. The main restriction is that the
associated finite displacement must be parallel in
map view. Although these models allow the understanding of many cases, they cannot represent the
complexity of a real three-dimensional case.

In the next section, we present a 3D model for
kinematic deformation of a sedimentary basin.
First, we explain the assumptions that sustain the

From: NIEUWLAND, D. A. (ed.) *New Insights into Structural Interpretation and Modelling.* Geological Society, London, Special Publications, **212**, 275–283. 0305-8719/03/$15

model, and how we express them in their mathematical form. Then we show the main results: (a) those in 2.5D that served to validate and to compare with existing model, (b) those in 3D to test the model on true tridimensional geometries. All the tests are performed for fault-bend-fold.

Principles of the model

The model proposed here can treat two mechanisms of deformation: vertical shear and flexural slip. We do not describe the equations of vertical shear, which are well known, but we propose a new model for the flexural slip mechanism. The modelling of the flexural slip is based on three main assumptions:

(1) the layers slip
(2) the thickness of each layer is preserved
(3) the area of the neutral surface of each layer is preserved.

Layer slip and preservation of thickness are the most often used assumptions in the literature (Suppe 1983; Waltham 1989, 1990), and they are relevant to the geological observations of the deformation of so-called competent layers (Ramsay & Huber 1987). The last assumption is more a mechanical one; it relies on the flexural mechanism and deals with the neutral surface of a layer, which is supposed to conserve its area through deformation (Ramsay & Huber 1987). Thus the motion of the neutral surface defines the motion of the layer. First the 2D model of the displacement of a neutral surface node is presented, then an extension to the 3D.

We now present the mathematical form of the model (Fig. 1a). In the 2D model, a polygonal line, cut by the bisectrix of each segment line angle, defines the basement of the basin. The layer is discretized with quadrilateral elements. For convenience, the neutral line is supposed to be the line that passes through the centre of the vertical edges.

Sliding support

The sliding support is defined with the bottom of the layer for the first layer of the hanging wall, and with the top of the previous layer for the upper layers. The support is then extended to the whole domain with the top of the other blocks. Each face of the layer is a draw segment, defined by equation:

$$\alpha x + \beta y + h = 0$$

where the coefficients α, β, and h are determined with the two points corresponding to the limits of the segment.

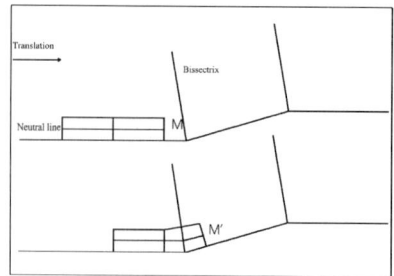

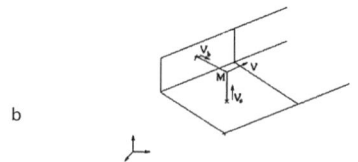

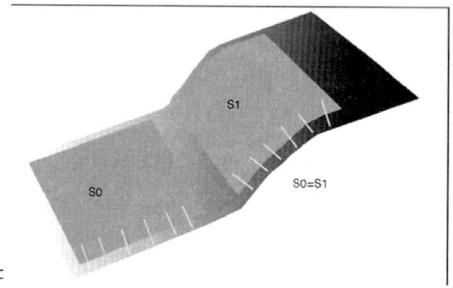

Fig. 1. (**a**) Diagram of the displacement of a node of the neutral line. (**b**) Definition of the displacement direction in 3D. (**c**) Diagram of the reconstruction of the vertical edges.

Bisectrix

As with Contreras & Suter's (1990) model, the sliding support is divided by the bisectrix of each intersecting segment. The coefficients of the bisectrix are determined by the definition of the segments pairs.

Curvilinear displacement

The displacement of the layer is achieved by the displacement of all the points of the neutral line. For M, a point of the neutral line, M' is the image point, by the transformation:

$$M' = \vec{t}_{\delta \vec{v}} \, (\mathbf{M})$$

where δ is the amount of displacement, and $\vec{v}$ the main direction.

Step of the displacement of a neutral line node

M_0 (x_0, y_0, z_0) is a node of the neutral line. Its displacement is parallel to the sliding support, and follows the draw D $(M, \vec{v}_h)$, where $\vec{v}_h$ is the orthogonal projection of $\vec{v}$ on the support:

$$D: \begin{cases} x_0 + v_{hx}t \\ y_0 + v_{hy}t \end{cases}$$

The point M_0 stay onto (D) until it intersects (D_b) one of the bisectrix of the support:

$$(D_b): \alpha_b x + \beta_b y + h_b = 0$$

The coordinates of $I = (D) \cap (D_b)$, intersection point of (D) and the bisectrix (D_b) are written:

$$I = \begin{cases} x_0 + v_{hx}t_{Ii} \\ y_0 + v_{hy}t_{Ii} \end{cases}$$

We deduced t_I with the following equation:

$$t_{Ii} = \frac{-(\alpha_b x_0 + \beta_b y_0 + h)}{\alpha_b v_{hx} + \beta_b v_{hy}}$$

We suppose $d = \|M_0 I\|$ to be the Euclidean distance between the two points I and M_0. The point M_0 has to displace of an amount of δ. We now have three possibilities:

1. $d > \delta$: the point M_0 has a displacement of δ along (D), and its new coordinates are:

$$M_0' = \begin{cases} x_0 + v_{hx}\delta \\ y_0 + v_{hy}\delta \end{cases}$$

2. $d = \delta$: the point M_0' is the same than point I.
3. $d < \delta$: M is displaced to I, but it still must displace a distance $\delta - d$. We repeat all the precedent operations, with initial point I and a new definition of $\vec{v}$ according to the sliding support.

Rebuild edges

The second step of the deformation is to rebuild the area of the layer. After translation of the neutral line, we have to rebuild the edges to restore the strained surface of the basin. The edges are rebuilt by a simple rigid rotation around the neutral line, whose area is preserved (Fig. 1c).

3D correction

The above equations are applied to the 2D cases, but they do not describe the lateral component of the displacement. Indeed, in 3D, we have to impose a new constraint on the displacement. The additional assumption we make in 3D is related to the conservation of the neutral lines, which is the lateral equivalent of the conservation of thickness: the width of the layer must be conserved. With this new assumption, the direction cannot be defined by a simple normal projection on the support. We define a laterally imposed surface, which allows the preservation of the neutral surface width (Fig. 1b).

The new direction $\vec{v}$ of the displacement is defined by:

$$\vec{v} = \vec{v}_s \Lambda \vec{v}_b$$

where $\vec{v}_s$ is the normal vector of the support, and $\vec{v}_b$ the normal vector of the lateral border (the imposed surface). Consequently the motion is parallel both to the basement of the layer and to the border we choose to impose as lateral reference.

Validation and results

The model was first validated in 2.5D and the results were compared to those of Wilkerson & Medwedeff's (1991) model. Similarly, we test the kinematic algorithm on two synthetic cases. The first test was made on a cylindrical basin, which is a succession of ten identical sections. It is 20 km long, 10 km wide, and 2 km thick (Fig. 2a). It is submitted to a displacement, which has a lateral variation that is defined by the relation for each time step:

$$\delta = 1000 \frac{(y_{max} - y)}{y_{max}}$$

The second test was made on an analytical basin, that has the same parameters as the first one, but instead of being cylindrical, it has a lateral variation of the strike of the ramp, from $10°$ to $25°$ (Fig. 2c). After displacement, 3 km for the first and 1 km for the second test, we observe good results for the deformation. We see similar shapes to those in Wilkerson & Medwedeff's model (Fig. 2b, d, e). We note a good geometric coherency, and the results obtained on a $60°$ fault confirm that we can handle more difficult geometries than the classical models, with strong dip for the ramp (Fig. 2f, g).

As the 2.5D results were satisfying, 3D tests have been performed on two similar cases. The purpose of these test cases is the validation on a 3D geometry, which is more complex than in 2.5D.

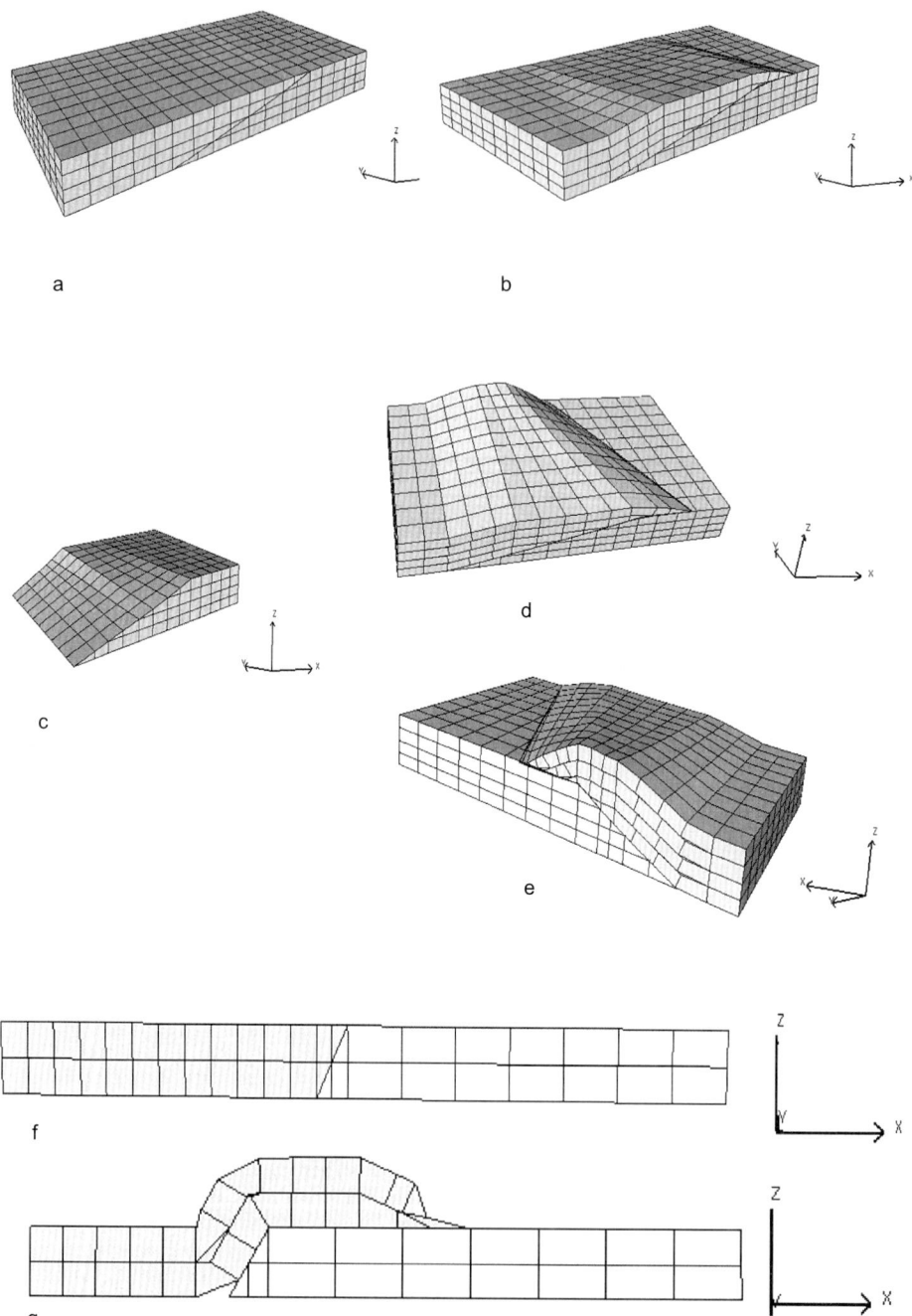

Fig. 2. 2.5D validation. (**a**, **b**) Initial and deformed configurations for a cylindrical block, deformed with shear displacement. (**c**, **d**, **e**) Initial and deformed configurations for a cylindrical block, with a variable ramp (from 10° to 25°). (**f**, **g**) Cross-sections for a basin with a 60° fault.

Two 3D cases were built both with the same hanging wall, and the same ramp. The only difference is the direction of the imposed surface that defines the 3D direction of the displacement. The first test case (Fig. 3a) has an imposed surface that is parallel to boundary displacement and the second test case (Fig. 4a) has an imposed surface that has a 5° angle with the boundary displacement.

The ramp of the basin is a curved ramp (Fig. 3b), with a 20° angle dip. The hanging wall at the

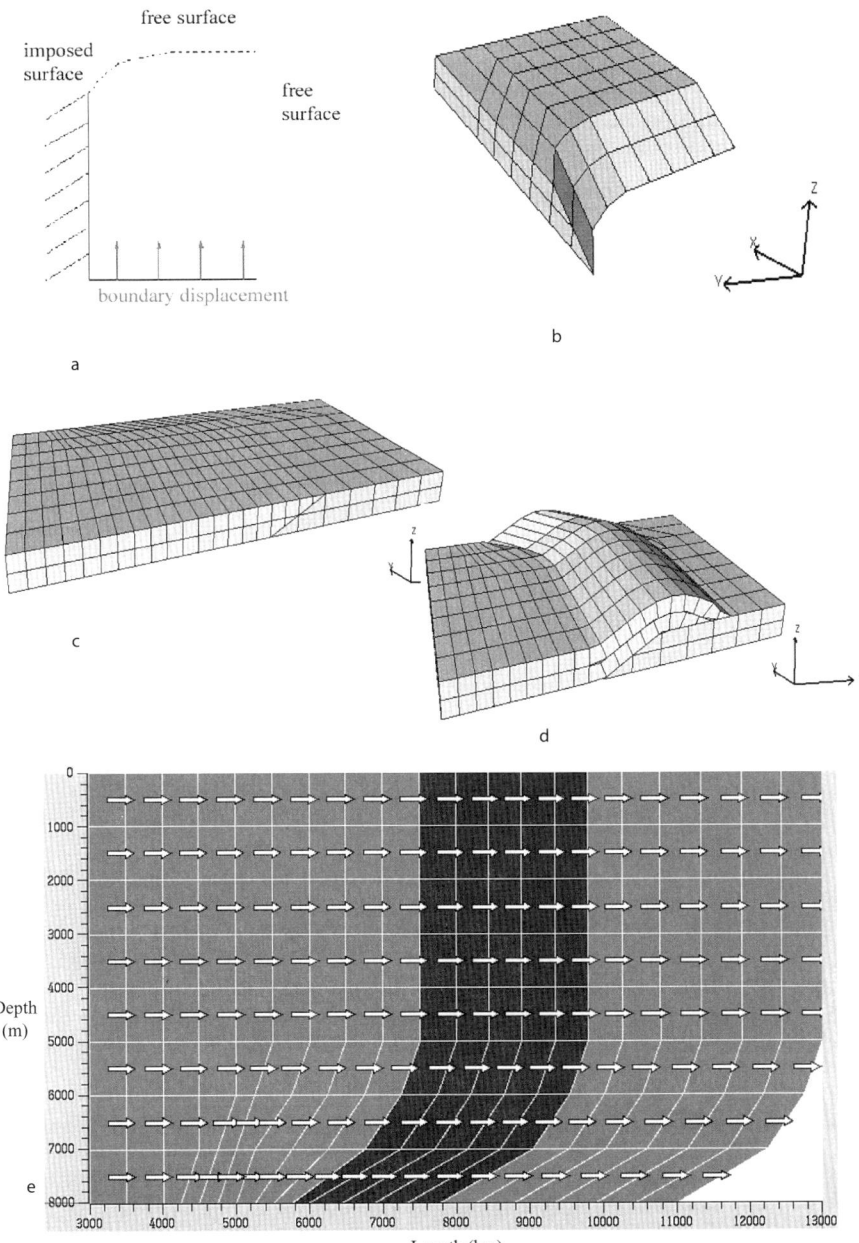

Fig. 3. First 3D validation test: lateral boundary parallel to the boundary displacement. (**a**) Diagram of the boundary conditions. (**b**) View of the curved ramp. (**c**) Initial state of the block. (**d**) Deformed configuration after 3000 m of displacement. (**e**) Map view of the projection onto a horizontal plane of the (x,y) components of the displacement of the nodes of the neutral surface.

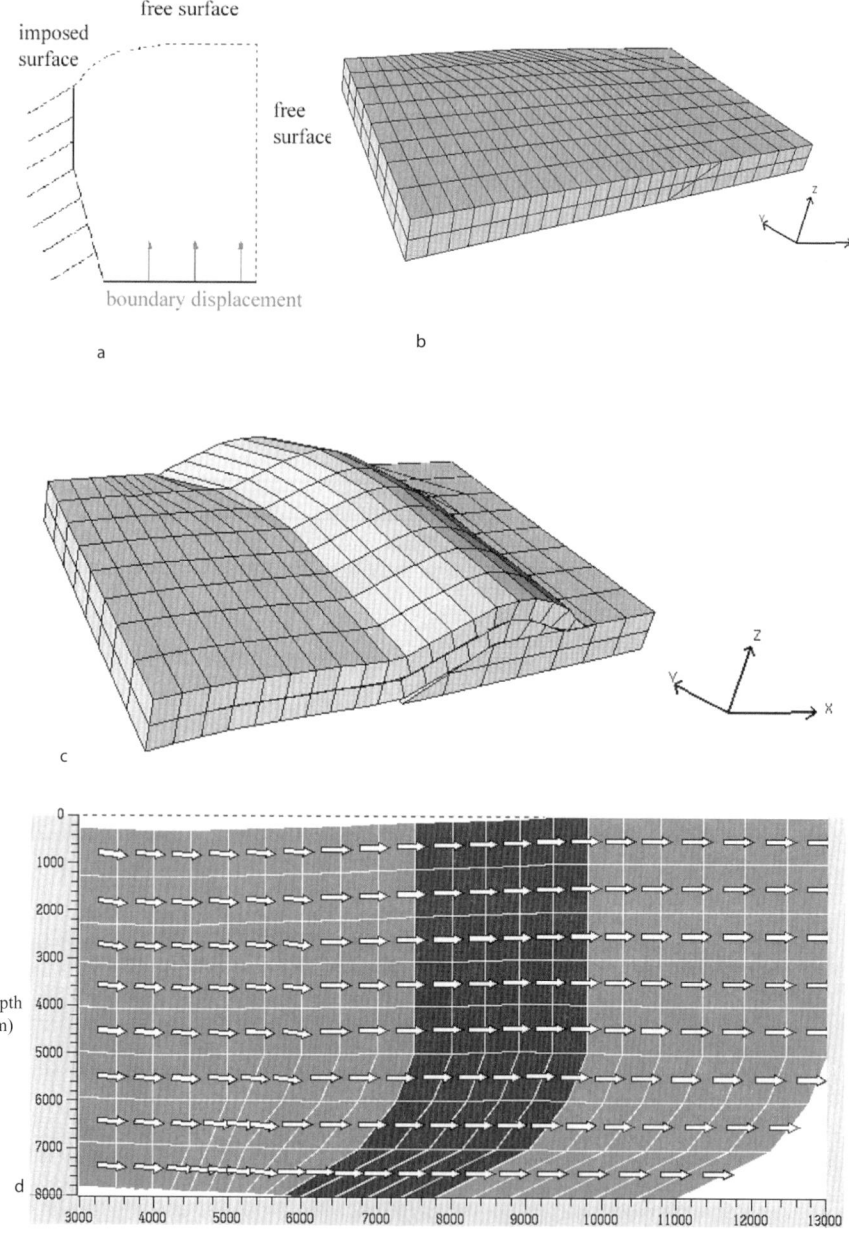

Fig. 4. Second 3D validation test: lateral boundary non-parallel to the boundary displacement. (**a**) Diagram of the boundary conditions. (**b**) Initial state of the block. (**c**) Deformed configuration after 3000 m of displacement. (**d**) Map view of the projection onto a horizontal plane of the (x, y) components of the displacement of the nodes of the neutral surface.

initial step (Fig. 3c) is 20 km long, 10 km wide and has a thickness of 1 km. The boundary condition is simple: we push at the back of the hanging wall, and we impose a 1 km displacement for each increment of time.

After 3 km of displacement, the basin presents a shape with a coherent geometry, and the edges, which are normal to the sliding support, show that we have a good flexural slip mechanism (Fig. 3d). If we look at the map in Figure 3e, which shows the (x, y) component of the displacement of the nodes of the neutral surface, we see all displacement vectors are parallel to the imposed surface, which we can also find in a 2.5D model. Nevertheless, a glance at the edges on the imposed border shows 3D rigid rotations, with lateral components due to the curvature of the ramp. So the decomposition of the displacement is in two steps: (1) motion of the neutral surface, (2) rebuilding of the edges with rigid rotations around the neutral surface. It allows 3D deformation even if the displacement is parallel.

For the second test case, the ramp is the same (Fig 3b), but the imposed lateral border has a 5° angle with the displacement boundary (Fig. 4a, b). This geometry is truly 3D and cannot be described by a 2.5D approach. The boundary condition is still simple: we impose a displacement of 1 km at the back of the hanging wall. After 3 km of displacement (three time steps), we observe a deformation that is geometrically coherent, and we notice that the width of the hanging wall is well preserved (Fig. 4c). This is the first indication of the deformation. The second, and most significant, is shown on the map of the component of the displacement of the neutral surface nodes: the components are not parallel and have variations in all space directions (Fig. 4d).

The kinematic model used here to deform the basin has no constraint on volume variation. In fact, the choice of a constant volume is a very strong assumption that is very difficult to realize with a direct algorithm. We have chosen to work with the neutral surface of each layer and tried to preserve it instead of volume. Nevertheless, if we do not have the conservation of the volume as a main constraint, we need at least to control its variations to verify that the results are not unrealistic. Table 1 presents the global volume variation of the hanging wall throughout deformation. The variations remain low and fully acceptable.

Still, we can observe a few volume variations, which are due to compression and elongation at the hinge ramp. Such volume changes are related to various parameters, the initial geometry of the layers and the fault surfaces being one of the most

important. However, for computatur limitation, it is not possible to reproduce exactly the natural geometry of the natural structures. So from this point of view, the modelling remains a crude approximation of natural behaviour. Introducing a more smooth variation along the fault surface could decrease the volume change. This will be tested and improved if necessary.

Several authors have pointed out the interest of introducing strain data in the balancing process (Woodward *et al.* 1986; Mitra 1994; Mac Naught & Mitra 1996; Von Winterfeld & Oncken 1995). This is theoretically possible with a discrete method but we did not try to implement such a constraint.

The kinematic model presented here was developed to be inserted late into basin modelling codes, as the structural part of the modelling. It is clear that there is more than one mechanism of deformation in the tectonic history of a basin. At least we can imagine that each layer can deform with its own mechanism. As an example of the versatility of the modelling code, we propose a model with two mechanisms of deformation: the bottom layer deforms with flexural slip and the top layer with vertical shear (Fig. 5a, b).

Conclusion

The model described here represents the next step of recent work performed in basin modelling at IFP (Schneider & Wolf 2000; Schneider *et al.* 2000). It was built to introduce more complex kinematics in the structural part as motions along 3D faults and 3D flexural slip. The great simplicity of its assumptions can be easily understood and implemented.

The geometries that can be treated are fully 3D, and are not restricted to cylindrical structures with parallel direction of displacement or built with topologically equivalent sections. One of the main strengths of the model is its capacity to apply different mechanisms of deformation to each layer of the basin (flexural slip, simple shear), and the layers are independent and treated individually. Another interesting point is the reversibility of the algorithm, which was one of the main goals to be achieved because it is to be inserted in a basin modelling code.

In the future, we now have to test the model on a real geological case, to see if the model is coherent with natural structures. This work will be of great help to pursue the study on internal deformations that take place in a layer throughout its deformation history.

Table 1. *Variation of volume (as a percentage) in the hanging wall between the initial and deformed state for each time step*

	T0	T1	T2	T3	T4
Test case with a parallel imposed surface	0	0.85	0.97	1.43	0.57
Test non-parallel imposed surface	0	0.61	1.63	1.80	1.34

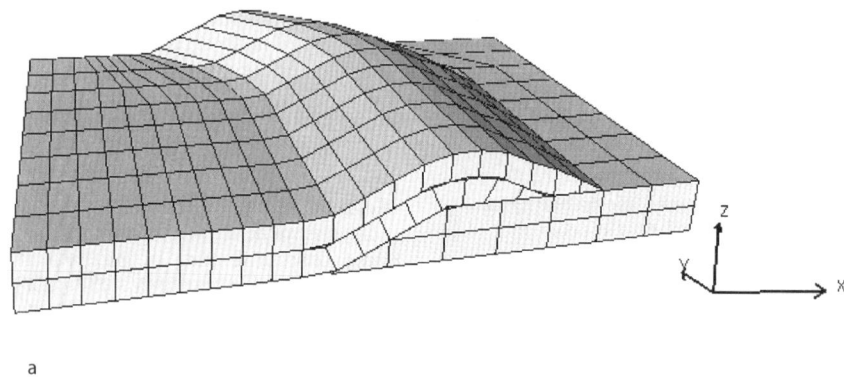

a

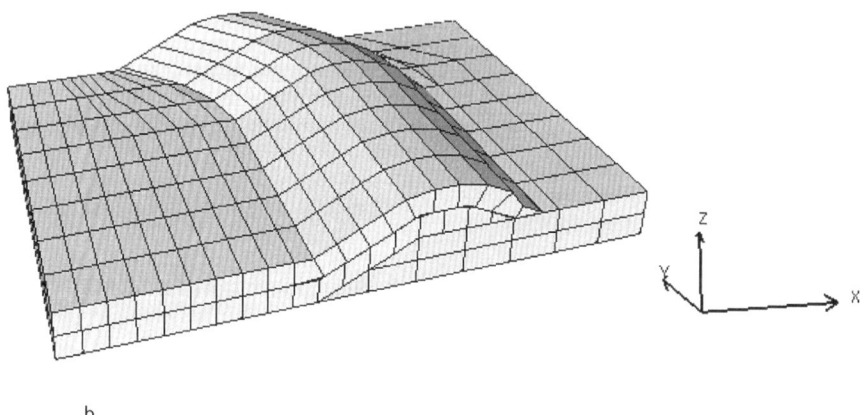

b

Fig. 5. Deformed basin, with combined flexural slip and vertical shear. (**a**) Flexural slip for the first layer and vertical shear for the second one. (**b**)Vertical shear for the first layer and flexural slip for the second one.

References

BARNICHON, J. D. 1998. *Finite element modelling in structural and petroleum geology*. PhD thesis, University de Liège.

BENNIS, C., VEZIEN, J. G. & IGLESIAS, G. 1991. Piecewise surface flattening for non-distorted texture mapping. *Computer Graphics,* **25**(4), 237–246.

CONTRERAS, J. & SUTER, M. 1990. Kinematic modelling of cross-sectional deformation sequences by computer simulation. *Journal of Geophysical Research,* **95**, 21913–21929.

CUNDALL, P. A. 1988. Formulation of a three-dimensional distinct element model-part I. A scheme to detect and represent contacts in a system composed of many polyhedral blocks. *International Journal of Rock Mechanics, Mineral Science & Geomechanical Abstracts,* **25**(3), 107–116.

DAHLSTROM, C. D. A. 1969. Balanced cross sections. *Canadian Journal of Earth Sciences,* **6**, 743–757.

DE PAOR, D. G. 1988a. Balanced section in thrust belts, Part 1: construction. *AAPG Bulletin,* **72**(1), 73–90.

DE PAOR, D. G. 1988b. Balanced section in thrust belts, Part 2: computerised line and area balancing. *Geobyte,* May, 33–37.

DIVIÈS, R. 1997. *FOLDIS: un modèle cinématique de bassins sédimentaires par éléments discrets associant plis, failles, érosion/sédimentation et compaction*. PhD thesis, Université Joseph Fourier de Grenoble.

EGAN, S. S., KANE, S., BUDDIN, T. S., WILLIAMS, G. D. & HODGETTS, D. 1999. Computer modelling and visualisation of the structural deformation caused by movement along geological faults. *Computers and Geosciences,* **25**, 283–297.

GRATIER, J. P. & GUILLIER, B. 1993. Compatibility constraints on folded and faulted strata and calculation of total displacement using computational restoration (UNFOLD program). *Journal of Structural Geology,* **15**(3–5), 391–402.

GRATIER, J. P., GUILLIER, B., DELORME, A. & ODONNE, F. 1991. Restoration and balance of a folded and faulted surface by best-fitting of finite elements: principles and applications. *Journal of Structural Geology,* **13**(1), 111–115.

HART, R., CUNDALL, P. A. & LEMOS, J. 1988. Formulation of a three dimensional distinct element model-part II. Mechanical calculations for motion and interaction of a system composed of many polyhedral blocks. *International Journal of Rock Mechanics, Mineral Science & Geomechanical Abstracts*, **25(3)**, 117–125.

HASSANI, R. 1994. *Modélisation numérique de la déformation des systèmes géologiques.* PhD thesis, Université de Montpellier II.

LAMOUREUX-VAR, V. 1997. *Modélisation de la compaction dans les bassins sédimentaires approche mécanique.* PhD thesis, Ecole polytechnique, Palaiseau.

MAC NAUGHT, M. A. & MITRA, G. 1996. The use of finite strain data in constructing a retrodeformable cross-section of the Meade thrust sheet, southeastern Idaho, U.S.A. *Journal of Structural Geology*, **18(5)**, 573–583.

MITRA, G. 1994. Strain variation in thrust sheets across the Sevier fold-and-thrust-belt (Idaho-Utah-Wyoming): implications for sections restoration and wedge taper evolution. *Journal of Structural Geology*, **16(4)**, 585–602.

NIÑO, F., CHÉRY, J. & GRATIER, J. P. 1998. Mechanical modelling of the Ventura basin: origin of the San Cayetano thrust fault and interaction with the Oak Ridge fault. *Tectonics*, **17**, 955–972.

RAMSAY, J. G. & HUBER, M. I. 1987. *The Techniques of Modern Structural Geology – Volume 2: Folds and Fractures.* Academic Press, New York.

SCHNEIDER, F. & WOLF, S. 2000. Quantitative HC potentiel evaluation using 3D basin modelling: application to Franklin structure, central graben, North sea, U.K. *Marine and Petroleum Geology.*

SCHNEIDER, F., POTDEVIN, J. L. & WOLF, S. 1996. Mechanical and chemical compaction model for sedimentary basin simulators. *Tectonophysics*, **263**, 307–317.

SCHNEIDER, F., WOLF, S., FAILLE, I. & POT, D. 2000. Un modèle de bassin 3D pour l'évaluation du potentiel pétrolier: application à l'offshore congolais. *Oil and Gas Science and Technology.*

SHAW, J. H., HOOK, S. C. & SUPPE, J. 1994. Structural trend analysis by axial surface mapping. *AAPG Bulletin*, **78**, 700–721.

SUPPE, J. 1983. Geometry and kinematic of fault-bend folding. *American Journal of Science*, **283**, 684–721.

VON WINTERFELD, C. & ONCKEN, O. 1995. Non-plane strain in section balancing: calculation of restoration parameters. *Journal of Structural Geology*, **17(3)**, 457–450.

WALTHAM, D. 1989. Finite difference modelling of hanging wall deformation. *Journal of Structural Geology*, **11**, 433–437.

WALTHAM, D. 1990. Finite difference modelling of sandbox analogues, compaction and detachment free deformation. *Journal of Structural Geology*, **12**, 375–381.

WILKERSON, M. S. & MEDWEDEFF, D. A. 1991. Geometrical modelling of fault-related folds: a pseudo-three-dimensional approach. *Journal of Structural Geology*, **13**, 801–812.

WOODWARD, N. B., GRAY, D. R. & SPEAR, D. B. 1986. Including strain data in balanced cross section. *Journal of Structural Geology*, **8(3–4)**, 313–324.

ZOETEMEIJER, R. 1992. *Tectonic modelling of forelands basins, thin skinned thrusting, syntectonic sedimentation and lithospheric flexure.* PhD thesis, Vrije Universiteit Amsterdam.

3D discrete kinematic modelling applied to extensional and compressional tectonics

T. CORNU[1], F. SCHNEIDER[1] & J.-P. GRATIER[2]

[1]*IFP, 1–4 avenue de Bois-Préau, 92500 Rueil-Malmaison, France (e-mail: tristan.cornu@ifp.fr; frederic.schneider@ifp.fr)*
Present address: Faculty of Earth and Life Sciences, Tectonics, Vrije Universiteit, De Boelelaan 1085, 1081 HV Amsterdam, The Netherlands (e-mail: tristan.cornu@falw.vu.nl)
[2]*LGIT, Observatoire de l'Université de Grenoble, IRIGM, BP 53, 38041 Grenoble, France (e-mail: Jean-Pierre.Gratier@obs.ujf-grenoble.fr)*

Abstract: The 3D simulation of coupled backward and forward deformation of geological layers is a new step in basin modelling. Although this problem could be addressed with either mechanical or kinematic approaches, the mechanical approach remains too complex to be addressed properly. The kinematic model described here allows a geologically valid path, which takes into account an incremental evolution in time. To obtain a better description of 3D geometries, the model uses a full hexaedric discretization and the discrete neutral surface of each layer is used when performing the flexural slip deformation. An application to a synthetic geological case is then proposed, to study the behaviour of the structure in compressional and extensional contexts.

Modelling the evolution and petroleum potential of sedimentary basins is a complex problem, involving two distinct steps: (1) the simulation of the tectonic deformation, and (2) the computation of the hydrocarbon generation and migration. Until now, most basin modelling tools were constructed to address only one of these problems (Schneider & Wolf in press; Zoetmeyer 1992). A first attempt to couple deformation and fluid flow simulations was done recently (Schneider *et al.* 2000) but it is still limited to 2D cases with relatively simple kinematic patterns (vertical shear mechanism). The model we propose here is a discrete model for 3D flexural slip deformation (or mixed flexural slip and vertical shear), where further computations to solve fluid flow simulations might be done by using the mesh of the deformed elements.

Two distinct approaches can be chosen to model the tectonic deformations that occur in a sedimentary basin: (1) a mechanical approach, or (2) a kinematic approach. The mechanical approach has already been tested on analytical or geological cases (Barnichon 1998; Niño *et al.* 1998; Erickson and Jamison 1995; Bourgeois 1997; Coussy 1995). However, these studies were done on 2D cases with simplified assumptions on the mechanical behaviour of the rocks. 3D mechanical modelling that would include all the parameters relevant to natural deformation has never yet been proposed.

This may be explained by the extreme complexity of the mathematical formulation and computer limitations on one hand, and on the other hand, by the complexity of the phenomenon at geological time and space scales (Ramsay & Huber 1987), which makes it difficult to find the adapted rheological laws and boundary conditions. Even a model like 3DEC (Cundall 1988; Hart *et al.* 1988) is limited by its restrictive hypothesis of incremental deformation within deformed blocks, which is not realistic for natural deformation of sedimentary basins.

To overcome the difficulty of the mechanical approach, geologists have instead focused on the kinematic approach (Dahlstrom 1969), which is the geometrical translation of mechanical assumptions. Kinematic modelling is a good alternative, which can be sufficiently representative of the natural processes. A discrete approach (Waltham 1989, 1990; Divies 1997) can also be used for further computation of thermal and fluid transfers, integration of rock attributes (i.e. porosity, permeability, thermal conductivity, etc.), and simulation of natural processes (i.e. sedimentation, erosion, and compaction). One limitation of the existing kinematic approaches is that the proposed models relate either to forward modelling (Suppe 1983; Gibbs 1983; De Paor 1988*a*, *b*; Contreras and Suter 1990) or to backward restoration (Moretti 1989; Gratier

From: NIEUWLAND, D. A. (ed.) *New Insights into Structural Interpretation and Modelling.* Geological Society, London, Special Publications, **212**, 285–294. 0305-8719/03/$15

et al. 1991; Gratier & Guillier 1993; Bennis *et al.* 1991). Therefore the modelling of tectonic deformation relies on two types of models to solve the problem of restoration and deformation (Egan *et al.* 1998). However, the main limitation is that most of the kinematic models are 2D, or 'pseudo-3D' at best (Wilkerson & Medwedeff 1991; Shaw *et al.* 1994). These pseudo-3D models extend the area conservation to volume conservation, but are limited to cylindrical cases, built with topologically equivalent 2D sections. The main restriction is that the associated finite displacement must be parallel in map view. To overcome these problems, we have developed and present here a discrete algorithm that can be used both for backward and forward modelling, and that can be applied to real 3D cases. The method is applied to an analytical sedimentary basin with a lateral termination, derived from real field structures.

The model

The assumptions used to describe the 3D flexural slip mechanism are first briefly presented. The model is supported by three main assumptions.

(1) Each layer of the basin is assumed to be independent, and the sliding between the layers is supposed to be perfect.
(2) As a general simplification, we assume that the thickness of the layer is preserved through the whole progressive deformation. Alternatively, thickness changes could be integrated if required.
(3) Because the flexural-slip mechanism would preserve the length of the neutral line of a layer, we assume that the area of the neutral surface is also kept constant in 3D.

Layer slip and preservation of the layer thickness are the most commonly used assumptions in literature (Suppe 1983; Waltham 1989, 1990), and they are consistent with the geological observations of the deformation of so-called competent layers (Ramsay & Huber 1987). The last assumption is a mechanical one, as it relies on the flexural-slip mechanism and deals with the neutral surface of a layer, which is supposed to conserve its area through deformation (Ramsay & Huber 1987). This is a powerful and useful assumption since further calculation of the progressive deformation will be greatly simplified by the use of a surface instead of a volume.

We take a discrete modelling approach here, and implementation of the algorithm is done in C++. The geological objects are defined as follows.

1. The basin defines the entire geological area. It contains all the tectonic portion of the studied domain within its geometric boundaries.

2. The faults are defined as the main zones of discontinuity within the domain. They allow the subdivision of the basin into a discrete number of subdomains, and they are defined as triangulated surfaces.
3. Each subdomain of the basin constitutes an independent block, which is bordered by the faults (footwall, hanging-wall). Their frontiers are defined either by a fault or by the boundary of the basin.
4. The layers are the simplified geometric representation of the lithologic beds. They are discretized with hexaedric elements with eight vertices, and they support the deformation algorithm.

Mathematical description

The first step of the modelling is to build a mesh: all the layers of each block are discretized in elements. The elements are height vertices hexaedric element, with six faces that are not always coplanar. After the definition of the geological domains, the motion of the basin is modelled through the displacement of each node of the neutral surface of the layers. The neutral surface is defined here as the median surface of the layer. It passes through the middle point of the 'vertical' edges of each hexaedric element (Fig. 1a).

Sliding support

The support of sliding is defined with faces of the element in each layer. For the first layer of the block, we use the base faces of the element, for the upper layer we use the roof faces of the previous layer, and for the lateral surface of sliding, we use the faces that coincide with the lateral imposed border. Each face is defined by four vertices, and we cut them into two triangles of three vertices. This allows us to define the surface of sliding as a C^1 piecewise surface, or plane:

$$(P): \alpha x + \beta y + \gamma z + h = 0$$

where α, β, γ, h are defined with the coordinates of the three vertices.

Bisector plane

The bisector planes define the intersection of two planes, and they will help us to preserve the distance through the displacement. They are defined with the help of the coordinates of the two planes they cut.

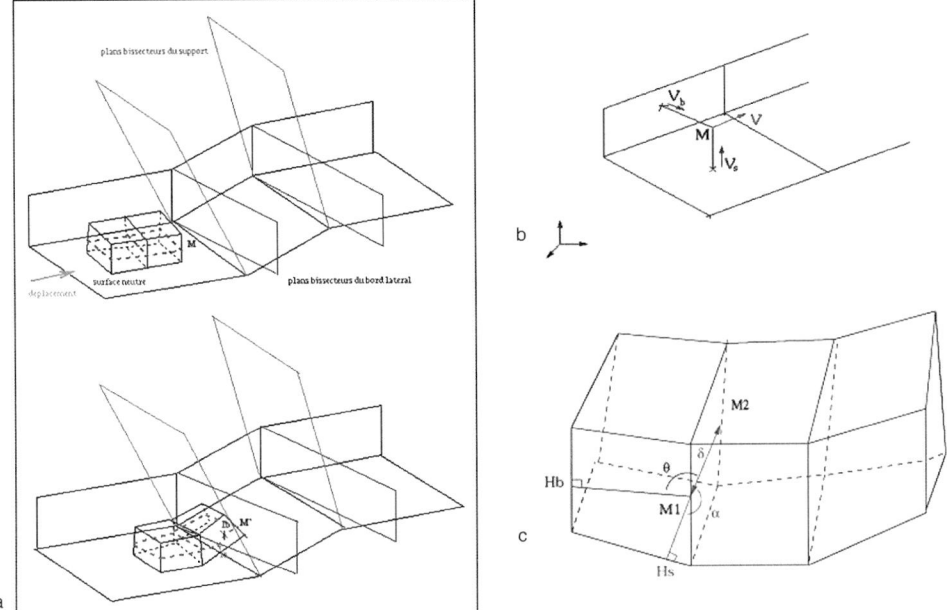

Fig. 1. (**a**) Diagram of the deformation of a layer with the algorithm based on neutral surface conservation; (**b**) diagram of the construction of the displacement direction for the bottom layer; (**c**) diagram of the construction of the displacement direction for the upper layers.

Curvilinear displacement

The movement of a neutral surface point is achieved with a curvilinear displacement $\vec{t}\,_\delta\vec{v}$, where δ is the distance of displacement, and $\vec{v}$ the direction of displacement. Due to the third space component, the direction of the displacement will depend not only on the basement, but also on the lateral border. We define $\vec{v}$ as:

$$\vec{v} = \vec{n}_s \wedge \vec{n}_b$$

where $\vec{n}_s$ is the normal vector of the support on which the layer slips, and $\vec{n}_b$ the normal vector of an imposed lateral surface (Fig. 1b).

The definition of $\vec{v}$ imposes to the displacement to be parallel both to the support and to the imposed lateral border. This first approximation provides a conservation of the width of the layer (the distance between each point of the neutral surfaces and the lateral boundary).

Displacement step of a neutral surface node

$M_0(x_0, y_{0,0})$ is a node of the neutral surface. It is displaced following the draw $D(M, \vec{v})$:

$$D: \begin{cases} x_0 + v_x t \\ y_0 + v_y t \\ z_0 + v_z t \end{cases}$$

The point M_0 follows (D) until it intersects one the bisector plane that split the domain. (P_{Bs}) is a bisector plane of the basement, (P_{Bb}) is a bisector plane of the lateral imposed border, defined by:

$$(P_{Bs}): \alpha_{bs}\,x + \beta_{bs}\,y + \gamma_{bs}\,z + h_{bs} = 0$$

$$(P_{Bb}): \alpha_{bb}\,x + \beta_{bb}\,y + \gamma_{bb}\,z + h_{bb} = 0$$

The coordinates of $I_i = (D) \cap (P_{Bi}, i = b, s)$, intersection point of (D) and the bisector plane (P_{Bi}) are written:

$$I_i = \begin{cases} x_0 + v_x t_{Ii} \\ y_0 + v_y t_{Ii} \\ z_0 + v_z t_{Ii} \end{cases}$$

The position of M_0 relative to the bisector planes is not known a priori, so we are obliged to calculate the two intersection points, with the following definition of t_{Ii}:

$$t_{Ii} = \frac{-(\alpha_{bi}x_0 + \beta_{bi}y_0 + \gamma_{bi}z_0 + h_{bi})}{\alpha_{bi}v_x + \beta_{bi}v_y + \gamma_{bi}v_z}$$

We suppose $d_i = \|MI_i\|$ to be the Euclidean distance between the two points I_i and M. The point M has to displace an amount of δ. Before the displacement, we must define which bisector plane is first cut by (D). It will be nearest the one with the smallest distance d_i. When the plane is determined, we have three cases:

(1) $d_i > \delta$: the point M has a displacement of δ along (D), and its new coordinates are:

$$M = \begin{cases} x_0 + v_x\delta \\ y_0 + v_y\delta \\ z_0 + v_z\delta \end{cases}$$

(2) $d_i = \delta$: the point M is the same than point I_i.
(3) $d_i < \delta$: M is displaced to I_i, but it still must displace a distance $\delta - d_i$. We repeat all the precedent operations, with initial point I_i and a new definition of $\vec{v}$ according to the basement and the lateral border.

Reconstruction of upper layers

The reconstruction of the upper layers tries to rebuild a geometry which conserves the angular and distance relationships of the previous step configuration. The complexity of the reconstruction comes from the parameters we have to define: the distance δ between two nodes of the neutral surface, and two angles α and θ relative to the basement and to the lateral border (Fig. 1c).

M_1 and M_2 are two successive points of the neutral surface: $\delta = \|M_1M_2\|$. H_s and H_b are the normal projection of M_1 upon the basement and upon the lateral imposed border:

$$\cos\alpha = \frac{M_1\vec{M}_2.\mathbf{M}_1\vec{H}_s}{\|\mathbf{M}_1\vec{M}_2.\mathbf{M}_1\vec{H}_s\|}$$

$$\cos\theta = \frac{M_1\vec{M}_2.\mathbf{M}_1\vec{H}_b}{\|\mathbf{M}_1\vec{M}_2.\mathbf{M}_1\vec{H}_b\|}$$

The direction of displacement $\vec{v}$ must now be defined according to α and θ. To do this, we place ourselves in a reference system $\Re(M_1, \vec{e}_1, \vec{e}_2, \vec{e}_3)$, where:

$$\vec{e}_1 = \frac{\mathbf{M}_1\vec{H}_s}{\|\mathbf{M}_1\vec{H}_s\|}$$

$$\vec{e}_2 = \frac{\mathbf{M}_1\vec{H}_b}{\|\mathbf{M}_1\vec{H}_b\|}$$

$$\vec{e}_3 = \frac{\mathbf{M}_1\vec{H}_s \wedge \mathbf{M}_1\vec{H}_b}{\|\mathbf{M}_1\vec{H}_s \wedge \mathbf{M}_1\vec{H}_b\|}$$

We can define $\vec{v}$ as:

$$\vec{v} = a\vec{e}_1 + \mathbf{b}\vec{e}_2 + \mathbf{c}\vec{e}_3$$

with:

$$\vec{v}.\vec{e}_1 = \cos\alpha$$
$$\vec{v}.\vec{e}_2 = \cos\theta$$
$$\|\vec{v}\| = 1 = \sqrt{a^2 + b^2 + c^2}$$

Volume reconstruction

After the displacement of all the nodes of the neutral surface of a layer, we have to restore the volume of the layer. To perform this restoration, we use the vertical edges. Each node of the neutral surface belongs by definition to a vertical edge. Thus, after displacement, we rebuilt the volume by a rigid rotation of the vertical edge.

Geological application

The test was done on a synthetic case. We constructed a geometric model (Fig. 2), which is similar to a field area in the Gulf of Mexico (Trudgill *et al.* 1999, fig. 2c; Rowan *et al.* 1999, fig. 31), or to Niger Delta structures (Crans *et al.* 1980). However, we are not trying to make any study of these areas, which are only named to help to figure what kind of application could be expected from the model. Such a structure contains both extensional and compressional structures, which result from gravity sliding processes. We assume the conservation of the global volume of the sediment (contained between the two main faults) during such a deformation. We also impose a lateral termination of the structure in order to model its lateral 3D evolution both in space and time. In this way, we can study the evolution of a part of a sedimentary basin ending along a non-vertical lateral ramp. With this experiment, we hope to get information on the 3D behaviour of the structure when it is subjected to gravity sliding. We will also study the local volume variations and the effect of the lateral ramp on the displacement field.

The limits (Fig. 2c) of the footwall are defined by a normal fault with a 20° dip at the rear, a reverse fault with a 20° dip at the front, and a lateral ramp with a 70° dip along one lateral boundary. The other lateral boundary is free. The hanging wall is a block of 11 286 elements. The lithologic sequence is made up of a composite sequence of eleven layers of various thicknesses, to see the behaviour of the algorithm with different thicknesses. The length of the model is 35 000 m, its width is 10 000 m, and its thickness is 4450 m (Fig.

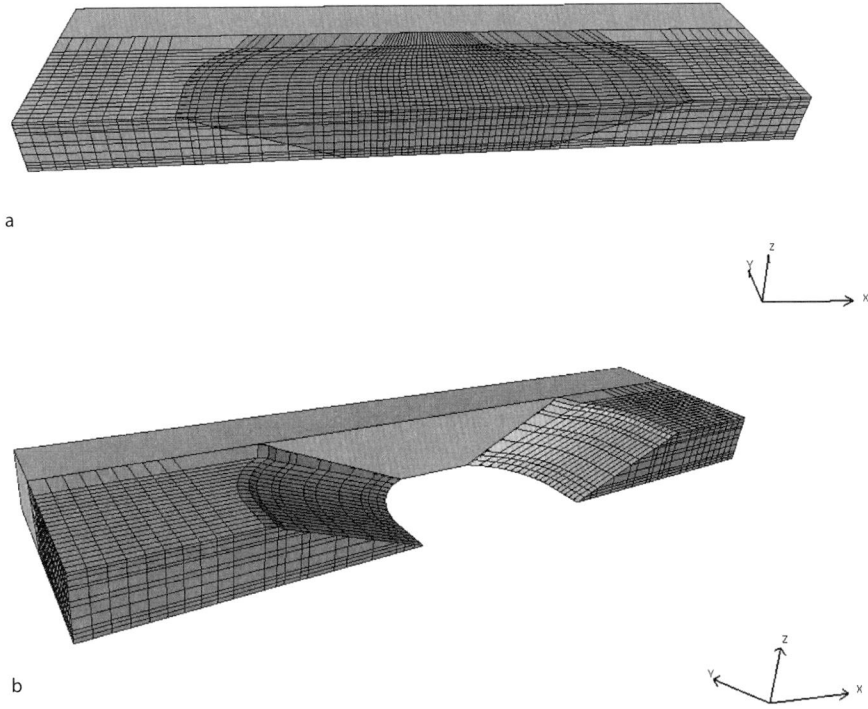

Fig. 2. (**a**) basin at the initial state; (**b**) description of the domain boundaries.

2b). The kinematic boundary condition is applied at the back of the hanging wall and imposes a displacement of 1000 m in the x direction at each time step. This means that all the elements of the first layer will move of 1000 m, and that those of the upper layers will be reconstructed according to the first layer deformation. All layers are supposed to deform by flexural slip, although it could also be possible to integrate other deformation mechanisms such as vertical shear for some layers. The footwall is supposed to be rigid, and is also discretized because its top helps to define the sliding support.

In Figure 3a–e we illustrate the evolution of the basin after four time steps and thus 4000 m of displacement. Below, we will outline our major observations.

Basin geometry

The resulting geometry remains consistent with the initial shape, and displays no anomalous or unexpected deformations like crossing edges. The global volume variation of the hanging wall after deformation is lower than 1% or 2%. This is an acceptable and realistic result. It implies that even if the volume conservation was not an independent constraint, the coupled assumptions made on the neutral surface and on the edge rebuilding were relevant.

Transport direction

In Figure 4a, the components of displacement for each node of the neutral surface are projected onto a horizontal plane. In that their direction varies widely in map view, we can conclude that our modelled deformation is fully 3D.

Boundary effects

The lateral displacements observed on the free vertical boundary (Fig. 4b, c) show the effect of a lateral ramp as a geometrical boundary condition. The displacement values are dependent on the direction and dip of the imposed lateral boundary as shown on Figure 4, and also on the thickness of the layers. They document the incidence of the geometry of a lateral termination on the kinematic evolution of the basin. Moreover, we suggest that lateral bedding slip or lateral flexural-slip deformation might act as evidence for a lateral termination, provided we could observe these features directly in the field.

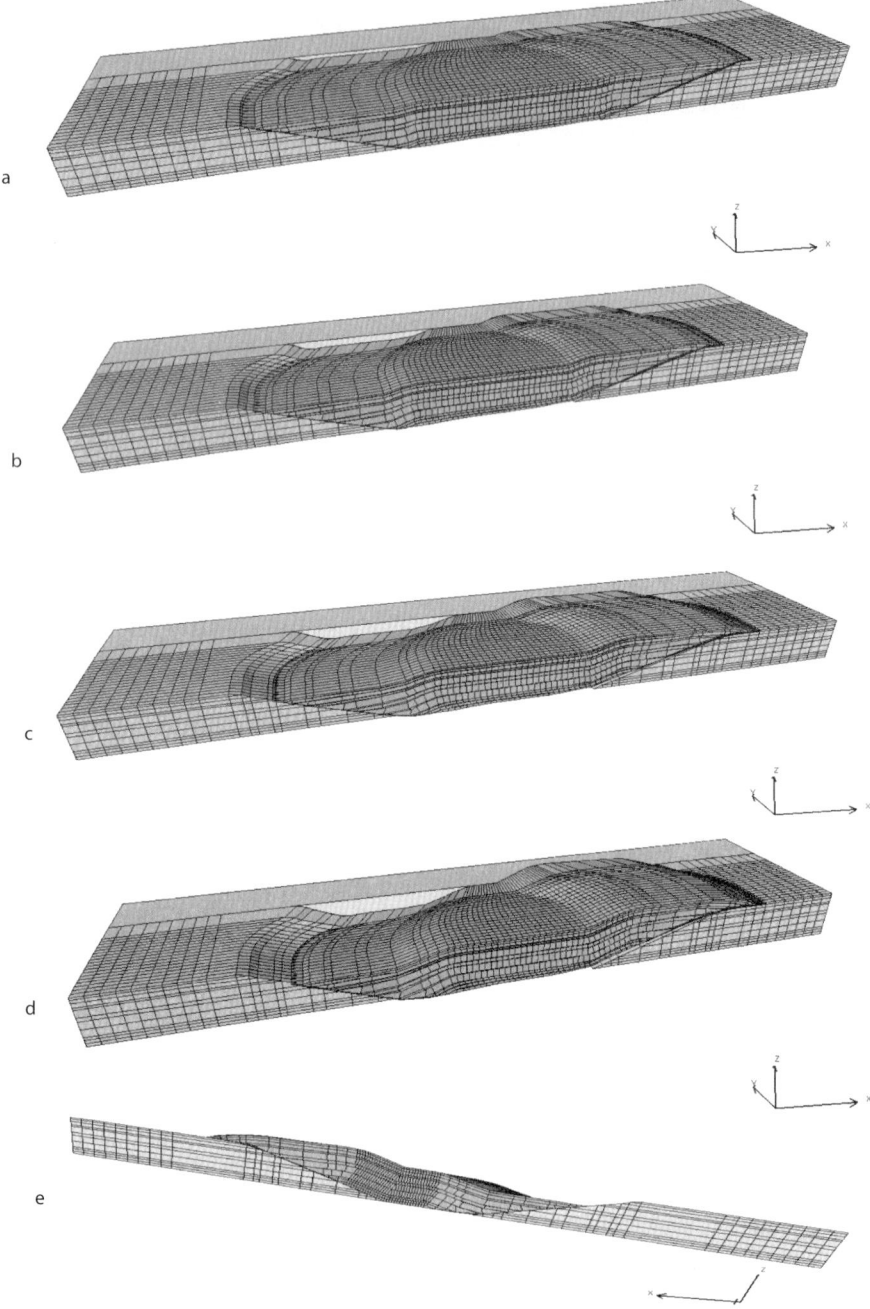

Fig. 3. (**a**) View of the deformed basin after 1000 m of displacement; (**b**) view of the deformed basin after 2000 m of displacement; (**c**) view of the deformed basin after 3000 m of displacement; (**d**) view of the deformed basin after 4000 m of displacement; (**e**) inverted view from the lateral imposed boundary of the deformed basin after 4000 m of displacement.

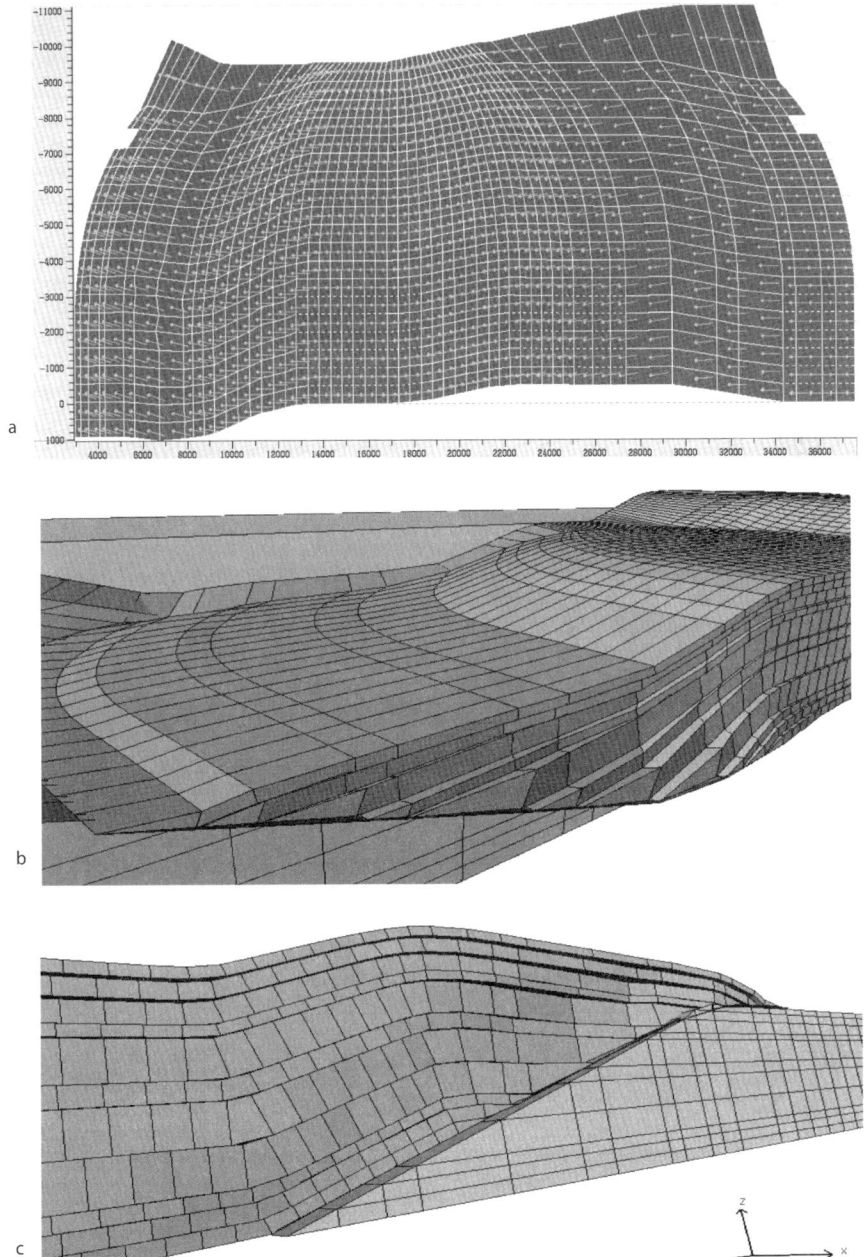

Fig. 4. (**a**) Map of the vertical projection on an horizontal plane of the neutral surface node displacement components, after 4000 m of displacement; (**b**) zoom of the lateral displacements observed on the normal fault after 4000 m of displacement; (**c**) zoom of the lateral displacements observed on the reverse fault after 4000 m of displacement.

Local volume variation

Here, the local volume variation will be designed by $\Delta V_i/V_i$, with $\Delta V_i = V_{iniial} - V_{final}$. According to literature, internal strain data must usually be taken into account in 2D during balancing (Woodward *et al.* 1986; Mitra 1994; Mac Naught & Mitra 1996), but also in 3D when deformation is no longer planar (Von Winterfeld & Oncken 1995). In our results, the local volume variations are clearly located in particular areas (Fig. 5a–d). Volume gain is observed for elements that cross the normal fault, whereas volume loss is observed for elements that cross the reverse fault. The geometric evolution of the elements can be tied to natural processes. If $\Delta V_i/V_i$ is positive, the internal processes are responsible for a decrease of the element overlaps, with a mechanism that could relate to compaction (for example by pressure-solution (Gratier 1993), with the occurrence of stylolites in field samples). On the other hand, if $\Delta V_i/V_i$ is negative, the internal deformation increases the amount of void space between the elements, and may be matched with micro-fracturing mechanisms such as open or sealed cracks in the field. These local variations are mainly caused by the geometric properties of the geological domain: basin morphology, fault architecture, blocks or layers. For example, volume variations are broadly correlated with the thickness of the layers. However, they still remain good indicators for the localization of highly strained zones. If we look at the evolution of $\Delta V_i/V_i$ within the basin, we see that the maximal negative values are located near the hinge zone of the fault-bend folds, where the curvature of the ramp is maximum. We also see a relative attenuation of the $\Delta V_i/V_i$ when getting farther from the curvature.

Normal fault geometry

The dip of the normal fault is clearly too low when compared with natural faults. We could have increased the dip of the faults; however, due to the discrete procedure, this dip must evolve progressively along the entire fault in the same way as for listric faults. Such a complex geometry was not tested here for simplicity.

Conclusion

The model presented here is a discrete kinematic model, which offers the opportunity to work both in extensional and compressional domains or coupled ones. This type of modelling is a first step in order to couple tectonic deformation modelling with the computation of fluid flow or hydrocarbon migration during progressive deformation. It is also a strong tool to study complex geological problems, even given that simplifications of the geometries may be required. It will ideally lead to a better understanding of the mechanism of internal strain, and to a better representation of localization of strain in three dimensions.

The quantitative values of the computed parameters, such as the local volume variation or the

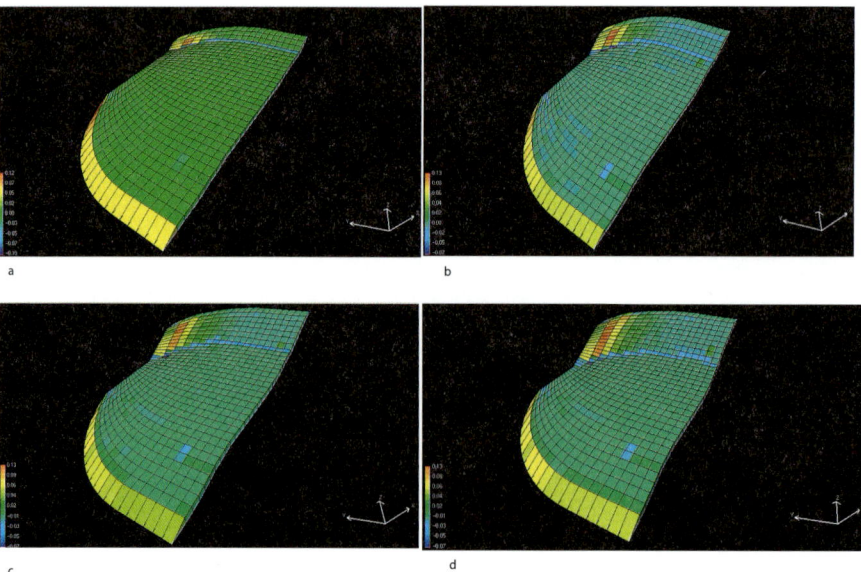

Fig. 5. The legend on the left of each picture corresponds to the $\Delta V_i/V_i$. (**a**) View of the local variation of volume on the first layer after 1000 m of displacement; (**b**) view of the local variation of volume on the first layer after 2000 m of displacement; (**c**) view of the local variation of volume on the first layer after 3000 m of displacement; (**d**) view of the local variation of volume on the first layer after 4000 m of displacement.

displacement directions, are strongly dependent on the geometry of the heterogeneous domains (faults, blocks and layers). Although the limitation of the computational scheme may restrict the numbers of layers, this is a feature common to all numerical models. Nevertheless, we believe that computed geometric and kinematic parameters should be used as semi-quantitative indicators of strain, thus allowing for comparison between 3D numerical modelling and field structures. In addition, systematic testing of the effect of the various parameters should provide general kinematic laws on the links between all these parameters, and guide further field study towards the most significant markers of the deformation (i.e. pressure-solution or fractures), both in the reservoir rock potential and seals.

References

BARNICHON, J. D. 1998. *Finite element modelling in structural and petroleum geology*. PhD thesis, Université de Liège.

BENNIS, C., VEZIEN, J. G. & IGLESIAS, G. 1991. Piecewise surface flattening for non-distorted texture mapping. *Computer Graphics*, **25(4)**, 237–246.

BOURGEOIS, E. 1997. *Mécanique des milieux poreux en tranformation finie: formulation des problèmes et méthodes de résolution*. PhD thesis, Ecole Nationale des Ponts et Chaussées.

CONTRERAS, J. & SUTER, M. 1990. Kinematic modelling of cross-sectional deformation sequences by computer simulation. *Journal of Geophysical Research*, **95**, 21913–21929.

CRANS, W., MANDL, G. & HAREMBOURE, J. 1980. On the theory of growth faulting: a geometrical delta model based on gravity sliding. *Journal of Petroleum Geology*, **2**, 265–307.

COUSSY, O. 1995. *Mechanics of Porous Continua*. Wiley, London.

CUNDALL, P. A. 1988. Formulation of a three-dimensional distinct element model-part I. A scheme to detect and represent contacts in a system composed of many polyhedral blocks. *International Journal of Rock Mechanics, Mineral Science & Geomechanical Abstracts*, **25(3)**, 107–116.

DAHLSTROM, C. D. A. 1969. Balanced cross sections. *Canadian Journal of Earth Sciences*, **6**, 743–757.

DE PAOR, D. G. 1988*a*. Balanced section in thrust belts, Part 1: construction. *AAPG Bulletin*, **72(1)**, 73–90.

DE PAOR, D. G. 1988*b*. Balanced section in thrust belts, Part 2: computerised line and area balancing. *Geobyte*, May, 33–37.

DIVIÈS, R. 1997. *FOLDIS: un modèle cinématique de bassins sédimentaires par éléments discrets associant plis, failles, érosion/sédimentation et compaction*. PhD thesis, Université Joseph Fourier de Grenoble.

EGAN, S. S. *et al.* 1998. Computer modelling and visualisation of the structural deformation caused by movement along geological faults. *Computers and Geosciences*, **25**, 283–297.

ERICKSON, S. G. & JAMISON, W. R. 1995. Viscous-plastic finite-element models of fault-bend folds. *Journal of Structural Geology*, **17**, 561–573.

GIBBS, A. D. 1983. Balanced cross-section construction from seismic sections in areas of extensional tectonics. *Journal of Structural Geology*, **5**, 153–160.

GRATIER, J. P. 1993. Le fluage des roches par dissolution-cristallisation sous contrainte, dans la crote supérieure. *Bulletin de la société géologique de France*, T. **164(2)**, 267–287.

GRATIER, J. P. & GUILLIER, B. 1993. Compatibility constraints on folded and faulted strata and calculation of total displacement using computationnal restoration (UNFOLD program). *Journal of Structural Geology*, **15(3–5)**, 391–402.

GRATIER, J. P., GUILLIER, B., DELORME, A. & ODONNE, F. 1991. Restoration and balance of a folded and faulted surface by best-fitting of finite elements: principles and applications. *Journal of Structural Geology*, **13(1)**, 111–115.

HART, R., CUNDALL, P. A. & LEMOS, J. 1988. Formulation of a three dimensional distinct element model-part II. Mechanical calculations for motion and interaction of a system composed of many polyhedral blocks. *International Journal of Rock Mechanics, Mineral Science & Geomechanical Abstracts*, **25(3)**, 117–125.

MAC NAUGHT, M. A. & MITRA, G. 1996. The use of finite strain data in constructing a retrodeformable cross-section of the Meade thrust sheet, southeastern Idaho, U.S.A. *Journal of Structural Geology*, **18(5)**, 573–583.

MITRA, G. 1994. Strain variation in thrust sheets across the Sevier fold-and-thrust-belt (Idaho-Utah-Wyoming): implications for sections restoration and wedge taper evolution. *Journal of Structural Geology*, **16(4)**, 585–602.

MORRETTI, I., WU, S. & BALLY, A. W. 1990. Computerized balanced cross-section LOCACE to reconstruct an allochtonous salt sheet, offshore Louisiana. *Marine and Petroleum Geology*, **7**, 371–377.

NIÑO, F., CHÉRY, J. & GRATIER, J. P. 1998. Mechanical modelling of the Ventura basin : origin of the San Cayetano thrust fault and interaction with the Oak Ridge fault. *Tectonics*, **17**, 955–972.

RAMSAY, J. G. & HUBER, M. I. 1987. *The Techniques of Modern Structural Geology - Volume 2 : Folds and Fractures*. Academic Press, New York.

ROWAN, M. G., JACKSON, M. P. A. & TRUDGILL, B. D. 1999. Salt-related fault families and fault welds in the northern Gulf of Mexico. *AAPG Bulletin*, **83**, 1454–1484.

SCHNEIDER, F. & WOLF, S. 2000. Quantitative HC potentiel evaluation using 3D basin modelling : application to Franklin structure, central graben, North sea, U.K. *Marine and Petroleum Geology*.

SCHNEIDER, F., WOLF, S., FAILLE, I. & POT, D. 2000. Un modèle de bassin 3D pour l'évaluation du potentiel pétrolier : application à l'offshore congolais. *Oil and Gas Science and Technology*.

SHAW, J. H., HOOK, S. C. & SUPPE, J. 1994. Structural trend analysis by axial surface mapping. *AAPG Bulletin*, **78**, 700–721.

SUPPE, J. 1983. Geometry and kinematic of fault-bend folding. *American Journal of Science*, **283**, 684–721.

TRUDGILL, B. D., ROWAN, M. G., FIDUK, J. C. *et al.* 1999. The perdido fold belt, Northwestern Deep Gulf of Mexico, part 1: structural geometry, evolution and regional implications. *AAPG Bulletin,* **83**, 88–113.

VON WINTERFELD, C. & ONCKEN, O. 1995. Non-plane strain in section balancing: calculation of restoration parameters. *Journal of Structural Geology,* **17(3)**, 457–450.

WALTHAM, D. 1989. Finite difference modelling of hanging wall deformation. *Journal of Structural Geology,* **11**, 433–437.

WALTHAM, D. 1990. Finite difference modelling of sandbox analogues, compaction and detachment free deformation. *Journal of Structural Geology,* **12**, 375–381.

WILKERSON, M. S. & MEDWEDEFF, D. A. 1991. Geometrical modelling of fault-related folds: a pseudo-three-dimensional approach. *Journal of Structural Geology,* **13**, 801–812.

WOODWARD, N. B., GRAY, D. R. & SPEAR, D. B. 1986. Including strain data in balanced cross section. *Journal of Structural Geology,* **8(3–4)**, 313–324.

ZOETEMEIJER, R. 1992. *Tectonic modelling of forelands basins, thin skinned thrusting, syntectonic sedimentation and lithospheric flexure.* PhD thesis, Vrije Universiteit Amsterdam.

Modelling the influence of tectonic compression on the *in situ* stress field with implications for seal integrity: the Haltenbanken area, offshore mid-Norway

T. SKAR[1] & F. BEEKMAN[2]

[1]*University of Bergen, Geological Institute, Allegaten 41, 5007 Bergen, Norway (e-mail: tore.skar@geo.uib.no)*
[2]*Tectonics/Structural Geology Group, Vrije Universiteit, De Boelelaan 1085, 1081 HV Amsterdam, The Netherlands (e-mail: fred.beekman@falw.vu.nl)*

Abstract: Exploration for hydrocarbons in overpressured domains is often considered high risk because of the possibility of seal failure and fluid leakage due to natural hydraulic fracturing. Several of the wells drilled in highly overpressured reservoirs on Haltenbanken, offshore mid-Norway, have proved to be devoid of hydrocarbons suggesting that ineffective seals are the cause of exploration failure. However, recent petroleum discoveries within this area demonstrate that fluid pressure is not the ultimate control on entrapment of hydrocarbons. We investigate the way in which far-field tectonic compression may have influenced the *in situ* stress conditions on Haltenbanken, and assess whether tectonic stresses also may facilitate local fracturing of the seal by reducing the retention capacity (minimum horizontal stress-fluid pressure).

We have approached the problem by applying a finite element model. The elasto-plastic model assumes two-dimensional plane-strain and is constrained from geological and geophysical data. The results show that: (1) contrasts in the rock's mechanical properties across discontinuities (e.g. sediment interfaces) cause rapid shifts in stress magnitudes; (2) the differences in stress magnitudes across such discontinuities can be subdued or enhanced under increased horizontal compression; and (3) structurally controlled variations in vertical displacements produce local concentrations of highs and lows in the stress field. The combined result of these three factors is that the magnitude of horizontal stress may vary quite considerably within spatially restricted areas. The implication of these predictions in terms of hydrocarbon preservation potential in highly overpressured regions is that rapid shifts in minimum horizontal stress magnitudes can reduce the retention capacity and therefore facilitate natural hydraulic fracturing.

An important concept in the exploration and production of hydrocarbons in deep sedimentary basins is that of pressure compartments (Bradley 1975). Pressure compartments are characterized by distinct pressure regimes, which affect the maturation, migration and trapping of hydrocarbons (Hunt 1990). Such compartments are typically bounded laterally and vertically by low-permeability barriers. Fault systems most often represent lateral barriers to fluid flow, whereas vertical fluid flow is restricted by the presence of low-permeability caprock lithologies.

The preservation potential for hydrocarbons in an overpressured (fluid pressure in excess of hydrostatic pressure) reservoir is a function of several factors, the most important being the relationship between the fluid pressure and *in situ* stress that controls hydraulic fracturing (e.g. Grauls 1998). Natural hydraulic fracturing occurs when fluid pressure reaches the mechanical strength of a rock, or its equivalent, the minimum horizontal stress. In a fluid dynamic system where lateral fluid flow out of a pressure compartment is restricted, overpressure and fluid movement are therefore regulated by the stress regime and retention capacity (σ_{Hmin} − fluid pressure) of the caprock (Caillet 1993; Gaarenstroom *et al.* 1993; Grauls 1998; Fig. 1). Thus, knowledge of spatial and temporal variations in the magnitude of horizontal stress is important in risk analysis when exploring for hydrocarbons in overpressured basins.

In this paper geomechanical modelling is applied

From: NIEUWLAND, D. A. (ed.) *New Insights into Structural Interpretation and Modelling.* Geological Society, London, Special Publications, **212**, 295–311. 0305-8719/03/$15

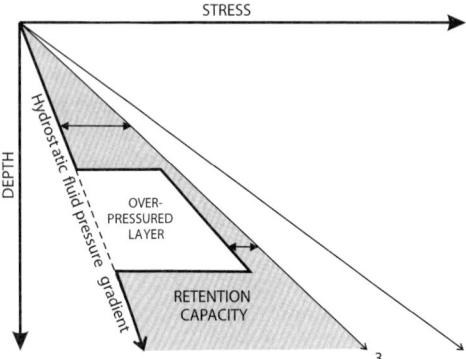

Fig. 1. Schematic illustration of the relationship between the fluid pressure, overburden pressure (1) and fracture pressure (3). The fracture pressure is determined from leak-off tests (LOT) and is an approximation of the minimum horizontal stress. The difference between the minimum horizontal stress and the fluid pressure is defined as the retention capacity.

to assess the influence of far-field tectonic compression on the local stress distribution in overpressured sedimentary rocks of Haltenbanken, offshore mid-Norway. A depth-converted regional geological cross-section that traverses part of the mid-Norwegian continental margin is utilized as a case study. In order to account for the present-day compressive stress field on the margin a slowly increasing horizontal load is applied at the edge of the model. The study area has a complex geological history where the basement structural grain is likely to have played an important role for the structural evolution (e.g. Gabrielsen *et al.* 1999; Pascal & Gabrielsen 2001). To keep the model simple, however, the influence of structural grain on the temporal stress and strain evolution is not considered.

One of the main goals of the modelling was to investigate whether tectonic stresses may facilitate local fracturing of the seal, for instance by locally reducing the retention capacity. In investigation of this effect and to obtain the best insight in the net/pure effect of tectonic stresses only in the local distribution of minimum horizontal stress, other possible mechanisms for fracturing must be excluded from the models. For that reason, pre-existing faults, fluid pressure and porosity are not included in the present model. This is acceptable because the aim of the modelling is not to accurately predict the magnitude of horizontal stresses in the subsurface, but rather to assess the most important factors that may contribute to local variations in the horizontal stress.

Geological setting

The study area is located on Haltenbanken (Fig. 2), which is part of a sedimentary basin situated on the mid-Norwegian margin, that has originated from multiple phases of extensional faulting, including Devonian collapse of the Caledonian Orogeny, late Permian stretching, and rifting episodes in the Mesozoic and early Cenozoic (e.g. Brekke & Riis 1987; Skogseid & Eldholm 1989; Eldholm *et al.* 1989; Skogseid *et al.* 1992; Blystad *et al.* 1995; Brekke *et al.* 1999; Doré *et al.* 1999; Gabrielsen *et al.* 1999; Reemst & Cloetingh 2000). The Triassic-Jurassic rifting may have affected the entire mid-Norwegian shelf and led to the formation of large Triassic and Jurassic basins (Gabrielsen & Robinson 1984; Bøen *et al.* 1984). The Jurassic-Cretaceous rifting episode structured the region by separating the stable Trøndelag Platform from the Møre and Vøring Basins, which were developed by subsequent thermal subsidence and deposition of a thick sequence of Cretaceous sediments (Fig. 3) (e.g. Blystad *et al.* 1995; Brekke *et al.* 1999; Doré *et al.* 1999). The last rifting episode started in the late Cretaceous and terminated with the opening of the northern part of the North Atlantic ocean during earliest Eocene (Skogseid *et al.* 1992).

The tectonic evolution after seafloor spreading between Norway and Greenland was strongly affected by a change in the regional stress regime from extensional to weakly compressional (Grunnaleite & Gabrielsen 1995; Doré & Lundin 1996; Vågnes *et al.* 1998; Gabrielsen *et al.* 1999). The post-late Eocene tectonic evolution of the mid-Norwegian margin was characterized by periods of regional uplift of Fennoscandia, growth of domes, arches and inverted structures in the Vøring and Møre basins and finally, by an increased rate of uplift of the continent and the deposition of thick clastic wedges in the adjacent offshore basins in close association with the Northern Hemisphere glaciations during late Neogene-Quaternary times (e.g. Stuevold & Eldholm 1996).

Today, the mid-Norwegian margin is in a compressive state of stress and is characterized by intermediate to low seismicity. Earthquake focal mechanism solutions give NW-SE oriented maximum horizontal stresses, with a dominance of strike-slip (mainly) and reverse faulting stress regimes (Bungum *et al.* 1991; Fejerskov *et al.* 1995; Lindholm *et al.* 1995*b*). This direction indicates that the stress field is the result of ridge push from the North Atlantic spreading zone. The same general orientation of the maximum horizontal stress direction also dominates in the North Sea basin (Lindholm *et al.* 1995*a*) showing the large-scale consistency of the stress field. Onshore *in situ*

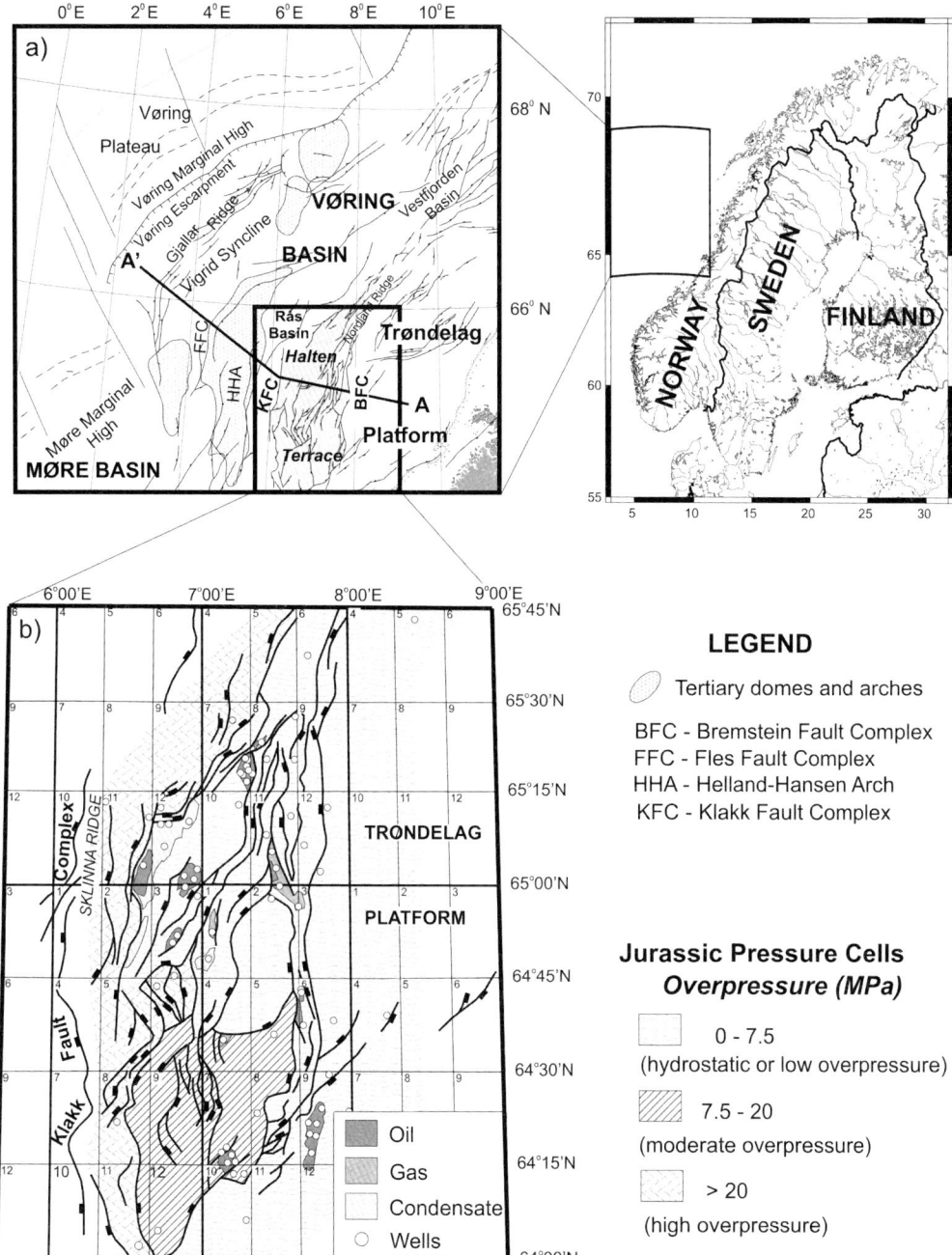

Fig. 2. (**a**) Map of the main structural elements within the study area (modified from Blystad *et al.* 1995). The black solid line indicates the location of the geological cross-section in Figure 3. (**b**) Jurassic pressure compartments and petroleum provinces on Haltenbanken (modified from Hermanrud *et al.* 1998).

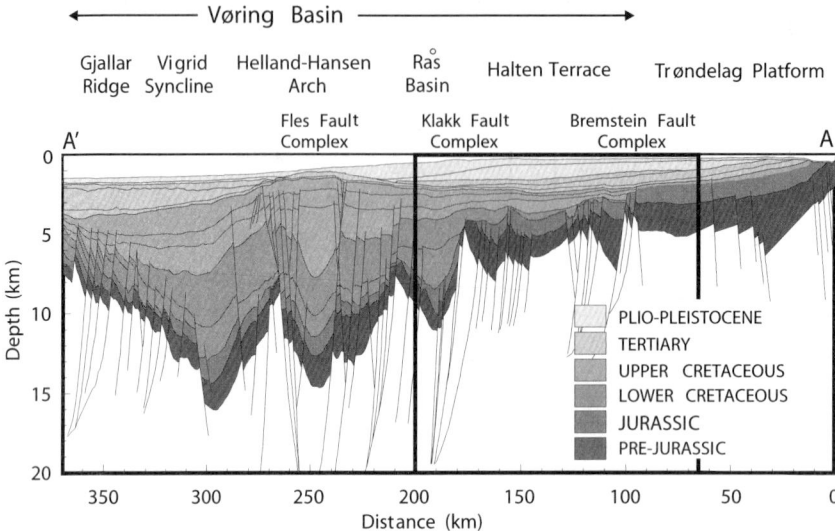

Fig. 3. Depth-converted geological profile across the Vøring Basin and Trøndelag Platform showing the main geological features. The Halten Terrace (the study area) is separated from the Trøndelag Platform by the Bremstein Fault Complex and from the Rås Basin in the west by the Klakk Fault Complex. The location of the cross-section is shown in Figure 2.

stress measurements obtained by overcoring (Fejerskov *et al.* 1995) and offshore bore-hole breakout directions (Borgerud & Svare 1995; Gölke 1996) confirm the existence of a NW-SE compressional σ_H.

The Haltenbanken area

Haltenbanken (including the Halten Terrace and Trøndelag Platform) is an important hydrocarbon province and has been extensively explored since the early 1980s. Successful drilling on a variety of structures has led to a good understanding of the sedimentary development and petroleum geology in this area (Heum *et al.* 1986; Cohen & Dunn 1987; Forbes *et al.* 1991; Ehrenberg *et al.* 1992; Koch & Heum 1995; Karlsen *et al.* 1995). Recently, the focus of exploration activity on the Norwegian Continental Shelf has shifted from 'shallow' water areas (Trøndelag Platform and Halten Terrace) to deep water areas (Vøring and Møre Basins) (Brekke *et al.* 1999). Exploration for hydrocarbons on the Halten Terrace remains, however, commercially attractive, especially on its western part where large accumulations of potentially overpressured hydrocarbons are thought to be present.

Sedimentology and petroleum geology

The stratigraphic framework of the Haltenbanken area was defined by Dalland *et al.* (1988) (Fig. 4). A thick Triassic sequence of alternating evaporites and claystones is overlain by a thick fluvio-deltaic and open-marine Jurassic sequence (Jacobsen & Van Veen 1984; Gjelberg *et al.* 1987; Dalland *et al.* 1988). The Båt Group comprises lower delta-plain deposits of the lower Jurassic Åre Formation and tidal, shoreface and offshore facies of the Tilje, Ror and Tofte Formations. The overlying middle to upper Jurassic Fangst and Viking Groups comprise tidal to increasingly open-marine facies of the Ile, Not, Garn, Melke and Spekk Formations.

Marine conditions continued to prevail during Cretaceous and Tertiary times. The Cretaceous sequence is up to 1700 m thick and comprises predominantly shales and siltstones of the Cromer Knoll and Shetland Groups. The Tertiary sedimentary record is composed of marine shales of heterogeneous composition. The highest rates of sedimentation occurred during Pliocene-Pleistocene times when the Haltenbanken area was covered by more than 1000 m of fine-grained deposits.

Laterally extensive marine sandstones of the Jurassic Tilje, Ile and, in particular, the Garn Formation are the best reservoir units in the area. Less

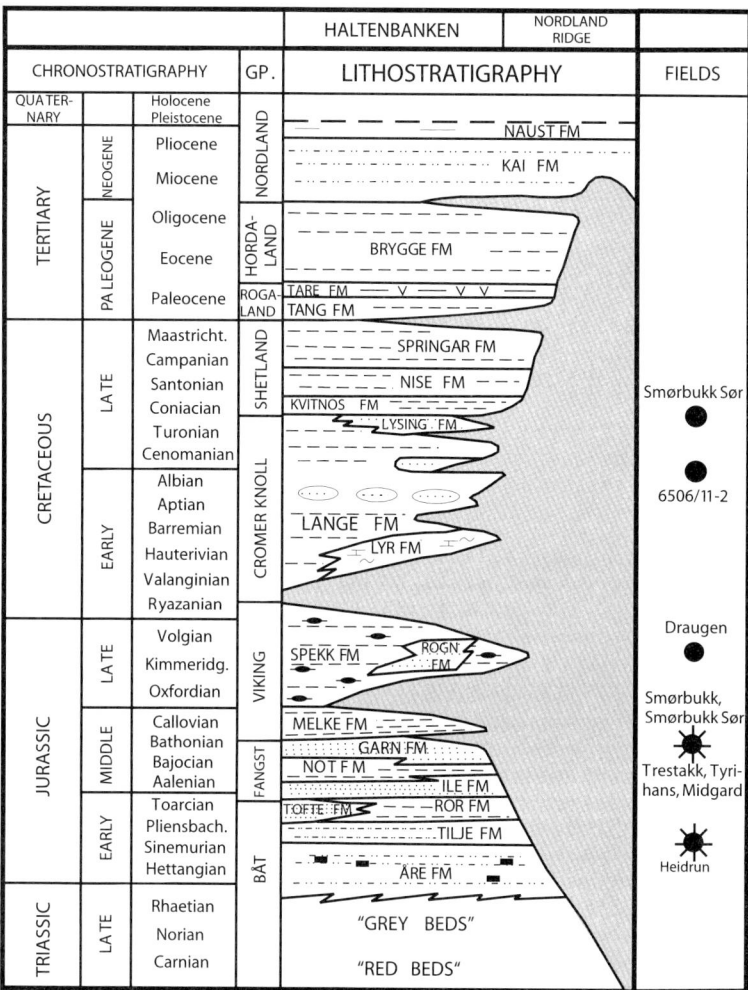

Fig. 4. Generalized lithostratigraphiy and oil/gas occurrences on Haltenbanken (modified from Dalland *et al.* 1988).

attractive Cretaceous plays are mainly confined to the sandstones of the Lange and Lysing Formations. The main source rocks are the Jurassic Åre and Spekk Formations, the latter being the time-equivalent of the Kimmeridge Clay of the North Sea.

Fluid pressure and in situ stress

Fluid pressures in the Jurassic reservoirs on Haltenbanken are characterized by pronounced spatial variations (Koch & Heum 1995; Hermanrud *et al.* 1998; Skar *et al.* 1999; Fig. 2). Reservoir fluids at depths exceeding 3 km have highly varying fluid pressure depending on the structural position on the margin. The highest fluid pressures are encountered on the western part of Haltenbanken where over-

pressure may exceed 35 MPa. Reservoir fluids on the Trøndelag Platform and on the central and eastern parts of the Halten Terrace are close to hydrostatic pressure. A transitional area with moderate fluid overpressure is encountered on the southern part of the terrace.

The high overpressures in the Jurassic reservoirs on the western part of Haltenbanken indicate that the fluids are entrapped in a relatively closed system with regards to fluid pressure. In such a system, overpressures and fluid movements are regulated by the stress regime and retention capacity of the caprock (Caillet 1993; Gaarenstroom *et al.* 1993). An estimate of the minimum horizontal stress of the caprock can be assessed by performing leak-off tests (LOT), which are conducted at each casing point prior to drilling ahead. Leak-off tests are typically taken in low-permeability lithologies (e.g. claystones and hard limestones), whereas fluid pressures are typically measured in sandstones.

LOT data from claystone formations on Halten-banken are shown in Figure 5a. The LOT data usually vary between 85% and 92% of the overburden, except at shallow depths and in the highly over-

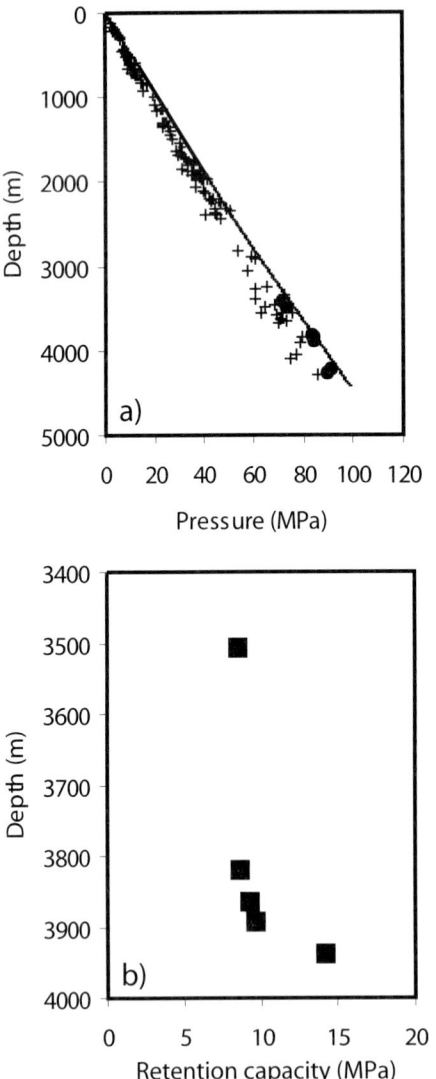

Fig. 5. (**a**) Relationship between leak-off test (LOT) data and overburden pressure (solid line). The cross symbols represent data from formations where the fluid pressures are close to hydrostatic pressure. The black dots represent data collected in formations with high fluid pressure. (**b**) Retention capacity in the upper Jurassic shales (caprocks) in the highly overpressured region on the western part of Haltenbanken.

pressured region. In the overpressured region, the leak-off pressures are close to the overburden pressure (Fig. 5a). Similar relationships between fluid pressure and minimum horizontal stress in highly overpressured formations are found in a number of basins worldwide (e.g. Breckels & van Eekelen 1982; Engelder & Fischer 1993).

The risk of seal failure increases when the retention capacity is less than 7 MPa as shown by Gaarenstroom et al.'s (1993) analysis of overpressured pre-Cretaceous rocks in the Central North Sea. Fluid pressures measured in the Jurassic reservoirs on Haltenbanken have been extrapolated to the top of the structures in order to relate the fluid pressure at the seal–reservoir interface to the local stresses. Retention capacities in caprocks in the overpressured region are as low as 8 MPa (Fig. 5b). In addition, the location of the measurements does not necessarily coincide with the up-dip position of the shallowest reservoirs, where natural hydraulic fracturing does take place. This implies that local variations in the minimum horizontal stress magnitudes may be of significance as a local decrease in horizontal stress can reduce the retention capacity and thereby facilitate fracturing.

Modelling approach

In the following section we investigate the influence of far-field tectonic compression on the stress distribution in the Mesozoic rocks and discuss some factors that may contribute to local variations in the local stress regime. We approached the problem by constructing a two-dimensional finite element model using the cross-section shown in Figure 3 as the geometrical geological constraint. The cross-section is oriented approximately NW-SE and therefore runs parallel to the orientation of the maximum horizontal stress on the mid-Norwegian margin. The numerical analysis assumes two-dimensional plane-strain and was carried out using the commercial finite element code Diana.

Model input and constraints

The modelled cross-section is 135 km long, has a height of 20 km and consists of basement and 15 sedimentary units ranging in age from pre-Jurassic to Quaternary. The Cretaceous-Quaternary sedimentary units are generally laterally continuous without any main faults offsetting the units. The pre-Cretaceous sequence, however, is faulted with vertical displacements ranging between a few tens of metres to almost 5 km along the Klakk Fault Complex. Shale is the dominant lithology with the highest sand content being encountered in the middle and lower Jurassic succession (see Fig. 4).

Material properties (Poisson's ratio, Young's modulus) assigned to the sedimentary layers and the basement are listed in Table 1. Due to a lack of published data on the rock mechanical properties of formations on the mid-Norwegian margin, the Young's modulus (E) of each sedimentary layer has been estimated from lithological data. The E-values have been calculated from (Zhang *et al.* 1995):

$$E = V_p^2 \, \rho \, (1 + \nu) \, (1-2\nu)/(1 - \nu)$$

where V_p is the velocity of the compressional wave (P-wave), ρ is the bulk density, and ν is Poisson's ratio. The bulk density data and velocity data used for estimation of E are constrained from an analysis of geophysical well log responses (Skar *et al.* 1998; Skar 2000). The calculated E values generally increase with depth, reflecting the increased stiffness of the rocks as a function of compaction. Poisson's ratio is assumed to be 0.20 for shale and 0.25 for sandstone unless measured data are available (Table 1). The calculated values for E are within the same order of magnitude as those used by Gölke (1996) for stress modelling on the mid-Norwegian margin.

The finite element model is based on the seismic cross-section, lithological data and the estimated material properties of the sedimentary layers (Fig. 6). The mesh is composed of almost 11 000 quadratic elements, with the highest density of elements in areas of interest.

To describe the stress and strain history of the rock, conventional rheological models for brittle rocks, isotropic linear elastic and elasto-plastic material behaviour are used. The yield condition, which specifies the state of stress at which rock failure is initiated, is described according to the Mohr-Coulomb criterion. In order to best study the net effect of far-field tectonic stresses the fault planes are not assigned any properties and no slip can therefore occur along the their surfaces.

Boundary conditions

The numerical predictions are sensitive to the assumptions of the initial boundary constraints, here defined as follows:

(i) the upper surface is free to deform horizontally and vertically;
(ii) the basal boundary is fixed in the vertical direction, but is allowed to move laterally;
(iii) the left edge of the model is fixed in the horizontal direction, but is allowed to move vertically;
(iv) the right edge of the model is allowed to move vertically, but is initially fixed in the horizontal direction to allow the geological structure to obtain an initial state of stress and strain associated with gravitationally derived body forces.

An assumption must be made on the *in situ* stress distribution in the subsurface before the modelled geological structure is deformed by horizontal compression. Ideally, information about the initial state is given by field measurements, but when these are not available, it is feasible to assess the *in situ* stress by considering a range of possible conditions and constraining factors (e.g. the system must be in equilibrium). In the model, the initial *in situ* vertical stresses are equal to the lithostatic

Table 1. *Material properties assigned to the sedimentary layers used in the numerical models*

Unit	Formation/Group	Bulk density (kg m^{-3})	Poisson's ratio	Young's modulus (GPa)
16	Naust	2100	0.2	8.96
15	Naust	2200	0.2	10.9
14	Naust	2300	0.2	13.4
13	Kai	2200	0.2	10.1
12	Brygge	1850	0.2	6.9
11	Tang/Tare	1900	0.2	6.2
10	Springar	2150	0.19	8.1
9	Nise	2300	0.2	14.6
8	Kvitnos	2400	0.2	18.2
7	Lange	2550	0.2	23.6
6	Lange	2600	0.2	30.1
5	Lange	2550	0.2	21.3
4	Spekk/Melke	2600	0.2	34
3	Fangst/Bat Gr.	2650	0.25	41.9
2	Triassic	2700	0.2	46.1
1	Basement	2800	0.3	50

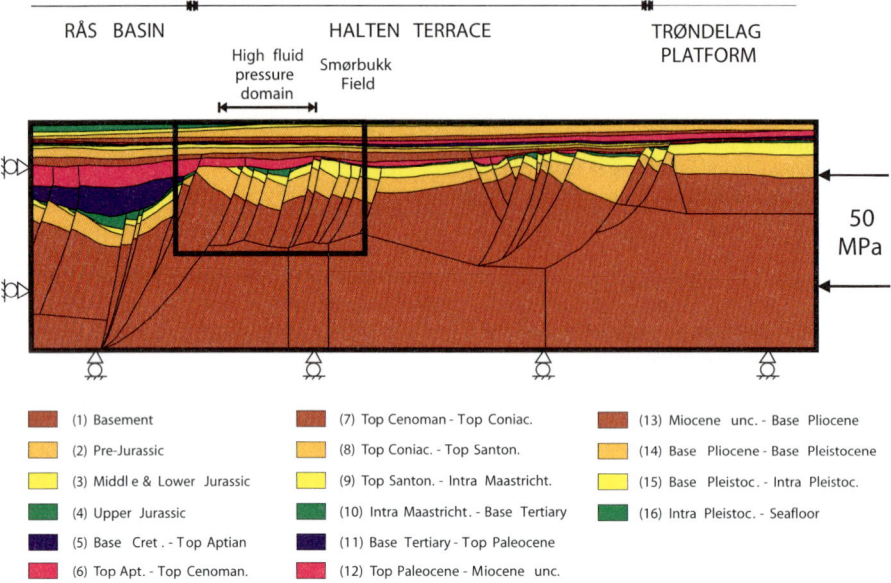

Fig. 6. The two-dimensional elasto-plastic model that was constructed to simulate the influence of far-field tectonic compression on the *in situ* stress field in the sedimentary fill. The model is based on the geological cross-section in Figure 2. The black box shows the location of the plots presented in Figures 10 and 11.

weight of the overburden: $\sigma_{zz} = \rho g z$, where g is gravitational acceleration, ρ is the mass density of the material, and z is the depth below the surface. The *in situ* horizontal stresses are given by the natural ratio between horizontal and vertical stress: $\sigma_{xx} = \sigma_{zz}(\nu/(1 - \nu))$ where ν is Poisson's ratio (e.g. Ranalli 1995). The formula is derived from the assumption that gravity is applied to an elastic mass of material in which lateral movement is prevented. This condition hardly ever applies in real geological systems due to repeated tectonic movements, material failure, erosion, sedimentation and remnant stress, e.g. associated with faulting. A compromise is therefore made by installing a set of stresses in the model, and running the model until an equilibrium state is obtained, which then serves as the initial *in situ* stress.

To study the mechanical behaviour of the upper crust under compression, the horizontal displacement constraint of the eastern edge of the model is removed and a stress of constant horizontal magnitude is applied at this edge, thereby generating a compressional intraplate stress field from 'far-field' plate tectonic boundary forces. The magnitude of the applied tectonic pressure is increased slowly from 0 to an arbitrary maximum of 100 MPa. Tectonic loading is applied in small incremental steps to closely track the load-history-dependent plastic strains and to gain a better convergence behaviour in the non-linear models. For the modelling results to remain valid, the model is not allowed to deform

to such an extent that the final geometry substantially deviates from the initial geometry.

The main advantage of using a constant stress magnitude as a boundary constraint is to better assess the role of rock heterogeneities and local discontinuities on the predicted *in situ* stress and strain. The predicted change in tectonic stress due to horizontal loading can easily be compared with the magnitude of the applied compressive stress and, thus, the role of lithology variations and dominant structural features can be assessed within different parts of the model. Some comments on the initial assumptions on the predicted stress and strain are presented later.

Model results

Gravitational body forces (vertical loading) and tectonic loading change the geometry and thereby the stress distribution in sedimentary basins. In order to assess factors that may influence the local stress distribution we need to consider aspects at different spatial scales.

The calculated results are displayed as two-dimensional contour plots showing the predicted vertical displacements and *in situ* stresses. Additional one-dimensional profiles are included at various locations along the two-dimensional profile to illustrate the predicted lateral and vertical variations. Results are shown for one load case with gravitational loading only and one case with the combined result of vertical loading and horizontal loading (in-plane compression of 50 MPa is applied at the edge of the model).

Sub-basin scale

The predicted variations in vertical displacement when only gravitational body forces are applied to the model are shown in Figure 7a. The magnitudes of the displacements are small, ranging from -0.5 m (subsidence) to 0.5 m (uplift), but pronounced spatial variations in signs and magnitudes are predicted. Uplift occurs in narrow zones that are separated in space by subsiding areas.

The predicted vertical deflections after applying a slowly increasing horizontal compression to the edge of the model are shown in Figure 7b. The pattern of vertical deflection as predicted for the gravitationally loaded model becomes progressively modified under increased tectonic loading. The calculated displacements are positive across the entire model because the basal boundary is fixed. However, lateral variations in the amount of displacement are indicated by the undulating nature of the contours (Fig. 7b).

The results show that the initial structural geometry on which the numerical model is based exerts an important control on the location of differential vertical displacements. The uplifted areas (uplift relative to the surrounding areas) occur in narrow zones and are spatially restricted to the crestal areas of the rotated fault blocks where top basement appear as a topographic high (e.g. Sklinna Ridge). In contrast, the largest subsidence

is concentrated where the basement is thin (e.g. Rås Basin) and at locations on the Halten Terrace where the top basement surface has the shape of a depression.

The vertical stresses induced by sediment loading reveal a gradual and rather uniform increase with depth (Fig. 8a). The minor variations in the stress magnitude at different locations reflect the rock's density structure. The estimates are largely in good agreement with those vertical stress estimates that are obtained by integration of the bulk density logs (Fig. 9). The predicted horizontal stress magnitudes also show an overall increase with depth, but the magnitudes display more spatial variability than those of the vertical stresses (Fig. 8c).

Adding a slowly increasing horizontal compression has only minor effect on the vertical stress distribution (Fig. 8b). The predicted increase in vertical stress magnitude is small (or negligible) compared to those stress magnitudes generated by sediment loading. The horizontal stress magnitudes, however, become amplified under increased horizontal compression. The most pronounced variations in horizontal stress magnitudes occur within the middle depth interval where the rocks are faulted (Fig. 8d).

Fault-block scale

Contour plots and one-dimensional vertical profiles are shown at the fault-block scale to better illustrate the factors that contribute to the spatial variations in *in situ* stress when the geological structure is

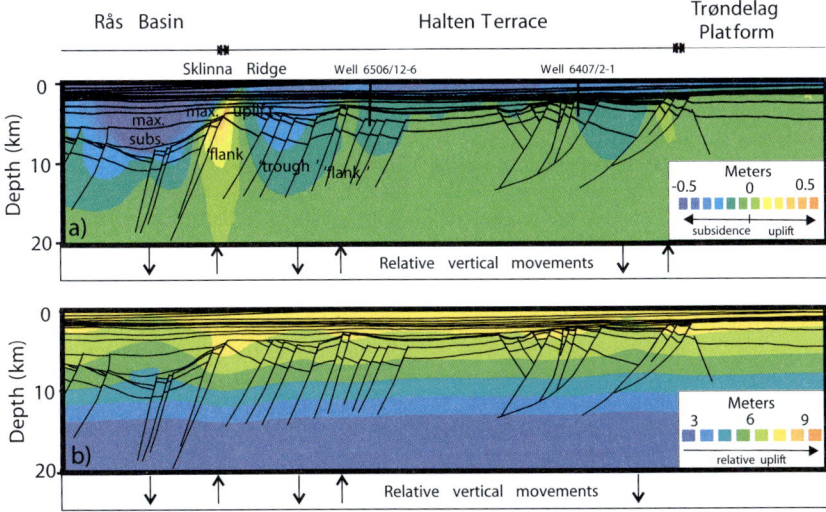

Fig. 7. Contour plots of the calculated vertical displacements in the elasto-plastic model for (**a**) only gravitational loading, and (**b**) with an additional horizontal in-plane compression of 50 MPa applied to the right side of the model.

Vertical stress

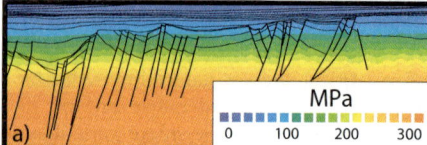
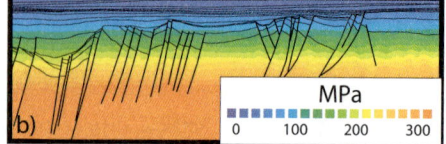

Horizontal stress

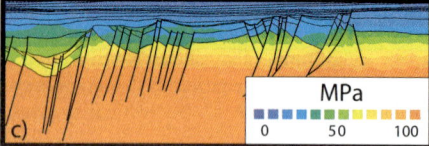
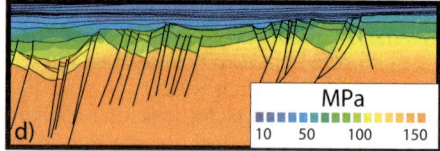

Fig. 8. Contour plots of the vertical and horizontal stress in the elasto-plastic model. (**a**) Vertical stress induced by gravitational loading; (**b**) vertical stress after 50 MPa of in-plane compression is applied to the right side of the model; (**c**) horizontal stress induced by gravitational loading; (**d**) horizontal stress after 50 MPa of in-plane compression is applied to the right side of the model.

subjected to an increase in far-field tectonic compression.

The predicted change in the horizontal stress in response to tectonic loading only (net tectonic effect) is shown in Figure 10. Note that the plots are contoured proportionally to the stress magnitude applied to the edge of the model. The modelling results reveal that the change in horizontal stress magnitudes is almost identical for 10 and 100 MPa tectonic loading. This demonstrates that the response of the rocks to increased horizontal compression is almost purely elastic.

In the previous section it was shown that the structural fabric (e.g. basement geometry) influenced the vertical deflection pattern (Fig. 7). It is therefore of interest to assess the way in which the stress magnitudes at specific structural locations change in response to an increase in tectonic loading. To allow for a visual comparison between the initial, gravitationally induced horizontal stress distribution and the stress distribution calculated after far-field tectonic compression is applied to the model, the plots in Figure 11 are contoured according to the level of horizontal compression applied to the edge of the model.

The model results show that the initial stress configuration only becomes slightly modified when compression is applied to the model (Fig. 11). A relationship exists, however, between the progressive change in stress magnitudes and the structural position (Fig. 11b). Relative stress maxima develop in the central part of the deep structure (trough), whereas relative stress lows appear at the crests of the rotated fault blocks (crest). This is caused by the vertical deflection pattern, producing bending-induced compressional and tensile stresses in the

troughs and crests, respectively. These bending stresses are superimposed on the compressional background stress field, thus producing local highs and lows in the stress field.

Figure 12 shows in more detail the horizontal stress distribution along two vertical profiles located in the trough and at the flank of the structure shown in Figure 11. The stress magnitude does not increase uniformly with depth, but changes abruptly at every stratigraphic interface. This is due to contrasts in the material properties and, in particular, the density and the elastic properties. The stress magnitudes generally become amplified under increased horizontal compression, but the magnitude of stress change within a specific formation depends on the actual depth, structural position and the magnitude of tectonic loading. The differences in stress magnitudes across the interfaces may either be enhanced (e.g. across interface between units 3 and 4) or reduced (e.g. across interface between units 2 and 3) in response to the increase in tectonic loading. This reflects the capability of the different rocks to transmit stresses.

Reservoir scale

One of our objectives was to assess the influence of far-field tectonic compression on the state of stress in the caprock in a faulted sequence. The predicted σ_{Hmin} in the Viking Group (caprock) on the western part of the Halten Terrace is characterized by pronounced lateral variations in magnitude, in particular across fault steps (Fig. 13a). In more detail, the difference in predicted stress magnitude at a specific depth may be in the order of megapascals depending on the applied magnitude

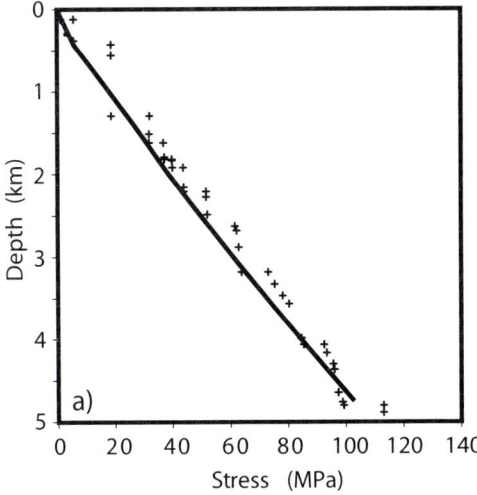

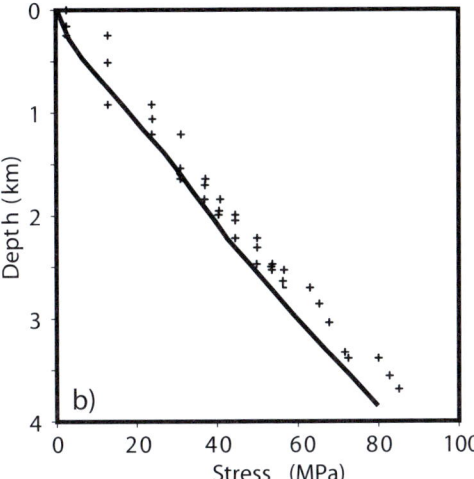

Fig. 9. Comparison of the predicted vertical stress (cross symbols) and the vertical stress as determined by integration of bulk density logs (solid lines) at the location of two wells. The location of the wells is shown in Figure 7: (**a**) well 6506/12–6; (**b**) well 6407/2–1.

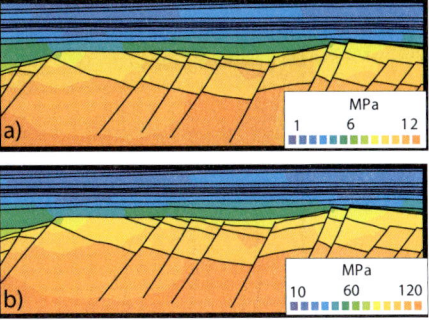

Fig. 10. Contour plots of the predicted change in horizontal stress due to an increase in horizontal compression of (**a**) 10 MPa and (**b**) 100 MPa. The colour codes used for contouring are the same in both plots, but the values in (b) are scaled proportionally relative to the magnitude of applied horizontal compression (multiplied by a factor of 10 compared to (a)). The location of the plots is shown in Figure 6.

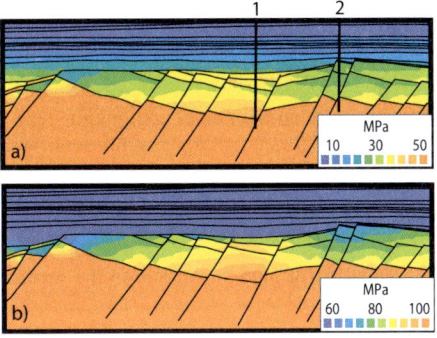

Fig. 11. Contour plots of the horizontal stress prediction on the western part of the Halten Terrace for (**a**) only gravitational loading and (**b**) with an additional horizontal in-plane compression of 50 MPa. Note that the colour codes used for contouring are the same in both plots, but the values in (b) are scaled linearly relative to the magnitude of applied compression (added 50 MPa compared to (a)). The location of the plots is shown in Figure 6.

Discussion

The elasto-plastic model applied in this study assumes two-dimensional plain-strain and is constrained by the use of geological and geophysical data. The model predictions show that the pre-existing basin geometry, the material properties and the loading conditions exert important controls on the calculated *in situ* stress. The model results are, however, based on several assumptions, and the use of other assumptions may result in predictions that differ from those presented in the previous sections. It is beyond the scope of this paper to present a comprehensive sensitivity analysis, but

of horizontal compression (Fig. 13b). The lateral variations cannot be attributed to variations in fluid pressure or active faulting because they are not included in the model. Neither can the variability be related to variations in rock properties as the shale layer has been assigned the same density value and the same elastic properties. Thus, the spatial variations must be the result of the differential vertical movements and the presence of faults causing rapid lateral shifts in rocks with different geomechanical properties.

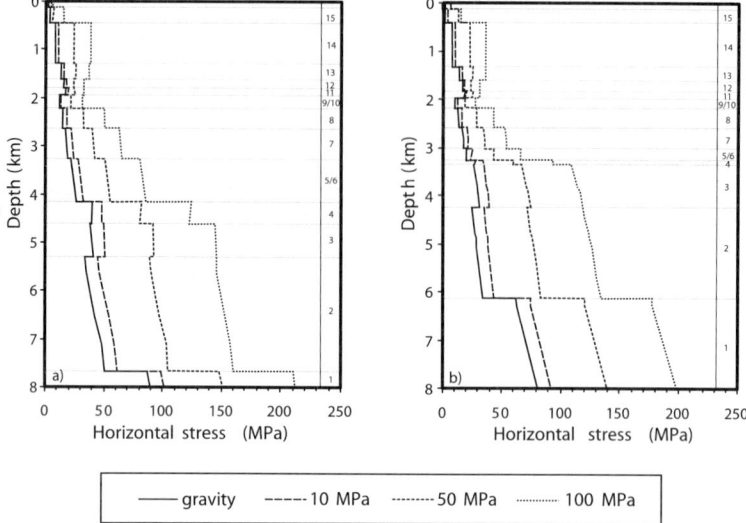

Fig. 12. Vertical profiles showing the effect of stratigraphic layering and in-plane horizontal compression of 10, 50 and 100 MPa on the predicted horizontal stress. (**a**) The central part of the trough (line 1 in Fig. 11) and (**b**) the structural flank (line 2 in Fig. 11). The index numbers refer to the sedimentary units as shown in Figure 6 and in Table 1.

some comments on the model assumptions and the applicability of the modelling results seem appropriate and are discussed below.

Model configuration

The finite element model is based on the geometry of the interpreted cross-section, representing the present-day geological structure of Haltenbanken and adjacent areas. The effect of applying a different basement structure and allowing reactivation of deep-seated faults could possibly affect the model predictions. As shown previously (Fig. 7), the relief of top basement controls the vertical displacement pattern that influences the stress predictions. Minor change in the top basement configuration is believed to be of secondary importance regarding the stress predictions. Although the top of the basement is difficult to interpret from seismic data, tilted fault blocks are observed in places. This causes the top basement to be an irregular surface and the thickness of the basement relative to the sediment thickness will vary across the model.

In the model, faults are not allowed to move, but their presence controls the geometry and architecture of the sedimentary units. Studies have shown that active faults can significantly modify the local *in situ* stress pattern (e.g. Bell *et al.* 1992; Addis *et al.* 1996). Furthermore, the role of major faults acting as buffer zones with respect to stress transfer has been emphasized by Pascal & Gabrielsen (2001). They show that the Møre Trøndelag Fault

Complex, a crustal weakness zone that separates the mid-Norwegian margin from the North Sea basin, may act as shield protecting the North Sea from Atlantic ridge push forces.

The Halten Terrace is bounded to the west by the Klakk Fault Complex (Fig. 2), a north-south striking fault zone with a displacement in the order of several kilometres. Some seismic activity has been recorded along the western segments of the fault complex (Bungum *et al.* 1991; Lindholm *et al.* 1996). Earthquake activity is, however, low on the Halten Terrace and the Trøndelag Platform, which may suggest that these areas are partly protected by the adjacent main fault complexes (e.g. Klakk Fault Complex and Møre-Trøndelag Fault Complex). A consequence of this is that the stress system on the Halten Terrace generally does not facilitate fault reactivation, and that our initial assumption of non-active faults on the Halten Terrace seems to be appropriate.

Geomechanical properties of the rocks

The Cretaceous, Jurassic and pre-Jurassic rocks are assigned density values and elastic properties appropriate for siliclastic rocks. The input parameters of the post-Triassic stratigraphic units are constrained from well data. Uncertainties regarding rapid changes in the rocks' properties caused by natural lithological heterogeneities cannot be accounted for in large-scale models such as the one presented here.

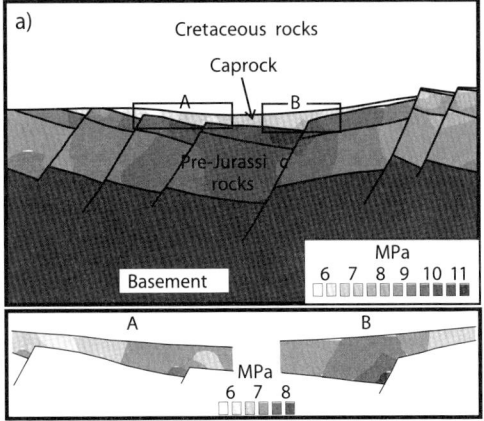

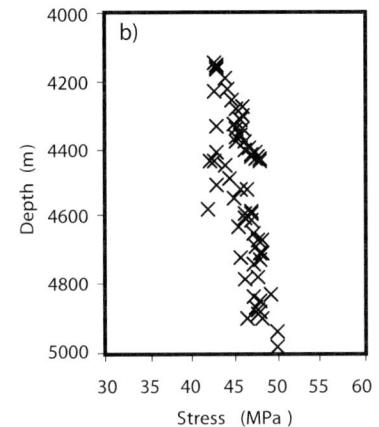

Fig. 13. (**a**) Contour plot of the minimum horizontal stress as predicted in the caprock on the western part of the Halten Terrace. (**b**) Predicted minimum horizontal stress values at each node within the caprock on the western part of Haltenbanken.

The pre-Jurassic stratigraphy is, however, less constrained as few wells have been drilled to top basement. It is known that layers of salt are present in the Triassic sequence. Studies have shown that the Triassic evaporites strongly influenced the structural development of the Haltenbanken area during Late Cretaceous and Cretaceous times (Jackson & Hastings 1986; Whitjack *et al.* 1989; Pascoe *et al.* 1999). Introducing such a ductile lithology in the model could probably act as a decoupling level between the reservoir beds and the strong crust below. Some of the overall stress would be released by ductile deformation. However, zones appearing as breakouts were recognized in the two salt layers in a well analysed by Borgerud & Svare (1995). These breakouts give a direction that fits to the regional horizontal principal stress direction.

Borgerud & Svare (1995) did not find evidence that the salt layers in the Triassic rocks act as decoupling levels.

Lateral boundary constraints

The predicted *in situ* stresses are dependent on the assumptions of the loading conditions. In this study a constant stress magnitude with depth was assumed. Using a different lateral boundary constraint where the horizontal load increases with depth would alter the results as would also different strain rates and horizontal shortening. Note, however, that although various stress indicators suggest that the mid-Norwegian margin is in a compressive state of stress, little information is available on changes in the far-field compressional stress magnitude with depth (see Pascal & Gabrielsen 2001). Applying a depth-dependent increase in horizontal stress magnitude or constant lateral strain as a boundary constraint would be as arbitrary as the one applied in the present analysis. A natural consequence of applying a boundary constraint where the increase in stress magnitude is depth-dependent is that the predicted horizontal stresses will show a more pronounced increase with depth and that the relative stress magnitudes within the layers will change. Applying constant lateral strain will result in the tectonic stress component varying with lithology (Young's modulus).

The aim of the modelling was to assess some influential factors that can contribute to local variations in the *in situ* stress field and not to accurately predict the magnitude of the horizontal stresses in the study area. As such, applying a different lateral constraint would change the relative magnitudes, but the effect of spatial variations in the rock's elastic properties would still be dominant for the distribution of the horizontal stresses. The same argument can be used regarding the assumption on the initial state of stress in the model. The magnitudes of the gravitationally induced horizontal stresses would be different, but the subsequent change in *in situ* stress due to far-field compression would remain the same.

Implications for top seal integrity in high fluid pressure regions

The highly overpressured reservoirs encountered in wells drilled prior to 1995 on Haltenbanken were mainly devoid of hydrocarbons although the Jurassic sequence is buried to similar depths and contains the same stratigraphic units as in the adjacent areas where reservoirs are filled with hydrocarbons (e.g. Ehrenberg *et al.* 1992). More recently, petroleum occurrences have been discovered in Jurassic reservoirs in the overpressured region, which demonstrates that fluid pressure is not the ultimate control for entrapment of hydrocarbons in this high fluid pressure domain. It is reasonable, however, to believe that ineffective seals are the cause of exploration failure where water-filled reservoirs occur in what seems to be well-defined closures, and where the reservoirs are also in direct contact with mature, organic-rich source rocks.

The minimum horizontal stress (σ_{Hmin}) magnitude is an approximate measure of the fluid pressure a seal can sustain before it will fracture. σ_{Hmin} increases typically with depth at a gradient higher than the hydrostatic pressure gradient, and consequently, the retention capacity increases with depth. The retention capacity is reduced by either an increase in the fluid pressure or a decrease in σ_{Hmin} or a combination of both. A fluid pressure increase can be caused by a number of mechanisms (for a review see Swarbrick & Osborne 1998), but on Haltenbanken it is believed that overpressuring occurred during Plio-Pleistocene times and was caused by hydrocarbon generation, rapid sedimentation and lateral pressure transfer (e.g. Kock & Heum 1995; Skar et al. 1999). The retention capacity in caprocks in the high fluid pressure domain may be as low as 8 MPa, which indicates that the seals are in the proximity of failure. A subtle increase in fluid pressure or a local decrease in the in situ σ_{Hmin} would facilitate fracturing. Natural fracturing is expected to occur at up-dip positions in a tilted reservoir where the fluid pressure exceeds the tensile strength of the caprock. However, it was shown that the structural fabric could cause a local decrease of several megapascals in the caprock (Fig. 13). Thus, the results suggest that natural fracturing may occur also at deeper locations and not only at the crest, depending on the local stresses. The predictions indicate that structurally controlled variations in σ_{Hmin} may be an important factor to consider regarding the preservation potential of hydrocarbons in deeply buried reservoirs when the retention capacity is low due to high fluid pressure.

A comparison with field data is ultimately required in order to assess whether the predictions are representative of the stresses observed in nature. In situ stress measurements are often limited to the locations of wells, and without high-density well control it may be difficult to recognize whether changes in stress orientation and magnitudes occur abruptly or gradually. Minimum horizontal stress data in caprocks (based on LOT) in adjacent hydrocarbon fields on Haltenbanken (Fig. 14) reveal large variations in magnitudes. The fluid pressures in these formations are close to hydrostatic pressure and the variability is therefore not likely to be caused by fluid pressure variations. More natural causes for such variability are lith-

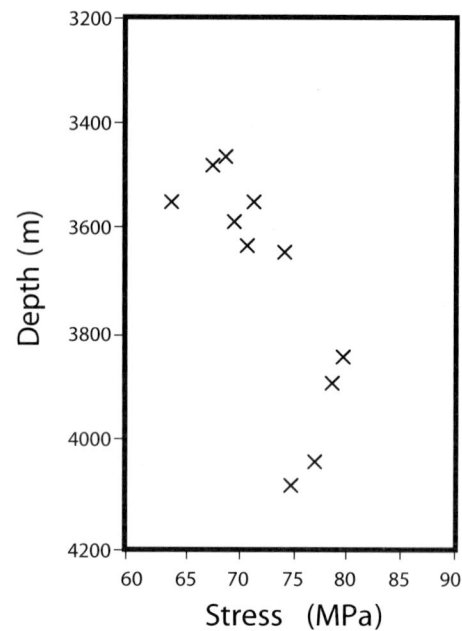

Fig. 14. Minimum horizontal stress values derived from LOT data in the caprock from nearby hydrocarbon fields.

ology variations and the quality of the measurements. However, the large spread of values from 65 to 80 MPa may suggest that the structural fabric in which these caprock formations occur has influenced the minimum horizontal stress magnitudes.

Although methods have been presented to constrain in situ stress and effective rock strength (e.g. Zoback & Peska 1995; Addis et al. 1996) the main uncertainty relates to the in situ stress and rock strength in between areas of well control. Numerical modelling techniques can be applied to assess changes in stress fields as they are able to investigate how various complex structures and discontinuities may affect the stress distribution within sedimentary basins (e.g. Sassi & Faure 1997; Beekman et al. 2000). As such, the predictive power of finite element tools must be acknowledged, but the results must be interpreted with care.

Conclusions

The model predictions must be interpreted qualitatively because of the intrinsic model simplifications. For example, the lateral boundary constraints, material properties in a heterogeneous medium, in situ stresses and the rocks' loading history will never be known completely. Furthermore, temperature, fluid flow and fluid pressure effects are not included. Given these modelling simplifications, the following conclusions can be drawn.

The model predicts pronounced spatial variations in horizontal stress, mainly as a result of (1) the structural configuration on which the model is based, (2) variations in the mechanical properties of the rocks, and (3) the magnitude of horizontal loading applied to the model. Rapid shifts in stress magnitudes occur across discontinuities (e.g. sediment interfaces) due to contrasts in the rocks' mechanical properties. The differences in stress magnitudes across such discontinuities can be subdued or enhanced under increased horizontal compression reflecting the different capability of rocks to transmit stresses. Furthermore, structurally controlled variations in vertical displacements produce bending-induced compressional and tensile stresses that are superimposed on the compressional background stress field, thus producing local concentrations of highs and lows in the stress field. Due to variance of the three above-mentioned factors the magnitude of horizontal stress varies significantly within spatially restricted areas.

With respect to hydrocarbon exploration in overpressured areas, our model results indicate that the minimum horizontal stress may be reduced by magnitudes in the order of megapascals in faulted regions, which may facilitate hydraulic fracturing given the conditions that the reservoir pore fluid pressures are sufficiently high.

The authors would like to thank Norsk Hydro for supplying sources of data and providing financial support for T.S during his PhD thesis. Discussions and useful comments from R. van Balen and H. Kooi are very much appreciated. The constructive comments of reviewers R. H. Gabrielsen and H. Doust have improved the paper. This is Netherlands Research School of Sedimentary Geology publication number 20030205.

References

ADDIS, M. A., LAST, N. C. & YASSIR, N. A. 1996. Estimation of horizontal stresses at depth in faulted regions and their relationship to pore pressure variations. *Society of Petroleum Engineers Formation Evaluation*, paper **11**, 11–18.

BEEKMAN, F., BADSI, M. & VAN WEES, J. -D. 2000. Faulting, fracturing and *in situ* stress prediction in the Ahnet Basin, Algeria; a finite element approach. *Tectonophysics*, **320**, 311–329.

BELL, J. S., CAILLET, G. & ADAMS, J. 1992. Attempts to detect open fractures and non-sealing faults with dipmeter logs. *In*: HURST, A., GRIFFITHS, C. M. & WORTHINGTON, P. F. (eds) *Geological Application of Wireline Logs II*. Geological Society, London, Special Publications, **65**, 211–220.

BLYSTAD, P., BREKKE, H., FÆRSETH, R. B., LARSEN, B. T., SKOGSEID, J. & TØRUDBAKKEN, B. 1995. *Structural elements of the Norwegian continental shelf, part II: the Norwegian Sea region*. Norwegian Petroleum Directorate, Stavanger, Bulletin **6**.

BØEN, F., EGGEN, S. & VOLLSET, J. 1984. Structures and basins of the margin from 62°–69° N and their development. *In*: SPENCER, A. M. *et al.* (eds) *Petroleum Geology of the North European Margin*. Norwegian Petroleum Society, Graham and Trotman, London, 3–28.

BORGERUD, L. K. & SVARE, E. 1995. *In situ* stress field on the mid-Norwegian margin, 62°–67° north. *In*: FEJERSKOV, M. & MYRVANG, A. M. (eds) *Proceedings of Workshop Rock Stresses in the North Sea*. Trondheim, Norway, 165–178.

BRADLEY, J. S. 1975. Abnormal formation pressure. *AAPG Bulletin*, **59**, 957–973.

BRECKELS, I. M. & VAN EEKELEN, H. A. M. 1982. Relationship between horizontal stresses and depth in sedimentary basins. *Journal of Petroleum Technology*, **34**, 2191–2199.

BREKKE, H. & RIIS, F. 1987. Tectonics and basin evolution of the Norwegian shelf between 62°N and 72°N. *Norsk Geologisk Tidsskrift*, **67**, 295–322.

BREKKE, H., DAHLGREN, S., NYLAND, B. & MAGNUS, C. 1999. The prospectivity of the Vøring and Møre basins on the Norwegian Sea continental margin. *In*: FLEET, A. J. & BOLDY, S. A. R. (eds) *Petroleum Geology of Northwest Europe: Proceedings of the 5th Conference*. Geological Society, London, 261–274.

BUNGUM, H., ALSAKER, A., KVAMME, L. B. & HANSEN, R. A. 1991. Seismicity and seismotectonics of Norway and nearby continental shelf areas. *Journal of Geophysical Research*, **96**, 2249–2265.

CAILLET, G. 1993. The caprock of the Snorre Field, Norway: a possible leakage by hydraulic fracturing. *Marine and Petroleum Geology*, **10**, 42–50.

COHEN, M. J. & DUNN, M. E. 1987. The hydrocarbon habitat of the Haltenbank-Traenabank area, offshore Mid-Norway. *In*: BROOKS, J. & GLENNIE, K. (eds) *Petroleum Geology of North West Europe*. Graham and Trotman, London, 1049–1063.

DALLAND, A., WORSLEY, D. & OFSTAD, K. 1988. *A lithostratigraphic scheme for the Mesozoic and Cenozoic succession offshore mid- and northern Norway*. Norwegian Petroleum Directorate, Bulletin **4**.

DORÉ, A. G. & LUNDIN, E. R. 1996. Cenozoic compressional structures on the NE Atlantic margin: nature, origin and potential significance for hydrocarbon exploration. *Petroleum Geoscience*, **2**, 299–311.

DORÉ, A. G., LUNDIN, E. R., JENSEN, L. N., BIRKELAND, Ø., ELIASSEN, P. E. & FICHLER, C. 1999. Principal tectonic events in the evolution of the northwest European Atlantic margin. *In*: FLEET, A. J. & BOLDY, S. A. R. (eds) *Petroleum Geology of Northwest Europe: Proceedings of the 5th Conference*. Geological Society, London, 41–61.

EHRENBERG, S. N., GJERSTAD, H. M. & HADLER-JACOBSEN, F. 1992. Smørbukk Field – a gas condensate fault trap in the Haltenbanken Province offshore mid-Norway. *In*: HARBURY, M. T. (ed.) *Giant Oil and Gas Fields of the Decade 1978–1988*. American Association of Petroleum Geologists, Bulletin Memoir, **54**, 323–348.

ELDHOLM, O., THIEDE, J. & TAYLOR, E. 1989. Evolution of the Vøring volcanic margin. *Proceedings ODP Scientific Results*, **104**, 1033–1065.

ENGELDER, T. & FISCHER, M. P. 1993. Influence of poro-

elastic behavior on the magnitude of minimum horizontal stress, S_h, in overpressured parts of sedimentary
basins. *Geology,* **22**, 949–952.

FEJERSKOV, M., MYRVANG, A. M., LINDHOLM, C. &
BUNGUM, H. 1995. *In situ* rock stress pattern on the
Norwegian continental shelf and mainland. *In:* FEJER-
SKOV, M. & MYRVANG, A. M. (eds) *Proceedings of
Workshop Rock Stresses in the North Sea.* Trondheim,
Norway, 191–201.

FORBES, P. L., UNGERER, P. M., KUHFUSS, A. B., RIIS,
F. & EGGEN, S. 1991. Compositional modeling of petroleum generation and expulsion: Trial application to
a local mass balance in the Smørbukk Sør Field, Haltenbanken area, Norway. *AAPG Bulletin,* **75**, 873–893.

GAARENSTROOM, L., TROMP, R. A. J., DE JONG, M. C. &
BRANDENBURG, A. M. 1993. Overpressures in the Central North Sea: implications for trap integrity and drilling safety. *In:* PARKER, J. R. (ed.) *Petroleum Geology
of Northwest Europe: Proceedings of the 4th Conference.* Geological Society, London, 1305–1313.

GABRIELSEN, R. H. & ROBINSON, C. 1984. Tectonic
inhomogeneities of the Kristiansund-Bodø Fault Complex, offshore mid-Norway. *In:* SPENCER, A. M. *et al.*
(eds) *Petroleum Geology of the North European Margin.* Norwegian Petroleum Society, Graham and Trotman, London, 397–406.

GABRIELSEN, R. H., ODINSEN, R. H. & GRUNNALEITE, I.
1999. Structuring of the Northern Viking Graben and
the Møre Basin; the influence of basement structural
grain, and the particular role of the Møre-Trøndelag
Fault Complex. *Marine and Petroleum Geology,* **16**,
443–465.

GJELBERG, J., DREYER, T., HØIE, A., TJELLAND, T. &
LILLENG, T. 1987. Late Triassic to mid-Jurassic sand-
body development on the Barents and mid-Norwegian
shelf. *In:* BROOKS, J. & GLENNIE, K. W. (eds) *Petroleum Geology of North West Europe.* Graham and
Trotman, London, 1107–1129.

GÖLKE, M. 1996. *Patterns of stress in sedimentary basins
and the dynamics of pull-apart basin formation.* PhD
thesis, Vrije Universiteit Amsterdam, The Netherlands.

GRAULS, D. 1998. Overpressure assessment using a minimum horizontal stress approach. *In:* MITCHELL, A. &
GRAULS, D. (eds) *Overpressures in Petroleum Exploration,* Proceedings Workshop, Pau 1998. Elf-EP editions, Pau, Bulletin du Centre Recherches Elf Exploration et Production, Memoire **22**, 137–147.

GRUNNALEITE, I. & GABRIELSEN, R. H. 1995. Structure
of the Møre Basin, mid-Norway continental margin.
Tectonophysics, **252**, 221–251.

HERMANRUD, C., WENSAAS, L, TEIGE, G. M. G., VIK, E.,
BOLÅS, H. M. N. & HANSEN, S. 1998. Shale porosities
from well logs on Haltenbanken (offshore mid-
Norway) show no influence of overpressuring. *In:*
LAW, B. E., ULMISHEK, G. F. & SLAVIN, V. I. (eds)
Abnormal Pressures in Hydrocarbon Environments.
American Association of Petroleum Geologists, Bull-
etin Memoir, **70**, 65–85.

HEUM, O. R., DALLAND, A. & MEISINGSET, K. K. 1986.
Habitat of hydrocarbons at Halten Terrace (PVT modelling as a predictive tool in hydrocarbon exploration).
In: SPENCER, A. M. *et al.* (eds) *Habitat of Hydrocarbons on the Norwegian Continental Shelf.* Norweg-

ian Petroleum Society, Graham & Trotman, London,
259–274.

HUNT, J. M. 1990. Generation and migration of petroleum from abnormally pressured fluid compartments.
AAPG Bulletin, **74**, 1–12.

JACKSON, J. S. & HASTINGS, D. S. 1986. The role of salt
movement in the tectonic history of Haltenbanken and
Trænabanken and its relationship to structural style. *In:*
Spencer, A. M. *et al.* (eds) *Habitat of Hydrocarbons
on the Norwegian Continental Shelf.* Norwegian Petroleum Society, Graham & Trotman, London, 241–
257.

JACOBSEN, V. W. & VAN VEEN, P. 1984. The Triassic
offshore Norway north of 62° N. *In:* SPENCER, A. M.
et al. (eds) *Petroleum Geology of the North European
Margin.* Norwegian Petroleum Society, Graham &
Trotman, London, 317–327.

KARLSEN, D. A., NYLAND, B., FLOOD, B., OHM, S. E.,
BREKKE, T., OLSEN, S. & BACKER-OWE, K. 1995. Pet-
roleum geochemistry of the Haltenbanken, Norwegian
continental shelf. *In:* CUBITT, J. M. & ENGLAND, W.
A. (eds) *The Geochemistry of Reservoirs.* Geological
Society, London, Special Publications, **86**, 203–256.

KOCH, J. O. & HEUM, O. R. 1995. Exploration trends of
the Halten Terrace. *In:* HANSLIEN, S. (ed.) *Petroleum
Exploration and Exploitation in Norway.* Norwegian
Petroleum Society, **4**, Elsevier, Amsterdam, 231–251.

LINDHOLM, C. D., BUNGUM, H., BRATLI, R. K., AADNØY,
B. S., DAHL, N., TØRUDBAKKEN, B. & ATAKAN, K.
1995a. Crustal stress in the northern North Sea as
inferred from borehole breakouts and earthquake focal
mechanisms. *Terra Nova,* **7**, 51–59.

LINDHOLM, C. D., BUNGUM, H., VILLAGRAN, M. &
HICKS, E. 1995b. Crustal stress and tectonics in
Norwegian regions determined from earthquake focal
mechanisms. *In:* FEJERSKOV, M. & MYRVANG, A. M.
(eds) *Proceedings Workshop Rock Stresses in the
North Sea.* Trondheim, Norway, 77–91.

LINDHOLM, C. D., BUNGUM, H., HICKS, E., FEJERSKOV,
M., LARSEN, B. T. & NØTTVEDT, A. 1996. *Seismicity
and Earthquake Focal Mechanisms North Atlantic
Area.* IBS-DNM, Norsar, NTNU & Norsk Hydro.

PASCAL, C. & GABRIELSEN, R. H. 2001. Numerical modelling of Cenozoic stress patterns in the mid-Norwegian margin and the northern North Sea. *Tectonics,* **20**,
585–599.

PASCOE, R., HOOPER, R., STORHAUG, K. & HARPER, H.
1999. Evolution of extensional styles at the southern
termination of the Nordland Ridge, mid-Norway: a
response to variations in coupling above Triassic salt.
In: FLEET, A. J. & BOLDY, S. A. R. (eds) *Petroleum
Geology of Northwest Europe: Proceedings of the 5th
Conference.* Geological Society, London, 83–90.

RANALLI, G. 1995. *Rheology of the Earth.* Chapman &
Hall, London.

REEMST, P. & CLOETINGH, S. 2000. Polyphase rift evolution of the Vøring Margin (mid-Norway); constraints
from forward tectonostratigraphic modeling. *Tectonics,*
19, 225–240.

SASSI, W. & FAURE, J. L. 1997. Role of faults and layer
interfaces on the spatial variations of stress regimes in
basins: inferences from numerical modelling. *Tectonophysics,* **266**, 101–119.

SKAR, T. 2000. *The influence of intraplate stresses on*

basin subsidence and fluid pressure evolution of Haltenbanken, mid-Norwegian margin. PhD thesis, Vrije
Universiteit Amsterdam, The Netherlands.

SKAR, T., VAN BALEN, R. T. & HANSEN, S. 1998. Overpressuring in Cretaceous shales on the Halten Terrace,
offshore mid-Norway: nature and causes. *In*: MITCH-
ELL, A. & GRAULS, D. (eds) *Overpressures in Petroleum Exploration.* Proceedings Workshop, Pau 1998.
Elf-EP editions, Pau, Bulletin du Centre Recherches
Elf Exploration et Production, Memoire **22**, 69–75.

SKAR, T., VAN BALEN, R. T., ARNESEN, L. & CLOET-
INGH, S. 1999. Origin of overpressures on the Halten
Terrace, offshore mid-Norway: the potential role of
mechanical compaction, pressure transfer and stress.
In: APLIN, A. C., FLEET, A. J. & MACQUAKER, J. H.
S. (eds) *Muds and Mudstones: Physical and Fluid Flow
Properties.* Geological Society, London, Special Publi-
cations, **158**, 137–156.

SKOGSEID, J. & ELDHOLM, O. 1989. Vøring Plateau continental margin: seismic interpretation, stratigraphy,
and vertical movements. *In*: ELDHOLM, O. THIEDE, J &
TAYLOR, E. (eds) *Proceedings of the Ocean Drilling
Programme, Scientific Results,* **104**, 993–1030.

SKOGSEID, J., PEDERSEN, T. & LARSEN, V. B. 1992.
Vøring Basin: subsidence and tectonic evolution. *In*:
LARSEN, R., BREKKE, M., LARSEN, B. T. & TAL-
LERAAS, E. (eds) *Structural and Tectonic Modelling
and its Application to Petroleum Geology.* Norwegian
Petroleum Society, **1**, Elsevier, Amsterdam, 55–82.

STUEVOLD, L. M. & ELDHOLM, O. 1996. Cenozoic uplift
of Fennoscandia inferred from a study of the mid-
Norwegian margin. *Global and Planetary Change,* **12**,
359–386.

SWARBRICK, R. E. & OSBORNE, M. J. 1998. Mechanisms
that generate abnormal pressures: an overview. *In*:
LAW, B. E., ULMISHEK, G. F. & SLAVIN, V. I. (eds)
Abnormal Pressures in Hydrocarbon Environments.
American Association of Petroleum Geologists, Bull-
etin Memoir, **70**, 13–34.

VÅGNES, E., GABRIELSEN, R. H. & HAREMO, P. 1998. Late
Cretaceous-Cenozoic intraplate contractional deformation at the Norwegian continental shelf: timing,
magnitude and regional implications. *Tectonophysics,*
300, 29–46.

WHITJACK, M. O., MEISLING, K. E. & RUSSEL, L. R.
1989. Forced folding and basement-detached normal
faulting in the Haltenbanken area, offshore Norway. *In*:
Tankard, A. J. & BALKWILL, H. R. (eds) *Extensional
Tectonics and Stratigraphy of the North Atlantic Margins.* Association of American Petroleum Geologists,
Memoir, **46**, 567–575.

ZHANG, Y. -Z., DUSSEAULT, M. B. & BRATLI, R. K. 1995.
Simulating stresses in overpressured zones using a
finite element approach. *In*: FEJERSKOV, M. & MYRV-
ANG, A. M. (eds) *Proceedings Workshop Rock Stresses
in the North Sea.* Trondheim, Norway, 108–126.

ZOBACK, M. D. & PESKA, P. 1995. *In situ* stress and rock
strength in the GBRN/DOE Pathfinder well, south Eug-
ene Island, Gulf of Mexico. Journal of Petroleum Technology, **47**, 582–585.

Integrated 3D geomechanical modelling for deep subsurface deformation: a case study of tectonic and human-induced deformation in the eastern Netherlands

J. D. VAN WEES[1], B. ORLIC[1], R. VAN EIJS[1], W. ZIJL[1], P. JONGERIUS[1],
G. J. SCHREPPERS[2], M. HENDRIKS[2] & T. CORNU[3]

[1]*TNO–NITG, Department of Geo-Energy, PO Box 80015, NL 3508 TA Utrecht, The Netherlands*
[2]*TNO Building and Construction*
[3]*Vrije Universiteit Amsterdam*

Abstract: Current advances in finite element computer codes and increase in computer power theoretically allow quantitative modelling of the geomechanical effects of hydrocarbon depletion from reservoirs. Here we show that it is technically possible to incorporate the full complexity of the 3D geological structure of a reservoir including faults into geomechanical models. In the workflow GOCAD is used for the structural modelling, integrated with DIANA for the geomechanical calculations.

A case study on the Roswinkel gas field in the eastern Netherlands illustrates the working methodology. The case study clearly shows the strong dependency of gas depletion deformation effects on the prevailing tectonic stress field. Our models for gas depletion predict a stabilization of the stress field (further away from failure) for reservoirs in compressive and strike-slip regimes. On the other hand extensional stress regimes will result in failure of the reservoir, in agreement with observed earthquakes, provided that (a) the reservoir material or existing faults are weak and (b) the state of stress is close to failure of the material.

The Roswinkel field, which is marked by a high abundance of earthquakes occurring after gas depletion started, is therefore most likely marked by an extensional tectonic regime and by geomechanically weak rock or pre-existing faults. According to base Tertiary fault displacements and the *World Stress Map*, the extension (minimum horizontal principal stress) is most likely NE-SW oriented.

Current advances in finite element computer codes and increase in computer power theoretically allow quantitative modelling of the geomechanical effects of hydrocarbon depletion from reservoirs.

Up to now the effects have been mainly studied in two dimensions (e.g. Orlic *et al.* 2001). Only a very few studies explore the effects of depletion in 3D (e.g. Kenter *et al.* 1998). It is noted that existing (3D) studies tend to oversimplify geological structures and the importance of pre-existing tectonic stresses. Geological structures are generally oversimplified because geomechanical software only is being used. As a consequence it is rather difficult to incorporate the full complexity of the geological structure, as derived from (3D) seismic interpretation and mapping software into the geomechanical

software. With respect to tectonic loading, most studies are performed adopting fixed lateral boundary conditions in which the ratio of the horizontal and vertical stress constant is chosen in a rather arbitrary way, often ignoring any effects of particular variations in tectonic stresses. This is not at all in agreement with the current improved understanding of the tectonic stress field (e.g. Zoback 1992; *World Stress Map* 2001), clearly showing that the ratio of horizontal and vertical stresses can vary significantly on a regional scale from normal faulting, to strike-slip and thrust faulting tectonic regimes.

A major aim of our study is to demonstrate that it is technically possible to incorporate the full complexity of the 3D geological structure of a res-

From: NIEUWLAND, D. A. (ed.) *New Insights into Structural Interpretation and Modelling.* Geological Society, London, Special Publications, **212**, 313–328. 0305-8719/03/$15

ervoir including faults, and furthermore that is it is important to incorporate the possible variations of the tectonic stress field in 3D.

First, the case study area around the Roswinkel gas field is introduced, describing its structural evolution and tectonic setting in the eastern Netherlands. The case study has been deliberately chosen for a reservoir for which pervasive (sub)surface deformation is observed through earthquakes and which is located in a complex tectonic setting.

Subsequently the working methodology for the modelling is described and demonstrated by the selected case study. The working methodology includes the use of GOCAD (2001) for the set-up of the 3D model geometry, DIANA (2000) for the geomechanical modelling, and specially developed software tools interfacing between both these software packages. Finally, the modelling results, in particular the effects of different tectonic stress scenarios and their 3D effects, are presented and discussed.

Case study area: Roswinkel field

The Roswinkel field is situated in the northeastern part of the Netherlands. This region is well known in the gas industry for its large Permian Rotliegend fields including the giant Slochteren field on the Groningen High. Besides these large fields, smaller occurrences can be found in younger deposits such as the Triassic Roswinkel field. The reservoir consists of clastic sediments of the Lower Germanic Trias Group, found at a depth of around 2400 m. These clastic fluvio-lacustrine sediments were deposited in the intracratonic Lower Saxony Basin spread out to Germany (Geluk & Röhling 1997) (Fig. 1).

The reservoir structure consists of an anticlinal structure above the Emmen Salt dome. Figure 2 shows the geological structure in the vicinity of the reservoir (TNO-NITG 2000).

The evolution of the Lower Saxony Basin is marked by various tectonic phases, in which halo×kinetic flow played an important role (Fig. 2). In Permian and Triassic times, a rather regionally uniform subsidence took place resulting in deposition of Permian sandstones and evaporites and subsequently deposition of the Triassic sediments. In Jurassic times more differentiated subsidence occurred because of extensional WSW-ENE trending faults soling in the Permian evaporites. The extensional faulting, which was most active during the Late Kimmerian (*c.* 150 Ma), caused halokinesis of the Permian evaporites resulting in WSW-ENE trending lows and highs.

The structures that formed were strongly reactivated in compression culminating in the Late Cretaceous. The compressional deformation resulted in

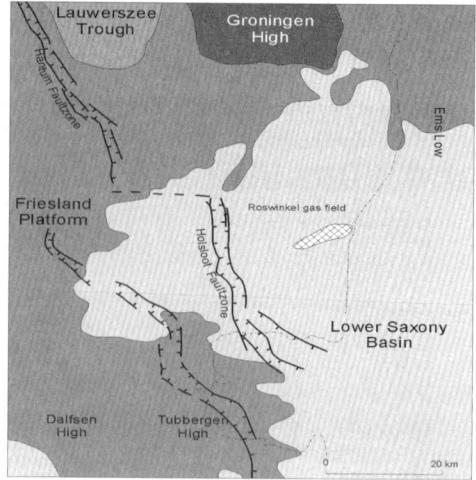

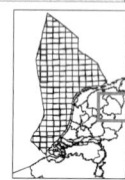

Fig. 1. Location of the Roswinkel field. Dark shaded area delineates gas-water contact. The reservoir is positioned in an anticlinal structure, delineated in Figures 2 and 3.

inversion of the Late Jurassic faults and generated WSW-ENE folds truncated by the base Tertiary unconformity (Fig. 2).

The base Tertiary structure shows NW-SE to north-south trending grabens in the area, which is in close agreement with the present-day NW-SE orientation of the largest horizontal stresses (*World Stress Map* 2001).

Detailed structural studies in the reservoir area are largely in agreement with the regional tectonic evolution. From 3D seismic studies, four groups of faults can be identified in the vicinity of the reservoir. The first two groups are situated in the Triassic deposits and consist of WSW-ENE trending normal and reverse faults, in close agreement with the regional phases of Late Kimmerian extension and Late Cretaceous inversion respectively. The third group occurs on a smaller scale and consists of en echelon patterns of SE-NW trending normal faults that can be on top of the anticline (Fig. 3). These are indicative for dextral wrenching most likely of Late Jurassic age. The fourth group of faults consists of WSW-ENE trending normal faults in the Late Cretaceous and Tertiary units in the crest of the anticline. These are believed to be local stress deviations resulting from the crestal collapse of the anticline.

The lithostratigraphic units in the area and the

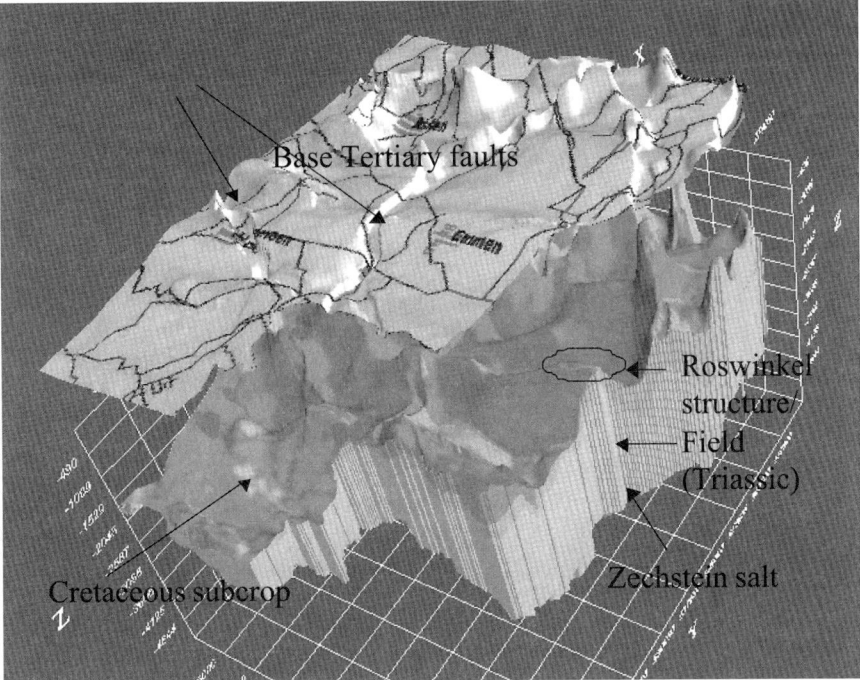

Fig. 2. Anticlinal structure of the Roswinkel field in Map Sheet 6 of the *Geological Atlas of the Deep Subsurface of The Netherlands* (TNO-NITG 2000).

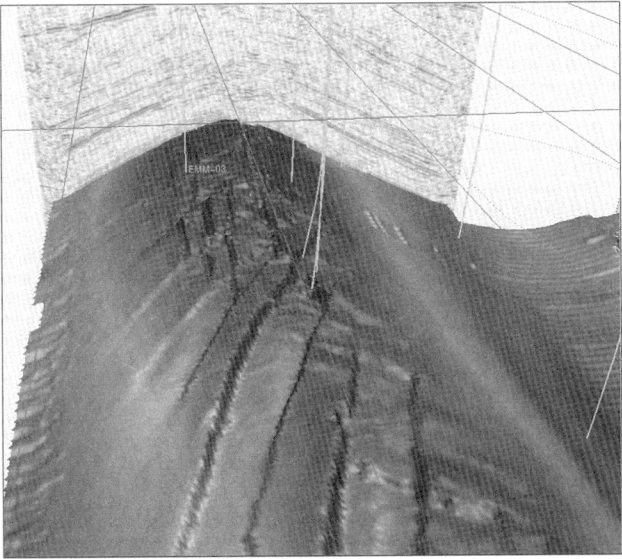

Fig. 3. 3D seismic interpretaton (looking to WSW) showing interpreted base of Jurassic sediments, highlighting NW-SE trending normal faults.

tectonic events can be found in Table 1 (Boogaert & Kouwe 1993).

Seismic events started to occur after 12 years of production, when the gas pressure had dropped severely. All epicentres are located above the crest of the field and magnitudes were in the range from 1 to 3.5 on the Richter scale. Depths of hypocentres have been indicated at or above reservoir depth; however, an uncertainty of ±500 m has to be kept in mind. Previous in-house research used 2D plane strain finite element models, combined with pressure information from reservoir models. The dis×advantage of these 2D models are the imperfect incorporation of the 3D stress field and of the 3D structural complexity. Using a 3D model, shear stresses can be imposed on the model and oblique slip on the faults can be modelled.

Modelling technique

In this section the modelling workflow is described using the Roswinkel field as a case study. The workflow integrates 3D structural mapping and 3D structural modelling into an integrated workflow for geomechanical modelling. This workflow is schematically depicted in Figure 4 and further described below.

Structural modelling with GOCAD

Different techniques can be used to represent, create and visualize shared earth models in the integrated 3D workflow for geomechanical modelling. Here, GOCAD (2001) is used for the construction of a boundary representation model based on geometries of mapped faults and horizons. A boundary representation model is defined by a set of triangulated irregular surfaces, which describe a model in terms of its bounding surfaces. Triangulated surfaces are well suited for representation of the complex structural geological surfaces and are flexible considering the density and locations of the nodes. Furthermore, a model represented by triangulated surfaces can readily be converted, i.e. meshed into a tetrahedral volume model, which yields the geometry for a finite element mesh. In analogy

Table 1. *Lithostratigraphic units and tectonic events in the study area*

Chronostratigraphy			N Area of the field Z	Tectonic phase	Orogeneisis
Cenozoic	Tertiary	Quarter.	Upper North Sea Group		Alpine
		Pliocene		Savic	
		Miocene		Pyrenean	
		Oligocene			
		Eocene		Laramean	
		Paleocene	Lower North Sea Group		
Mesozoic	Cretac.	Late	Chalk Group	Sub Hercian	
		Early	Riinland Group	Austrian Late Kimmerian II	
	Jurassic	Late	Niedersaksen Group	Late Kimmerian	
		Middle		Mid Kimmerian	
		Early	Altena Group		
	Triassic	Late	Upper Germanic Trias Group	Early Kimmerian	
		Middle			
		Early	Lower Germanic Trias Group	Hardegsen	
Palaeozoic	Permian	Late	Zechstein Group	Saalian	Varistian
		Early	Lower Rotliegend Group		
	Carbon.	Late	Limburg Group	Asturian	

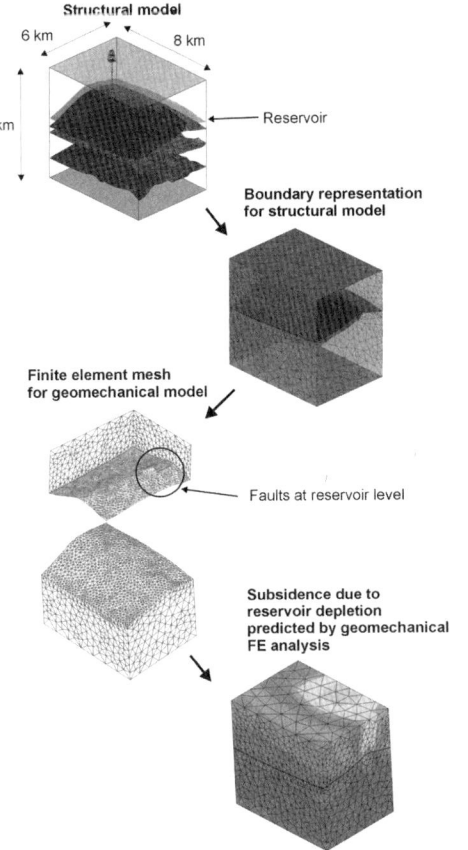

Structural model

6 km   8 km

7 km

Reservoir

Boundary representation
for structural model

Finite element mesh
for geomechanical model

Faults at reservoir level

Subsidence due to
reservoir depletion
predicted by geomechanical
FE analysis

Fig. 4. Integrated methodology. First, the structural model is translated to a boundary representation in GOCAD. Second, the model is transformed into a 3D geomechanical mesh of tetrahedral.

with triangulated surfaces, tetrahedral volume elements and tetrahedral meshes are suitable for representation of complex geological bodies. Tetrahedral meshes can be fully constrained by the bounding triangulated surfaces of a model and easily refined locally in specific areas of interest.

The main steps in structural modelling are schematically presented in Figure 5. Inputs for the modelling are in ZYCOR mapped depth surfaces of differentiated geological units and mapped fault surfaces (Fig. 5, step 1). These interpreted surfaces are usually available as gridded surfaces and first have to be converted to triangulated surfaces.

Additional editing of surfaces, for which GOCAD provides a versatile set of tools, is required in order to develop a consistent 3D structural model of the gas field (Fig. 5, step 2). Geometric operations that may have to be carried out on surfaces commonly involve filling the gaps in surfaces by interpolation, extending surfaces by

extrapolation, clipping surfaces and finding surface intersections. As a result of mutual intersections and clipping, the regular partitioning of the surfaces may be altered, due to incorporation of the intersection curves. At the intersection curve many additional vertices, and therefore triangles, are introduced in the surface. Afterwards the edited surfaces may need to be extended to the boundaries of the volumetric box that bounds the domain of interest (Fig. 4, top). It is convenient to orientate the volumetric box parallel to directions of the three principal stresses of the regional stress field. The three principal stresses of the initial stress tensor in the models are then perpendicular to the model boundaries, which reduces negative influence of boundary conditions in geomechanical modelling.

The edited surfaces are clipped against the boundaries of the volumetric box to make sure a volumetric consistency is preserved. However, automatic building of the boundary representation model from the edited surfaces partially fails in GOCAD. This is due to the insertion of vertices at intersection curves, which result in zero-length curve segments and associated newly inserted zero-area triangles. At these edges connections between surface patches cannot be determined, resulting in undefined volume. Coinciding vertices can be filtered out, which rectifies the problem with zero-length segments and zero-area triangles. However, near-zero-length segments and near-zero-area triangles are still preserved in the structural model. As a result, some triangles and tetrahedra in finite element mesh will have elongated and distorted shapes, which are not acceptable in a good quality mesh for finite element calculations.

The above-mentioned problem can be overcome by resampling all border line segments, which bound model surfaces, to regular intervals (Fig. 5, step 3). The interval length can be used to control the density of vertices along the line segments, which in turn determines the density of triangles in the surface patches, which in turn determines the density of the tetrahedron mesh.

The procedure for building and resampling border line segments comprises the following steps.

(1) All border line segments of the surface patches are first isolated from the structural model in GOCAD.

(2) The borders are subsequently subdivided into individual keylines. Each keyline is stored into a separate curve object.

(3) The endpoints of the keylines correspond to keypoints, which are stored as point objects. The location of keypoints will not be altered during resampling.

(4) For the resampling every keyline is copied

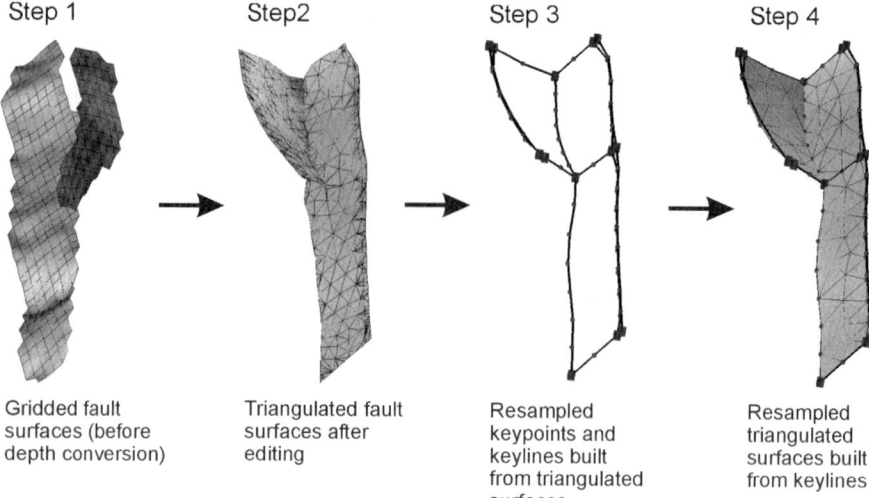

Fig. 5. Structural modelling of fault surfaces in GOCAD.

and a resampled midline is generated. A resampled keyline, which consists of a number of segments of equal length, approximates the original border line. The interval length can be specified by the user.

The resampled keylines provide a starting point for generation of resampled triangulated surfaces, which approximate the original surfaces of the structural model (Fig. 5, step 4). For the purpose of triangulation, the curves have to be grouped into boundaries of regions, which are subsequently triangulated. The newly created resampled surfaces may still need to be constrained by the internal vertices of regions. This operation can readily be accomplished in GOCAD by modifying a surface object by a point data set. The resulting surface patches marked by quality triangles define a boundary representation model, which is topologically and volumetrically consistent.

Meshing

For generation of a quality finite element mesh from a boundary representation model, the preprocessor of DIANA (2001) can be used. This preprocessor is capable of generating good quality tetrahedral meshes from 3D general bodies, i.e. bodies bounded by a set of planar triangles. The quality and density of a tetrahedral mesh is generally determined by the quality of the surface patches in the boundary representation. In the case considered here, the boundary representation model, bounded by good quality surface patches, proved to be a solid basis for generation of a good quality tetrahedral mesh.

Properties

The geomechanical units were assumed to be homogeneous with respect to their geomechanical parameters. A Mohr-Coulomb material model was used for the reservoir model unit and for the faults. Other differentiated geomechanical units were assumed to be pure elastic (Table 2).

Interface elements

In finite element analysis the behaviour of faults can adequately be modelled by interface elements, which can slip if their shear strength, described by the Mohr-Coulomb friction model, has been exceeded. In order to investigate the impact of depletion on faults at the reservoir level, we have incorporated two wedge-forming faults into the finite element model. For a tetrahedral mesh like the one considered here, triangular prismatic interface elements have to be inserted along fault surfaces between existing tetrahedral elements. This operation is, however, not trivial as the insertion of interface elements alters the topology of the existing tetrahedral mesh and deteriorates its quality, bearing in mind a near-zero-thickness of faults. For this reason we developed an algorithm for the insertion of interface elements into an existing finite element mesh and applied it to this case study.

Boundary conditions

Structural boundary conditions were defined by imposing displacement constraints along the model

Table 2. *Geomechanical parameters for the model units and the faults*

Lithology	Model unit	ρ (kg/m^3)	E (MPa)	ν	c (MPa)	ϕ (°)	ψ (°)		
Various	OVER-BURDEN	2650	20000	0.25					
Sandstone	RESERVOIR	2650	10000	0.25	10	35			
Various	BASEROCK	2650	30000	0.25					
Structural features		Model unit	Dn (MPa)			Ds (MPa/m)	c (MPa/m)	ϕ (°)	ψ (°)
Fault		FAULTS	2000			2000	0	25	

ρ, density of rock mass; E, Young's modulus; ν, Poisson's ratio; c, cohesion; ϕ, angle of internal friction; ψ, angle of dilatancy; Dn, normal stiffness; Ds, shear stiffness

boundaries, i.e. the faces of the volumetric box defining the study domain. No displacements were allowed along the bottom model boundary, while the top boundary was free to move in any direction. Each of the lateral model boundaries was constrained in the direction perpendicular to the lateral boundary under consideration, i.e. displacements were allowed to take place only in the plane of the lateral boundary under consideration.

Tectonic stress and initial loading conditions

Initial loading conditions determine the initial state of stress in the geomechanical model, which is characteristic for the period before gas production. The *in situ* state of stress in the gas field under study was not measured. However, the prevailing stress regime can be estimated from the current insights of the intraplate stresses of the area. *World Stress Map* (2001) data and earlier studies for the southern North Sea area (e.g. Van Wees & Cloetingh 1996), indicate that the stress field is marked by largest principal horizontal stresses oriented NW-SE. Inspection of the regional fault displacement patterns of the base Tertiary in the *Geological Atlas of the Deep Subsurface of The Netherlands* (TNO-NITG 2000; Figs 1 & 2) shows multi-kilometre-scale normal faults trending N-S to WNW-ESE with displacements of a few hundreds of metres. Markedly older east-west trending faults have not been reactivated since the end of the Cretaceous, being sealed by the base Tertiary unconformity. As further analysed in a palaeostress analysis on fault reactivation by G. Worum *et al.* (unpublished work) these observations indicate a stress field with the smallest principal stress ($\sigma3$) oriented approximately NE-SW, and its maximum principal stress ($\sigma1$) oriented vertically. This is in close agreement with the regional NW-SE princi-

pal horizontal stress being the intermediate principal stress ($\sigma2$).

The axes of the geomechanical model, aligned according to the reservoir structure, are oriented approximately parallel to the directions of the principal stresses of the above inferred regional stress field. The model y-axis points in a NNW direction (with an azimuth of 341 degrees) and the model x-axis in a ESE direction (Fig. 6).

For the above described stress field the initial stress field can conveniently be approximated by the initial stress tensor with the three principal stresses perpendicular to the model boundaries. Consequently, the alignment deviates by some 15° in an approximately NW-SE direction of the stress field. However, we believe this is well inside the range of uncertainty of the stress field and sufficient to calculate the typical response.

In the modelling we did not adopt the above described stress conditions as the only stress scenario for the model. Actually we have assessed the effects of a depleting reservoir for a wide range of characteristic stress regimes (Fig. 6), to illustrate the different (3D) effects. Five different stress regimes were considered: radial extension (case 1), isotropic stress regime (case 2), radial compression (case 3), strike-slip stress regime (case 4) and normal faulting (case 5). Cases 1 to 3 correspond to cases in which the principal horizontal stresses are equal in magnitude. Case 5 corresponds to the most likely regional stress field, in agreement with patterns of the base Tertiary faulting. Case 4 is marked by the same orientation of horizontal stresses with maximum horizontal stress approximately NW-SE; however, the horizontal stress magnitudes correspond to a strike-slip scenario, in which the vertical overburden pressure corresponds to the intermediate principal stress ($\sigma2$).

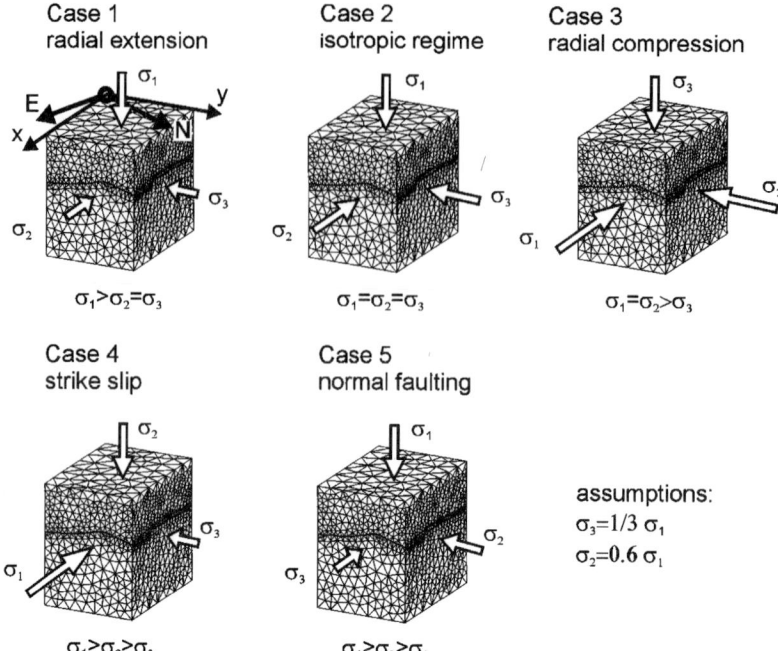

Fig. 6. Initial tectonic loading conditions of the model prior to gas depletion. σ_1, σ_2, σ_3 denote largest, intermediate and smallest principal stresses. N and E are orientations of geographical north and east; x and y represent coordinate axes of the model. The model y-axis points in a NNW direction (with an azimuth of 341°).

Magnitude of stresses and application into the model

The principal effective stress acting in a vertical direction is equal in all loading cases. This stress is calculated by combining:

(1) the weight of the sediments, which is calculated by applying the gravity acceleration on the model;
(2) the overpressure in the reservoir, estimated from reservoir simulations;
(3) the hydrostatic pressure in other model units, except in the reservoir;
(4) the pore pressure along the faults, specified as reservoir-dependent pore pressure; in parts of faults and bedding planes intersecting the reservoir, the pore pressure was equal to the pressure in the reservoir.

Two principal stresses acting in horizontal directions are derived from the principal stress acting in a vertical direction. The following assumptions about the magnitudes of the three principal stresses have been made: the ratio between the minor principal stress (σ_3) and the major principal stress (σ_1) was kept constant, $\sigma_3/\sigma_1 = 1/3$; and the intermediate principal stress was assumed to be equal to $\sigma_2 = 0.6\,\sigma_1$. The 1/3 ratio actually agrees with the

tectonic loading scenario which allows failure to occur at small deviations of the existing stress field (e.g. Van Wees & Stephenson 1995).

For load cases 1 to 3 we use the K_0 procedure in DIANA to impose the initial state of stress onto the model. This procedure initiates the horizontal effective stress (σ'_h) into the model from the calculated vertical effective stress (σ'_v), keeping the ratio between the two stresses constant and equal to a user-specified value for the lateral stress ratio ($K_0 = \sigma'_h/\sigma'_v$). The K_0 procedure suffices for load cases 1 to 3, in which the two principal horizontal stresses are equal. However, this procedure it is not applicable to initiate the stresses for load cases 4 and 5, in which the three principal stresses are different in magnitude. For these two load cases we used the following procedure. First we calculate and output the effective vertical stress caused by the gravity loading. From the effective vertical stress, which is one of the principal stresses, we calculate the other two, non-equal, principal stresses, acting in a horizontal plane. We do so for each element in the model taking into account the adapted assumptions about the principal stress ratios. From calculated horizontal principal stresses we define a new load, which we will apply on the model in order to initialize the 3D stress field at the beginning of geomechanical finite element analysis. The initial loading of the model is then carried out by using the pre-stress load type. As in DIANA the stress components for the pre-stress load type are referenced to the global (NE) coordinate system; the two principal horizontal stresses, pointing in the x- and y-directions of the model coordinate system, must be transformed to the global (NE) system.

During initialization of the effective stresses in the model, deformation of the model was not allowed. This was required in order to preserve the initial model geometry.

Modelling reservoir depletion

Withdrawal of gas from the reservoir was modelled by decreasing the fluid pressure in the reservoir in gradual steps. Reservoir simulation was used to derive changes in pore pressure for a given period of gas production. Full depletion in the crest amounted to 30 MPa.

Modelling results

Modelling results provide prediction of stress evolution in the model during reservoir depletion with the associated pattern of displacements and deformation. Various attributes and presentation techniques are available to present and analyse the results of a finite element analysis. As the case study presented here is used to illustrate the 3D workflow for geomechanical modelling, we choose two characteristic attributes to demonstrate the result of geomechanical finite element analysis: the proximity to failure and the stress path. The results are presented for the reservoir part of the model only, where the expected change in the natural initial stresses is the highest.

The first attribute which will be analysed is the proximity to failure, which represents the ratio between the shear strength and the shear stress at the point under consideration in the model (Fig. 7).

The shear stress and the shear strength are defined considering that the Mohr-Coulomb material model was used to define the behaviour of the reservoir and other model units. For other material models the proximity to failure could have been defined differently, for example as a ratio between the normal effective stress and the deviatoric stress.

In the Mohr-Coulomb material model the failure criterion is defined by taking into account two principal effective stresses, the major stress (σ_1') and the minor stress (σ_3'). The shear strength of the material is determined with two shear strength parameters, namely the cohesion and the angle of internal friction. For the reservoir material, two sets of values were used for the cohesion (c) and the angle of internal failure (ϕ): $c = 10$ MPa and $\phi = 35°$; and $c = 5$ MPa and $\phi = 20°$. The higher values of the shear strength parameters characterize the more competent, brittle part of the reservoir rock while the lower values characterize the less competent part with the clay-like materials.

The maps of the reservoir showing the proximity to failure, calculated for five considered stress regimes, are presented in Figure 8. Several colour-coded classes of the proximity to failure are used to map the calculated variation of this attribute over the reservoir. Such maps with a few colour-coded classes are deemed suitable to represent the general variability of an attribute. For easy visualization of the effects of depletion on the reservoir, the proximity to failure is presented for the initial state of stress in the model, before depletion (Fig. 8, left-hand columns) and for the final state of stress, when the full depletion has been reached (Fig. 8, middle columns). The total change in magnitude of this parameter is shown in the right-hand columns of Figure 8 to assess whether depletion favours the failure of the rock mass in the reservoir or stabilizes the rock material.

Changes in the effective stress during depletion of the Roswinkel gas field are shown also on the $\sigma' - \tau$ diagrams in the form of the effective stress

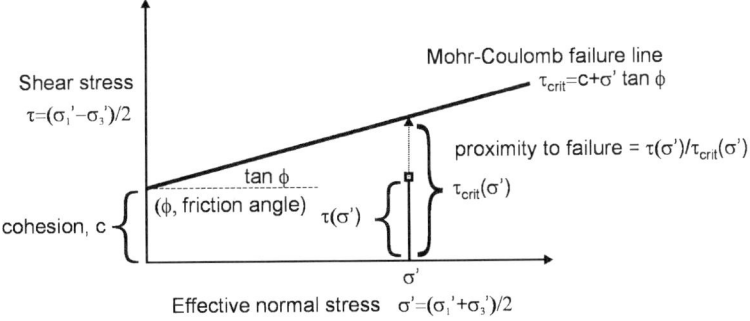

Fig. 7. Schematic diagram showing derivation of the proximity-to-failure parameter.

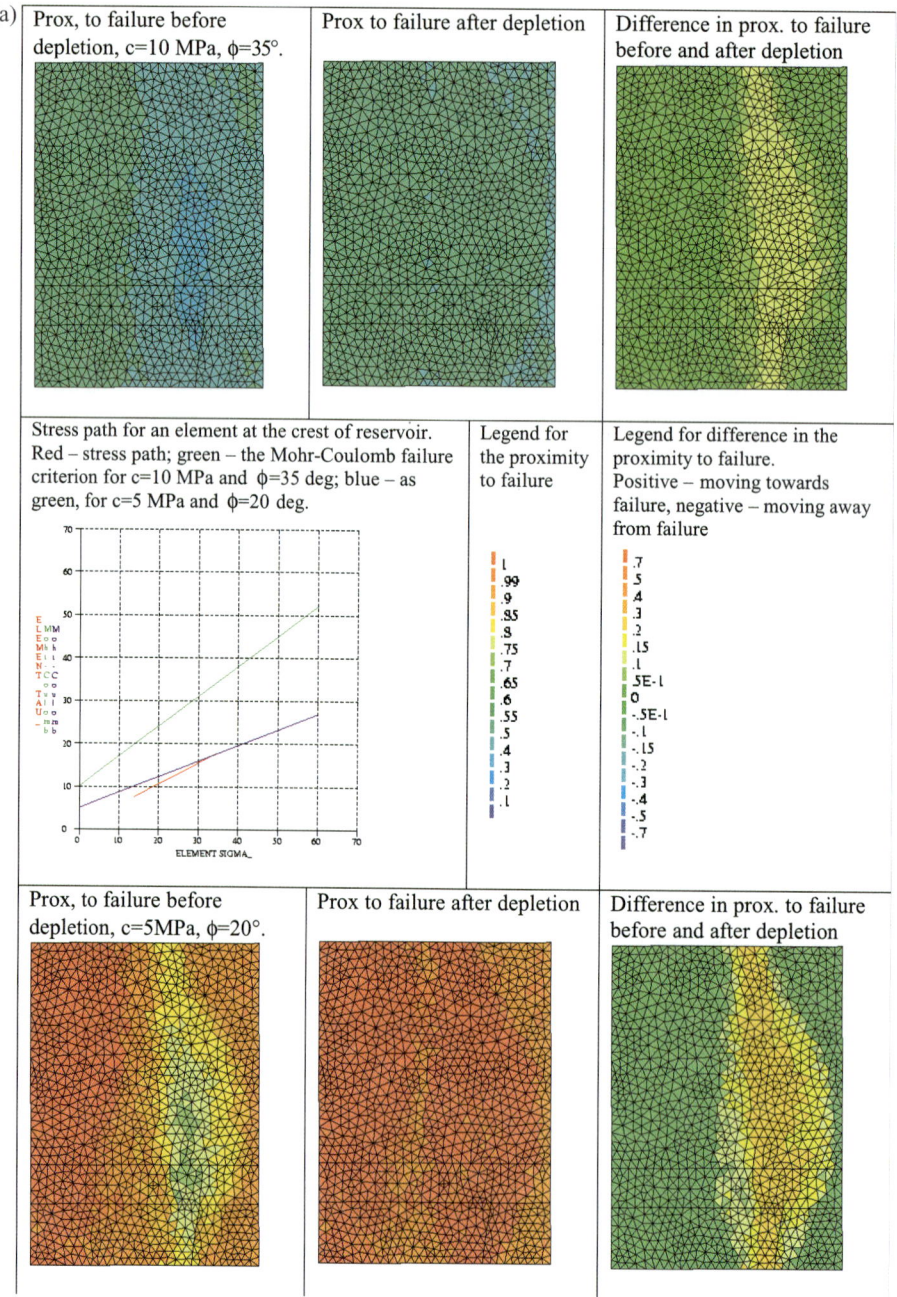

Fig. 8. Stress diagrams with top view of the reservoir showing proximity to failure and stress path diagrams during reservoir depletion for (**a**) case 1, radial extension.

paths. The effective stress path is defined by the two parameters as follows:

$$\sigma' = \frac{1}{2} (\sigma'_1 + \sigma'_3)$$

$$\tau = \frac{1}{2} (\sigma'_1 - \sigma'_3)$$

where σ' = the normal, i.e. isotropic, effective

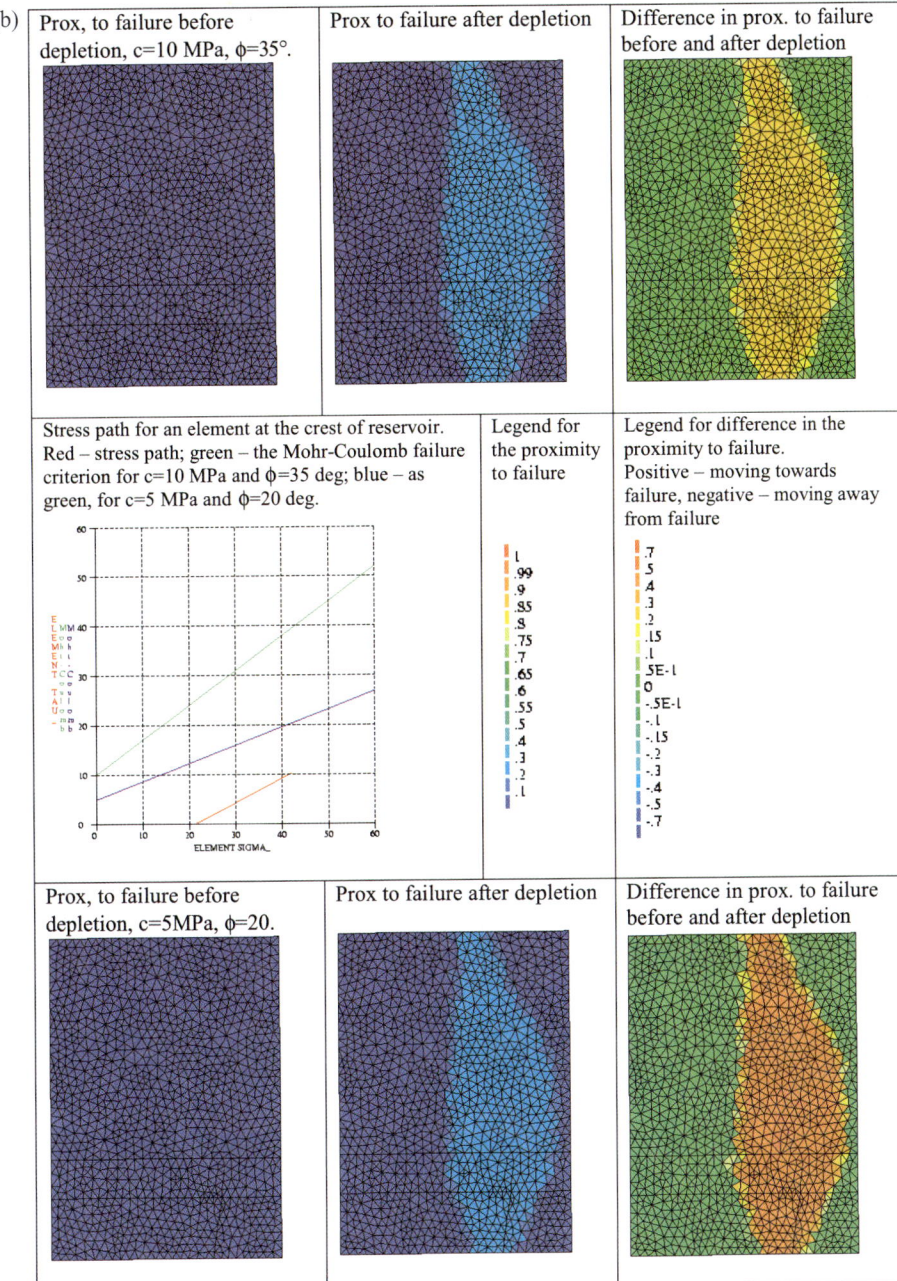

Fig. 8. Stress diagrams with top view of the reservoir showing proximity to failure and stress path diagrams during reservoir depletion for (**b**) case 2, hydrostatic stress regime.

stress, τ = the shear, i.e. deviatoric, stress, σ_1' = major principal effective stress and σ_3' = minor principal effective stress.

Development of the stresses during reservoir depletion is analysed for a characteristic element located at the crestal part of the reservoir, where the expected depletion is the highest. The stress path diagrams depict the failure criterion for the

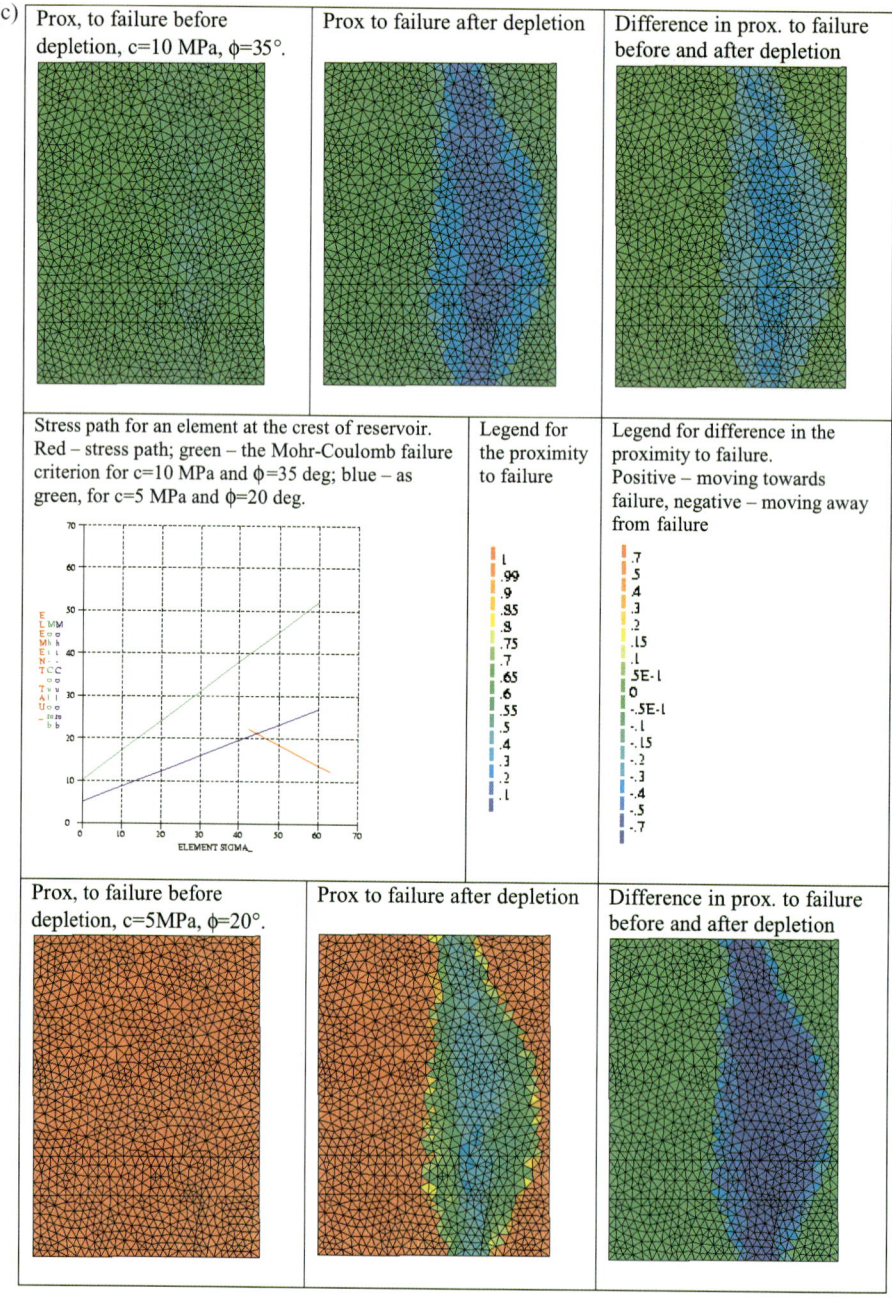

Fig. 8. Stress diagrams with top view of the reservoir showing proximity to failure and stress path diagrams during reservoir depletion for (**c**) case 3, radial compression.

two sets of shear strength parameters used for the reservoir material.

A characteristic of the initial state of stress in the model is that the rock material in a part of the reservoir filled in by the gas is further away from the failure than the rest of the reservoir. The overpressure reduces the effective stress in the gas part of the reservoir, as explained by Biot's theory (Biot

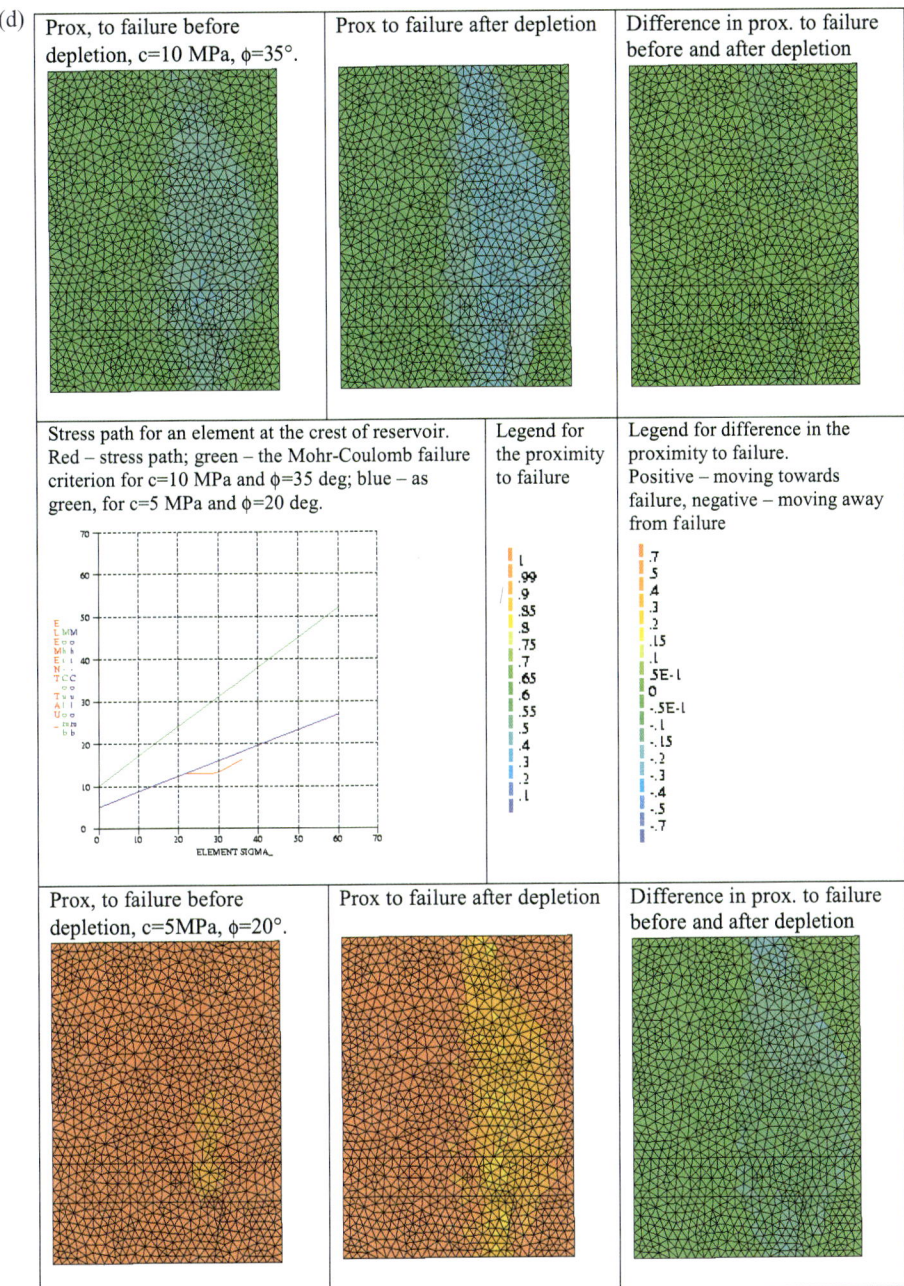

Fig. 8. Stress diagrams with top view of the reservoir showing proximity to failure and stress path diagrams during reservoir depletion for (**d**) case 4, strike-slip.

1941) and Terzaghi's principle of effective stress (Terzaghi & Peck 1962). An apparent exception is an isotropic stress regime (Fig. 8b), with equal principal stresses and a zero shear stress.

Depending on the initial stress regime, depletion of the reservoir has a different effect on the material in the reservoir. Depletion will generally cause an increase in the effective stresses through-

(e)

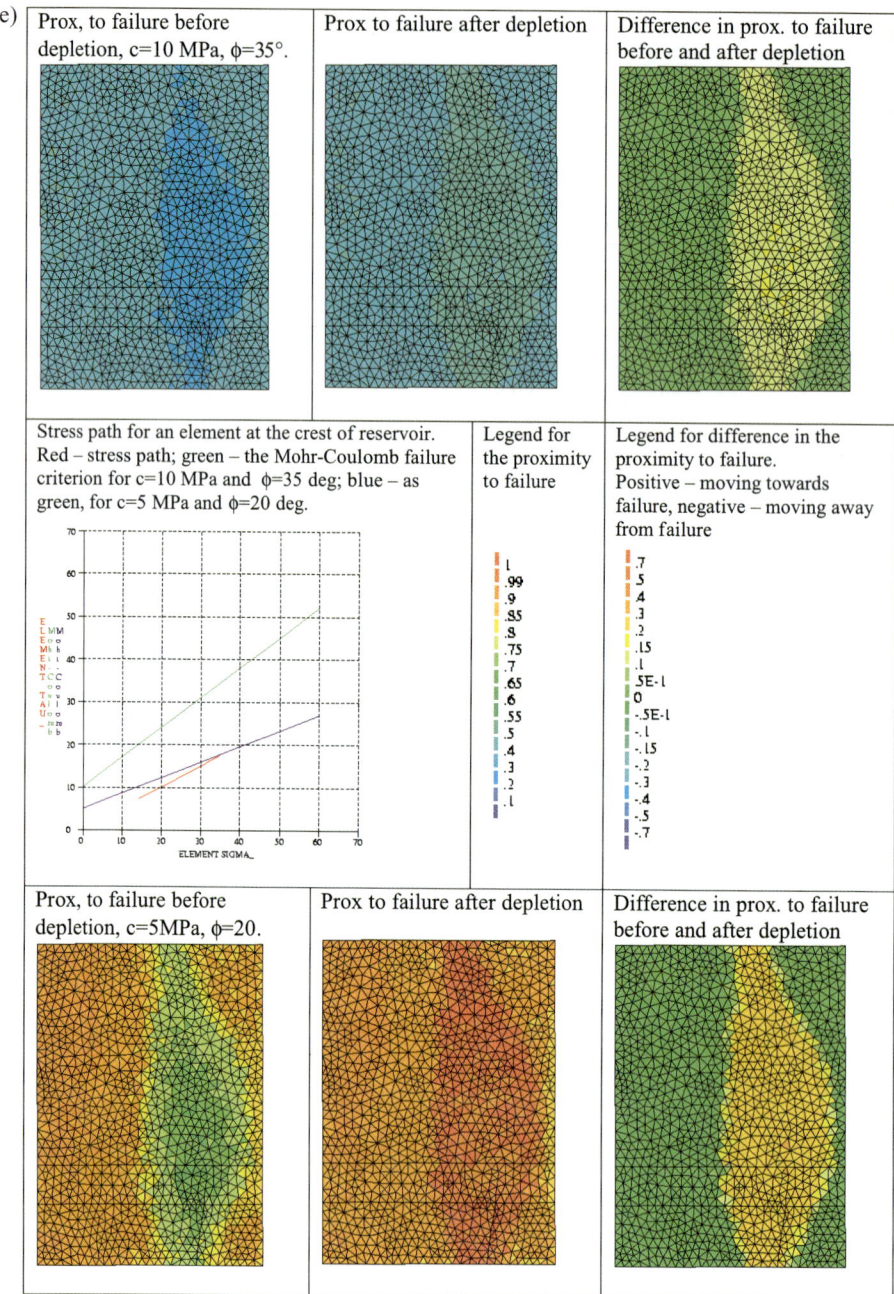

Prox, to failure before depletion, c=10 MPa, φ=35°.	Prox to failure after depletion	Difference in prox. to failure before and after depletion

Stress path for an element at the crest of reservoir. Red – stress path; green – the Mohr-Coulomb failure criterion for c=10 MPa and φ=35 deg; blue – as green, for c=5 MPa and φ=20 deg.	Legend for the proximity to failure	Legend for difference in the proximity to failure. Positive – moving towards failure, negative – moving away from failure

Prox, to failure before depletion, c=5MPa, φ=20.	Prox to failure after depletion	Difference in prox. to failure before and after depletion

Fig. 8. Stress diagrams with top view of the reservoir showing proximity to failure and stress path diagrams during reservoir depletion for (**e**) case 5, normal faulting.

out the reservoir. The overall effects of this stress increase on the rock material are to a large extent dependent on the initial stress regime in the model.

In the case of radial extension (Fig. 8a) and the case of normal faulting (Fig. 8e) depletion moves the material towards the failure, as is clearly visible for the weaker reservoir material, characterized by a set of lower material parameters ($c = 5$ MPa, $\phi =$

20°). Increase in the vertical effective stress during depletion is generally higher than the increase in the horizontal effective stresses due to a Poisson's ratio of $\nu < 0.5$. Consequently, the shear stress increases and the material approaches the state of failure.

In the case of radial compression, depletion moves the material away from the failure and improves its stability (Fig. 8c). The evolution of stresses during depletion is here more complex and the stress rotation occurs (Orlic *et al.* 2001). For a combination of parameters which we used to define the initial state of stress in the model and to simulate gas production, the overall effects of depletion on the rock material are favourable.

In the case of strike-slip, the effects of depletion are relatively small (Fig. 8d). This is due to the small changes in the shear stress due to depletion, as in the strike-slip stress regime both the maximum principal stress and the minimum principal stress are acting in a horizontal plane. In a depleting reservoir the rate of change of these two principal stresses, controlled by Poisson's ratio, will be approximately the same and no significant increase in shear stress will occur.

In all the cases fault movements on the implemented faults did not occur during the gas depletion. This is attributed to relatively small changes in the shear stress field, insufficiently large for the faults which are not aligned to the stress field in a preferential direction for failure.

Summarizing, it is concluded that only in the case of radial extension or normal faulting can the shear stress change result in a relative increase in shear stresses which can cause sufficient geomechanical failure to produce earthquakes. Given the moderate slope of the shear stress increase, failure can occur only in weak reservoir rock or weak faults aligned in a preferential direction for failure, both marked by a low friction angle of $c.$ 20°.

In our case study none of the faults adopted became reactivated.

Conclusion and discussion

The integrated workflow for 3D geomechanical modelling facilitates the construction of faulted structural models and their evolution to geomechanical finite element models. With a case study of a depleting gas reservoir we demonstrated the work process and the use of tools for structural modelling and finite element geomechanical modelling. Due to their flexibility to represent spatially varying degrees of geometric complexity, we have chosen triangulated surfaces and tetrahedral volumetric meshes to represent geological surfaces and volumes.

In the workflow we rely on advanced modelling capabilities of GOCAD to generate structural and property models. The key link between the GOCAD shared earth models and the quality tetrahedral meshes are the boundary representation models in GOCAD, which can be exchanged between GOCAD and DIANA and meshed by DIANA's pre-processor.

The case study on the Roswinkel field clearly shows the strong dependency of gas depletion deformation effects on the prevailing tectonic stress field. Our model for gas depletion predicts a stabilization of the stress field (further away from failure) for reservoirs in compressive and strike-slip regimes. On the other hand reservoirs in extensional stress regimes will result in failure of the reservoir, in agreement with observed earthquakes, provided that (a) the reservoir material or existing faults are weak and (b) the state of stress is close to failure of the material.

In the Netherlands only a few gas reservoirs are marked by the occurrence of earthquakes. Our results indicate that the absence of earthquakes in these areas should be attributed primarily to the prevailing stress field being less favourable for bringing the reservoir to a state of failure and stronger reservoir rock properties. In this respect it should be noted that the Roswinkel field is located on top of a salt dome and its structure is detached from basement rocks. Therefore it is easily possible that extensional stresses are locally magnified and closer to failure than in other areas.

The presented model has a relatively coarse geometry with rather uniform geomechanical properties. In its present form it is not very capable of predicting in detail secondary stress and strain effects such as arching on top of the reservoir (e.g. Kenter *et al.* 1998). Ongoing work is focused on the incorporation of property models and appropriate upscaling, capable of representing the spatial variability of geomechanical rock mass properties, into the workflow. The theoretical work on the upscaling of properties, related to the homogenization of Hooke's law for linear elastic material behaviour, has been successfully completed (Zijl *et al.* 2002) and its implementation in the workflow is currently under way.

References

BIOT, M. A. 1941. General theory of three-dimensional consolidation. *Journal of Applied Physics,* **12**, 155–164.

BOOGAERT, H. A. & KOUWE, W. F. P. (compilers) 1993. *Stratigraphic nomenclature of the Netherlands.* Revision and update by RGD and NOGEPA-mededelingen Rijks Geologische Dienst, **50**.

DIANA. 2000. *DIANA 7.2, Program and User's Manuals* (CD-ROM). TNO Building and Construction Research.

GELUK, M. C. & RÖHLING H. -G. 1997. High-resolution

sequence stratigraphy of the Lower Triassic 'Buntsandstein' in the Netherlands and northwestern Germany. *Geologie en Mijnbouw*, **76**, 227–246.

GOCAD. 2001. *GOCAD 2.0, Program and User's Manuals.* T-Surf & National School of Geology (ENSG), Nancy.

KENTER, C. J., BLANTON, T. L. & SCHREPPERS, G. J. 1998. *Compaction Study for Shearwater Field.* SPE/ISRM Eurock, Trondheim, SPE/ISRM paper **47280**, 63–68.

ORLIC, B., VAN EIJS, R. & SCHEFFERS, B. 2001. Geomechanical modelling of the induced seismicity for a gas field. *63rd European Association of Geoscientists and Engineers Meeting*, Amsterdam, paper P 604.

TERZAGHI, K. & PECK, R. 1962. *Soil Mechanics in Engineering Practice* (12th edition). John Wiley & Sons, New York.

TNO-NITG. 2000. *Geological Atlas of the Deep Subsurface of The Netherlands.* Explanatory notes to mapsheet VI: Veendam-Hoogeveen.

VAN WEES, J. D. & CLOETINGH, S. 1996. 3D flexure and intraplate compression in the North Sea Basin. *Tectonophysics*, **266**, 243–359.

VAN WEES, J. D. & STEPHENSON, R. 1995. Quantitative modellig of basin and rheological evolution of the Iberian Basin (Central Spain): implications for lithosphere dynamics of intraplate extension and inversion. *Tectonophysics*, **252**, 163–178.

World Stress Map. 2001. Internet source: http://www-wsm.physik.uni-karlsruhe.de/index.html

ZIJL, W., HENDRIKS, M. & HART. M. (2002). Numerical homogenization of the rigidity tensor in Hooke's law using the displacement-based conformal-nodal finite element method. *Mathematical Geology*, **34**, 291–332.

ZOBACK, M. L. 1992. First and second-order patterns of stress in the lithosphere: the World Stress Map Project. *Journal of Gephyisical Research*, **97**(B8), 11703–11728.

Index

Note: Page references in *italics* refer to Figures and Tables